Science & Technology in Fact and Fiction

Science & Technology in Fact and Fiction™

A Guide to Children's Books

□ □ □

DayAnn M. Kennedy
Stella S. Spangler
Mary Ann Vanderwerf

R. R. BOWKER
New York

Published by R. R. Bowker, a division of Reed Publishing (USA) Inc.

Printed and bound in the United States of America

Library of Congress Cataloging-in-Publication Data

Kennedy, DayAnn M.
Science & technology in fact and fiction : a guide to children's books / DayAnn M. Kennedy, Stella S. Spangler, and Mary Ann Vanderwerf.
p. cm.
Includes bibliographical references.
ISBN 0-8352-2708-1
1. Science—juvenile literature—Bibliography. 2. Technology—Juvenile literature—Bibliography. 3. Children's literature—Bibliography. 4. Bibliography—Best books—Science. I. Spangler, Stella S. II. Vanderwerf, Mary Ann. III. Title. IV. Title: Science and technology in fact and fiction.
Z7401.K46 1990 [Q163]
016.5—dc20 89-27374
CIP

ISBN 0-8352-2708-1

Contents

Foreword

Since this book is about science and technology, one should perhaps begin by trying to distinguish between the two. Science may be regarded as the discipline that attempts to describe the reality of the world around us, including the nature of living organisms, by rational means. Technology, however, attempts to exploit the fruits of science in order to attain human goals. In short, science is to be thought of as related to knowledge, while technology is concerned with the utilization of knowledge for, one hopes, the betterment of the human condition.

Four hundred years ago one could not possibly have anticipated the enormous strides that science and technology were destined to make in the ensuing centuries. Even as recently as 100 years ago, who would have predicted the great revolutions in science and technology which the twentieth century held in store for us? Thus, the theories of relativity and quantum mechanics, our perception of the universe around us, both in the large and in the small, the nature of the structure of matter, molecular biology and our understanding of life processes changed forever the way we look at the world around us and, at the same time, irrevocably established the rational mode of inquiry, the essential element of the scientific method, as preferred above all others.

Technological applications, a mixed blessing at best, followed quickly on the heels of the more basic scientific discoveries. For example, progress in communications, transportation, space exploration, and electronics, the invention and rapid development of the digital computer, improved methods for the diagnosis and treatment of disease, the use of the atom as the source of limitless amounts of energy were among the more notable accomplishments; on the dark side, however, were the successful development of intercontinental missiles armed with nuclear warheads and atomic, chemical, and biological means of mass destruction. The pollution of the environment was also a result of the recent scientific and technological advances. These were only a few of the consequences of the scientific revolutions of the twentieth century.

Then too, the fruits of technology itself fed back into and facilitated the ever increasingly rapid advance of science so that today we are racing ahead at breakneck speed to a future filled with uncertainty.

Thus, it is clear that the spectacular advances of science and technology in this century and the current trends hold enormous promise for good and an equally great threat to our very survival. The promise is that the fruits of science will be used for the benefit and well being of mankind, leading to the never ending improvement in the quality of life for everyone; the threat is that the fruits of science will be used for destructive purposes, leading to consequences ranging from the irreversible pollution of the environment to the destruction of human life as a result of a nuclear holocaust.

Which road do we take and how do we decide? Clearly, in a society such as ours it is an informed public that makes the crucial decisions; and how better to inform public opinion than to point the way to the relevant information? It seems to me that this was precisely the goal of the authors in writing this book. They have attempted to provide the information concerned with the availability of the vast storehouse of scientific knowledge that currently exists. It is an important goal. One hopes that they have succeeded, but only time can tell.

Herbert A. Hauptman
1985 Nobel Laureate in Chemistry

Preface

Historically and currently science is defined as the basic rules of order in the universe, revealing how things are put together, and technology is defined as the art of using scientific knowledge to develop tools or devices. This volume contains annotations about children's books on science and technology, vital subjects of our present-day world. Children's literature pertaining to these topics reveals our present technological society and hints at future developments; even an amusing futuristic concept or fantastic story line may contain a thread of probability. Multimedia exposure and the availability of electronic games and toys have acquainted most children with the basic elements of scientific knowledge and information. For example, when children play games of space flight, they want to know how to build a rocket, where they as passengers will go, how long the journey will take, what provisions are necessary, what to do when they arrive, and other significant details. Older children want to build working models and perform experiments in response to their scientific and technological inquiries. Students need specific answers to specific questions and also feel the need and desire to expand their present knowledge as society becomes more technologically oriented. Every year an increasing number of librarians, teachers, and parents report that young people of all ages are requesting more books about science and technology.

Because of this perceived need, we have gone in search of both fictional and nonfictional books for boys and girls that will satisfy their curiosity as well as build a broader scientific and technological knowledge base about space, aeronautics, computers, mechanics, robots, and other subjects. We have also looked for books about scientific wonders that exist today or may occur in the future. This investigation involves our research in both public and school libraries and book stores for books of this nature. Our search was further aided by the guidance and assistance of librarians. Children's interests, as well as teachers', librarians', and parents' observations and communications with children, were considered. Furthermore, professional reference sources, such as *The Horn Book*, *The Reading Teacher*, and *School Library Journal*, along with children's literature textbooks—including *Literature and the Child* by Bernice E. Cullinan (Harcourt, 1989) and *Children and Books* by Zena Sutherland and May Hill Arbuthnot (Scott, Foresman, 1986), as well as lists published by the Children's Book Council and American Association for the Advancement of Science—were consulted. Attendance and presentations at regional, national, and international conferences during the past several years were a significant factor in our search for books on these topics. Books from "The Reading Rainbow," which features and reviews children's books on public television, were also included. Both fictional and nonfictional books having themes that relate to scientific and technological topics have been selected. Some futuristic topics have been included since they suggest scientific and technological advances. Literature reviewed for this volume was chosen on the basis of appeal, timeliness, and authenticity for books of nonfiction, and appeal, originality, and believability for books of fiction.

Many children's books today are both fictional and informational, complicating discrete categorization. Within what can be thought of as a fictional framework, much scientific information can, incidentally or intentionally, be transmitted more or less as a backdrop to the story. A typical example would be *Regards to the Man in the Moon* by Ezra Jack Keats (Four Winds, 1981). Louie, with his parents' help, fashions a spacecraft made from junk. On his way through outer space, as he leaves Earth behind, he passes planets that are recognizable as Saturn and Jupiter. Excitement is created during a "rock storm" where the spacecraft hurtles through the asteroid belt.

Other interesting and unusual books have been selected that provide recreation and some scientific information. For instance, manipulative books, i.e., books having moving parts, such as *Robot* by Jan Pieńkowski, are included because they have become popular and more durable due to improvements in paper engineering. A book such as *Make a Bigger Puddle, Make a Smaller Worm* by Marion Walter may not be found in a library because an accessory, a small metallic mirror included in a pocket on the inside cover, could be easily lost. However, we have included a book such as this because parents and other caregivers, as well as children, will enjoy using this book. In two instances the older editions of books and the new, revised editions were both included because these books provide an opportunity to compare former knowledge with current available information. The books are Franklyn M. Branley's *What the Moon Is Like* (1963 and 1986) and Roma Gans's *Icebergs* (1964) and *Danger—Icebergs!* (1987).

The specific sciences selected for this volume include the physical sciences and earth sciences. Books involving the life sciences have been excluded because attention to all of these categories would suggest a separate volume.

Our reported age- and grade-level determinations are the result of years of experience as professionals in the fields of teaching and children's literature. Thus, our suggested age and grade levels in some instances differ from those indicated by publishers. For the most part, the age ranges in this volume are from 0–11; we have used the designation of 0 for infants who will enjoy hearing the book being read aloud by a caring individual, and because of the current emphasis on "lap reading" and reading aloud. In addition, some books span the age and grade ranges of the middle school and the secondary schools. These books would be of interest to readers at several stages of their reading development.

All entries include both a summary and an authoritative evaluation. The summary reviews the content of the book and is often detailed. The evaluations consist of the following features as they apply to a given entry:

1. overall literary quality
2. attention to scientific and technological detail and accuracy
3. clarity, style, and use of language
4. artistry and appropriateness of illustration
5. coordination of text and illustration
6. ability of author to convince audience of possible futuristic events
7. appropriateness of topic for intended audience
8. scope of reference materials (e.g., glossary)
9. unusual format: manipulative, cloth, cardboard, sensory, and the like.

Illustrations, where they occur, may be discussed in either the summary or the evaluation, or in both parts of the entry. For example, in a wordless picture book, the illustrations obviously tell the story and reveal the plot; therefore, they are discussed in both the summary and the evaluation. In most cases, however, the illustrations are dealt with in the evaluations.

Detailed bibliographic information is cited for each entry. We have provided the name of the author(s), title, illustrator(s), publisher and year of publication, number of pages and ISBN(s) for hardbound or library binding (LB) and paperback (pap.) editions. No ISBN has been provided for out-of-print (o.p.) works. If a given book is part of a series, that information has been provided. In addition, we have indicated those books that have won awards.

Occasionally, older or out-of-print books were included if the books continued to be maintained in both school and public libraries or if the book contained relevant subject matter. In some of these books, the information was basic and unchanging over time, for instance, books that dealt with mathematics, such as *Names, Sets and Numbers* by Jeanne Bendick (Watts, 1971), or one on temperature, *Hot and Cold and In Between* by Robert Froman (Grosset & Dunlap, 1971).

A list of professional references that identifies interesting articles and chapter readings that pertain to the subject matter in this volume has been provided, and precedes the indexes. There are five indexes: Author, Title, Illustrator, Subject, and Readability.

The Fry Readability, which purports to estimate the level of difficulty at which a book is written, is given in the Readability Index for the books for which it may be applicable. The readability is calculated on a graph using an average number of sentences per 100 words and an average number of syllables per 100 words evaluated from a 300-word-book sample. This calculation yields a level of reading corresponding to grade placement. It should be remembered that a calculated grade level indication may be limiting to readers whose ability causes them to read above or below an expected level of difficulty, or whose interest in a particular subject matter transcends grade placement. Either situation could be due to a variety of personal or environmental factors and in these cases teachers can adapt to childrens' needs. An NA (Not Applicable) indication for Fry Readability means that the book does not contain a sufficient number of words to do a readability evaluation.

This volume will help librarians, teachers, parents, and other caregivers to become familiar with books children will enjoy and that will answer their questions and stimulate their curiosity. Recently we received a letter in which a parent had written that his child "has been reading *Berenstain Bear's Science Fair* with me with great enthusiasm, telling me related things he learned in school. Of course he wants to do all the experiments!" We hope this volume will encourage adults to explore books involving science and technology with children because these worthwhile topics are a natural part of a child's curiosity, often overlooked, and are necessary for living in today's fast-moving technological and scientific society.

□ □ □

The authors appreciate the contributions of the following people: Amy McNaughton, Margaret Mooar, and Margaret Verver, and also Barbara A. Frank and Shirley Long. We are also thankful for the generous services of the Buffalo and Erie County Public Library System of the State of New York. We also acknowledge the help of the Windermere Boulevard Elementary School and the firm of Brody & Weiss. We are most grateful to our husbands, Thomas P. Kenney, Robert A. Spangler, and Malcolm D. Brutman, and to our families and friends, for their assistance and support.

DayAnn M. Kennedy
State University of New York
at Buffalo

Stella S. Spangler
Daemen College, Amherst,
New York

Mary Ann Vanderwerf
Amherst Central Schools,
Amherst, New York

Science: Fiction

Within many children's books of fiction there are scientific explanations about natural phenomena. Information about the changes of seasons, patterns of weather, cycle of day and night, and phases of the moon are often integral parts of story plots. The following books acquaint readers not only with astronomy and meteorology but also with geology, mathematics, and paleontology. Through the use of imaginary story backgrounds, descriptions are presented about the Earth's internal structure, optical illusions and mathematical concepts of combinations and probabilities, and ancient life forms of the past through fossil remains. Furthermore, the marvels of the space age are seen in adventures that deal with zero gravity and the possibility of extraterrestrial life and space colonies. Some books of poetry that refer to many of these topics have also been included.

□ □ □

Adoff, Arnold. *Tornado!* Illus. by Ronald Himler. Delacorte, 1976, o.p. Unp. Age levels: 4–7. Grade levels: PreK–2.

SUMMARY: This book relates, in verse form, the tragedy of a tornado in Xenia, Ohio, in spring of 1974. In the United States, tornados can occur when the cool air from Canada meets the warm air of the southern states; the air masses meet high in the sky but usually descend in the central states. The strong winds bring rain, hailstones, and often the strong, whirling, devastation of the tornado.

The family pictured in the book listens to the weather reports on the radio and seeks a safe place in the basement to wait out the storm. They unplug the appliances and the television, and use a transistor radio to get news and weather information. After the funnel passes, the family inspects the damage. They are safe and the house has received little damage.

Sleep comes hard for the still nervous family but school is held the next day. Some other families weren't as lucky. The children see one home torn away from its foundation even though the table is still set for dinner. The people of Xenia rebuild their town but remember the devastation of the tornado with the certainty that more will occur in the future.

EVALUATION: The utter frustration of a family being unable to prevent a tornado is told in poetic form. The parents express their own vulnerability by telling the children that it is all right to be afraid if one is in real danger of being injured. After the storm the family is thankful, and checks to see if relatives and friends are safe or need help. Hailstones the size of tennis balls are collected for the freezer, since power is down. Police and ambulance sirens scream through the night as the injured are taken to hospitals. The path of the tornado was traced by its path of destruction and it was figured to have lasted about 80 seconds, a brief time to leave such devastation.

This act of nature is effectively illustrated in charcoal sketches, which capture the dark, menacing clouds and swirling funnels. The text is printed without the use of punctuation or capital letters. The poetry is somber, adequately conveying the horror of the event. The reader learns how tornados are formed and is told which states experience them. In the beginning of the book, the author includes a dictionary's definition for tornado in English and in Spanish. *Tornado* educates the young reader in a poetic and artistic style.

□ □ □

Allison, Diane Worfolk. *In Window Eight, the Moon Is Late.* Illus. by author. Little, Brown, 1988. Unp. ISBN: 0-316-03435-5. Age levels: 6–9. Grade levels: 1–4.

SUMMARY: This rhyming story relates the experiences of Ann, a young girl, on the first day of summer. The attention of the reader is quickly drawn to the warm, comfortable extended family setting. As Ann goes through the day's activities, she becomes conscious of nature as it is seen through the windows of her house.

When she is asked to go to the dark and forbidding cellar to bring up clean clothes, she is comforted by the sight of familiar garden growth seen through window number one. Other windows display a blue sky, a weeping tree, and two swallows. As twilight falls, black tree branches appear to grimace through an upstairs window. At night, a lit attic makes a window become a reflecting mirror, and Ann needs to cup her hands in order to see stars and fireflies outside. Finally, the moon makes a late appearance outside Ann's bedroom window. Windows nine and ten are Ann's eyes, shut in sleep, but they become windows that open up a fantasy dream.

EVALUATION: The poet-illustrator conveys aspects of a summer's day and also gives the reader a counting book in rhythmic sentences. The security of the extended family situation is reinforced through soft and glowing colors. An easy progression occurs as windows revealing the scenery of a typical summer's day give way to windows of the mind. There is a successful juxtaposition of light and dark, fear and security, and realism and fantasy.

□ □ □

Anno, Mitsumasa. *Anno's Counting Book.* Illus. by author. Crowell, 1977. Unp. ISBN: LB 0-690-01288-8; pap. 0-06-443123-1. Age levels: 3–7. Grade levels: PreK–2.

SUMMARY: In sequential order, a rural landscape reveals numbers from 0 to 12. Each number is represented on a two-page spread. The first illustration is an empty and stark snow scene with a river. A prominent 0 is displayed in the margin on the right, and 10 empty blocks are seen in the margin on the left. The countryside changes and accumulates trees, animals, buildings, and many other things as the numbers increase. The number 4 is introduced in a springlike setting where 4 fish are swimming in the river, 4 geese fly overhead, and 4 trees are abloom, while children and adults engage in various activities. One guesses that this spread represents the month of April. Four colored blocks fill 4 of the 10 places in the left-hand margin, and a very large 4 stands out in the right-hand margin.

The number 11, representing the month of November, portrays 11 trees that have dropped their leaves, and 11 birds, presumably migrating since they fly in the opposite direction from the geese observed in April. A solitary barrel stands apart from 10 barrels that have been stacked and grouped together close to a building. In the left margin, the column of 10 blocks is now filled, and another column next to it has been started with one block.

Snow is falling on the holiday scene of number 12. A group of people surround the Christmas tree outside the church where the clock on the steeple shows it's midnight. Twelve reindeer soar across the winter sky. A year has gone by, the seasons have changed, mathematical concepts have been portrayed, and counting opportunities have been provided. At the end is an "About Numbers" page for adults, describing how in the Stone Age record-keeping was probably done with pebbles, which eventually led to more convenient ways of dealing with numbers.

EVALUATION: This imaginative, wordless counting book is marvelously illustrated by the author-illustrator. The colorful two-page spreads, with their plentiful counting situations, contain delightful touches. For instance, a rainbow with seven colors is depicted in July, and eleven December evergreens are covered with snow on a distant hillside while the twelfth one is festooned with Christmas decorations.

Mathematical relationships, such as one-to-one correspondences, groups, and sets, are effectively illustrated in pastoral settings with natural, everyday situations. The changing of seasons is subtly imparted in soft watercolors, while the hands of the clock on the steeple of the church point to another passage of time. Even the margins, with cardinal numbers and groupings of blocks, have mathematical information. This is an exceptional counting book that both children and adults will find fun.

□ □ □

Aragon, Jane Chelsea. *Winter Harvest.* Illus. by Leslie Baker. Little, Brown, 1989. Unp. ISBN: 0-316-04937-9. Age levels: 0–6. Grade levels: PreK–1.

SUMMARY: The young girl in this story takes grain, apples, and corn to the salt lick in the clearing beyond the trees in order to feed deer in the dead of winter, when woodland animals have trouble finding food. The deer, who come close to her house only at night, eat the treats she has brought, then prance around in the moonlight. In the morning, her mother feeds the birds and Papa shovels a path from the door while the deer, hidden among the trees, quietly watch. We know the deer are friendly and unafraid because as they pass by the cottage, they look through the window, and the little girl looks back. The spare text is more accurately described as poetry rather than prose, and it complements the sequential illustrations that convey all the events of the story.

EVALUATION: Winter as seen in *Winter Harvest* may generate real longing in readers who have moved away to sunny climes. This exquisite book has luminous artistic qualities, both verbal and visual. The little family's close-to-nature life could not be portrayed with greater warmth or more tender beauty. After reading the poetic text and enjoying the feast for the eyes displayed on every two-page spread, one is inclined to flip to the beginning and go through it all over again. Although the book has been assigned an appropriate age and grade level, all children will surely love it.

□ □ □

Archbold, Tim. *The Race*. Illus. by author. Henry Holt, 1988. Unp. ISBN: 0-8050-0954-X. Age levels: 0–6. Grade levels: PreK–1.

SUMMARY: A boy, a dog, and a homemade go-cart are the ingredients in this book for very young readers. The boy's enormous dog wears a bewildered look that indicates he can't wait for the grand adventure described in *The Race* to be over. The boy, however, is single-minded about the perilous journey, and has a determined expression that leads the reader to believe that he knows just what he is doing and where he is going. As the text reads "ready, set . . ." the eager boy, wearing a helmet, and the unbelievably clumsy dog crowd aboard a go-cart, a fragile-looking vehicle with tiny wheels. The magic word *go* sends the go-cart down the hill, leaving the dog virtually in midair. After landing in a dump, the boy shoves the go-cart and his dog through the mud and over a bump. Curious onlookers watch the rickety go-cart race under a bridge and around a bend in the road. A wheel flies off, but undaunted, the boy pushes the go-cart—complete with dog—up a flight of steps; next we see it turning left, then right. It goes down (while the dog goes up) as the text tells the reader to hold on tight. Ahead a clothesline—the finish line—awaits the happy-looking boy and dog who, as winners, receive a fine dinner.

EVALUATION: The cartoon-style line drawings are hilarious in this delightful book, which convincingly uses vocabulary associated with space and position in a story context. The clumsy, placid dog never knows what is going on but loyally allows himself to be stuffed into a go-cart, which the reader can guess will collapse or meet with inevitable disaster. The plot is fun, the print is bold and bright, and the concepts are clearly presented. Although it was originally published in the United Kingdom, this book is really universal in its appeal.

□ □ □

Asch, Frank. *Mooncake*. Illus. by author. Prentice-Hall, 1983. Unp. ISBN: 0-13-601013-X; pap. 0-13-601048-2. Age levels: 0–7. Grade levels: PreK–2.

SUMMARY: Since they feel hungry, Bear and Little Bird yearn for a piece of the moon to eat. They attempt to hit the moon with a spoon that's attached to an arrow, and they fail. Then they decide to build a rocket ship from material they buy at a junkyard. Their project is still incomplete at the end of summer, when Little Bird must migrate with the flock.

Bear finishes building the rocket ship in the fall, alone. He climbs inside and falls asleep during countdown. Sometime during winter, Bear wakes up and thinks he has reached his destination. Having never seen snow, he scoops some up and makes a mooncake to eat.

Feeling adventurous, he decides to explore his surroundings. He is quickly frightened by his own paw prints, which he thinks were made by a moon monster. He seeks the safety of his rocket ship, and again he falls asleep during countdown.

In the spring, Little Bird returns and wakes his friend. Bear tells him about the delicious mooncake treat he savored on his trip.

EVALUATION: This trip to the moon is made in a bright red rocket ship that never leaves the ground. The words *rocket ship, takeoff,* and *countdown* are effectively used in the text to aid in lifting young minds beyond Earth's bounds.

Determination and imagination allow Bear to attain his goal of reaching the moon and tasting it. The illustrations are soft, and Bear appears gentle and cuddly within serene surroundings.

The seasons are woven into this story, and the author-illustrator subtly uses winter to fool Bear by creating a moonscape with snow. Young readers, however, will not be deceived and will be able to enjoy this rather stationary but highly creative journey.

□ □ □

Asch, Frank. *Moongame.* Illus. by author. Prentice-Hall, 1987. Unp. ISBN: 0-13-600503-9; pap. 0-13-601055-5. Age levels: 0–6. Grade levels: PreK–1.

SUMMARY: Two animals, Bird and Bear, are the main characters in this book about the moon. After Bird shows Bear how to play hide-and-seek, they play all day. When the sun sets, Bear wants to play the same game with the moon. He hides in a tree trunk, then comes out, looks at the moon and says that the moon found him. When it's the moon's turn to hide, a breeze hides it behind a cloud, but when Bear finishes counting to ten, the moon doesn't reappear. After searching everywhere, Bear becomes worried and asks Bird, who in turn asks the other forest animals to help. Eventually Bear exclaims that he gives up; then another breeze blows and the moon reappears from behind a cloud, making everyone very happy.

The text is fairly long but illustrations dominate, either taking up an entire page or most of a page. Few colors are used in the illustrations but there is contrast, which is especially evident and necessary when Bear searches for the moon.

EVALUATION: The book should appeal to very young children who enjoy hide-and-seek. Schoolchildren could be led in a discussion about the moon and phases of the moon after hearing or reading the book. Clouds, stars (which are also pictured) and the wind could be included in a science discussion. This book could also be a catalyst for a number or counting lesson, since numbers 1 through 10 are in the text as part of the game.

The illustrations are cute and childlike, with very simple lines and little shading. The coloring is solid; for instance, the bear is a single shade of brown, with a lighter color only at the edge, almost as an outline.

□ □ □

Barrett, Judi. *Cloudy with a Chance of Meatballs.* Illus. by Ron Barrett. Macmillan, 1978. Unp. ISBN: 0-689-30647-4; pap. 0-689-70749-5. Age levels: 0–8. Grade levels: PreK–3.

SUMMARY: Through a tall tale that Grandpa tells at bedtime, two children find out how the people in the town of Chewandswallow get their food at mealtime: the weather brings it to them. It rains juice or soup, snows mashed potatoes, or the wind blows in hamburgers, for example. Everything runs smoothly for awhile. People wrap some goodies for between-meal snacks, and all the pets and animals are fed, too. The Sanitation Department even takes care of the extras.

However, one day the weather took a turn for the worse; the townspeople only had spaghetti, cheese, or over-cooked broccoli to eat that day. Storms of pancakes came another time, and people sneezed terribly from the salt and pepper wind. Finally they had to abandon the town, which they did in a very unusual manner. The difficult part was that in their new homeland they had to purchase food in a supermarket; it seemed strange that food came in boxes, packages, bottles, and cans.

EVALUATION: Almost everyone enjoys hearing a tall tale, and this one could provide an exciting introduction to weather, or simply be fun reading. Part of the book is written in weather-report form, and many weather terms are used throughout. Children could count and name weather words or phrases, and compare the tale with actual weather reports or predictions. Weather idioms known and used today, such as "raining cats and dogs," could be compared to descriptions in the book. The author also plays on words. Children could do a play-on-words newspaper article with weather terms as another activity involving the book.

The illustrations are humorous, especially if the details are not ignored, and would certainly delight the reader. They have a soft touch of color added over the black-and-white penline-type drawings.

□ □ □

Baum, Arline, and Joseph Baum. *Opt: An Illusionary Tale.* Illus. by authors. Viking, 1987. 32 pp. ISBN: 0-670-80870-9; pap. 0-14-050573-3. Age levels: 8–11. Grade levels: 3–6.

SUMMARY: The reader is taken through a magical kingdom of optical illusions. Vocabulary associated with castles, such as *Great Hall* and *Castle Guard,* is used to introduce the illusions. Each page is devoted to at least one visual riddle. Questions require the reader to study colors, lines, and shapes in a variety of situations. Visual surprises abound, as a special guest is anticipated at a special event, a birthday party for the prince. The awaited guest, a dragon, appears with gifts and adds more visual excitement. At the party, an optical illusion gift needs the participation of the reader to find out the age of the prince.

The last seven pages of the book indicate that some illusions have more than one explanation. The illusions featured on each page are explained in text and illustration in a more sophisticated fashion than the preceding pages and cover suggest. A two-page activity section that tells children how to make their own illusions concludes the book.

EVALUATION: This book can entertain children who want to concentrate only on a few pages of this intriguing exercise of visual play. Although the vocabulary and the characters appear to relate to a castle of medieval times, a real story does not emerge. This artificial device fails to provide a continuous thread; however, it is probably unnecessary as the book can be enjoyed a page at a time.

Some of the pages require a high degree of visual discrimination and this can make them too challenging for some children. Consequently, the reading ages suggested on the book jacket, 3–8, appear to be somewhat inappropriate. The colors are bright, the characters are cute, and an adult sharing this book with a child will probably enjoy it, too. The text consists of a question and answer format that involves the reader throughout. A jester at the bottom of each page enthusiastically directs the reader to continue to the final section, which contains the answers. There is much to fool and delight the eye in this book.

□ □ □

Baylor, Byrd. *If You Are a Hunter of Fossils.* Illus. by Peter Parnall. Scribner, 1980. Unp. ISBN: 0-684-16419-1; pap. 0-689-70773-8. Age levels: 7–11. Grade levels: 2–6.

SUMMARY: While collecting fossils to take home, the author imagines what life was like in ancient times—from animal life, such as a trilobite, 5 billion years old, to a seed fern in shale, or sponges in rocks, to chalky limestone. Several scientific words are included in this fiction book. The illustrator had previously illustrated three Caldecott Honor books.

EVALUATION: This "Reading Rainbow" story easily shifts back and forth from past to present, and at times the two seem to blend. Even the format of the text—one-, two-, or three-word lines, with some double columns—makes the story seem like it happened long ago. It has a poetic but believable quality because the author writes in the first person, and she effectively draws the reader into her world. The illustrations complement the text; earth and water tones set the mood.

□ □ □

Berenstain, Stan, and Jan Berenstain. *The Berenstain Bears in the Dark.* Illus. by authors. Random House, 1982. 32 pp. ISBN: LB 0-394-95443-2; pap. 0-394-85443-8. Age levels: 5–8. Grade levels: K–3. Series: First Time Books.

SUMMARY: Brother Bear chooses a mystery book from the library and begins to read it to Sister. When the story about the dark, mysterious cave begins to frighten her, Papa Bear puts an end to the reading and the children go to bed. The children can hear the soft wind and other night sounds. Sister Bear begins to see cave creatures in the chest of drawers. When Brother Bear lets out a wailing noise to scare his sister, both parents come in and try to calm down the commotion.

The next day Father Bear takes Sister into the attic to look for something. Meanwhile, he explains that her imagination is playing tricks on her, and she should be careful not to let it control her; most everyone is afraid of the dark at one time or another. Sister realizes that the things she sees are not creatures but shadows cast on the wall by old furniture.

When Brother discovers that his sister wants to hear the end of the mystery, he can hardly believe it. She is somewhat disappointed to find out that the wailing noise was only the wind blowing across the cave. However, both children go to bed with the night-light on, which Father Bear found in the attic.

EVALUATION: Being afraid of the dark, and finding out that noises or shadows can be explained, is an experience to which many children can relate. The night-light or reflected light from the moon can cast shadows in an unlit room.

This book contains a considerable amount of text, and some of the vocabulary may have to be explained to younger children. The illustrations at times capture the story within the story.

□ □ □

Bourgeois, Paulette. *Franklin in the Dark.* Illus. by Brenda Clark. Scholastic, 1987. Unp. ISBN: pap. 0-590-40631-0. Age levels: 3–5. Grade levels: PreK–K.

SUMMARY: Franklin the little turtle is afraid of small, dark places. This fear extends to his shell, and he resorts to dragging it behind him rather than crawling in to rest. His mother shines a flashlight into the shell to reassure him that there are no creepy things inside, but Franklin is not convinced. He sets out on his own, dragging his shell behind him, to seek help from the other animals.

When he tells the duck his problem, she immediately identifies because at times she is afraid of very deep water; she offers him the use of her water wings. Franklin tells her this wouldn't help him as he is afraid of the dark, not water.

By and by, he meets a lion who is afraid of loud noises, a bird who is afraid of flying too high, and a polar bear afraid of icy, cold nights. All offer their personal "crutch" for help but, of course, none can solve Franklin's problem.

Finally, Franklin returns home to his anxious mother, who tells him she was afraid he had gotten lost. When Franklin realizes his mother can be frightened, too, he sets aside his own fears and crawls inside his shell to sleep—but with a night-light.

EVALUATION: This tender story is sure to be enjoyed by each child who identifies with Franklin's problem. At the beginning of the story, Franklin appears to be a very competent, capable young turtle. He can count, zip, and button; the fact that he can count backward adds a comic touch. However, when it grows dark, he becomes frightened. Children love flashlights, and it is a sensitive parent who tries to coax a youngster to bed by showing the child that there is nothing to be afraid of. This tactic doesn't work with Franklin. As he searches for help, each animal, from the small bird to the mighty lion, confesses that he or she is also afraid sometimes and needs some security. Since this book is best enjoyed by small children, Franklin could have encountered three animals rather than four. The point is sufficiently made with three, and the number *three* has always held magic in fables, folk tales, and children's stories.

The writer and illustrator should have used pond animals exclusively: one would not find a polar bear and a tiger in the same climate, and this may confuse the child. Illustrations of sweet-faced animals were drawn in the background, but some nocturnal animals might have been included in the night pictures. The use of the night-light is a timely touch that all children will find comforting.

□ □ □

Briggs, Raymond. *The Snowman*. Illus. by author. Random House, 1979. Unp. ISBN: 0-394-93973-5; pap. 0-394-88466-3. Age levels: 0–6. Grade levels: PreK–1. Awards: American Library Association Notable Book, 1978; Boston Globe-Horn Book Award, 1978.

SUMMARY: In this charming wordless picture book, a little boy wakes up one morning to discover that snow has fallen during the night. In the nine tiny framed pictures on the first page, we follow the boy as he quickly dresses himself and says good-bye to his mother. At the bottom of the page the boy rushes outside in a single colorful strip. His facial expression reveals that his mind is full of plans. The next few pages tell us that he has waited for a heavy snowfall so he could build a snowman. And it's a jolly snowman he builds, complete with a carrot nose, lumps of coal for eyes, buttons, and a scarf and hat his mother let him borrow. Later the boy goes inside, but every illustration lets the reader know that the snowman never leaves his mind. The boy goes to bed, but we see that in his dream he joins the snowman for many wonderful adventures, both in- and outdoors. When the snowman observes that the sun is coming up, he guides the boy back to his bedroom and hugs him good-bye. When the boy wakes up the next morning he races to the window only to find that his friend the snowman has become a lump of melting snow on the lawn.

EVALUATION: The award-winning art in this book proves the argument that a story can be effectively told without the use of type. The colors are soft, the winter scenes are totally convincing, and the family relationships and the boy's wonderful friendship with the snowman are warm and comfortable. Little children will want to tell the story in their own words, making the book an excellent vehicle for the promotion of creative thought and language. Humor abounds, and an adult sharing this book with a child will probably have a generous share of smiles and chuckles. This book received *School Library Journal*'s Best Book of the Year designation in 1978.

□ □ □

Burton, Virginia Lee. *The Little House*. Illus. by author. Houghton Mifflin, 1942. 40 pp. ISBN: LB 0-395-18156-9; pap. 0-395-25938-X. Age levels: 4–7. Grade levels: PreK–2. Awards: Caldecott Medal, 1943.

SUMMARY: In this 1943 Caldecott Award-winning story, a man builds a lovely little house in the country. He builds it strong so that it will last for generations. At night the house sees the lights of the big city far away and she wonders what it would be like to live there.

Years go by and horse-drawn carriages and sleighs make way for automobiles and a road is built near the little house. Other houses, gas stations, and roadside stands spread over the land; soon the little house is part of a city. Elevated trains, subways, and skyscrapers are built around her. No one lives in or takes care of her any more. Although the windows are cracked and the paint peeling, her structure is still sound underneath.

The great-great-granddaughter of the builder discovers the house and has it moved to the country and placed on a hilltop. The house is repaired and painted, and once again enjoys the quiet and peace of the country, and never dreams of the city again.

EVALUATION: The reader first meets the little house as she stands proudly on a hill in the country. During the five generations covered, the technological advancements of humanity are brought to her very doorstep. The progress is carefully sequenced so the reader understands that some of the advancements occurred within a single decade. In one illustration, for example, a gas station is built as automobiles begin to travel the road. In the next illustration, the hills are dotted with new houses, telephone poles, and a steady stream of cars and trucks belching gray smoke. Of the four apple trees that once flanked the house, only one remains as a reminder of things as they once were. The young reader learns that progress takes its toll on the environment—as in the form of air pollution.

Burton's illustrations have a folk quality about them. The lack of shading and perspective enhances the text. In the earlier pages of the book, the house and land glow radiantly in the summer sun. However, as the dirty city rises up, gray becomes the predominate color.

The author's message is subtle, but the options and trade-offs choosing various life-styles and environments are best explained by an adult. The end papers are worth noting: the automotive, communication, and environmental changes of this century are drawn chronologically with the little house centered in each of the 18 frames.

□ □ □

Carle, Eric. *Papa, Please Get the Moon for Me.* Illus. by author. Picture Book Studio, 1986. Unp. ISBN: 0-88708-026-X. Age levels: 0–7. Grade levels: PreK–2.

SUMMARY: A little girl wants her father to get the moon for her to play with. Father gets a ladder of extraordinary length, shown through two double-page fold-outs. In a single vertical fold-out page, Father balances the ladder on top of a mountain, in order to reach the full moon. In another gigantic fold-out Father climbs close to the moon, and the moon tells him that he'll play with his daughter when he's smaller. Later, Father brings the moon—which has become crescent shaped—to his daughter. She plays with the moon, which continues to change shape, until it disappears. One night the moon reappears in the sky and becomes increasingly larger until it becomes a full moon again.

EVALUATION: In this fantasy, vivid, childlike illustrations highlight the nighttime sky. The characters recognize the great distance from the earth to the moon, and the fold-out pages emphasize this distance. Clever structure and imaginative artwork combine to create a playful interpretation of the phases of the moon.

□ □ □

Clifton, Lucille. *The Boy Who Didn't Believe in Spring*. Illus. by Brinton Turkle. Dutton, 1973. Unp. ISBN: 0-525-27145-7; pap. 0-525-4365-7. Age levels: 3–6. Grade levels: PreK–1.

SUMMARY: A little boy from the inner city named King Shabazz didn't believe in spring. His mother spoke of the crops coming up and his teacher told the class about the birds. Though the days are getting longer and warmer, King doesn't realize these to be signs of spring. He and his friend Tony take off on a quest to find spring. As they walk aimlessly through the neighborhood, they pass the school, the grocery store, a church, and a restaurant. They venture out of their neighborhood to an area where they find a garbage dump surrounded by tall apartment buildings. An abandoned car holds the promise of spring—a nest of bright blue robins' eggs and near it, a colorful patch of crocuses.

EVALUATION: The author uses black dialogue in the story. This book does not explain how the seasons change or why birds, crops, and crocuses are symbols of this annual event. It could have been written so that the mother, the teacher, or a neighbor would give King and Tony this information.

The boys leave home without permission, but this goes unpunished. Indeed, it is in a sense rewarded by their sighting the robin's nest and flowers. This is not an appropriate lesson to be conveyed to children the ages of three to six, for whom this book is intended.

The illustrations, however, are a pleasure. The inner city landscape provides a rich visual backdrop for the two cute boys, one black and one white. A mailbox and a fire hydrant can be spotted, and the street signs can be "read" by the preschooler. The illustration of the dump has a menacing quality so it is a pleasant surprise to discover that in the grayest, dirtiest place in town, spring can be found (although the adult may wonder how crocus grow in this shaded area when there are none in the sunnier locations in the boys' own neighborhood). In the scene where the boys walk away from the dump, Turkle has painted larger-than-life flowers in the background to evoke a dreamlike quality. This nice little story has been written to entertain rather than instruct.

□ □ □

Coats, Laura Jane. *Marcella and the Moon.* Illus. by author. Macmillan, 1986. Unp. ISBN: 0-02-719050-1. Age levels: 5–7. Grade levels: K–2.

SUMMARY: Marcella the duck is a dedicated artist who enjoys painting the moon. In fact, she likes it so much that she refuses the chance to swim with her duck friends. They tease her because she won't join them. Marcella takes her moon painting seriously; she is never bored since the moon is always changing. Studying her subject keeps Marcella busy producing a variety of moon paintings.

One night there is consternation among her duck friends when the moon fails to appear. They do not know where the moon has gone, and they seek Marcella for answers. As she leisurely relaxes in her hammock, Marcella reassures her concerned friends that the moon will reappear. She has gained important knowledge while oberving the moon's cycle.

Friendship flourishes among the ducks. Marcella finds time to join the others in swimming, and her friends start to appreciate her paintings.

EVALUATION: Accurate scientific information on the phases of the moon is cleverly conveyed through a sequence of pictures Marcella hangs on a clothesline next to her easel: her artwork displays the waxing and waning of the moon. Marcella is intelligent and artistic, and is accepted for both these qualities. The reader learns that art can be a hobby as well as a conveyance for scientific knowledge.

The text flows in an easy manner. The illustrations of the nighttime scenes, which could be difficult to portray, are softly done with no extraneous detail to distract from the subject. Furthermore, the illustrations reinforce the scientific concepts.

□ □ □

Cole, Brock. *The Winter Wren.* Illus. by author. Farrar, Straus, 1984. Unp. ISBN: 0-374-38454-1; pap. 0-374-48408-2. Age levels: 3–7. Grade levels: PreK–2.

SUMMARY: One year, the winter seemed endless. Simon's family could not grow crops, and soon the villagers would be out of food. Simon goes looking for Spring and overhears two crows chattering about Spring being asleep at Winter's farm. Simon sets out with his little sister Meg to wake up Spring. He brings a sack of meal and an apple for food on their journey.

Simon walks with his sister on his back until he comes upon an old man sowing a field with sleet. When the old man, Winter, sees the two children, he throws ice at them. His evil magic turns Meg into a Winter Wren, who then tells Simon how to trick Winter. Simon does as he is told. Where Winter sowed sleet, Simon sows wheat meal taken from his sack. Where Winter cut the buds off the apple trees, Simon throws his apple into the grove and a large apple tree springs up. Angry Winter runs into his house with Simon at his heels. Inside, Simon sees no sign of Winter but instead finds Meg sleeping like a princess in a bed of flowers. As the two walk back to the village together, green grass and flowers roll before them. Simon had awakened Spring at Winter's farm, and all of the villagers plant lettuce and potatoes in the warm earth.

EVALUATION: This book has a fairy-tale quality. Simon, Meg, and their mother live in what appears to be an old European village with stone fences and gnarled trees surrounding Tudor-style cottages. The family and indeed the whole village is poor and down to their last bit of porridge and milk when Simon sets out to wake up Spring.

The main characters are like the folktale characters of old. Simon is treated as a careless boy, yet he is given the opportunity to save the village from starvation by tricking Winter. Meg is seen as weak and helpless until she is changed into the magic wren and eventually a special princess.

The book is illustrated with elegantly detailed watercolors, which give further evidence of Old World charm. The text is formal and understated. Simon's mother calls him "daft" and "befuddled fool." The symbols of spring are aesthetically pleasing to twentieth-century man, but Simon's family and neighbors needed the long, sunny days and warm nights for their survival.

□ □ □

Coleridge, Sara. *January Brings the Snow.* Illus. by Jenni Oliver. Dial, 1986. Unp. ISBN: LB 0-8037-0314-7. Age levels: 4–8. Grade levels: PreK–3.

SUMMARY: The months of the year are described in this poem written in the 1800s by Sara Coleridge, daughter of the renowned English poet, Samuel Taylor Coleridge. For this publication, it is noted at the end of the book that only the September stanza was altered; however, the original wording is not included. Each month is illustrated on a two-page spread containing two lines of rhyme.

The chill of ice and snow is conveyed in the winter months' verse. December is festive with a decorated Christmas tree, and filled stockings hang from the mantel of a fireplace. Gusty breezes announce the arrival of the spring months with an abundance of flowers and the birth of lambs. In an April garden, the reader sees a budding tree and a primrose path leading to a gazebo where a shaggy dog sits.

The rain sometimes cools the hot summer months that abound with plentiful crops. "Sheaves of corn" are mentioned in the August verse; however, a crop of wheat is illustrated. A note at the end of the book explains that wheat was actually harvested, but the reference to corn is a result of the English laws that require all grain to be called corn. Autumn brings falling leaves; harvests of fruit and nuts are to be gathered. Cows are seen in September grazing under laden apple trees. The year is traversed in country home and farm scenes.

EVALUATION: This poem is beautifully illustrated starting from the title page where two twisted, flowering wisteria vines frame the title. Colorful and delicate watercolor paintings poignantly extend the monthly verses. No children are seen, yet their presence is felt: in the January two-page spread children's mittens, scarves, and boots are strewn on a porch near a sled and toboggan. In the July illustration, half-eaten cookies, a doll, and a teddy bear suggest that children were in this scene before the reader arrived. The change of seasons is subtly portrayed in soft watercolors, enhancing the lyric quality of Coleridge's poem.

□ □ □

Davis, Hubert. *A January Fog Will Freeze a Hog*. Illus. by John Wallner. Crown, 1977, o.p. Unp. Age levels: 5–9. Grade levels: K–4.

SUMMARY: Before radio or television weather reports existed, people tried to predict the weather by observing patterns in the sky, people and animal behavior, or some other phenomena. This collection of 30 weather sayings is based on these early weather forecasts. The sayings in this book are somewhat humorous and vary in their reliability, from being good predictors to having no bearing whatsoever on the weather. Included in the notes in the back of the book are the origins of the sayings, and where in the world or within our own country the sayings have been used.

EVALUATION: Young children should enjoy the references to animals or signs in the sky in this easy-reading book. The appreciation of the sayings is not likely to be marred by comparison to scientific weather reports. Older readers will enjoy learning the history behind the sayings, and assessing their validity since third and fourth graders are usually very concerned with fact and fiction. The illustrations are elaborate black-and-white drawings that add beauty to the text.

□ □ □

DeWeese, Gene. *The Dandelion Caper*. Putnam, 1986. 159 pp. ISBN: 0-399-21326-0; pap. 0-440-40202-6. Age levels: 9–13. Grade levels: 4–8.

SUMMARY: Calvin and Kathy have already had experiences with visitors from outer space, which is why the old Diefenbacher place is so intriguing to them—they are on the lookout for more. Here they encounter Dandelion, a big yellow talking cat who is really a space visitor. With Dandelion acting as their guide and protector, the children experience invisibility gadgets, green lizards, and shrinking spaceships. Best of all, the cat is a friend of Calvin's Uncle Harold, who 50 years earlier had encountered space visitors. Together, they search out the mysterious intruders in Calvin's house and investigate the kidnapping of his mother and brother. A battle occurs, leaving Calvin and Kathy greater believers in outer space visitors than they were before the Dandelion Caper.

EVALUATION: This book humorously and imaginatively presents space visitors with faulty equipment and less-than-average abilities. The text is wonderful with its "off the track" comments, typical of a child trying to tell a story. Unusual terminology or situations are simply defined. Each chapter is headed by a catchy quote from the chapter. The jacket illustration suggests the wonderful fun in this story. *The Dandelion Caper* is a sequel to *Black Suits from Outer Space* (Putnam, 1985).

□ □ □

Euvremer, Teryl. *Sun's Up*. Illus. by author. Crown, 1987. Unp. ISBN: LB 0-517-56432-7. Age levels: 0–7. Grade levels: PreK–2.

SUMMARY: This wordless pictorial fantasy pretends to progress from sunrise to sunset; however, seasonal changes occur as well. Despite its recent publication, a fair amount of sexist and ageist stratification exists in this book. Among the human characters, the man goes to the office while Grandmother stays at home and bakes. Mother is also at home, decorating the house for a party. Children are present but in minor, insignificant roles.

Additionally, the sun is male and the moon is female. The sun wears clothes—including a tie—and carries a briefcase. He works at providing energy for growth, health, and light, and paints the proverbial rainbow. There is elaborate activity throughout. For example, there is a dramatic clashing between the sun and a thunderhead, while the earth below teems with busy people and animals. The objects and pretty colors that surround them look soft and suggest variety in texture.

EVALUATION: The scientific aspects of the book appear to be inaccurate and consequently can be confusing to young children. For example, many seasonal changes occur within a one-day period.

Children will certainly enjoy examining the fine detail in every picture; there are seek-and-find opportunities, surprises, and strikingly colored images of earth and sky.

The personification of the sun and moon allows children to exercise imagination and artistic expression. This is the type of story that encourages children to extend their fantasies beyond the time frame of the book, through the introduction of the female moon on a final, mysterious two-page spread.

□ □ □

Fast and Slow. **Illus. by Peter North. Grosset & Dunlop, 1977. Unp. ISBN: LB 0-448-13380-6. Age levels: 4–6. Grade levels: PreK–1. Series: Opposites.**

SUMMARY: The Opposites series uses pairs of words to introduce concepts of size, space, sound, speed, and position. This picture book for young children uses the words *fast* and *slow,* and many full-color pictures to show relative speed. One two-page spread pictures a hot-air balloon moving slowing through space opposite a pair of rockets zooming through clouds across the sky. Another page shows an elderly lady slowing being led by a little girl as she walks on a slippery surface opposite five ice skaters racing on a frozen pond. On a third spread, people leisurely row on a lily pond opposite racing sailboats and a catamaran. The last pages ask the reader questions about what is fast and what is slow in the pictures, and also what the children pictured can do that is fast and slow.

EVALUATION: The illustrations throughout the book are absolutely beautiful; rarely does one see a "concept" book that exhibits the artistic detail found here. Unfortunately, the concepts of fast and slow are not always pictured as distinctly as they might be. For example, the truck and cars shown on a highway don't appear to be going at different speeds. A road sign on the slow side of the page is supposed to indicate that the vehicles are going up a hill; however, young children will not be able to interpret the sign. Furthermore, the "fast" side shows only that the two cars are passing the truck. This is a driving strategy, and does not necessarily indicate that the cars are going appreciably faster than the truck. In this and several other pictures, particularly one showing fast and slow rides at a fair, a greater degree of both visual discrimination and experience are necessary for children to understand that opposite speeds are being presented.

□ □ □

Florian, Douglas. *A Winter Day.* Illus. by author. Greenwillow, 1987. Unp. ISBN: LB 0-688-07352-2. Age levels: 2–6. Grade levels: PreK–1.

SUMMARY: The passage of a winter's day is portrayed from its gray beginning to its dark ending in poetic form. A loving family is presented, and indoor and outdoor winter activities are pictured. The reader is taken to a country setting where snow is falling during the morning hours; smoke rises from the chimneys of the homes in the landscape. A family finishes a pancake breakfast in a kitchen warmed by a black wood-burning cooking stove. Preparing to go outside and enjoy some winter fun, the children dress warmly, including putting on headgear.

The snow has stopped when the children are seen skating, sledding, and making a snowman. Soon the children are back indoors warming their toes before the fire, and steaming soup is on the table. Late in the day they help Father bring in wood from the woodpile. Night falls on this cold white world. There's a full moon in the sky, and lights glow in the windows of homes.

EVALUATION: A cold wintry day is depicted with much color and warmth in this poem for young children. Colors subtly change as the day progresses from a gray, snowless morning to a night lit by a full moon overlooking a snow-covered landscape. A great deal of warmth permeates the pages of this book: there is the warmth of family, of comradery, of a fire, of clothing, and the warmth of working together.

Even though there are no more than three words on a page, the text is effective in delivering this seasonal message in rhyme. As the children bundle up to go outside, the text reads, "Cover your heads." Since much heat is lost from one's head in winter, the instructive line inserted in this poem will be appreciated by parents. The full-page childlike illustrations are very pleasing, imaginative, and beautifully capture the mood of a winter day.

□ □ □

Gibbons, Gail. *Sun Up, Sun Down.* Illus. by author. Harcourt, 1983. Unp. ISBN: 0-15-282781-1; pap. 0-15-282782-X. Age levels: 0–7. Grade levels: PreK–2.

SUMMARY: A small child wakes up as the sun comes in through her window from the east. As she narrates the book in the first person, she notices and experiences many things about the sun: it gives light and makes patterns, and the energy and power from the sun enables things to grow. She notices how high the sun is in the sky, allowing for ample playtime in summer; shadows, or the lack of them, tell her what time of day it is. She finds out how far away from Earth the sun is, what it consists of, and what it would be like if there were no sun by questioning her parents. As the sun begins to set in the west, clouds appear and it rains. From her parents, the girl learns what causes rain and rainbows, and that when the sun sets here it is shining on the other side of the world. The book ends with a page of facts about the sun and its energy.

EVALUATION: Much information about the sun, its characteristics and its effects, is contained in this short science picture book. Illustrations dominate each page, and the text ranges from approximately two to four sentences per page. Although written from the point of view of a child, this book contains vocabulary such as *separating, prism, vapor,* and *different,* which may pose reading difficulties for younger children. The illustrations contain simple lines and have bold color—mostly red, blue, yellow, and green. Scientific concepts are extremely simple in presentation, yet accurate. This book is an excellent introduction to the sun for younger children.

□ □ □

Goennel, Heidi. *Seasons*. Illus. by author. Little, Brown, 1986. Unp. ISBN: 0-316-31836-1. Age levels: 0–6. Grade levels: PreK–1.

SUMMARY: An assortment of activities and feelings associated with the various seasons are vividly shown, beginning with spring and ending with winter. The first illustrated page depicts four seasons and foreshadows the concept of seasonal continuity and repetition. The simple, large print text utilizes the first person singular—"I." Urban and rural settings are included; children are shown decorating with fresh flowers, sailing toy boats, selecting pumpkins, and enjoying the snow. The last page of text reminds the reader that seasons are cyclical.

EVALUATION: Since there are few science books for young children, this book about seasons makes a worthwhile contribution. The activities associated with the seasons are plausible and will appeal to children. The artwork, although it is bright and bold, comes across as flat. The illustrations are childlike, but lack outline and perspective. The children are faceless and devoid of personality, but this ensures universality. The concept of the continuity of seasons fostered on the first page via illustration is reinforced on the last page through the text.

□ □ □

Goldsmith, Howard. *Invasion: 2200 A.D.* Doubleday, 1979, o.p. 137 pp. Age levels: 10–15. Grade levels: 5–10.

SUMMARY: This story features the Tuks, ratlike creatures reminiscent of the characters in Robert C. O'Brien's *Mrs. Frisby and the Rats of NIMH* (Atheneum, 1971). Paul Bennett and his working farm family become embroiled in a struggle to save their home planet, Earth, from takeover by forces which are bent on mind control. Paul has many enemies. Because he is a farmer, his interests have historically been in opposition to the interests of the Domes, belligerent city dwellers whose urban concerns conflict with Paul's attempts to sustain life through a time-honored agrarian model. The scientist Rog, an alien from another planet that has been infiltrated by the Tuks, is at first perceived as an enemy, but it is actually he who convinces Paul and his farm workers to join forces with the hostile Domes to ward off the Tuks, who clearly want to systematically diminish the intelligence of Earth beings, rendering them helpless and subservient.

EVALUATION: Set in the twenty-third century, this fast-paced adventure keeps the reader enthralled from beginning to end. The setting is futuristically interesting; it consists of a city, run by robots and computers, which survived a twenty-second century nuclear war. Freedom and liberty were destroyed in the city. The farms, collectively called the Grange, maintained liberty and independence. The story, actually an allegory, conveys the message that whether life is maintained naturally or mechanistically, the preservation of human intelligence is a common goal. The Tuks effectively represent the forces of evil, seeking to dominate both city and farm. The Tuks are defeated because Dome and Grange—forced to cooperate—are stronger than the Tuks, who are defeated and sent out "beyond the rim of time." Even though it is not totally free of contrivance, the book is just plain good fiction.

□ □ □

Gundersheimer, Karen. *Happy Winter*. Illus. by author. Harper & Row, 1982. 37 pp. ISBN: LB 0-06-022173-9; pap. 0-06-443151-7. Age levels: 4–8. Grade levels: PreK–3.

SUMMARY: The special charm of wintry days is enjoyed by two sisters as they take part in a variety of outdoor and indoor activities. One morning, frost etches designs on the window panes; outside a blanket of snow that has fallen in the night covers the rural landscape. After a warm breakfast, the children put on clothing to protect themselves against the biting wind. They play in the snow and leave snowprints and writing on the soft wintry tablet.

On a bitter cold day with driving sleet, the children stay indoors and use their imaginations in creative play. They also help their mother bake a fudge cake. During another day when falling snow keeps the sisters house-bound, they make presents for their mother's forthcoming birthday. Reading fairy tales, spooky stories, and nonsense rhymes is another indoor pastime. Bubbles in a warm bath are reminders of how snow looks perched in a tree. Finally, the mother's winter lullaby closes this story in rhyme.

EVALUATION: The marvel of winter is the subject of this delightful book. Changing winter scenes reflect some aspect of nature as the characters appreciate their surroundings. The full-color illustrations will appeal to children. The text is a lively poem that includes a rhythmic get-dressed song and a lullaby. Sprinkled throughout the text, isolated drawings depict the words next to which they are placed and offer help in word recognition. A recipe for a fudge cake is included, and carries a note to the child that adult assistance is necessary. The children's joyful experiences are unified by the smooth blend of text and illustrations. Winter's harsh weather is softened in this comforting and pleasant read-aloud or read-alone book.

□ □ □

Hague, Kathleen. *Numbears: A Counting Book.* Illus. by Michael Hague. Henry Holt, 1986. Unp. ISBN: 0-03-007194-1. Age levels: 0–3. Grade levels: PreK–K.

SUMMARY: This counting book features bears and the numbers one to twelve. Meghan is happy alone (suggesting number one), Sam buttons two buttons on his coat, Brittany has two friends for tea (making three), Kathleen counts four blue eggs, Mikey builds five snowmen, Alison collects six shells, Cori takes a walk with seven ducks, Elisa rocks eight dolls, Anthony ties nine balloons to his rented spaceship, Mary sees ten lambs, Kevin conducts an orchestra in which the players are eleven bunnies, and Heath has trouble sitting in his bedroom because twelve balls crowd his space.

EVALUATION: Hague's breathtaking art dominates this beautiful picture book for the very young. The Numbears are adorable, and the adult who shares this book with a child will no doubt go back to it again and again for the sheer thrill of the illustrations. However, it is not an appropriate concept book to teach numbers. Numerals do not play a prominent part, and when they do appear they are arranged haphazardly. On the cover, for example, the numeral 3 is shown turned backward. The cover, title page, and final two-page spread all display numerals 1 through 12, but these are sometimes shown out of order and crudely drawn, to mimic the chaotic way younger children may write numerals when they are just beginning; unfortunately, children may want to imitate the way the numerals appear here. The body of the book is playful in its featuring of the charming Numbears, but on every page the opportunity to display a large, clear, unmistakable representation of the appropriate numeral is missed. A subtly placed numeral *does* appear on every page, but it is always a challenge to find it, even though the search can be fun. Two particular illustrations may worry or confuse: Kathleen Bear is pictured standing on an open windowsill, and the ten lambs appearing in cloud formations are a challenge even for the discerning adult, who already knows what real lambs are supposed to look like.

□ □ □

Hayward, Linda. *Windy Day Puppy.* Illus. by Lucinda McQueen. Grosset & Dunlap, 1985. Unp. ISBN: 0-448-10453-9. Age levels: 0–6. Grade levels: PreK–1.

SUMMARY: As the book begins, the Windy Day Puppy looks outside to see the leaves stirring on the ground. When Grandfather Dog asks the Windy Day Puppy if he would like to play a game or take a walk in the woods, Puppy chooses the walk because he is eager to see what happens when the wind is blowing outside. Together they discover that the air is cool. The first thing they observe is clothes blowing on a clothesline. Next they see a pretty bird dodging blowing leaves. Puppy has a good time walking on logs and chasing leaves that have blown into the water. Then Grandfather Dog's hat blows off his head; the hat circles in the wind and sails far away. Windy Day Puppy finds the hat covering a turtle. Before they go home, Puppy and Grandfather play hide-and-seek. Back at the house, the tired puppy climbs in his bed and takes a nap; the windy day weather has exhausted him.

EVALUATION: The blowing wind is visible in the woods as Grandfather and Windy Day Puppy watch colorful leaves blow. The illustrations are cute and childlike, and the reader sees only smiling faces in this story, from beginning to end. One page that may be slightly confusing shows a disproportionately large bird compared to the animal characters featured in this story. There is a warm, comfortable relationship between Grandfather Dog and Puppy.

A real plus for parents is that, although Grandfather wants to continue playing, Puppy is so tired from being outside that he puts himself to bed for an afternoon nap. The book also shows that Puppy can both dress and undress himself. This board book is spiral-bound with plastic, leading one to wonder how durable it might be after many readings.

□ □ □

Holl, Adelaide. *Moon Mouse*. Illus. by Cyndy Szekeres. Random House, 1969, o.p. 36 pp. Age levels: 0–7. Grade levels: PreK–2.

SUMMARY: In this story, Mother Mouse tells Arthur that he is now old enough to stay up at night. Arthur notes that the night is cool and still. He thinks that night is very nice. All at once, Arthur notices the full moon and excitedly points it out to his mother. Then, as children do, Arthur wants to know where the moon is, what it is for, and what is is made of. Mother Mouse replies that the moon is far away, high in the sky, that it shines to give us light, and admits that she does not know what the moon is made of. She says she has heard it is made of cheese, but doesn't believe this is true. Arthur continues to think about the moon, and one night he sets out to look for it by himself. After walking a long distance and noticing many things under the starry sky, he spies the moon at the top of a tall building. Arthur tries to reach to the moon by climbing up a fire escape; he finds an open window and goes inside. When he spots an enormous round yellow cheese, he exclaims that this is the moon. He nibbles away, running in and out of the craters (holes), until he is full. When he returns home, he tells Mother Mouse that he found the moon and it *is* made of cheese. After many rainy, cloudy nights, the moon reappears as a small, thin slice. Arthur tells his mother that he has eaten the rest. She smiles and tells him it's a good thing that he didn't eat it all.

EVALUATION: In the beginning of this appealing book, Mother Mouse tells Arthur that the moon gives light at night. When Arthur is allowed to go out after dark, he uses his powers of observation to notice temperature, stillness, stars, the changing moon, and the cloudy sky. He takes quite literally his mother's reference to the moon being made of cheese and never really finds out the difference in this story. The author and artist convincingly picture Arthur's concept of the moon as a round yellow cheese. The simple black, white, and yellow artwork effectively details Arthur's nighttime explorations. The story moves along well, with the reader wishing it did not end as soon as it does.

□ □ □

Hughes, Shirley. *Up and Up*. Illus. by author. Lothrop, 1986. Unp. ISBN: 0-688-06261-X. Age levels: 8–10. Grade levels: 3–5.

SUMMARY: In this wordless picture book, a very young child observes birds in flight and then attempts to fly herself. She even makes herself wings out of paper and sews them onto her clothing. After an unsuccessful attempt at flying, she then tries to fly with a bunch of balloons that break when they touch trees' branches, thwarting her plan. After a realistic beginning, the book shifts into a fantasy mode. A deliveryman brings her what appears to be an enormous gift-wrapped chocolate egg, approximately four times her size. After unwrapping it, she lifts the top half and climbs inside, then breaks through as a bird would break its shell. Next she eats the entire chocolate egg, after which magical things happen. She begins to bounce through the air, taking bigger and bigger steps, finally walking on walls and ceilings, defying the laws of gravity. Her adventures in flight take her all over the city of London. People try to catch her, and an elderly man in a hot-air balloon almost succeeds. She flies away from him and punctures his balloon, after which they both fall to Earth, ending the fantasy. The next few frames show her waving to all the people she encountered, and returning home to her family. She sits down to eat an egg, and the expression on her face indicates that it reminds her of her adventure in flight.

EVALUATION: This rather long book is an artist's view of a child's adventure. In some ways the book is probably inappropriate for children because negative behavior modeling dominates much of the book. Except for several frames at the beginning and the end of the book, the child is aggravating, rude, disobedient, thoughtless, disruptive, and mean. It is suggested that this behavior is connected with substance consumption, probably the caffeine in the chocolate egg. During her flight she knocks over a produce stand, disrupts a classroom, breaks a television antenna, and punctures an elderly man's hot-air balloon so that it deflates, among other things.

Also, the illustrations do not differentiate between realism and fantasy. Some of the pages are crowded, such as the single page containing 14 separate images. The lack of color and fine detail sometimes render it difficult to determine what is happening in the frame. Overall, the artwork appears to be too sophisticated for the age range of the intended audience.

□ □ □

Hutchins, Pat. *The Wind Blew.* Illus. by author. Macmillan, 1974. Unp. ISBN: 0-02-745910-1; pap. Penguin 0-14-050236-X. Age levels: 0–7. Grade levels: PreK–2. Awards: Kate Greenaway Medal, 1974.

SUMMARY: The sun, peaking out from beneath a cloud, holds little promise on the day that a gale-force wind blows across the countryside. A balding man loses his umbrella, which has turned inside out, and a balloon escapes from little Priscilla's hand. At a wedding the groom loses his top hat, and a little boy's kite flies away. As they chase their possessions, these characters move closer and closer to the city.

A lady loses a shirt from her clothesline. Another woman sees her handkerchief fly away, while a lawyer's wig is caught up in the wind. Joining the parade is a postman whose letters scatter into the air. A uniformed guard tries to catch an escaped flag. Then a pair of twins lose their scarves, and finally a newspaper blows out of the hands of a bystander. The turbulent wind tosses the lost items around and drops them helter-skelter on the surprised characters. After reclaiming their possessions, they watch as the gusty wind, which has now moved out to sea, fills the sails of a boat. The sailor is delighted as the boat quickly heads toward the horizon.

EVALUATION: This award-winning cumulative tale dramatically shows the power of wind. The blustery wind is portrayed as a force of nature that can not only cause havoc but also be beneficial. The use of texture in brick, stone, and leaf as well as in clothing patterns makes the illustrations very interesting. The rhymed text is clever and appropriate for the predictable situations. As the wind gains force and the characters chase their possessions, the scene effectively moves from countryside to city. There are incidental counting opportunities throughout.

People in the community, who might have been strangers before this windy day, are brought together by the wind. All the illustrations are two-page spreads that contain humor as well as fast-paced action. This story will amuse and entertain children and the adults who may share this book.

□ □ □

Janosch. *Dear Snowman.* Illus. by author. World, 1970, o.p. Unp. Age levels: 3–8. Grade levels: PreK–3.

SUMMARY: Johnny the snowman is a village favorite. As the people stand around and talk Johnny hears all the news. He also receives poetic weather predictions from his friends, the seven ravens. The only people Johnny doesn't like are firemen because he worries when they are around. His favorite thing is the 8:22 train to the city, because she informs him about the outside world and the changing seasons.

A little girl, Lisa, loves Johnny the snowman best. She receives her unusual birthday wish of having the beloved snowman visit her at home. Unfortunately, Lisa builds a fire in the stove in Johnny's bedroom; in the morning Johnny has disappeared. She is consoled by her father, who tells her that Johnny might come back on the train next year with the first snowfall.

EVALUATION: A snowman is a pleasant association children have about the winter season. This particular snowman is a favorite of the townspeople and particularly beloved by Lisa, who assumes he can survive in a human environment. As a consequence of her friendship with Johnny, Lisa learns about temperature changes and seasons. The illustrations are bold, bright, and childlike, and the catchy poetic weather report can lead to a discussion of weather rhymes, such as the popular saying "red sky at night, sailors' delight. . . ." The gentle but necessary loss of the snowman provides a soft ending for the story.

□ □ □

Jupo, Frank. *A Day Around the World.* Illus. by author. Abelard-Schuman, 1968, o.p. Unp. Age levels: 8–10. Grade levels: 3–5.

SUMMARY: Steve and Mitsuko live in opposite parts of the world; while one experiences daytime, it is nighttime for the other. This is how the author introduces the concepts of the Earth's rotation, and from here he goes on to time zones and a trip around the world. In either one- or two-hour intervals, Jupo places the reader in a certain part of the world and describes what a child who lives there might be doing, thereby informing the reader what life is like in other countries. Simple sketches portray the country and events and a map of time zones is included.

EVALUATION: The author uses a natural tie-in between the rotation of the Earth and time zones and the time zones map shows every hour of the day. However, sometimes there is one hour to the next time zone and at other times two hours to the next time zone: it is possible that this may be confusing to younger children. It might have been helpful to use a small insert of the time zone map to illustrate each time zone move. In that way a child would more easily relate to such uncommon places as Bulshire, Iran, or Kargalinsk, U.S.S.R., along with more recognizable places such as New York and London.

The text is generally easy reading for younger children but some vocabulary, such as *astride* or *bullock carts,* may pose difficulty. Bullock carts are shown in an illustration; however, they are so small that it is difficult to determine what they are. There is no ending to the story of Steve and

Mitsuko that begins the book; the book ends abruptly after the last time zone is discussed.

□ □ □

Karl, Jean E. *Beloved Benjamin Is Waiting*. Dutton, 1978. 150 pp. ISBN: 0-525-26372-1. Age levels: 10–15. Grade levels: 5–10.

SUMMARY: In this example of science fantasy, we learn that after Lucinda's parents abandon her, she fends for herself in the safety of an abandoned caretaker's house inside the gates of a cemetery. She is befriended by a space being who takes on the form of the discarded statue of a dead little boy, Benjamin. In the library, she finds records about the old cemetery and learns more about the statue and her new home. The library also becomes a source for materials to teach the space being about life on Earth and its surrounding solar system. In the process, Lucinda, too, is learning how to cope with her life: her contact with the space being and her research help her to eventually solve her problems.

EVALUATION: This is a simple fantasy, in a realistic setting, which is very sensitively presented. It shows the strength and determination of a young girl as she cares for herself in the wake of the heartlessness of others. Karl creates an unusual friendship that is believable and holds readers' interest throughout the story.

□ □ □

Kessler, Ethel, and Leonard Kessler. *All for Fall*. Illus. by Leonard Kessler. Parents Magazine Press, 1974, o.p. Unp. Age levels: 3–6. Grade levels: PreK–1.

SUMMARY: This story, told in verse form, is about the changes in weather from late summer to early winter. We sense the change in the colors of the grass and particularly trees as their leaves turn from green to red and orange until finally they turn brown, die, and fall in the cool autumn wind. We see yellow in the harvest of corn as well as in the school bus that signals the end of summer fun. The birds and geese fly to a warmer climate, while the mammals gather food and begin to rest for hibernation. The days grow shorter as we lead up to the first days of winter, which have the fewest hours of daylight.

EVALUATION: The breezy rhythm and poetic style of *All for Fall* will enchant the preschooler. For the child who is beginning to read, the repetition of words and frequent naming of colors will make this an "easy reader." Many words describing green, and the words *yellow, bright, dark, light, blue,* and *gray,* also describe the colors of early fall. On other pages, colors in the illustrations provide contextual clues to help children with the color names.

The text tells of the gradual change from summer to winter. The days grow shorter and the weather turns colder; animals, plants, and people all prepare themselves for the cold snowy months ahead. Leonard Kessler chose to show the gradual change in the seasons by muting the colors in the illustrations from one page to the next. This process is successfully used to subtly show the young child how the colors of seasons change from the bright hues of summer, through the reds and oranges of fall, to the stark white, gray, and brown of winter. The endpapers open to summer outdoor activity and close on the barren landscape of autumn.

□ □ □

Lapham, Sarah, reteller. *Max Chases the Moon.* Illus. by S. Rainaud. Derrydale, 1985. Unp. ISBN: 0-517-49029-3. Age levels: 0–6. Grade levels: PreK–1.

SUMMARY: Max, a mouse, waits until nightfall to set out for a new adventure. Neither the lure of city lights nor the glowing stars in the sky attract him. It is the moon he wants, because he thinks it's made of cheese. He plans to have an ample supply for the winter months.

Spotting the moon above a flower, he carefully and slowly climbs to the top. When he tries to grab the moon, he discovers to his dismay that the moon is no longer over the flower. It is out of reach, and dawn is approaching. Deciding to outrun the moon, he plans to capture it at the distant hill. Again the moon outsmarts him by disappearing. Max returns home and rationalizes that because winter is a long way off, he has plenty of time to catch the moon.

EVALUATION: This delightful book tells about a whimsical game between a mouse and the moon. The adventurous mouse is endearing as it tries to snatch the moon for its selfish reasons. The moon escapes by appearing to move across the sky; its facial features seem to indicate that it enjoys the chase. Both characters are charmingly portrayed in a bucolic setting. The simple text is lively and captures the spirit of the race. The colorful illustrations subtly connect the sky and ground scenes. This little story provides an interesting look at the fascination that the moon holds for many.

□ □ □

Lee, Jeanne M. *Legend of the Milky Way.* Illus. by author. Henry Holt, 1982. Unp. ISBN: 0-8050-0217-0. Age levels: 6–10. Grade levels: 1–5.

SUMMARY: This Chinese legend describes a small boy who tends a buffalo and who matures into a young man. When he plays the flute, it is heard by the weaver princess in the heavens. She follows the sound to Earth, meets the young man, and marries him. Before he dies, the buffalo, who had become friends with the boy, tells his master to make a cloak of his hide after he dies, and predicts that the cloak can perform miracles for him. In the meantime, the weaver princess' mother orders guards to bring the princess back.

When the young man finds his wife missing, he dons the cloak and is miraculously lifted toward the heavens. When the mother sees him coming, she draws a silver river with her silver pin across the heavens to stop him. Then she turns the young man and the princess into stars, one on each side of the river. When the princess will not stop crying, her father persuades his wife to allow the young couple to visit each other once a year, on the seventh day of the seventh month (on the Chinese calendar). Blackbirds form a bridge across the silver river so that the princess can cross.

In the Prologue, the constellations Lyra and Aquila are drawn as well as the stars Vega and Altair, which represent the young man and the princess. The birds and the silver river are also pointed out.

EVALUATION: The illustrations are outstanding, done in simple line cuts with beautifully blended or contrasted colors. They reflect Chinese art in tone as well as subject. The legend could be told to children, with reference to the book's illustrations, and the tie-in with the sky map is easy to discern. This is a lovely "Reading Rainbow" book to read with children when doing a unit related to space.

□ □ □

Leonard, Marcia. *Find That Puppy*. Illus. by Maxie Chambliss. Bantam, 1988. Unp. ISBN: pap. 0-553-05429-5. Age levels: 3–6. Grade levels: PreK–1. Series: Show Me.

SUMMARY: Annie and Joe take a walk in the park with their puppy, Bo. The puppy runs off and the children must look for him in the busy park. The story is told in verse and as each page is turned, a different area of the park is depicted, such as the pond, the tennis court, and the playground. In each picture, the reader is asked to find certain objects or people involved in certain activities.

Bo appears on each page but is usually obscured by a larger object. The reader is asked to count specific objects on some pages and point to pond creatures on another. The puppy is found at the end and all is forgiven; Bo has led Annie and Joe on a merry chase, giving them an adventure to remember.

EVALUATION: There is much to be learned in this simple, playful story book. Each page is loaded with action so that visual memory must be used to find Bo. Clues are given in the text and the able child can quickly deduce where the puppy is. For example, Bo takes off sniffing the air and disturbs a family's picnic. Where is Bo? Why, in the bushes, of course, with a stolen hot dog. Also on this page the reader is asked to pick from the array of food "something sour, something sweet."

On the page where the playground is pictured, Bo is attracted to the sound of children playing and laughing. Here the young reader is asked to find the "boy sliding down and the girl climbing up."

The illustrations are an important part of this story. Activity abounds on each page, making the search for Bo difficult. Readers become involved in the story as they are asked to follow the directions and find not only the lost puppy but several other sight clues that enhance skill in reading. Young children will learn how to figure out a mystery by seeking clues and using their five senses.

□ □ □

Lightner, A. M. *Star Dog*. McGraw-Hill, 1973, o.p. 179 pp. Age levels: 10–14. Grade levels: 5–9.

SUMMARY: Holton has visions of breeding his dog, Mitzi, with a purebred and selling the pups at a profit, but Mitzi is too frisky and never cooperates. One night Holton's friend Willy excitedly tells him that a flying saucer has landed in the field. As the boys dash out so does the dog, thus beginning the adventure.

Returning from the field, they come across a dog with six legs, which has been killed by a car. They take it home and eventually to the university for an examination. Mitzi bears a pup, Rov; however, Mitzi is later hit by a car and dies. Holton and Rov develop a special friendship and have the ability to communicate telepathically. After a battle between good and evil forces to gain control of Rov, both boys and the dog are picked up by a spaceship. This ship is looking for the dog of an alien boy. When they find that Rov is the son of that dog, they use mental telepathy to return the boys and dog to Earth.

EVALUATION: This is a delightfully written, fast-paced variation of the boy and his dog story. There is simple, direct conversation though it is not always believable. Each of the 15 chapters has a short statement heading, and an epilogue follows. This book has all the elements for a great deal of fun without being heavy reading on the alien theme.

□ □ □

Locker, Thomas. *Sailing with the Wind*. Illus. by author. Dial, 1986. Unp. ISBN: LB 0-8037-0312-0. Age levels: 5–8. Grade levels: K–3.

SUMMARY: This is the story of Elizabeth, a little girl who lives with her parents in a secluded area by the river. Her family receives word that her Uncle Jack, a sailor who works on ships and travels around the world, is coming to visit. Elizabeth climbs to the top of a hill to watch for Uncle Jack's sailboat.

When he finally arrives, he spends time with Elizabeth and her parents. He tells them of the sights he has seen: the Great Wall of China, strange animals and birds, and tombs and palaces. He asks Elizabeth's parents if he can take her on a sailing trip down the river to the ocean.

Elizabeth has never seen the ocean or been on a sailboat, so she looks forward to the one-day trip. They pack a picnic lunch, and set their alarm clocks for six o'clock. As they sail down the river the next morning, there is hardly any light or wind. As the sun rises, the wind picks up and they are on their way.

As they get to the mouth of the river, the weather changes dramatically. A violent wind tosses the tiny boat and they have to turn back. It rains constantly on their return trip, and Elizabeth is kept busy bailing. They eventually get home safely, and Uncle Jack has to return to his ship. He promises Elizabeth he will return to share another adventure with her.

EVALUATION: Although the text is simple and straightforward, the art dominates this book. The illustrations are oil paintings reproduced in such a way that they do not lose their delicate luster in the printing process. The outdoor scenes are extremely luminous; the author-illustrator has truly captured the shimmering reflection of the moon on still water and the hazy, hot summer sun high in the sky. Indeed, the characters are minute in comparison to the dominant weather and landscape.

The land surrounds the small boat as the two sail down the river, but there is no land in sight as they reach the waves and swells of the ocean. They look vulnerable as they are tossed about under the gray sky. But as the sun sets that evening, the sky is painted in hues of orange and yellow signaling a "sailor's delight" for the next day.

Elizabeth clearly adores her Uncle Jack. She is so isolated that she is clearly starved for adventure. The author understands human nature and family relations. The parents are reluctant to allow their daughter to take the trip but finally agree. Elizabeth takes off her life jacket at one point, and is surprised that Uncle Jack doesn't object. For all the parents' doubts and Elizabeth's fears, the uncle proves an excellent sailor as he maneuvers the small boat around the rocks in the storm.

This beautiful "Reading Rainbow" story can be read by a middle-grade child or read to a younger one. Adults are sure to enjoy the quality of the illustrations, as the originals are befitting an art gallery.

□ □ □

McCloskey, Robert. *Time of Wonder.* Illus. by author. Viking, 1957. 62 pp. ISBN: 0-670-71512-3; pap. 0-14-050201-7. Age levels: 4–10. Grade levels: PreK–5. Awards: Caldecott Medal, 1958.

SUMMARY: The family in the story spends summers living on an island off the coast of Maine. Their daily activities and amusements are recounted. One day a storm is brewing, which is a common occurrence in late summer, and the vacationers, fishermen, and animals prepare for it. The storm comes and goes, leaving fallen timber and a changed terrain. At the end of the season, the family packs to return to the mainland for the opening of school.

EVALUATION: The author writes in a descriptive, almost poetic fashion, and watercolor paintings help to enhance the text. As the fog descends, the accompanying picture evokes a moist, boggy haze we can almost touch and feel.

McCloskey recounts the preparations the fishermen and lobstermen must make to protect their boats, traps, and homes from the storm: boats are docked quickly, and gear is stowed and tied down. Food, supplies, and fuel are purchased. As the storm hits, the family seems to have a routine as stories are read, songs are sung, and flashlights are used for security. The storm has lifted the next morning, with fallen timber and a treasure of crushed clam shells—which the children imagine had once been played with by Native American children long ago.

Information on how the Earth was formed is included as the book relates how the rocks at the ocean's shore were formed by glaciers that moved across the Earth thousands of years ago. Now strong winds and heavy rain make changes on the terrain. The reader knows that man can protect and defend himself from the elements, but may wonder where the animals go. A timelessly beautiful and informative work.

□ □ □

McKee, David. *The Day the Tide Went Out . . . And Out . . . And Out . . . And Out . . . And Out . . . And Out.* Illus. by author. Harper & Row, 1976. Unp. ISBN: LB 0-200-00160-4. Age levels: 3–6. Grade levels: PreK–1.

SUMMARY: When the tide goes out, the jungle animals play on the beach and make sandcastles. One large animal, called a beachkeeper, flattens the sandcastles in order to keep the beach tidy. While the beachkeeper sleeps at midday, the animals build a sandcastle on his back. When he awakens, he shakes off the sandcastle, walks to the shore, and lies down. The water of the incoming tide washes the remaining sand from his back.

One day, the animals, angry with the beachkeeper for destroying their castles, ask the sea to help them with their plan. The following day, the animals wait for the beachkeeper to fall asleep. Then they build sandcastles on the beach as well as on the beachkeeper's back. The beachkeeper goes to the water and waits for the tide to come in. But the ocean retreats; each time the beachkeeper gets up and moves closer to the edge of the sea, but the sea will not wash the sand off the beachkeeper's back.

Meanwhile the animals build many tall sandcastles on the shore. When they climb to the top they cannot see the ocean because it has receded. Since the beachkeeper cannot remove the sandcastle from its back, it decides to keep it there.

The animals, angry because their plan failed to trick the beachkeeper, retreat to the jungle. The ocean stays where it is. People begin to come from miles around to see the sandcastles and wonder how they got there.

EVALUATION: This *pourquoi* book, telling how the camel got its hump, has been written and illustrated for the preschooler. The tale is told simply, with only a few lines of text on each page.

The illustrations have a primitive folk art quality. The animals have bright eyes and smiling faces, and even the lions and tigers aren't the least bit ferocious. All the creatures are painted in a rainbow of bright colors sure to enchant the young audience.

The beachkeeper is actually a camel with no hump. He is a sober creature intent on spoiling the rolicking beach fun of the other animals. At the end he is willing to settle for living with the hump on his back.

The tale is sure to prompt a discussion about tides, as the animals wait for the sea to come in and out each day. The silvery sliver of the moon hangs in the sky above the ocean, suggesting its role in the changing of the tides, although no explanation is given in this simple story book.

On the last page, children are pictured frolicking on the inviting sand dunes. The next time the readers visit the ocean and play in the dunes, they'll know how the dunes got there—the animals built them.

□ □ □

McNulty, Faith. *How to Dig a Hole to the Other Side of the World.* Illus. by Marc Simont. Harper & Row, 1979. 32 pp. ISBN: LB 0-06-024148-9. Age levels: 4–8. Grade levels: PreK–3.

SUMMARY: McNulty's tongue-in-cheek blueprint for an imaginary geological adventure suggests that the reader will have a greater chance of success journeying to the other side of the world if a friend comes along to help. What follows is a descriptive catalog of substances to be found between the Earth's surface and its core, from loam and topsoil to oil, basalt, magma, and iron. Attention is given to the proper clothing and equipment needed to protect oneself against the elements of water, heat, and so on.

After digging through the outer and inner cores of the Earth, the child is finally at the center where there is weightlessness. The reader is told not to stay at the Earth's core very long, but to hurry home to tell everyone all about what went on. After the child stays a short time, the trip continues, but to the opposite side from where the trip began, and the Earth's many layers are discussed again. This trip begins in the United States and ends in the Indian Ocean.

EVALUATION: This is a science book intended for little children that may cause adults to laugh aloud, due to its cleverness and charm. Useful information is sandwiched between deliciously droll messages; for example, if one finds dinosaur bones, they must be dusted off and saved. The reading level is quite easy but the information is surprisingly detailed. Colorful pictures add to the author's humor by depicting a person drilling in several kinds of special suits through different types of rock.

□ □ □

Maestro, Betsy, and Giulio Maestro. *Through the Year with Harriet.* Illus. by Giulio Maestro. Crown, 1985. Unp. ISBN: LB 0-517-55613-8. Age levels: 0–7. Grade levels: PreK–2.

SUMMARY: The seasons roll by as Harriet, an elephant, awaits her next birthday. Month after month she is involved with seasonal activities, such as building a snow elephant, ice-skating, planting seeds, playing baseball, swimming, raking leaves, and throwing a football. Harriet also takes note of the changes that take place in nature during the year. In January, for instance, she sees that it is dark outside when she eats her dinner; however, in March, while washing the dinner dishes, she notices that it has become lighter outside. In June there is daylight until her bedtime.

There are also different seasonal temperatures. Harriet feels the cold winter air as she ice-skates, experiences the warmth of the spring sun on her back as she rides her bike, seeks the shade in order to read during the summer heat, and notices the changing colors of autumn leaves as the weather becomes chilly. Halloween and Christmas are celebrated before her favorite day, her birthday, arrives again in January.

EVALUATION: This picture book involves the senses as a complete year goes by. The illustrations are whimsical, such as Harriet's snake friend ice-skating on one skate and wearing a matching stocking and stocking hat. Food is frequently and vividly pictured and one can almost smell the steaming hot cocoa topped with whipped cream, or taste the ice cream that Harriet savors. Her fluffy winter slippers have tactile qualities. At play, Harriet and her friends are nearly audible.

Sensory experiences also extend to the seasons. Seeing the length of daylight change, feeling the difference in the air temperature, wearing appropriate clothing, and participating in a variety of seasonal activities, holiday events, and sports are some of the highlights in this concept book for young children.

Utilization of a birthday as the device to carry the text and illustrations through an entire year is a clever approach for this age group. The text is brief and clear, and the illustrations are bold and delightful.

□ □ □

Marshall, James. *Merry Christmas, Space Case*. Illus. by author. Dial, 1986. Unp. ISBN: LB 0-8037-0216-7. Age levels: 4–8. Grade levels: PreK–3.

SUMMARY: Buddy McGee's friend, the thing from outer space who visited on Halloween, has promised to return at Christmas. Since the McGee family plans to spend Christmas at Grannie's house, Buddy leaves directions for his friend to find him there. He tells his understanding grandmother, a scientist, about the thing, and she humors him. However, when he tells her neighbors, twin boys, about the forthcoming visit, they threaten to harm him if he is lying.

Meanwhile, zillions of miles away, the thing is enjoying a party when it suddenly remembers its promise. It zooms to Earth and lands on a log. Surrounded by alligators, the thing knows a mistake has been made and heads north. It encounters a few more surprises before it finds the note at Buddy's house and then flies to Grannie's. Just as the nasty twins are about to harm Buddy, the thing makes its appearance and transforms the bullies into snowmen. After being forced by the thing to apologize to Buddy, the humbled twins are returned to their former state. Grannie meets Buddy's unusual friend, and they discuss science. She wants it to take her for a ride, too, but only after Buddy has circled over the neighborhood and buzzed those mean twins.

EVALUATION: This is Marshall's sequel to *Space Case* (Dial, 1980), and it is full of action and humor. The interactions with the thing from outer space actually seem believable. The excitement of the Christmas holiday is doubled with the anticipated visit of a special friend from far away. The holiday spirit is dampened somewhat by the unbelieving and malicious twins, who are depicted with buckteeth and angry faces. The loving and sympathetic grandmother is portrayed in a nontraditional role that's pleasing and fun. The text is lively and mainly conversational, and the illustrations are colorful, amusing, and boldly executed. Young readers will find year-round pleasure in this book that takes place on a major holiday.

□ □ □

Mathews, Louise. *Bunches and Bunches of Bunnies*. Illus. by Jeni Bassett. Putnam, 1978. Unp. ISBN: 0-396-07601-7; pap. Scholastic 0-590-41880-7. Age levels: 5–8. Grade levels: K–3.

SUMMARY: This book illustrates multiplication concepts using rabbits in various predicaments. The multiplication facts presented include 1 × 1, 2 × 2, 3 × 3, and so on through 12 × 12. The situations are comical, and grow in humor as the number of rabbits increases.

EVALUATION: This book gives more weight to the enjoyment of learning than it does to the actual learning of the facts of operations. Younger children will enjoy watching the number of rabbits grow on the successive pages, and may even count the bunnies on each page. The illustrations clearly show what is happening, and each rabbit is appropriately represented by totaling the correct product on the corresponding page. Although not a school math text or drill for facts, this book presents multiplication in a pleasurable way, and gives the beginning math student a positive outlook toward multiplication and increasingly complex operations.

□ □ □

Mendoza, George. *The Marcel Marceau Counting Book*. Photos. Doubleday, 1971, o.p. Unp. Age levels: 3–5. Grade levels: PreK–K.

SUMMARY: This is a picture book that combines the humor of pantomime with practice in learning numbers 1 through 10. At the opening of the book, the reader is invited to count the ways Marcel Marceau can make believe. Marceau is pictured wearing his familiar white pants, striped shirt, and short gray jacket. He wears a different hat on each page, and he is seen miming activities associated with each hat.

"1 is a cowboy," reads the first page. We see Marceau wearing a cowboy hat, his legs in a gallop with one arm holding the imaginary reins of a horse, the other high overhead as if spinning a lasso. On the next page, the cowboy hat is in a see-through cube and this time Marceau is seen wearing a straw hat and appears to be digging. "2 is a farmer."

The numerals count up to 20, with each represented by a different hat and set of gestures to help the reader guess the number. On the last page, Marceau is wearing the battered top hat and red flower of his famous character, Mr. Bip.

EVALUATION: The book combines a child's desire for make believe with his ability to count. Both Marceau and the hats are white, black, gray and a touch of red against a flat beige surface for contrast. The clever child will want to guess the occupation Marceau is portraying before the answer is read to him. Most of the hats are well chosen; however, a few may be unrecognizable to the very young child for whom this book is intended. The cap of a rogue and the top hat of a gentlemen may need explanation, for example.

The numerals are clearly printed in five, narrow parallel lines. The hats are displayed so the child can count them and easily make the relationship between numeral and set of objects. This simple, colorful book is sure to be a hit with preschoolers. Their ability to read numerals and count and an interest in dressing up make this book a success.

□ □ □

Miller, Jane. *Seasons on the Farm.* Photos. Prentice-Hall, 1986. Unp. ISBN: LB 0-13-797275-X. Age levels: 4–7. Grade levels: PreK–2.

SUMMARY: This book is a photographic catalog of the activities taking place on a farm throughout the seasons. It opens on spring, featuring new beginnings with the birth of lambs, chicks and goslings hatching, and a young colt grazing near its mother. Flowers are blooming, and ducks seek new sources of food.

The summer scenes reveal a variety of farm machines at work. Grass is cut to make fodder for the animals to eat in the winter. The combine harvester cuts and threshes the wheat and barley that have ripened. A baler rolls up the straw for convenient storage. However, strawberries are picked by hand. This year a second set of piglets are born during the summer.

In autumn, the farmer picks apples and the trees begin to change color. There are also the births of many calves, and birds prepare for their winter journey to warmer climes. Horses, as well as machinery, are shown plowing autumn fields. At this time, farms also prepare for next year's growth.

When winter comes, some animals grow heavy coats to keep them warm. Others come inside to live in the barn. The farmer's hard work provides food for these cold months.

EVALUATION: Many colorful and very appealing photographs portray seasonal farm landscapes. The cyclical nature of seasons holds promises that young children can easily look forward to. The author-photographer invites the reader to participate in a counting activity when she suggests that baby pigs in a particular photograph are counted. Realism is enhanced when the author includes a bleak and muddy winter scene. The impressive machinery tells the reader that much heavy work is done on a farm. The sparse text is appropriate throughout and leaves the reader with the expectation that spring is around the corner.

□ □ □

Mori, Tuyosi. *Socrates and the Three Little Pigs.* Illus. by Mitsumasa Anno. Philomel, 1986. 44 pp. ISBN: 0-399-21310-4. Age levels: 9–17. Grade levels: 4–12.

SUMMARY: Socrates the wolf is a philosopher and naturally spends a lot of time thinking, which aggravates his fat wife, Xanthippe. She has no patience for thinking and is hungry most of the time. Although Socrates thinks about the three little happy pigs, he knows that he must think about Xanthippe's dinner. The philosopher gets an idea about how he can catch one of the pigs. However, the pigs live in five houses and Socrates begins his mathematical calculations on how he might capture one of the three pigs while it sleeps in one of the five houses. He begins by wondering where the first little pig could be; illustrations show the one pig in five different houses. This is repeated for the second pig, and other arrangements are added such as two pigs in one house. After the third pig is introduced, the possibilities become even more complex, so the author introduces a new way to look at the problem using a tree with limbs, branches, and twigs. This is done with the help of Socrates' friend Frog, who is a mathematician. The possibilities are described by a tree with five thick limbs, representing the five houses for the first pig; five branches at the end of each limb, which are the five houses for the second pig; and five twigs at the end of the branches representing five choices for houses for the third pig, so that the choices are five times five times five, or 125 possibilities. Following this is another representational illustration in which the pigs are actually placed inside the houses showing the 125 combinations.

Thus the reader is not only introduced to but involved in combinatorial analysis. The author explores combinations, permutations, and probabilities, showing readers how it is possible to make choices by using a mathematical system approach. A note to older readers and parents explains some of the mathematics used in the story, and suggests things readers could do on certain pages. Suggestions are also given to young children for working in a systematic mathematical manner. A brief biographical sketch of the author and illustrator follows.

In the end, Socrates convinces his wife that they shouldn't eat the nice little pigs anyway.

EVALUATION: Children and adults alike will enjoy these illustrations and text either to gain an understanding of the mathematical concepts presented or just for entertainment. This book contains very sophisticated mathematical ideas and the basis of many computer operations that are difficult to understand; however, with the help of the thoroughly well-thought out illustrations, even a younger reader can begin to comprehend.

□ □ □

Novak, Matt. *Claude and Sun*. Illus. by author. Bradbury, 1987. Unp. ISBN: 0-02-768151-3. Age levels: 0–6. Grade levels: PreK–1.

SUMMARY: Diminutive Claude, impressive in his beard and straw hat, spends a full day with his friend, Sun. They both tend the garden and have fun together. Sun helps Claude find berries in the woods and guides him through a hidden passage. A change in weather reveals a rain spot where there are many children under umbrellas. The children welcome Claude and Sun, and they all enjoy playing hide-and-seek until late in the day. Then the sun, through lengthening shadows, points the way home where Claude and Sun end the day and bid each other a good night.

EVALUATION: A day's sun cycle is accurately and playfully represented. The whimsical character of Claude, who looks like Van Gogh, creates interest for young children whereas the artwork appeals to all ages. The vivid illustrations could be used to acquaint readers with the pointillism style of painting. Opening sunflowers in the beginning of the book and closing sunflowers at the end effectively emphasize the day's cycle. On the next to the last two-page spread, the impression of horizon, created by a colorful quilt and the setting sun, evokes a feeling of relaxation as the book takes the reader to the day's end.

□ □ □

Offerman, Lynn. *The Truck That Drove All Night.* Illus. by Patti Boyd. Western, 1986, o.p. Unp. Age levels: 0–6. Grade levels: PreK–1. Series: First Little Golden Books.

SUMMARY: This book in rhyme reveals that when the sun goes down, Charlie and his truck Old Roller Mack wake up and get ready to go to work. While children and bunnies are sleeping, they drive all night through a rain storm. Just as the sun comes up, Charlie and Old Roller Mack arrive at a dock on the river. Next to the river is a little building where Charlie goes to have breakfast while Old Roller Mack takes a nap.

EVALUATION: This little book uses personification of animals—Charlie is a dog who owns a panda bear—and inanimate objects—Old Roller Mack is a truck whose headlights are eyes and whose grillwork is a smiling mouth—to tell the story. On Old Roller Mack's side and on Charlie's clothes there is a logo resembling a cupcake, so the reader can infer that bakery goods are inside but this is never stated. Actually there is minimal text, and what there is does not constitute a story line. However, it is fun to look at and little ones will like it. The main concept conveyed is that work goes on at night when many people are asleep.

□ □ □

Pearson, Susan. *My Favorite Time of Year.* Illus. by John Wallner. Harper & Row, 1988. Unp. ISBN: LB 0-06-024682-0. Age levels: 4–8. Grade levels: PreK–3.

SUMMARY: With her family, a young girl, Kelly, experiences the uniqueness of each passing season and finds that all seasons have special qualities. In the autumn, the leaves of maple trees are turning red, and the leaves of elms are becoming yellow. Preparing for cold weather, the neighbors cover their rose beds with straw. Geese, in V-formation, begin their migration, and the whirling wind sets the falling leaves into motion.

Winter arrives with a morning frost, and Kelly, clad in heavy clothing, can see her breath in the air. A heavy, gray sky brings a covering of snow, and the family enjoys winter activities. Much later the snow starts to melt and is seen only in patches here and there. As crocuses begin to blossom, spring showers also arrive. Budding trees burst forth, geese return, and the landscape turns a deep green. Summer appears abruptly. Light clothing is needed, vegetables are picked from the garden, and fireflies are caught and placed in jars to make lanterns. The family enjoys the beach, and at night Kelly hears the songs of the frogs and crickets. Throughout the seasons, Kelly and her family observe different weather patterns and participate in seasonal activities.

EVALUATION: The characteristics of each season are beautifully portrayed in this picture book. Special details, such as the migration and return of the geese and the colors of autumn leaves, describe seasonal highlights and the wonders of nature. The descriptive and lively text has a poetic quality. The colorful illustrations are lovely and are positioned in such a manner that readers might feel they are looking at a family album.

The aerial perspective of two illustrations and the accompanying text are exceptional. In one illustration Kelly and some family members are seen from the tree's point of view. The mother remarks that the tree must see them like baby birds. Another illustration shows Kelly's family from above, "like ants in a line," walking at the beach with their footprints behind them in the sand. The author's voice and illustrator's eye combine to produce a sensitive and informative book about the seasons.

□ □ □

Peters, Lisa Westberg. *The Sun, the Wind and the Rain.* Illus. by Ted Rand. Henry Holt, 1988. 32 pp. ISBN: 0-8050-0699-0. Age levels: 5–9. Grade levels: K–4.

SUMMARY: Elizabeth creates a mountain of wet sand on the beach. Elizabeth's sand mountain offers a basis for comparison to a real mountain and the way it has evolved. The book informs about the way in which temperature and the elements of nature contribute to mountain formation and change. For example, water creates streams in Elizabeth's miniature mountain in much the same way as a river gouges a valley in the great mountain seen in the distance behind her. After a rain, Elizabeth's mountain peak becomes rounder and flatter. In earth time, rocks turn to sand and jagged peaks become flat layers of sandstone. The end of the book suggests that the process is constantly repeated.

EVALUATION: The comparison of a small child's sand mountain to the formation of a real mountain is an unusual technique. Spectacular art enhances the scientific information and will be enjoyed by adults as well as children. Rand uses two-page spreads throughout, and usually pictures the real mountain on the left, but carries the image across the page to where Elizabeth and her sand mountain are shown as an insert. In this manner the images and continuity of land and seascape are not interrupted. One minor distraction is the emphasis on the effects of wind and rain, and there is little mention of the sun, even though *sun* is the first word of the title. The author's background and research in geology lend authenticity to this excellent book.

□ □ □

Prelutsky, Jack. ***It's Snowing! It's Snowing!*** **Illus. by Jeanne Titherington. Greenwillow, 1984. 48 pp. ISBN: LB 0-688-01513-1. Age levels: 6–10. Grade levels: 1–5.**

SUMMARY: This book contains 16 poems by the author, all pertaining to winter. The elements and effects of the weather, shorter daylight hours, and a winter shadow cast on the snow all show an awareness of the scientific aspects of winter. A child experiences the air as a silvery blur because it is snowing, finding not much time to play because the days are short; he feels the nipping winter winds, the sharp coldness of the air, and other noticeable differences between this season and others.

The poems range in length from one to four pages; however, there are as few as four lines on one page and the type is large. Illustrations are in black and white and vary in size, some encompassing an entire page. The last page includes short biographical sketches of both the author and illustrator.

EVALUATION: The author is a popular one with young readers, and his descriptions and vivid imagery in these poems would appeal to an audience, whether they are read to or if the poems were read alone. Through his writings, young people can develop an awareness of, or vicariously experience, many of the elements of the winter season. Also, the author has written some of these poems in the first person, so the reader can feel a part of the experience.

The illustrations are somewhat subdued, so as not to dominate the poetry. All titles are shown in blue ink, the only color used.

□ □ □

Radin, Ruth Yaffe. *A Winter Place.* Illus. by Mattie Lou O'Kelley. Little, Brown, 1982. 31 pp. ISBN: LB 0-316-73218-4; pap. 0-316-73219-2. Age levels: 4–10. Grade levels: PreK–5.

SUMMARY: On a winter afternoon, a family begins a trip to a special place a distance from their farm low in the hills. There is no snow on the ground as they follow the road past a burned-off mountain, a rushing river, and a town with filigreed buildings. As the family continues on their uphill walk, another town—with homes that lean against each other—is crossed. Soon only trees occupy the snow-covered winter wonderland that emerges. In a clearing, a frozen lake holds skaters and ice fishermen; the family joins in the winter recreation. Bonfires on the shore warm the winter enthusiasts. As the sun sinks in the sky, the family leaves the snowy hilltop and begins the familiar walk home. At night, while the family is settled in their cozy environment, snow is falling outside.

EVALUATION: This simple winter tale is idyllic and exquisite. The primitive, rich-toned paintings are lovely; panoramic scenes are captured with bright colors, and snowy woodlands are gorgeously depicted. The winter vistas reflect the colder temperatures as the family moves progressively higher in altitude during their walk into the hills. Even the houses lean against each other, as if huddled together for warmth. The text is lyrical, descriptive, and brief. Format and style combine to create the beautiful winter mood. This exceptional book is a "Reading Rainbow" selection that will enrich readers.

□ □ □

Rayner, Mary. *The Rain Cloud.* Illus. by author. Atheneum, 1980, o.p. Unp. Age levels: 3–6. Grade levels: PreK–1.

SUMMARY: On a warm May day, a rain cloud travels over the sea gathering moisture until it is ready to burst. When it reaches land, the cloud discovers that rain will disappoint many people. The children at the beach will not be able to complete their sand castles, the wash hanging on the line will have to be taken in, a baby sleeping outside in a carriage will be disturbed, and a picnic outing and a painter's work will be spoiled.

The considerate cloud attempts to find the right place to release its rain. It finally arrives over a tilled field where a farmer is pleased to see the dark cloud. Soon the farmland and countryside are drenched with rain, and the cloud at last feels comfortable. That year the farmer's wheat is the best ever.

EVALUATION: In this story, a very simple explanation for the formation of a rain cloud and the release of its moisture as rain is given. The use of a rain cloud as the main character in this book for very young children is uncommon, yet interesting and effective. Aerial views from the cloud's perspective provide sweeping panoramic vistas. In the scenes from the ground, there are components that carry the eye upward. This visual link in the illustrations keeps the rain cloud in the forefront. Within the text, the personification of the rain cloud allows its understanding and sympathetic nature to unfold. Although it darkens and looms overhead, it is not perceived as dangerous. In fact, the beneficial part of rain is skillfully noted at the end of this book. Watercolor illustrations subtly note the advent of precipitation.

□ □ □

Rich, Gibson. *Firegirl.* Illus. by Charlotte Purrington Farley. Feminist Press, 1972, o.p. Unp. Age levels: 6–8. Grade levels: 1–3. Series: Children's.

SUMMARY: Brenda, an eight-year-old girl, proves to her family, the firemen at the station, and herself that girls can be fire fighters too.

Brenda has always been interested in fire engines, and she tries to help at a fire down her street but gets in the way. She is told only boys can be firemen. This makes Brenda more determined than ever to learn all about fire safety and regulations. At school and on a class trip to the fire station, she holds on to her dream of becoming a fire fighter even against firemen's disbelief of a girl doing the job.

She shows her determination and saves a rabbit by climbing up the ladder to an attic window at a fire. Thus she saves the day; Brenda is awarded the title of "firegirl" at the station and is given special duties after school hours.

The insides of the front and back covers of this book display fire fighting equipment, labeled for the reader who wants to learn more.

EVALUATION: A feminist issue is dealt with in a sensitive way. Brenda's feelings are portrayed realistically, and many young readers will identify with her when she is told "that's for boys only." The treatment and discussion of a girl's viewpoint and feelings is well done.

Brenda shows courage, bravery, and determination even against adult opinions and derision. Unfortunately she also shows bad manners and stupidity in the way she conducts herself at the fire station. It is hoped that readers will not want to imitate Brenda as she hides on a fire engine as it races to a fire, or as she takes and uses equipment without permission. The author strives to show that girls can have the same goals and opportunities as boys, but then shows girls as impulsive risk takers.

The mostly black-and-white illustrations, which include touches of red here and there, and text go well together. They reinforce the text, as in the fire safety poster within the illustration of a classroom. An open-minded attitude is also shown in the multicultural and multiracial emphasis of the illustrations.

The text is easy to read and understand for the young reader as it deals with the familiar such as the home, feelings, and school. The text and illustrations are arranged in an appealing fashion.

Fire safety is taught in the story line and illustrations, and on the inside covers, making this book attractive to both boys and girls.

□ □ □

Rylant, Cynthia. *This Year's Garden*. Illus. by Mary Szilagyi. Bradbury, 1984. Unp. ISBN: 0-02-777970-X; pap. 0-689-71122-0. Age levels: 0–6. Grade levels: PreK–1.

SUMMARY: At the end of winter, a large family walks down to the garden to take stock. As the adults discuss what to plant in the spring, the children anticipate the fun they will have working in the garden. Soon, however, the children realize they must wait for warmer weather and gentle rains before they can garden. Finally, the day comes when they can work, and plant the seeds. As summer begins to yield a crop, animals and insects are described as they begin to inhabit the garden. Harvesting soon follows, and they begin preparing the food for canning. Finally, when the garden is empty, the family sits with the neighbors to compare notes on who had the best luck with their garden this year. Once again everyone anticipates spring.

EVALUATION: The seasons of the year are beautifully portrayed in both text and illustrations. In few words the author captures the language the adults or children might say or think. In some instances incomplete sentences are used. This in no way detracts from story and in fact reflects the way the character would think or speak. The entire story has a sense of family unity through the repeated use of the word *we* as well as scenes of everyone working or playing together. The illustrations are colorful, simple, and almost Impressionist in style. The frontispiece includes all the hand tools needed for working a garden. The connection between growth and death cycle of the seasons may not be understood by young children.

□ □ □

Shulevitz, Uri. *Dawn*. Illus. by author. Farrar, Straus, 1974. Unp. ISBN: 0-374-41689-3; pap. 0-317-59335-8. Age levels: 6–10. Grade levels: 1–5. Awards: International Board on Books for Young People, 1974.

SUMMARY: A grandfather and his grandson experience the beauty of dawn while camping outdoors. On this special trip together, they choose a spot under a tree and beside a lake to spend the night. Darkness and stillness envelop them as they sleep under their blankets.

Slowly a little light creeps into this peaceful and quiet scene, and a breeze causes ripples on the lake's surface. The sounds of a frog and bird are heard. More light breaks through the dark blues and grays of the night, and the campground begins to brighten with light green colors as the grandfather awakens his grandson. They break camp and set off in their rowboat to continue their trip. When they reach the middle of the lake, dawn's light suddenly covers the landscape in brilliant blues, greens, and yellows. A rosiness is added as the sun begins to appear in the sky. Together the grandfather and his grandson witness a colorful beginning of the day.

EVALUATION: This extraordinary picture book is a fantastic presentation of the splendor of dawn. The author-illustrator's watercolors are phenomenal in capturing a wondrous time of day. The natural setting is presented in exceptional color contrasts and shadings. Some interesting illustrations display vistas that are really reflections in the lake. The text is brief and has a poetic quality. Motion and sounds are subtly captured in the simple text.

The natural phenomenon of the rising sun heralds the beginning of a day that is bound to be memorable. The experience of viewing together the early light of the sun will bond the grandfather and his grandson together in a special way. Readers, too, will experience a spectacular visual treat.

□ □ □

Shulevitz, Uri. *Rain Rain Rivers*. Illus. by author. Farrar, Straus, 1969. Unp. ISBN: 0-374-36171-1; pap. 0-374-46195-3. Age levels: 3–6. Grade levels: PreK–1. Awards: American Library Association Notable Book, 1969.

SUMMARY: The theme of this book for young children is what happens when it rains. A young girl sits in her bedroom in the attic. She hears the rain on the roof and smiles; she can imagine all the fun she will have in the morning. She will sail boats into the large puddles and jump over the small puddles. Her friends will join her in playing barefoot in the squishy mud.

Meanwhile, the animals also enjoy the rain. Frogs stop croaking and splash and dive into the pond. Birds bathe and drink from the newly formed puddles and gutters flowing with water. The plants and grass receive moisture to help them grow.

As it pours, water flows from streams to brooks and down from the rocks to the rivers and lakes. Ocean waves bound and swell with water that the sun evaporates and then brings back to the earth in the form of rain once again. The reader discovers that all of nature benefits from the rain.

EVALUATION: The story begins and ends in the little girl's bedroom. Her memories and imagination are sparked by the sound of the rain on the windows of her bedroom. She imagines all the wonderful things the rain brings to her neighborhood and to the world. Rain is not a negative, fun-spoiling event, but rather a vehicle to enchance fun as well as benefit nature.

The illustrations are in blue and yellow paints that blend and bleed to produce green—the color of life. Black ink highlights important objects. Across each page, pale diagonal slashes designate the falling rain. The drops of rain grow from puddles in the city street to flowing streams and finally the huge waves of the ocean. Images of people protecting themselves from the rain with umbrellas contrast with frogs frolicking in the refreshing downpour.

The preschooler learns through verse and illustrations how rain brings benefits and pleasure to all living things, as well as sustaining and generating life. The young child will easily identify with the daydreaming in which the girl engages. Rain drops on her window as images of rainy day fun dance in her imagination.

Spier, Peter. *Peter Spier's Rain*. Illus. by author. Doubleday, 1982. Unp. ISBN: LB 0-385-15485-2; pap. 0-385-24105-4. Age levels: 0–11. Grade levels: PreK–6. Awards: American Library Association Notable Book, 1982.

SUMMARY: This detailed wordless picture book involves the reader in a series of activities experienced on a rainy day. The variety, size, and placement of illustrations carry the reader through a rain shower and a violent thunderstorm to the dawn of a calm, sunny morning. A warm family feeling pervades the story when mother assists the children in putting on rain gear to have fun outside. They splash in puddles, watch cars spray water in all directions, and observe brilliant raindrops sparkling on a spider web. When a dark, threatening storm approaches, the children run inside to the comfort of their cozy home. The storm continues through the night.

EVALUATION: The renowned author Peter Spier presents rain through sensational illustrations in this "Reading Rainbow" selection. On a two-page spread, raindrops hit the pavement in circular patterns that are emphasized through changes of light and shadow.

The juxtaposition of huge raindrops and tiny children's legs creates a humorous and effective focus on the rain itself. The children's activity continues throughout but is cleverly subordinated to the power of nature.

□ □ □

Sussman, Susan. *Hippo Thunder*. Illus. by John C. Wallner. Whitman, 1982, o.p. Unp. Age levels: 0–7. Grade levels: PreK–2.

SUMMARY: Thunderstorms are the subject of this small book for tots. A young child is frightened by the sound of thunder in the night. The family tells the child many different tales about the overhead noise; the young child's fears are not alleviated by the folklore that thunder is caused by angels bowling or snoring stars.

It is Daddy who explains that the distance of the approaching storm can be estimated by the lightning and thunder. He teaches his young child to count by saying "one hippopotamus, two hippopotamus . . . ," between a flash of lightning and the clap of thunder to determine how close the storm is.

Daddy counts hippopotamuses and provides reassurance until the storm arrives overhead; then they count together as the storm recedes. Finally, tired Dad convinces the child to tell him in the morning how many hippopotamuses he counted during the night. Sleep overcomes the child by the time the number 20 is reached.

EVALUATION: This entertaining book informs younger children about lightning, thunder, and calculating a storm's distance. During the thunderstorm, father and child count hippopotamuses between the flash of lightning and the crash of thunder in order to gauge the storm's distance. At the count of five hippopotamuses, the storm is approximately a mile away. The brief and clear text cleverly indicates that sound travels at about one mile in five seconds.

Helmeted hippopotamuses with spears cavort and create havoc in humorous and dramatic illustrations. This book not only portrays the comfort that a parent can provide a child who is scared by nature's forces, but also gives counting exercises while supplying some scientific information about the sound that thunder makes.

□ □ □

Szilagyi, Mary. *Thunderstorm*. Illus. by author. Bradbury, 1985. Unp. ISBN: 0-02-788580-1. Age levels: 3–6. Grade levels: PreK–1.

SUMMARY: This book is about a storm and what happens before and after one occurs. It starts with a little girl playing peacefully in the summer sun. Her dog is sleeping nearby as she moves sand from her sandbox to her wagon. The stillness of the day is broken by the rustling of tree leaves and the warning calls of a bird. The little girl, sensitive to sounds of a coming storm, pauses, listens, and then hears the crackle of thunder.

She drops her shovel and runs into the house with her dog to seek comfort from her mother. They watch the storm together from the window, safely inside the house. The thunder and lightning subsides, and the little girl and her mother read stories while a heavy rain falls outside.

After the storm, the little girl and her dog run outside to once again play in the summer sun.

EVALUATION: The author, who lives in Ohio and is well aware of the fear and destruction a sudden storm can bring, manages to convey the quiet before the storm. Animals signal the coming storm with their calls. The wind—one moment still, the next stirring the trees—tells the little girl to beware. She runs into the house where her mother provides the reassurance she needs. After she is comforted, she gives a comforting hug to her cowering dog in a manner characteristic of a girl this age.

The book is illustrated in bright pastels, which have the grainy, streaky appearance of crayons. This style enhances its relevance to the primary-age group. Large tears well up in the sad eyes of the little girl as she becomes frightened. The dog attempts to hide under the couch, but his hind legs and tail stick out. The bolts of lightning brighten the sky with flashes of light and huge drops of rain splash onto objects scattered at the first sound of the storm: all these are visual symbols with which a young child can easily identify.

This is beautifully written and illustrated book, just right for the preschooler. The young child will love being told of the solitary play of the little girl, her fear of the storm, and the protective feeling she has for something smaller than her, the dog. The reader will find comfort in the shared experience that security comes from Mother, who is nearby to reassure that after the dark, noisy storm, the sun comes out again.

□ □ □

Thurber, James. *Many Moons*. Illus. by Louis Slobodkin. Harcourt, 1971. Unp. ISBN: 0-15-251873-8; pap. 0-15-656980-9. Age levels: 8–11. Grade levels: 3–6. Awards: Caldecott Medal, 1944.

SUMMARY: Princess Lenore is sick and it will take the moon to make her well again. The Lord High Chamberlain, who is first given the responsibility of bringing the moon to the palace, informs the princess and her father, the king, that it is made of copper and is located 35,000 miles away, making it impossible to get. The Royal Wizard is sent for next; he replies that although he can perform magic of all kinds, he simply can't bring the princess the moon, which is twice as big as the palace, made of green cheese, and 150,000 miles away. The Royal Mathematician is next because he knows many scientific facts, but he reports that the moon is 300,000 miles away and insists that it is pasted on the sky and therefore immovable. Only the Court Jester can help now. He asks the princess what she thinks the moon is made of and when she says gold, he has the Royal Goldsmith make a gold moon for her to wear on a chain around her neck. Her explanation of why the real moon still shines in the sky is full of childlike logic and imagination.

EVALUATION: Thurber's tongue-in-cheek humor is full of imaginative wit and wisdom in this tale about the elusive moon and how we describe its characteristics. The illustrations in pen and ink with color wash are dainty and somewhat nebulous. This suits the charming story, which implicitly suggests that the reader "think into" the plot his or her own wildly exaggerated no-tions about how these characters think of the moon. Thurber's special inventive language imprint is all over this story; actually the words are fun for readers of all ages. The book is a classic and will never be outdated. (*Many Moons* would be a natural for dramatization.)

□ □ □

Tresselt, Alvin. *Autumn Harvest.* Illus. by Roger Duvoisin. Lothrop, 1951. Unp. ISBN: LB 0-688-51155-4. Age levels: 4–8. Grade levels: PreK–3.

SUMMARY: In this picture book the child reads of the seasonal changes in the weather from late summer to autumn. Farmers know how to listen to the animals to tell when the first frost will come: katydids chirp, birds chatter, while squirrels and chipmunks silently gather nuts for winter storage.

Children walk home from school, crunching through the dry leaves, and find milkweed pods to split open. Once home, they see the reaper, thresher, and tractor in the farmer's fields harvesting wheat and gathering apples. High overhead they hear geese honk as they fly south in a V formation. Short autumn days give way to a cold, crisp Halloween night and finally to the warm gathering of a large family Thanksgiving.

EVALUATION: The book carries a 1951 copyright; however, all the information is current. The song of the katydids still heralds the coming of autumn, along with the changing colors of the leaves and the southern flight of the birds. In today's society, a child living in the city may miss such seasonal cues as the changing colors of the leaves, busy squirrels, fields of goldenrod and mums. But even the urban child will be aware of the cooler nights, shorter days, and frost on the ground.

The author has written in beautiful prose, yet each sentence has been chosen to instruct as well as entertain. The text is found on the left page and bordered with the fruit or foliage about which Tresselt has written. The right page bears a more complete pictorial description of the written information—katydids, threshing machines, and seed-eating rodents and birds reinforce the information the author gives in the text.

The author has written a restful, almost poetic story while the illustrator has disturbed the mood with hard-edged pictures painted with fiery reds, electric blues, and neon oranges. The illustrations—rather crude pen and ink drawings with a garish paint wash—do not complement the text.

□ □ □

Williams, Jay, and Raymond Abrashkin. *Danny Dunn and the Smallifying Machine.* Illus. by Paul Sagsoorian. Simon & Schuster, 1977, o.p. 117 pp. Age levels: 9–12. Grade levels: 4–7.

SUMMARY: In this story, Danny Dunn and his friends Joe and Irene set out on a new scientific adventure. The three investigate a barn where they suspect that Professor Bullfinch's newest experiment is under surveillance by spies. Danny and his friends find evidence that some kind of spaceship is involved. They find a platform. When they step on it and push a button the gravity changes; they spin around and tumble to the ground. The professor is with them and when they see a gigantic daddy longlegs, he explains that the spider is normal in size but that all four humans have shrunk. Changing back to their normal size proves quite complicated. Eventually they create a cobweb parachute to facilitate their escape, and they discover that the "spy" who has been trying to learn the secret of the smallifying machine is really a legitimate government representative.

This Danny Dunn adventure is one in a group of books that have been popular for several years.

EVALUATION: This clever science fantasy succeeds in convincing middle-grade readers to believe its implausible story because it is so realistically written. For the first few chapters, the people and events are predictable and even ordinary. The element of fantasy creeps in gradually, and when the transition to fantasy occurs, the reader is ready to accept it. The authors include a great deal of scientific information, weaving it into the plot in a surprisingly uncontrived manner. Many readers will identify with the curiosity Danny and his friends have about scientific phenomena.

□ □ □

Wyndham, Lee. *The Winter Child.* Illus. by Yaroslava. Parents Magazine Press, 1970, o.p. Unp. Age levels: 3–6. Grade levels: PreK–1.

SUMMARY: Maria and Sergei are a kind, hardworking Russian peasant couple. Maria is an excellent housekeeper and cook, and the house is always filled with the aroma of fresh-baked goods. Sergei takes excellent care of his fine pigs, hens, and ducks and his cow gives the richest milk around. It is a happy home except that they have no child to share their happiness. As this folktale unfolds, the couple is rewarded for their kindness and hard work—they fashion a snow daughter who comes to life. They are radiantly happy together for the entire long, Russian winter until the snow daughter melts in the warm spring sun. The next year they make *Snegurochka* (snow daughter) again and once again she comes to life.

EVALUATION: This wonderful old tale will remind the reader of *Frosty the Snowman.* The author gives the readers clues as to the young girl's fragility and the clever child should guess the cause of her demise. The snow daughter refuses hot foods and drinks, and chooses to sleep outdoors. She protects herself from the strong rays of the sun, yet dances among the snowflakes when it snows. Maria and Sergei adapt to her needs and prepare cold foods and build a porch outside in which she can sleep and play.

The rich text is sprinkled with Russian vocabulary (printed in italic type). Because it is used contextually, it is easy for the reader to find the meaning. *The Winter Child* is sure to aid young children in their understanding of the change of seasons. The crisp, clear illustrations add much to the cold, wintery feeling. A paint-on-glass technique was used by the illustrator with great success, giving a folk look to this tale.

□ □ □

Zolotow, Charlotte. *Summer Is . . .* Illus. by Ruth Lercher Bornstein. Crowell, 1967. Unp. ISBN: LB 0-690-04304-X. Age levels: 3–5. Grade levels: PreK–K.

SUMMARY: Many of the sights and sounds of the four seasons are recounted in this picture book. For example, "Summer is" singing birds, flowers, and lemonade, and the sound of lawn mowers on a hot afternoon. "Fall is" blowing leaves, wearing heavy clothes, and selling fruits and vegetables on roadside stands.

In winter the sound of snow shovels scraping on the driveway can be heard, and the days are short and the nights are long. Logs burning in fireplaces mixed with the scent of freshly cut pine can be smelled. The winter sky is pink and the moon comes up early. In spring, the sight and smell of hyacinths, crocuses, and tulips abound. Pussy willows and forsythia bloom, saying that winter is over and it is now time for nature to waken.

EVALUATION: On each page, a matronly, warm figure and spirited preschool-age child enjoy the splendors of each season. Indeed, the adult seems to be enjoying these experiences as if for the first time through the eyes and emotions of her young companion. The household pets also frolic among dry leaves or stick their noses into newly sprouted spring flowers.

The author reduces the events of each season to its simplest symbol in order to maintain the attention of the preschooler. Bornstein's illustrations also add much to the poetic quality of the text. They are soft, sherbety pastels. The adult figure is that of a woman but she seems more mature than the child's mother; perhaps she represents the child's grandmother or babysitter. The sex of the child cannot clearly be determined, which further adds to the universality of the characters.

This lovely book will delight and educate the preschooler. Each page bears pictorial examples of the highlights of each season. One or two sentences on each page are just enough to hold the youngster's attention to the end. The seasons are introduced in sequence (summer is first) so as not to cause confusion. The book completes the cycle with spring, in which the Earth warms in answer its promise of renewal.

□ □ □

Zolotow, Charlotte. *When the Wind Stops.* Illus. by Howard Knotts. Harper & Row, 1987. 32 pp. ISBN: LB 0-06-026972-3. Age levels: 0–7. Grade levels: PreK–2.

SUMMARY: As a little boy approaches his bedtime, he notices that the sky is getting darker and the sun is fading away. Sorry to see the day end, he recalls some of the things he did that day. When his mother comes to tuck him in, he begins to ask many questions. He queries his mother as to why the day must end, where the sun goes, where the wind goes when it stops, and where clouds go when they go across the sky. In her answers, the mother gives simple scientific answers, which help the boy to conclude in the end that nothing really ends, but everything goes on. His mother reassures him by telling him that although this day has ended, the moon that he now sees will begin night somewhere else and he can look forward to a new day.

The black-and-white illustrations dominate this picture book and depict the text. One to seven lines of text are on each page and the print is rather large.

EVALUATION: Charlotte Zolotow is a popular and skillful author for children. In this book, she holds the interest of the reader through the questions of a child—to which the young reader can relate—and the answers of the mother, which allow the child to think, reason, and deduce answers independently. With sensitive words—and without saying "Yes, you're right"—the young boy/reader is assured that the deduction was indeed correct. Although written in prose, the text has a poetic quality through the flow of words, cadence, and descriptive choices.

The book itself may reassure the young reader who comes to the end of a good day and wonders if another one will occur. This may be a particularly appropriate book to use at bedtime or at the end of the school day in the classroom. This book could be a catalyst for finding out more factual information about the sun, moon, weather, waves, and other topics mentioned. The illustrations are quiet and restful, giving the reader the sense that it is the end of the day.

Science: Nonfiction

Did dinosaurs exist? What is a black hole? Why do volcanoes erupt? Is there life on other planets? How far away is the sun? Answers to these and many other questions that children ask about the world of science can be found in this section. Books about electricity, molecular structure, solar energy, maps and globes, and geometry supply readers with simple explanations and interesting facts. In order to understand the scientific method and stimulate direct practical experiences, some books feature experiments to be undertaken by children. Biographies of famous scientists, such as Madame Curie and Leonardo da Vinci, are also provided.

□ □ □

Adler, David. *Roman Numerals*. Illus. by Byron Barton. Crowell, 1977. 33 pp. ISBN: LB 0-690-01302-7. Age levels: 6–9. Grade levels: 1–4. Series: Young Math.

SUMMARY: Adler tackles the topic of Roman numerals for young children. Most people do not recognize Egyptian or Babylonian numbers when they see them because they are no longer used. However, Roman numerals are still used today on some buildings to indicate when the buildings were completed; occasionally they are used in books to denote chapters. Children will realize that some clocks use them. The symbols for one, five, and ten are shown, and it is explained that combinations in the sequencing of these symbols can denote other numbers. When Roman numerals are written with the largest value first addition is indicated, but when a smaller-valued symbol is before a larger-valued symbol subtraction is intended. No more than three of the same symbols can be used in a row.

EVALUATION: A delight to read, *Roman Numerals* presents ways Roman numerals are used in our society and explains a logical way of interpreting them. The project of clipping cardboard squares of different sizes to match the size of Roman numerals is an excellent way to get children involved in the analysis and comprehension of Roman numerals. The illustrations are nice, but not necessary to the understanding of text since the text is self-explanatory. For students interested in Roman numerals, or needing extra help with them, this is an excellent resource in development of concepts. Because it asks many questions without providing the corresponding answers, ideally it should be read and worked through with parental or teacher guidance.

□ □ □

Alexander, Alison, and Susie Bower. *Science Magic*. Illus. by Carolyn Scrace. Simon & Schuster, 1987. 46 pp. ISBN: 0-13-795311-9; pap. 0-13-795022-5. Age levels: 5–9. Grade levels: K–4.

SUMMARY: The simple experiments in this book can be done using everyday objects found around the house. Some of the experiments include making a fountain, blowing up a balloon without blowing into it, designing a kaleidoscope and a periscope, turning a glass filled with water upside down, and making simple musical instruments. The directions are clear, and the illustrations are presented in such a way that the child experimenter won't try to perform the steps in the experiments out of order. At the corners of pages, set off by a contrasting color, the reader finds scientific principles explaining what is taking place when a given experiment works.

EVALUATION: This colorful activity book will delight children and stimulate their inquiring minds. Many of the experiments are messy (parents need to realize this and be prepared in advance), all of them are fun, and the authors have made generally good choices for their intended age range of children from ages four to eight. However, that estimation excludes nine-year-old fourth graders, who would love it. In addition, age four is a little young to begin the experiments, although it is suggested that a parent help. The very first experiment involves using hot water—hot *tap* water might have been specified in the interest of safety. (Although the introduction says boiling water is never used, this information needs to go on the page with the experiment as well.)

□ □ □

Aliki. *Digging Up Dinosaurs*. Illus. by author. Crowell, 1981. 34 pp. ISBN: LB 0-690-04099-7; pap. 0-06-445016-7. Age levels: 4–8. Grade levels: PreK–3. Series: Let's-Read-and-Find-Out Science.

SUMMARY: A museum visit begins this book, which contains information about dinosaurs and fossils. There are labeled dinosaur skeletons, models of various kinds, and visitors who comment on specific details about dinosaurs. The main character is a little girl who guides the reader page by page through the museum. Knowledge about dinosaurs, such as when they lived and what they ate, became available about 200 years ago. New information continues to be found and catalogued by experts: geologists, paleontologists, draftsmen, photographers, excavators, and museum specialists. A one-page data sheet about the iguanodon reveals details gathered from many finds throughout the world. Fiberglass skeletons of dinosaurs are made using molds from actual bones so that many museums can have dinosaur displays.

EVALUATION: *Digging Up Dinosaurs,* a "Reading Rainbow" selection, is full of pertinent information about dinosaurs and fossils presented in clever ways. Details are imparted in text, on museum labels, and through cartoon balloons above the cast of characters in this book. Black-and-white illustrations and the use of earth tones complement and enhance the topic without distracting from the text. The text is simple and clear, and reads like a basic reference for children. The step-by-step process of fossil study reveals many facts and also shows a fascinating scientific occupation.

□ □ □

Aliki. *Dinosaur Bones.* Illus. by author. Crowell, 1988. 32 pp. ISBN: LB 0-690-04550-6. Age levels: 4–8. Grade levels: PreK–3. Series: Let's-Read-and-Find-Out Science.

SUMMARY: In this Let's-Read-and-Find-Out Science book, the reader discovers how scientists study fossils and bones to determine how many millions of years ago the dinosaurs lived. Using rocks and rock layers to tell the age of fossils and bones is explained through the text with the help of charts and diagrams of the eras and periods of rock.

Scientists do not know many things about dinosaurs such as their color, the noises they made, or whether they were warm- or cold-blooded creatures, but they can follow the evolution of dinosaurs through the Triassic, Jurassic, and Cretaceous periods. At the back of the book is a chart of the known dinosaurs, indicating where they were found and where they lived.

The author also points out additional sources of information about dinosaurs.

EVALUATION: An enormous amount of information is presented here as simply as possible. The text and illustrations work together to make the information sensible to the young reader.

The artistry of the author is clear in his ability to keep readers enthralled not only with dinosaurs, but with the science of studying ancient bones and fossils as well.

Children will require a few readings to absorb all the difficult terms and names, but the information is presented in a manner that is commendable. The author firmly believes in giving children correct information in a matter-of-fact manner, and in using the proper names and terms instead of watering down the information.

Readers will enjoy the colorful illustrations as well as the informative text. Anyone interested in learning about dinosaurs should be sure to check out Aliki's books in this series.

□ □ □

Aliki. *Dinosaurs Are Different.* Illus. by author. Crowell, 1985. 32 pp. ISBN: LB 0-690-04458-5; pap. 0-06-445056-2. Age levels: 4–8. Grade levels: PreK–3. Series: Let's-Read-and-Find-Out Science.

SUMMARY: How scientists divide the many types of dinosaurs into different orders, according to their special characteristics, is described in this book. The dinosaurs are classified by lizard or bird hip structure and plant- or meat-eating habits. Once the various orders and suborders of dinosaurs are explained, each dinosaur is shown with its classification given at the top of the page.

Narrated by a young boy, this book carefully explains the different characteristics of various dinosaurs so the reader will understand how the classification system works. Included at the end of the book is a herbivore/carnivore chart of the orders of dinosaurs.

EVALUATION: Dinosaur buffs will love this information-packed book; they can enjoy the wonderfully detailed illustrations as much as the easy-to-read text. Even the children pictured in the illustrations show their fascination with dinosaurs in their curious faces.

Names of dinosaurs are given in italic type, highlighting the correct name with the description given. The text is clear but bombards the reader with big words and labels for dinosaurs and vast information about them, thereby making the illustrations and diagrams more crucial as they simplify, organize, and clarify the information.

The author's understanding of her audience is apparent in her illustrations, which portray children as they really are: egocentric, vivacious, energetic, and curious—even when they are on a trip to the museum. In doing so the author promotes an impression of museums that is more interesting than the stuffy, old buildings many children perceive them to be.

□ □ □

Aliki. *My Visit to the Dinosaurs*. Illus. by author. Crowell, 1985. 32 pp. ISBN: LB 0-690-04423-2; pap. 0-06-445020-1. Age levels: 4–8. Grade levels: PreK–3. Series: Let's-Read-and-Find-Out Science.

SUMMARY: In this book, readers take a tour through a museum of natural history. A little boy narrates the story and is obviously fascinated with the topic of dinosaurs and his visit to the museum. Readers are introduced to 14 kinds of dinosaurs, learning something about their physical characteristics, habits, and habitats. After an introduction to Apatosaurus, readers get a glimpse of the work of archaeologists and paleontologists.

The book begins and ends with skeletal figures of dinosaurs, but the bulk of the book contains illustrations of completely formed dinosaurs. A page or two is devoted to each kind of dinosaur, with a brief description.

The climax of the book is the meeting of Tyrannosaurus rex, the fiercest dinosaur of them all.

EVALUATION: Just paging through the book will capture the attention of any young reader who is even remotely interested in dinosaurs. A lot of information is packed into this appealing and readable little book; sentences are generally short and to the point.

Apart from the speculation about what Ornithomimus may have eaten, information is presented as fact. The detailed illustrations are especially helpful in making these long-extinct creatures come alive. The full color drawings are spectacular in detail and loaded with information, reinforcing the text with signs, labels, diagrams, and charts as they are used as captions of the museum displays. One could learn just as much about dinosaurs reading only the pictures. Either way, there is plenty to see just as in a real visit to a museum.

□ □ □

Althouse, Rosemary, and Cecil Main. *Air*. Photos. Teachers College Press, 1975, o.p. 30 pp. Age levels: 3–5. Grade levels: PreK–K. Series: Science Experiences for Young Children.

SUMMARY: Properties of air explained in simple terms for young children form the focus of this book. To help children understand the concept that air takes up space and can be found anywhere, six activities are suggested. Activity 1: Using plastic bags, children compare them when they are empty and when they contain things. When air is trapped in a bag it is bulgy just like the bags with things in them. Even though it cannot be seen or felt, the air is in the bag. The odors children may say they smell are *in* the air but they are told that pure air has no smell. Activity 2: Children learn that air is real by using air from a plastic bag to blow into a toy party horn. Activity 3: Each child receives a balloon to play with outside. Children discover that air was put into the balloon and when it burst, the air inside the balloon went into the air outside. Activity 4: Children experiment with blowing bubbles, discovering that the more air that is blown into a bubble, the bigger the bubble becomes. When the bubbles pop, air is removed from inside. Activity 5: Plastic glasses pushed straight down underwater do not fill up with water until tilted because air takes up space and prevents water from getting in. Activity 6: By immersing different objects under water, children can observe which objects produce the most bubbles, signifying the presence of air.

EVALUATION: Very practically written for pre-kindergarten and kindergarten children, *Air* details how to develop basic concepts about air. Each activity has a guided, hands-on approach listing the concept to be developed as well as the activity, materials, procedures, and helpful suggestions. Parents or teachers might prepare themselves by reading the activity before conducting it with the children. While this is written for presentation to young children, it is not illustrated or written to be read by them; it prepares the instructor to help children learn as the activities are performed.

□ □ □

Anno, Masaichiro, and Mitsumasa Anno. *Anno's Mysterious Multiplying Jar*. Illus. by Mitsumasa Anno. Philomel, 1983. Unp. ISBN: 0-399-20951-4. Age levels: 3–8. Grade levels: PreK–3.

SUMMARY: The story begins as an imaginative fairy tale. A lid is taken off a jar; water is found inside. The reader is asked to imagine that the water is the sea. On the sea is 1 island. On the island are 2 countries. On each country are 3 mountains, and so on. Eventually the reader ends up with 10 ginger jars that are within 9 boxes. The authors explain that not only has the reader counted to 10 but to 3,628,800. In order to show how this occurred, dots are used on the pages that follow this statement, and the mathematical property of factorials is revealed. The story is reviewed and the reader is able to visualize the dots as they are multiplied. A few simple examples are included so that the reader can experiment on his own.

EVALUATION: What begins as a simple lap book ends with a rather sophisticated concept. A young child will enjoy the pictures and be able to count farm animals, birds, and people on some pages. The terms *more than* and *less than* are taught as well as the principles of addition and subtraction. The interesting illustrations of flags, towers, roofs, trees, and so on also provide the opportunity to discuss *the same as* and *equal.*

The older child is sure to be intrigued with factorials. A knowledge of multiplication is necessary in order to understand the concept. Although a few concrete examples are given, a clever child could think of many more. This kind of literature is a useful addition to the "hands-on" approach to math currently utilized in most math programs.

The authors draw readers into the story by asking them to use their imagination and pretend the water in the jar is the sea. Later the authors depart from the tale when they teach a lesson in factorials. The illustrations are breathtaking. The ink and water color paintings are so detailed that one needs to look long and hard in order to take in each object. The reader will sight some objects, people, and animals repeated in illustrations from previous pages. All in all this is a great book for the preschooler who is learning to count. However, this book can also be enjoyed by the third- or fourth-grader who has a knowledge of multiplication.

□ □ □

Arnold, Caroline. *Charts and Graphs: Fun, Facts, and Activities.* Illus. by Penny Carter. Watts, 1984. 32 pp. ISBN: LB 0-531-04719-9. Age levels: 7–11. Grade levels: 2–6. Series: Easy-Read Geography.

SUMMARY: After noting the use and availability of charts and graphs in daily life, the author gives directions for making pictographs, bar graphs, and pie charts, a family tree, and other mathematical projects. It is explained that bar graphs aid in comparing things, such as the relative sizes of major cities in the world or the means of transportation used by children to travel to school. Some mathematical information is furnished in pictographs that reveal the monthly rainfall in southern California and the daily count of lunch boxes brought to school during a week. Mileage charts and time lines are some of the other diagrammatic representations that are clarified. The names of living family members and ancestors are organized in the form of a family tree. Activities are included for all charts and graphs that are included in the book. Directions are accompanied with a list of materials needed for each activity. Along with an index and table of contents, there is a glossary with 13 entries.

EVALUATION: The book's format is appealing and organized not only to impart information about charts and graphs but also to stimulate hands-on activities. Colorful illustrations, photographs, and diagrams are numerous and enlarge the text. The activities, with simple and clear directions, are practical, fun, and easy to do. Concepts and skills are learned that are useful in social studies as well as in mathematics. Older students will find this book handy for quick and easy reference purposes.

□ □ □

Arnov, Boris. *Water: Experiments to Understand It.* Illus. by Giulio Maestro. Lothrop, 1980, o.p. 63 pp. Age levels: 9–12. Grade levels: 4–7.

SUMMARY: Arnov begins his description of what water is and how it works by asserting that people generally take water for granted. He suggests that by getting to know and understand more about water, people may develop a greater appreciation for it. This book contains many experiments that will help students become aware of the properties of water. The properties discovered in this way are then given practical applications in the relevant sections at the end of each chapter. Some of these properties include density, expansion, overturn, evaporation, heat storage, pressure, buoyancy, how water interacts with gravity, how it dissolves plant and other substances, and how it interacts with light.

EVALUATION: This is an excellent book for experimenting with water. Arnov has arranged each chapter in orderly structure. Each chapter begins with an idea or premise; then the experiment is broken into categories of materials, procedures, and conclusions. Finally the relevance of this property of water to everyday life is explored. Its step-by-step approach is easy to follow and the language is understandable. The reader is brought to an awareness of the many properties of water and thus this book develops a greater appreciation for someone too many people take for granted. The illustrations play an integral part in helping to clarify the experiments and relevance of the properties of water. The book is filled with an appropriate amount of scientific detail.

□ □ □

Arvetis, Chris, and Carole Palmer. *Why Is It Dark?* Illus. by James Buckley. Rand McNally, 1984. Unp. ISBN: 0-317-65112-9. Age levels: 0–8. Grade levels: PreK–3. Series: Just Ask.

SUMMARY: In this book, the authors explain the concepts of day and night, the Earth as a planet—its orbit, revolution, and rotation—and the moon's orbit and reflection of the sun. Presented in a way that very young children can understand, animals—with a wise owl as the teacher—act out the concepts using a ball for the earth, a smaller ball for the moon, and a flashlight for the sun. The bright illustrations almost appear animated as the animals follow the owl's directions and finally realize and understand her explanations.

EVALUATION: The book is a combination of *Bambi*like illustrations and words in cartoon rectangular balloons. At first glance it appears that a child could read the book alone; however, some difficult nonscientific words are interspersed throughout, such as *fascinating, imagine, disappear,* and *interesting*. Some scientific words such as *rotate, orbit,* and *reflects* may also be difficult for young children to read. With adult assistance, however, the text can be clarified and the scientific explanations are very simply and clearly presented. Bold type is used for important concepts, such as *planet, orbit, rotate,* and *reflect*.

The concepts are told and then role-played by the animals. For example, when the owl is discussing the Earth's orbit around the sun, she has the animals use a ball for the earth and a flashlight for the sun: a large path is made by walking the ball around the sun. It is explained that it takes 365 days for the earth to accomplish this task. For reinforcement, other animals skip around the same path, repeating or adding ideas to the same concept. The authors use this teaching technique for the remainder of the concepts as well. In two instances an inset illustration is used for further clarification.

The illustrations, showing the animals' happy and positive attitudes about learning, may help to draw and maintain the readers' attention and possibly make them eager to know more about the concepts discussed. There are several books in the Just Ask series, including some about rain, rainbows, and volcanos.

□ □ □

Asimov, Isaac. *How Did We Find Out about Black Holes?* Illus. by David Wool. Walker, 1978. 64 pp. ISBN: LB 0-8027-6336-5. Age levels: 10–17. Grade levels: 5–12.

SUMMARY: Asimov's book explains black holes and how they were revealed to the scientific world. Friedrich Bessel, a German astronomer living in the 1800s, discovered—with the use of a primitive telescope—a star that couldn't be seen with the naked eye. While charting the paths of the stars he noted that certain stars appeared to revolve around an entity much as the earth revolves around its sun. Since Bessel's telescope yielded inexact information, he assumed that some of the objects he observed were dead stars. Years later, data obtained using more sophisticated telescopes revealed that the "dead" stars were small and hot; they were called white dwarfs because of their size and white-hot appearance. By this time scientists had determined that the hotter a star is, the more quickly it burns to extinction. These and other facts about stars, neutron stars, pulsars, and supernovas, and the gravitational pull affecting the escape velocity of stars and the tidal effects that cause what are referred to as black holes, are discussed in considerable detail.

EVALUATION: Excellently written and presented, *How Did We Find Out about Black Holes?* provides considerable detail in a way that the reader can easily comprehend. The language is simple. Terms and concepts are explained clearly, with the pronunciation of technical terms given in parentheses. There are few pictures in the book and most do not add greatly to the text, as the text seems self-explanatory. Overall, the book presents fascinating information, such as its explanation of the tidal effect, in an enlightening way. Readers should enjoy learning of these discoveries as much as if they were the scientist making these discoveries for the first time.

□ □ □

Asimov, Isaac. *Realm of Numbers*. Illus. by Robert Belmore. Fawcett, 1982. 202 pp. ISBN: pap. 0-449-24399-0. Age levels: 10–15. Grade levels: 5–10.

SUMMARY: The use of numbers is necessary in all aspects of life, from simple mathematics to complicated scientific equations. Many people are put off by mathematics because it looks too complicated or uninteresting. Asimov begins this book with very basic uses of numbers by counting on fingers, and discusses possible steps primitives took to increase number count when the fingers ran out. The abacus and Roman numerals are discussed as solutions to increased numerals. The hows and whys of mathematical procedures are discussed: multiplication, division, fractions, decimals, negative and positive numbers, and to infinity. Asimov shows how they are integrated for use in clocks, weights, distance, and direction. There is no end to what numbers can do, or what can be done with numbers.

EVALUATION: This prolific writer achieves his goal of explaining mathematics with simple examples, bits of humor, and many references to history. Points are clearly explained, giving many examples and correlations to other aspects of life. The meaning and origin of terms are explained. Though this text is 30 years old, the basic foundation of mathematics remains the same.

Therefore, this book should still be considered a source for understanding numbers. An index is included.

□ □ □

Asimov, Isaac. *Stars*. Illus. by Herb Herrick and Mike Gordon. Follett, 1968, o.p. 32 pp. Age levels: 7–11. Grade levels: 2–6. Series: Beginning Science.

SUMMARY: This book flows from the simple concept of the sun to other stars, constellatoins, and the First Magnitude (the 20 brightest stars), continuing to our galaxy and finally the universe. The author then discusses stars and constellations in the Northern and Southern Hemispheres, and why the position of the stars in the sky appears to change during the seasons. Mention is made of how stars have been viewed by astronomers throughout the centuries and the characteristics that they noted. Asimov ends by telling how star distances are measured from the Earth. A vocabulary list is included with a page citation indicating where the word first appears. A list of several activities for children to do to understand more about this topic is included.

EVALUATION: This book is authored by a notable science writer; it has a large, easy-to-read, informative text that defines many terms used in astronomy. The dot-to-dot illustrations of the constellations might help the stargazer find the different star shapes in the night sky. The star maps provide a broader view of the constellations. Asimov facilitates learning by proceeding from simple to more complex concepts in an orderly fashion.

□ □ □

Asimov, Isaac. *What Makes the Sun Shine?* Illus. by Marc Brown. Little, Brown, 1971, o.p. 57 pp. Age levels: 7–11. Grade levels: 2–6.

SUMMARY: The sun and its functions, including energy and motion, are compared to a continuously exploding hydrogen bomb. Gravity is described in detail, but in a simplified way. The author uses comparisons to explain ideas, such as the waves that radiate from an electric iron when in use or not in use. Explanations about atoms and molecules offer useful information for the reader interested in the sun and its properties.

EVALUATION: The drawings are intentionally childlike; however, in many cases they cloud the basic scientific concept discussed. For example, on pages 10 and 11 a swirl pattern, which appears to serve no useful purpose, obscures the relative distances among planets. On page 12, the art seems to be particularly lacking in value because of intervening background lines. The distortion of clouds, building, and horizon on page 15 might confuse readers trying to understand the principle in the accompanying text. The sequence of a thrown baseball breaking a window is shown on three consecutive pages, disturbing the continuity of both idea and image. The concepts covered in the text are clear and informative. The large print is helpful, as are the index and glossary, which include page references.

□ □ □

Bendick, Jeanne. *Names, Sets and Numbers.* Illus. by author. Watts, 1971, o.p. 65 pp. Age levels: 5–8. Grade levels: K–3.

SUMMARY: Readers are introduced to set theory in this beginner book, which explains why it is easier to talk about objects if names are used to clarify them in categories. Empty sets and subsets are discussed, as is the concept of estimation. For example, children learn that it is all right to qualify numbers with words such as *about* when referring to objects that cannot be counted exactly, such as stars or raindrops.

EVALUATION: It is difficult to imagine how an author could make set theory clear to children between five and eight years of age, but Bendick appears to have accomplished it. The book is repetitive in its presentation of the concepts of names and sets, but does not belabor the point. The repetition clarifies the concepts and gives readers practice experimenting with them. The end of the book contains an index and answer key for each question in the book. The illustrations are well integrated with the text and aid in the presentation of ideas and examples. This book is recommended to get children off to a good start understanding set theory.

□ □ □

Bendick, Jeanne. *Solids, Liquids and Gases.* Illus. by author. Watts, 1974, o.p. 72 pp. Age levels: 8–11. Grade levels: 3–6. Series: Science Experiences.

SUMMARY: States of matter are highlighted in this book for children in middle elementary grades. Solids are explained as being things that take up space and have a specific shape of their own. Solids may be the same size, but not necessarily weigh the same. Solids can also be counted and measured. Liquids take up space. You can put something directly into a liquid, which you cannot do to a solid. Liquids take the shape of the container they are in. Some liquids mix together so they cannot be separated, and others do not mix. Gases usually cannot be seen. They spread out to take up as much space as possible. Gases all have weight although most are light. One can feel gases when they move. Gases mix with other gases and are hard to unmix. Many things in the world are combinations of solids, liquids, and/or gases; for example, a person is a combination of all three. Things are always changing and some things change from one state to another. Even when things go away or seem like they are used up, they are really just changing to something else.

EVALUATION: In this thorough examination of the states of matter, the reader discovers the properties of each set through a Socratic questioning style. Through beautifully blended questions and answers, the author pushes the reader toward self-discovery. The most thought-provoking questions are listed at the end of the book with hints for solving some of the questions. The illustrations are cleverly and often humorously presented, and help clarify experiments and concepts. An index is located at the end of the book. Overall, the book gives wonderfully detailed scientific information on the three basic concepts of solids, liquids, and gases.

□ □ □

Berenstain, Stan, and Jan Berenstain. *The Berenstain Bears' Science Fair.* Illus. by authors. Random House, 1977. 64 pp. ISBN: LB 0-394-93294-3; pap. 0-394-86603-7. Age levels: 4–7. Grade levels: PreK–2.

SUMMARY: With the help of his neighbor, Actual Factual Bear, Papa Bear gives a science lesson to the cubs in preparation for the Science Fair. Machines, matter, and energy are explained and illustrated. Actual Factual Bear explains that energy is important because it results in motion or change. We get our energy from food; our body burns food to make energy. Machines and vehicles get their energy from matter in the form of coal, oil, and natural gas. Matter is burned to make fuel that makes engines work.

A few experiments are explained and illustrated in the book. In one example, Mother Bear shows that matter can change its form from solid to liquid to gas: ice cubes are solid in the freezer, but will melt and turn into water if left at room temperature. If allowed to boil, water changes to steam. At the close of the book, each of the Bear cubs has a project to take to the Science Fair and share with others in the neighborhood.

EVALUATION: Actual Factual Bear is a welcome addition to the Science Fair. He wears glasses, carries a briefcase, and uses charts and diagrams to support what poor Papa Bear has attempted to explain. The book is well organized, with a short list on the title page of some of the topics presented so one can choose to read only the section that supports the science subject being taught. The simple experiments are illustrated and explained very well, and would be easy to duplicate at home or school.

The reader may choose to read the two or three sentences on each page and/or include the labels for the illustrations to help the child understand the concept. For example, on the pages about matter, solids we can find in our environment are illustrated and labeled. This should prompt children to name other solids around them.

As a lap book, *The Berenstain Bears' Science Fair* is a fine way to introduce and explain the subjects of matter, energy, and machines to preschoolers. The school-age child, however, may find the silly humor and simplistic format too unsophisticated.

□ □ □

Bramwell, Martyn. *Glaciers and Ice Caps.* Illus.; photos. Watts, 1986. 32 pp. ISBN: LB 0-531-10178-9. Age levels: 8–11. Grade levels: 3–6. Series: Earth Science Library.

SUMMARY: The opening page shatters the idea that the Arctic and Antarctica are alike except for their location at opposite poles. Antarctica is land covered by a large ice sheet. The Arctic is mostly ocean surrounded by land; the Arctic Ocean is covered by glaciers. In the frozen areas of land close to the Arctic, moss and lichen grow in the summer. With the heavy snowfall, ice forms and is packed into a hard mass. Because of the dense weight of the glacier mass, it begins to move slowly down mountainsides. On steep slopes the ice breaks causing deep cracks in the rocks and soil beneath. Rocks underneath the glacier are moved along with it, and erode the valleys below. In more temperate areas where water is trickling underneath it, the glacier moves more swiftly: glaciers may move anywhere from a few inches to 100 feet per day. Winters around glaciers are usually quiet, except for an occasional cracking of ice breaking under stress. In the summer, a constant trickling or rushing of water is heard everywhere. Glaciers change the shape and texture of the landscape. The classic pyramid-shaped Matterhorn was formed by glacier heads on all sides of the mountain.

Antarctica doesn't accumulate much more snow than the Arctic. Usually the wind just picks up the snow and blows it around. Where animal life does exist in this area, it is being protected by bans on weapons tests and military devices.

EVALUATION: Although this book is geared toward the intermediate grades, it does not assume that the reader already has much of a background on glaciers and ice caps. At the same time it does not provide so much information that the reader would be lost in detail. Fascinating pictures enhance the text while providing examples of the text. Aerial photographs of glaciers display examples of terms being defined. The pictures and photographs cover more than what is explained in the text. This book appears to be a good solid introduction to glaciers for upper-elementary-age readers.

□ □ □

Branley, Franklyn M. *Air Is All Around You.* rev. ed. Illus. by Holly Keller. Crowell, 1986. 32 pp. ISBN: LB 0-690-04503-4; pap. 0-06-445048-1. Age levels: 5–8. Grade levels: K–3. Series: Let's-Read-and-Find-Out Science.

SUMMARY: The book sets out to explain that everywhere there is air, even though it cannot be seen, smelled, or felt. The first experiment proves that there is air in a glass: by putting a paper towel into a tumbler, turning the glass upside down, and sticking it straight into the water, the napkin will stay dry because of the air pocket. However, if you tilt the glass slightly, it will allow the air to move out and the water to move in, so the napkin will get wet.

Air covers the earth like an orange peel—air is heavy, although it may not seem so because it is spread out. The air in an average room weighs 75 pounds; all of the air on Earth weighs a total of 5 quadrillion tons. Air is dissolved in water so fish can breathe underwater. After a glass of water has sat for an hour you can see air bubbles on the inside of the glass. Spaceships take air with them in order to keep astronauts alive. The spaceships themselves fly above the air that surrounds the Earth, while airplanes and balloons actually fly *in* the air. Air *is* all around us!

EVALUATION: Geared toward younger readers, this book aids in the comprehension that air is everywhere even though it may not always be sensed. The two experiments in this book are practical and easy to do. The pictures are colorful and enticing, though the text is fairly self-explanatory. One feature that detracts from the ease of reading involves the use of black print on a dark blue background near the middle of the book. This dark on dark situation is hard on the eyes. Readability seems appropriate for primary grades, and technological details are appropriately limited.

□ □ □

Branley, Franklyn M. *A Book of Flying Saucers for You.* Illus. by Leonard Kessler. Crowell, 1973. 72 pp. ISBN: LB 0-690-15189-6. Age levels: 7–10. Grade levels: 2–5.

SUMMARY: Many reports of "seen" flying saucers are briefly and vividly recounted, along with the illustrator's conceptions of how they appeared to the observer. Some reports are almost believable, such as a pilot's description of having to veer out of the way of a UFO to avoid a collision; others are bizarre, such as a report of a flying platform that glowed, with timbers running across it. Ancient reports dating back to biblical times and around the years 1270 and 1290 are also related. Investigations have shown that some sightings were hoaxes or planned for reasons such as publicity; others are unexplained and remain a mystery. The author cautions that our eyes are not always reliable and supports this with an exercise in the book. He also explains mirages, and how and why people see them. Branley discusses discharges of electricity in the atmosphere, ball lightning, and how planets and shooting stars can be mistakenly identified as UFOs. The northern lights, reflected light that moves, and man-made satellites and balloons have often been misidentified as UFOs. Branley states the differences between UFOs, IFOs (identified flying objects, such as weather balloons), and flying saucers. Who studies and investigates reports of UFOs and the reports published on these scientific studies are discussed. Branley explains why flying saucers and ships from other planets are not possible. The future possibility of visits to planets outside our solar system—and what we might see—ends this book on flying saucers.

EVALUATION: The author, an astronomer, explains what has been sighted, why some sightings remain a mystery, and why others can be explained in a very understandable way. Scientific evidence and explanations are cited wherever possible, although he still points out fascinating aspects of some assumptions made about UFOs. The illustrations add humor at times, and are mostly black and white. Large print and easy vocabulary make this a very readable book for an age group with a natural curiosity about UFOs.

□ □ □

Branley, Franklyn M. *Comets*. Illus. by Giulio Maestro. Crowell, 1984. 32 pp. ISBN: LB 0-690-04415-4; pap. 0-694-00199-6. Age levels: 5–8. Grade levels: K–3. Series: Let's-Read-and-Find-Out Science.

SUMMARY: *Comets* tells young children some basic information about our solar system by relating features of comets to familiar objects and actions. For example, comets are like dirty snowballs made of dust, stone, and ice. When the sun heats the comet, some of it is changed to gases giving it a head and a tail. The orbit of a comet around the sun is elliptical, like a flattened circle. Most of the time the comet is far away from the sun. A comet is always moving fast but because it is so far away it cannot actually be seen to move.

Comets are formed by particles clumping together and eventually being pulled by the gravity from Jupiter and the sun, neither of which can hold them so they sail away.

At one time people were afraid of comets, thinking they were warnings that disaster was coming. The word *disaster* means evil star.

Halley's Comet was named after the man who predicted that this particular comet would return every 76 years. It was seen in January and March 1986.

Julius Caesar saw it in 87 B.C. when he was 13 years old, and Mark Twain was born and died in years that Halley's Comet returned. Halley's comet is getting dimmer and dimmer, and will return again in 2062.

EVALUATION: Because Halley's Comet returned in 1986, this book was published in time to capitalize on renewed interest. Even though Halley's Comet has come and gone, *Comets* still provides a fascinating look at the source and composition of comets. The book is very readable for young children, providing thorough explanations in detail without bogging the reader down. The illustrations are well integrated with the text and add to the explanations by example. One of the most useful pictures is that of the orbit of a comet crossing the orbits of other planets. Despite the fact that readers may have missed Halley's Comet, the book promises that many other comets will come before Halley's Comet returns in 2062.

□ □ □

Branley, Franklyn M. *Flash, Crash, Rumble, and Roll.* Illus. by Barbara Emberley and Ed Emberley. Crowell, 1985. 32 pp. ISBN: LB 0-690-04425-9; pap. 0-06-445012-0. Age levels: 4–8. Grade levels: PreK–3. Series: Let's-Read-and-Find-Out Science.

SUMMARY: This book explains why and how thunderstorms happen and the safety measures to be observed when lightning strikes. The author explains how it is possible for a thunderstorm to occur even on a day that begins calm and sunny. Warm air moves up from the Earth to the clouds bringing moisture with it. The clouds begin to grow and the water vapor cools turning into water droplets and also ice crystals. The colder air in a cloud moves down while the warmer air rises. The air becomes unstable, so airplanes avoid thunderclouds. Also, each water droplet carries electricity, which may jump from the top to the bottom of a cloud creating lightning that may at times be a mile long.

Thunder is heard after lightning is seen because sound waves travel slower than light waves. By counting seconds after a flash of lightning, a person can estimate the distance of a storm. The storm is one mile away if five seconds elapse between seeing lightning and hearing thunder. Thunder is noisy but not harmful. Lightning, however, can be dangerous; people should stay out of water, seek shelter indoors, and stay away from wires, pipes, and metal things. Outside, metal fences and trees should be avoided; keeping close to the ground or staying in a car is recommended. The author points out that knowledge about thunder and lightning and awareness of safety measures will eradicate fears about stormy weather.

EVALUATION: Thunderstorms are naturally a topic of high interest to younger children, and books about thunderstorms intrigue them. This book, with its fantastic illustrations, is no exception. The illustrations are very colorful and dynamic in portraying the dangers that may be encountered during a thunderstorm and the safety precautions required. The amount of scientific information is appropriate for PreK through third graders, with older children benefiting more from the concepts presented. Overall, the book is very well written, with many facts given in a meaningful and interesting way.

□ □ □

Branley, Franklyn M. *Gravity Is a Mystery.* rev. ed. Illus. by Don Madden. Crowell, 1986. 32 pp. ISBN: LB 0-690-04527-1; pap. 0-06-445057-0. Age levels: 4–8. Grade levels: PreK–3. Series: Let's-Read-and-Find-Out Science.

SUMMARY: Beginning with the proposition that it is possible to dig a hole through the center of the Earth and come out at the Indian Ocean, the author suggests that you should next jump into this hole. Going past the center of the Earth, you would eventually slow down before reaching the ocean floor. Your body would swing back and forth several times until you finally stopped at the center of the Earth, where the center of gravity is. Gravity is the force that pulls things to the center. The heavier the object, the harder gravity pulls down on it, thus the more the object weighs. The moon also has gravity, but not as much as the Earth. Therefore, one would not weigh as much on the moon. Some planets have more gravity than the Earth and some have less; the sun has a lot of gravity.

EVALUATION: This delightful book begins with the idea that gravity is mysterious because we cannot see it, even though we know it exists. As it takes one through explanations of the qualities of gravity, the book again repeats its conclusion that gravity is a mystery. The illustrations aid in the concept of gravity. The facial expressions of the boy who finds himself in the center of the Earth, and his weight on each of the different planets—a light 8 pounds on Pluto ranging to the extreme 16,470 pounds on the sun—are examples of how the illustrations add humor that can be enjoyed by old and young alike. This is a simple but informative and appropriate science book for children.

□ □ □

Branley, Franklyn M. *Hurricane Watch.* Illus. by Giulio Maestro. Crowell, 1985. 32 pp. ISBN: LB 0-690-04471-2; pap. 0-06-445062-7. Age levels: 4–8. Grade levels: PreK–3. Series: Let's-Read-and-Find-Out Science.

SUMMARY: This book details some of the characteristics of hurricanes and provides an awareness of the safety measures to be taken in anticipation of this violent weather. When air is moving slowly, it is felt as a light breeze. If it is blowing more than 74 miles per hour and covers nearly 500 miles, it is called a hurricane. Most hurricanes occur in August through October and often begin in the Atlantic Ocean, the Gulf of Mexico, or the Caribbean Sea. When conditions are such that a tropical storm is created, winds may exceed 100 or 200 miles per hour. Some hurricanes reach the coast and cause damage. They may pull trees out from their roots or move entire houses and cars. Next to a map that shows the hurricanes' paths, three destructive hurricanes are listed along with the year they struck, the location of the worst hit areas, and the highest wind speed reached.

Once a hurricane has started, it cannot be stopped. The eye of a hurricane is calm, and the sun often shines through it. Planes fly into hurricanes to measure how fast the wind is blowing, and also to measure air temperature, the amount of water in the clouds, and air pressure. Scientists have tried to spread dry ice in the clouds, but it has not prevented hurricanes. Weather satellites aid in tracking hurricanes and warning people about impending danger. People are advised to take some necessary precautions when they are unable to leave an area when a hurricane is approaching, such as filling a bathtub with water in case pipes are broken; keeping a flashlight, candles, and matches nearby in case of power failure; and listening to a transistor radio for current weather conditions and advice.

EVALUATION: The natural force of a hurricane and its potential dangers provoke wonder and amazement in young readers. Despite the disturbing evidence of catastrophes caused by hurricanes, it is fascinating to discover specific information on this powerful force of nature. The author and the illustrator combine their talents to present their topic in an exciting manner.

Information is presented simply and clearly in the text and is often repeated a second time elsewhere in the book. This repetition facilitates a good, basic foundation in the study of hurricanes. The illustrations vividly depict the causes and tragic results of hurricanes. One illustration in particular—the passage of an airplane through storm clouds—dramatically shows the vast contrast between the enormity of the hurricane and the tiny dot of the airplane. This informative book offers exciting reading.

□ □ □

Branley, Franklyn M. *Is There Life in Outer Space?* Illus. by Don Madden. Crowell, 1984. 32 pp. ISBN: LB 0-690-04375-9; pap. 0-06-445049-X. Age levels: 4–8. Grade levels: PreK–3. Series: Let's-Read-and-Find-Out Science.

SUMMARY: Earth teems with life, and throughout time people's imaginations have run rampant about the possibility of life on other planets. On occasion, false sightings of aliens have been reported, and in one instance curiosity turned to widespread fear when a radio play announced the arrival of Martians.

The excitement of the lunar landing in 1969 was accompanied by the anticipation that some form of life might be discovered beyond Earth, but no signs of life were found on the moon. Later, space probes to Mars showed no life in the photographs taken or in the soil that was tested. Pictures returned from the probes to Mercury and Venus also indicated that life does not exist there. Earth seems to be the only planet in our solar system that sustains living things.

Only further scientific and technological advances will help us determine if there are plants and animals beyond our solar system.

EVALUATION: Young readers should find the illustrations of imaginary moon creatures and Martians amusing in this "Reading Rainbow" book. Actual lunar landing and space probe photographs are informative. Simple and clear language describes the facts and fiction about life beyond Earth. By using the latest scientific knowledge available to date, answers are supplied for readers who seek specific information about the existence of life on planets other than Earth.

□ □ □

Branley, Franklyn M. *Journey into a Black Hole*. Illus. by Marc Simont. Crowell, 1986. 32 pp. ISBN: LB 0-690-04544-1; pap. 0-06-445075-9. Age levels: 5–8. Grade levels: K–3. Series: Let's-Read-and-Find-Out Science.

SUMMARY: In this book, a black hole is described in language that's easy for young children to understand. The author explains what can occur when the hot gases of a star cool and the star collapses. Its gases can draw together very tightly and form a black hole from which no light escapes.

Although scientists have not seen black holes, they have determined that they exist because of the following observations: a star that is near a black hole changes position as the strong gravity of the black hole pulls on it. Also, X rays that are produced by black holes have been found by X-ray telescopes.

The author then takes the reader on a distant and imaginary trip to a black hole. He explains the effect of gravity as the traveler nears the black hole and finally enters it. The reader is cautioned not to accept an invitation to visit a black hole because it would be a one-way trip.

EVALUATION: This book presents a difficult scientific concept to young readers in an easy, step-by-step method. Branley uses his expertise as an astronomer to clarify an astronomical theory with text that is informative, understandable, and at times playful. Simont's illustrations add to the clarification of a complex occurrence when some stars die out and collapse. This book is concise, yet challenging. In the classroom, Branley's book and Jane Yolen's

Commander Toad and the Big Black Hole (Coward, 1983), an easy-to-read and humorous story about black holes, could be used for comparative purposes.

□ □ □

Branley, Franklyn M. *Light and Darkness.* Illus. by Reynold Ruffins. Crowell, 1975. 33 pp. ISBN: LB 0-690-01122-9. Age levels: 4–8. Grade levels: PreK–3. Series: Let's-Read-and-Find-Out Science.

SUMMARY: Light and darkness are of interest to every young child. This book explains day and night in very simple terms and can be used to explain these phenomena to toddlers or older children. It reveals that because the sun shines on the earth by day, light is discernible and at night the moon and electricity provide all the light.

Reflected light, as contrasted with direct light, is explained. Objects that reflect light, such as the moon, are mentioned and the author, while not attempting to explain the speed of light scientifically, convinces young readers that light travels very fast. Readers are told that without light, they could see nothing at all.

EVALUATION: Overall the book is well written, providing a detailed account of light and darkness and how they are related. It is sequentially presented from the most to the least obvious, from day and light to night and dark, to the more sophisticated conclusion that light is everywhere. The illustrations aid understanding of the concepts, for example, reflecting light is illustrated by a bouncing ball compared to bouncing sun rays. In looking around it is obvious that most things reflect light rather than generate it. This principle is illustrated by the artistic shading of trees, buildings, and hair color. This is an appropriate book for young children to explore the properties of light and darkness.

□ □ □

Branley, Franklyn M. *Measure with Metric.* Illus. by Loretta Lustig. Crowell, 1975. 33 pp. ISBN: LB 0-690-01117-2. Age levels: 5–9. Grade levels: K–4. Series: Young Math.

SUMMARY: Branley uses familiar situations and objects to acquaint children with metric measurement. For example, he explains that when the reader was born, his or her length was measured in inches, and his or her weight measured in pounds. Almost everywhere else in the world, people are measured in centimeters for length and grams for weight. These examples of measurement are part of the metric system for determining length and weight.

It is explained that *centi* means 100, as children will recognize from knowing that there are 100 cents in a dollar. Longer distances are measured in kilometers. A kilometer is 1,000 meters in length or distance. Readers are told that a gram weighs about the same as ten paper matches or six toothpicks. If someone weighs 40,000 grams, that person weighs 40 kilograms. Liquid measurements in the metric system are reported in liters. The author concludes that eventually everyone might be using the metric system.

EVALUATION: At the time this book was published, there was much talk of the United States possibly converting to the metric system. Presently that conversion does not seem imminent. Some math text books for young children no longer include chapters on the metric system, so the interest of children today in this title would probably not be very high. While the book makes a few base comparisons between the English and the metric systems, it barely highlights conversions within the metric system itself. Also, it seems to only dabble in the topic and does not provide an inclusive overview of the metric system, which may leave readers more confused than when they started. Illustrations are large and brightly colored; they aid concept development on a small scale. While the author's style seems geared toward younger readers, comparisons of size, weight, and distance require higher cognitive abilities. This book may leave the reader with the impression that the metric system is foreign and thereby probably difficult to adjust to, which most likely is not the case.

□ □ □

Branley, Franklyn M. *The Planets in Our Solar System.* Illus. by Don Madden. Crowell, 1987. 32 pp. ISBN: LB 0-690-04581-6; pap. 0-06-445064-3. Age levels: 4–8. Grade levels: PreK–3. Series: Let's-Read-and-Find-Out Science.

SUMMARY: The nine planets in our solar system are featured in this science book. *Sol* means sun, and the sun is the most important and dominant part of our solar system. Mercury, Venus, Mars, Jupiter, and Saturn can be seen without a telescope and look like bright stars in the sky. Because the moon goes around our planet, it is often called a satellite of the Earth. Satellites circle other planets, too, and thus are also a part of our solar system. Asteroids are large chunks of rocks that orbit the sun. Comets are dirty iceballs, and meteorites or shooting stars are rocks that have fallen to Earth. Therefore, asteroids, comets, and meteorites are also a part of our solar system. Although Mercury is the closest planet to the sun, it is still millions of miles away from it. Uranus is over a billion miles away. Neptune and Pluto, the coldest planets, are the farthest from the sun. Mercury and Venus are the hottest and closest to the sun. Plants and animals cannot live on any other planets in our system except Earth. Earth is a mid-sized planet. In our solar system there are four planets larger and four smaller than Earth. The biggest planet is Jupiter and the smallest is Pluto. Directions for a mobile and a model of the solar system are included in the book.

EVALUATION: Illustrated by both drawings and photographs, this book provides an appealing look at our solar system. Each photograph has an illustrated character with a cartoon bubble, explaining the photographs. Also, the pictures are integrated well into the text, actually playing an important role in fulfilling explanations. The suggested projects of mobile and model would be excellent activities for children to deepen their comprehension of the subject matter. The book provides some practical tips for sky watching and brings the big, difficult-to-comprehend concept of the solar system to terms easily understood by young readers. Enjoyable to read, this book presents up-to-date information in a pleasurable manner.

□ □ □

Branley, Franklyn M. *Rain and Hail.* Illus. by Harriett Barton. Crowell, 1983. 39 pp. ISBN: LB 0-690-04353-8. Age levels: 4–8. Grade levels: PreK–3. Series: Let's-Read-and-Find-Out Science.

SUMMARY: Branley presents information about precipitation in language even young readers can understand. He explains that rain comes from clouds, and clouds are made of billions of tiny water droplets that come from water vapor that is in the air. Heat causes water to evaporate, and one can see one's breath on a cold day because water vapor is condensing and changing from a gas to a liquid. Water vapor condenses to form tiny droplets that make clouds when the air is very dense. Differences in cloud formations and what causes them are covered, as are the conditions under which rain, ice, and hailstones occur. The book explains the evaporation-precipitation cycle in sufficient detail for readers to understand.

EVALUATION: After stating his initial theme, Branley backtracks to expound on the parts of his theme, after citing several examples at each phase. He cleverly repeats important concepts while building in new information with each repetition. The end of the book recaps the water cycle. Colorful illustrations add to the enjoyment of reading the text. The book is well written for young audiences.

□ □ □

Branley, Franklyn M. *Saturn: The Spectacular Planet.* Illus. by Leonard Kessler. Crowell, 1983. 57 pp. ISBN: pap. 0-06-446056-8. Age levels: 9–12. Grade levels: 4–7.

SUMMARY: This book is entirely devoted to man's knowledge about Saturn, dating back to the time when the planet could be seen by the naked human eye to today's sophisticated instruments and spacecraft, which have revealed even more information. Following some historical background, the author devotes chapters to specific aspects of Saturn, such as its rotation and revolution, temperature, density and gravity, its rings and satellites, and theories about the planet.

Many black-and-white photographs and drawings accompany the text. A page of "Facts about Saturn" and an index appear at the end of the book.

EVALUATION: Branley is an authority on astronomy and has written many books for children in this field. He relates well to young people in his writings, and this book gives much information about Saturn in an easy-to-read and understandable manner. Branley's presentation of theories, particularly about Saturn's rings, are matter-of-fact, clear, and simply stated.

Almost anything a child might want to know about Saturn can be found either in the text or in the helpful page of facts about Saturn.

□ □ □

Branley, Franklyn M. *The Sky Is Full of Stars*. Illus. by Felicia Bond. Crowell, 1981. 34 pp. ISBN: LB 0-690-04123-3; pap. 0-06-445002-3. Age levels: 5–9. Grade levels: K–4. Series: Let's-Read-and-Find-Out Science.

SUMMARY: *The Sky Is Full of Stars* explains that many stars can be seen at night, particularly by using a telescope. The sun, however, is one star that can't be seen at night, even with the use of a telescope. Initially stars look alike, but careful scrutiny reveals that some are bigger than others, and some are different colors—white, red, blue, or yellow. Different stars are seen at different times of the year, such as the Milky Way (a blend of billions of stars), which is best viewed in the summer. Eighty-eight star pictures or constellations have been charted by astronomers; stargazers must use their imaginations to see the constellations.

EVALUATION: The pictures of constellations in the book give the reader clues as to what pictures to look for in the night sky. The author also provides information about the best months and times for seeing certain constellations. The illustrations are helpful in showing how much imagination is required to see the constellations. Actually, few constellations are "pictures" in themselves; many details must be filled in by the imagination, and this book helps to explain that. This book encourages readers to carefully observe the differences in size and color of stars. Also noteworthy is the use of photographs of the sky with the constellations drawn in place to provide accurate formations of constellations. An experiment is also given in which one may become familiar with some of the constellations. Although readable, the information in the book may be difficult to transfer to the real sky;

however, this book does a superb job in attempting to make that transition possible.

□ □ □

Branley, Franklyn M. *Snow Is Falling*. Illus. by Holly Keller. Crowell, 1986. 32 pp. ISBN: LB 0-690-04548-4; pap. 0-06-445058-9. Age levels: 4–8. Grade levels: PreK–3. Series: Let's-Read-and-Find-Out Science.

SUMMARY: The ordinary results of falling snow are recited in this primary book for children. The reader sees snow falling silently and it can be seen at night by looking at the street light. Everything that snow covers begins to look white. Pictures of magnified snowflakes show that they look different although all snowflakes have six sides. Various uses of snow are also described. Snow is fun to play in, and it is also good for plants and animals. Snow melts in the spring and waters the plants. Snow keeps animals warm by acting like a blanket. Eskimos use snow to build their houses. Snow can make life hard when it becomes a blizzard or when it melts quickly and causes a flood.

EVALUATION: This book appears appropriate for the younger child. It covers the basic concepts of a snowfall, snowflake, snowstorm, and spring thaw. The illustrations are colorful and match the text. Students in the second and third grades could read the book on their own. An experiment is suggested using two thermometers outside: one thermometer is buried in the snow, and the other hangs from a tree. After an hour one could look at both thermometers and see that the snow acts like a blanket for plants and animals, because the buried thermometer would have a warmer temperature. Technological details seem appropriate for the ascribed age level.

□ □ □

Branley, Franklyn M. *Space Colony: Frontier of the 21st Century*. Illus. by Leonard D. Dank. Lodestar, 1982. 102 pp. ISBN: 0-525-66741-5. Age levels: 10–14. Grade levels: 5–9.

SUMMARY: Branley asks what life will be like in outer space and how man will live. He provides a very imaginative view of life in space colonies via the hypothetical journey of a couple into space. The couple sends letters back to Earth describing their life. The colonies range from those closest to Earth (low orbit), which act as satellites, to the farther colonies, which manufacture products for use on Earth (e.g., crystals for electrical use). The couple explains that major space colonies will be completely self-contained, sealed environments. The shape of colonies is discussed, taking into consideration such aspects as how air, light, and gravity will be used. They talk about the division of residential and work space, and separate climate needs. Work beyond the colonies and how the colonists will live, build, and transport products is also examined. The use of a power satellite, with its transport, and the negative and positive effects on the Earth of this satellite are examined. Coping with disaster is addressed, and the quick thought and action necessary to repair damages is discussed. This story projects life on space colonies as a very comfortable, rewarding existence under ideal conditions. Branley makes it something to look forward to.

EVALUATION: This book is entertaining and realistic. It is very detailed; but things are simply explained, especially the design and workings of the colony. Of particular interest is the discussion of living conditions—air, water, and food chain—back on Earth, and the effects of these elements on earth dwellers versus the effect of the same elements on life in the colonies. Illustrations are very clear and quite informative, adding to an understanding of the text. Measurements are in metric (some converted, some not). There are many cross-references within the text. Technical terms are defined. A further reading list is included, as is an index.

□ □ □

Branley, Franklyn M. *The Sun: Our Nearest Star.* Illus. by Don Madden. Crowell, 1988. 31 pp. ISBN: LB 0-690-04678-2; pap. 0-06-445073-2. Age levels: 4–8. Grade levels: PreK–3. Series: Let's-Read-and-Find-Out Science.

SUMMARY: In this book basic information is given about our nearest and most important star. A description of its characteristics and uses is given; for example, the daylight star provides light and energy for plant and animal life on earth. Solar energy is introduced through examples of its use. The energy found in natural resources is shown to have come from the sun. Young readers are given an illustrated experiment to follow, showing the necessity of sunlight and its effects on plant life.

EVALUATION: The text is straightforward and presented in an easy manner for the young reader. The colorful illustrations clarify and reinforce the text appropriately. For example, the distance between the Earth and the sun, expressed in large numbers in the text, is also illustrated in a clearly labeled diagram so a young child can understand the distances by size comparison.

The use of bright, warm colors captures the reader's interest in the text and gives a feeling of the warmth of our important and nearest star.

Children will enjoy performing their own experiment, as illustrated in the book, to prove that a plant needs light although they may not have the patience to wait for a seed to sprout and grow.

A character is used throughout the text to illustrate the everyday benefits of the sun. The text is well written and does not contain so much detail that young children will be put off by this basic introduction to the sun.

□ □ □

Branley, Franklyn M. *Sunshine Makes the Seasons.* rev. ed. Illus. by Giulio Maestro. Crowell, 1985. 32 pp. ISBN: LB 0-690-04482-8; pap. 0-06-445019-8. Age levels: 4–8. Grade levels: PreK–3. Series: Let's-Read-and-Find-Out Science.

SUMMARY: This book explains that sunshine makes the earth warmer, and that it is warmer in the summer, in part, because there are more hours of sunlight. To show the reasons for the seasonal changes an experiment is suggested. The reader is told to stick a pencil through an orange, and draw a line halfway around the middle. A pin should then be put somewhere in the top half. When the pencil is held straight up and down and a flashlight (representing the sun) is turned on it, the orange is lighted evenly from pole to pole. The author tells the reader that if the Earth were exactly like the orange, there would be no seasons. But when the orange is turned slightly on its axis (the pencil), it is apparent that the North Pole would have a long winter night and be cold, and then in summer a long day. The reader learns that when it is summer in the north, it is winter in the south, but at the equator, the temperature and length of the days stay about the same year round. Thus it is shown that seasons are a result of the earth being tilted on its axis.

EVALUATION: *Sunshine Makes the Seasons* draws the interest of the reader by its very title. Many children are naturally curious about the seasons and why they change. In this enjoyable book, Branley is able to summarize clearly two reasons for the change of seasons. He first explains that more hours of sunlight leads to warmer temperatures. The second explanation is more detailed with its fun hands-on experiment illustrating how the earth, when tilted, changes seasons. This example is central to the understanding of the concept, as the illustrations alone may be difficult for children to understand. Also, showing that the axis of the Earth cannot be straight up and down if there are seasons helps prevent the experimenter from making that mistake. Although the illustrations would be inadequate without the experiment, they do add much in the way of a visual explanation of our earth and its rotation and revolution.

□ □ □

Branley, Franklyn M. *Volcanoes.* Illus. by Marc Simont. Crowell, 1985. 32 pp. ISBN: LB 0-690-04431-3; pap. 0-06-445059-7. Age levels: 4–8. Grade levels: PreK–3. Series: Let's-Read-and-Find-Out Science.

SUMMARY: Almost 2,000 years ago, Mt. Vesuvius in Italy blew up; it is still spouting out steam and ash, but not as much as it once did. In 1815 when Mt. Tambora in Indonesia erupted, the ash blew around the Earth and blocked out the sun. The year 1816 was called the year without a summer because June had snow, and July and August had frosts. Mt. St. Helens in Washington blew in 1980, but geologists anticipated it because of the rumblings it made. The Earth's crust is broken into sections called plates that are always moving. Earthquakes occur when the plates move together or under each other. Inactive volcanoes will probably never erupt again. Geologists are always monitoring the earth for changes in order to give ample warning to people before a volcano erupts.

EVALUATION: Using descriptive drawings, the illustrator brings to light the causes and functions of a volcano. Several actual volcanic eruptions are cited, which brings validity to the information as we are told of devastating natural events. The book gives a solid scientific detail with simple vocabulary (for example, "blows its top" rather than "erupt") without oversimplification of the topic. After journeying to many volcanoes, we are reassured by the concluding message: the chances are minimal that a volcano would appear in one's backyard since geologists are always watching for changes in the earth.

□ □ □

Branley, Franklyn M. *Weight and Weightlessness*. Illus. by Graham Booth. Crowell, 1971. 33 pp. ISBN: LB 0-690-87329-8. Age levels: 5–9. Grade levels: K–4. Series: Let's-Read-and-Find-Out Science.

SUMMARY: This account for very young readers explains that although one would weigh about the same on any scale on Earth, in a spaceship orbiting the Earth, one would be weightless. Astronauts must fasten themselves to chairs and beds so they do not float around, and outside of the ship they must be fastened so they don't float away. Weight on Earth is a result of gravity. While a person or an object is being pulled down through gravity, a corresponding push upward occurs. If one were on a scale and suddenly fell through into a great hole, there would be no push upward because a state of weightlessness would have occurred. The author explains that every 5 miles a spaceship in orbit is pulled down to 16 feet. If a ship is moving 5 miles a second, it never comes down to Earth because the Earth is curved, and everything remains weightless while in this orbit. A spaceship must slow down and change its path when the astronauts want to return to Earth. The force of gravity is revealed using the image of a spaceship, since children frequently have a lively interest in space travel.

EVALUATION: *Weight and Weightlessness* deals with a topic taken for granted and explains it. Weighing oneself on a scale is easy to do, but it is because of gravity that we have weight. Astronauts are out of range of the Earth's gravitational pull and so do not have weight. Branley does an excellent job of explaining how orbiting the Earth at five miles per second can keep the ship at a steady height without its falling down to the Earth. This book is very readable, with remarkable explanations of the concept of weight and weightlessness that are neither too complex nor too simple. The book is illustrated with drawings and diagrams that enhance the text and help explain how the push up balances the pull down, as well as the constant orbit of a spaceship being systematically pulled down as it matchs the curve of the earth. Also enticing is the conclusion that says that astronauts get tired of being weightless. Nevertheless an impression is left that it would be fun to try weightlessness for a while.

□ □ □

Branley, Franklyn M. *What Makes Day and Night.* rev. ed. Illus. by Arthur Dorros. Crowell, 1986. 31 pp. ISBN: LB 0-690-04524-7; pap. 0-06-445050-3. Age levels: 4–8. Grade levels: PreK–3. Series: Let's-Read-and-Find-Out Science.

SUMMARY: Through pictures, experiments, and examples, Branley provides a simple explanation of how the rotation of the Earth causes night and day. An example is given that the Earth is like a big ball that spins smoothly at 1,000 miles per hour. Because it spins smoothly at a constant speed, we cannot feel it moving. The Earth rotates once every 24 hours. One half of it is lit by the sun. Sunrises and sunsets are a result of the Earth turning, and so we move from day to night. An explanation of day and night on the moon is also given. The explanation states that the moon moves so slowly, it has two weeks of daylight and then two weeks of darkness. In contrast the Earth has an average of 12 hours of daylight and then 12 hours of darkness.

EVALUATION: Branley covers a complex question very thoroughly in his explanation of how the rotation of the Earth causes night and day. Younger children often have a difficult time comprehending this concept. They accept the nature of night and day, but lack the understanding of its causes. The hands-on experiment with globe and flashlight (which represents the sun) and the experiment with the child acting as the Earth and turning in front of the flashlight are excellent ideas to involve a child in active learning. The text is written with vocabulary appropriate for primary grades. The illustrations seem essential to understanding the causes of night and day.

□ □ □

Branley, Franklyn M. *What the Moon Is Like.* Illus. by Vladimir Bobri. Crowell, 1963, o.p. Unp. Age levels: 0–8. Grade levels: PreK–3. Series: Let's-Read-and-Find-Out Science.

SUMMARY: The many assumptions about the nature of the moon contained in this book are naturally limited by the knowledge available at the time of its publication. Since the moon had not yet been explored, the information is a mixture of fantasy and speculation. Make-believe situations highlight the narrative, featuring fictional astronauts who explore mountains, valleys, craters, and waterless seas.

The author personalizes concepts by addressing them directly to the reader using second person singular. For example, he states that *you* would be able to leap over a house, if the moon had houses. A cutaway section of a rocket ship represents the possibility of (then forthcoming) lunar explorations. Further knowledge about the moon is anticipated as a consequence of humankind's quest to understand the universe.

EVALUATION: An intriguing topic is presented in a childlike manner by an interesting combination of artwork and content. Many illustrations appear to be collages and several pages utilize light print on dark backgrounds. Shades of purple and aqua are interspersed with black, gray, and white, suggesting the lifeless character of the moon. The vocabulary avoids complex scientific terminology. Yet the term *lunar sea* is mentioned and explained, while the more significant concept of gravity is mentioned but the word *gravity* itself is omitted. Scientific generalizations made throughout the book appear plausible. However, in the cutaway section of the rocket ship, children might observe and question the fact that the astronauts are not displaying weightlessness. This 1963 book about the moon was revised in 1986; it might be an interesting classroom exercise to compare these two editions written by the astronomer Franklyn M. Branley.

□ □ □

Branley, Franklyn M. *What the Moon Is Like.* rev. ed. Illus. by True Kelley. Crowell, 1986. 32 pp. ISBN: LB 0-690-04512-3; pap. 0-06-445052-X. Age levels: 0–8. Grade levels: PreK–3. Series: Let's-Read-and-Find-Out Science.

SUMMARY: Information about the appearance and characteristics of the moon is presented in this science book, which is an updated and reillustrated revision of an earlier edition. In the view from Earth, the appearance of the moon suggests different interpretations to an observer. For example, one might see a man's face, a rabbit, or even Jack and Jill and their pail. On closer view, however, astronauts report that the surface of the moon is covered with dust and includes hills, valleys, and cliffs.

The moon has changed little over a very long period of time because there is no wind, air, or water. In order to travel on the moon, the astronauts used a mooncar with wide tires to maneuver over the fine dust. In their exploration, astronauts found no evidence of current or previous life.

In addition to relating the astronauts' experiences on the moon, the moon's gravity, the temperature extremes, and the long cycles of day and night are covered. A comparison is made of the sky as it is seen from the Earth and from the moon and reasons given for the difference in the sky's color. There is also an explanation of why stars can be seen in the daytime from the moon. The prediction that a moon colony may be built in the future is made.

EVALUATION: While the photographs, drawings, and cartoons attempt to explain what the moon is like, the utilization of several art media in this book is unusual for a children's book of this short length, and may lead to misinterpretation of the material. For children who are accustomed to a vertical thermometer, the horizontal red and blue model that indicates the extreme temperatures on the moon might need explanation. More importantly, however, the thermometer is not labeled centigrade or Fahrenheit; although Fahrenheit is specified in the text, it might not be associated with the picture. In addition, mention of a colorless moon is followed by the description of the moon as being grayish brown.

Analogies are employed that are effective and easily understood by children, such as the surface of the moon is as dry as a desert, and the statement that some moon rocks are as large as houses.

□ □ □

Broekel, Ray. *Experiments with Light*. Photos. Childrens, 1986. 46 pp. ISBN: LB 0-516-01278-9; pap. 0-516-41278-7. Age levels: 8–10. Grade levels: 3–5. Series: New True.

SUMMARY: Many facts about the origin and characteristics of light are featured in this book intended for children in the elementary grades. For example, Broekel explains that light comes from the sun and other stars, from fire, hot metal, and electricity, and points out that without light human beings could not see colors. He explains luminosity, absorption, reflection, light energy, and refraction, light's conversion to chemical and heat energy, and the speed at which light travels. Children will be interested to know that moonlight is really sunlight, since the moon gives off no light at all. Translucency and opacity are discussed, and color separation is demonstrated when the author explains the process through the use of a prism.

EVALUATION: The photographs parallel the text in clarifying concepts and providing examples. The information is clearly presented and up to date, providing younger readers with a basic study on light. Several experiments are suggested that give practical examples of the characteristics of light. A few of the experiments need adult guidance in selecting appropriate materials, but most are simple enough for the readers to perform alone. A "Words You Should Know" list is included at the back of the book along with an index.

The writing style and vocabulary allow for independent reading and understanding by most third to fifth grade students.

□ □ □

Carona, Philip B. *Crystals.* Illus. by Phillip D. Willette. Follett, 1971, o.p. 30 pp. Age levels: 7–9. Grade levels: 2–4. Series: Beginning Science.

SUMMARY: Crystal formation is discussed in this beginning science book for young children. They learn that crystals can be found in rocks and minerals and that they differ in color, size, and shape (salt crystals are tiny cubes). Carona explains that the three states of matter are solids, liquids, and gases, and that most solids change to liquids and then gases when heated. Matter is made of atoms that join to make molecules, which are always in motion. The warmer a substance is, the faster the molecules move and the farther apart they become. When molecules in solids arrange themselves in an orderly manner, they form crystals. Some molecules cannot arrange themselves in order and so are referred to as crystalline. The book also explains that there are many types of minerals. The *crystalline lattice* is the term describing the arrangement of molecules and atoms.

Crystals exist in steel as iron crystals, but gem crystals are valued for their beauty, the most prized being the diamond. Crystals are used in communicating by radio, television, and telephone. Glass crystals that can withstand a lot of heat without cracking are used on rockets and missiles. Light is shone on a ruby, which becomes energized and generates bursts of laser light. Crystallographers are scientists who study crystals.

EVALUATION: Although informative, this book lacks the spark needed to hook and maintain the interest of young readers. It presents too much information too quickly for students who are starting to explore crystals. The illustrations are well rendered and clarify some textual information. The captions adjacent to the illustrations provide additional scientific detail. The book is supposedly written for seven and eight year olds to read by themselves, but there are many technical words—for example, *sublimation, molecules, atoms,* and *amorphous combinations*—which they may find difficult.

□ □ □

Catchpole, Clive. *Deserts.* Illus. by Brian McIntyre. Dial, 1984. Unp. ISBN: 0-8037-0035-0; pap. 0-8037-0037-7. Age levels: 6–10. Grade levels: 1–5.

SUMMARY: In this picture book for young children, the author explains desert life by showing how deserts affect living things. He reveals that deserts are places that receive very little rain, but are not necessarily dry, hot, and sandy. There are deserts in frozen climates, too, and some deserts are rocky or salty. Even though deserts are not suitable for many living things, a variety of plants and animals live in deserts. Facts about camels will interest youngsters; camels can travel about 30 miles per day and live for 17 days without water, and the large hump on the camel's back stores reserved food. Cacti have prickly spines and thick skins to protect the retention of water. Beetles also have a thick covering to prevent water loss. Lizards and snakes live in the desert, too, in spite of the fact that some deserts receive rain only once a year. Often this is enough for special desert flowers to grow, bloom, and produce seeds.

EVALUATION: The illustrations in this book are exceptionally handsome. They are beautifully colored with an extremely lifelike look. The green plants look alive; the sky is a bold blue and is realistic. The illustrator has effectively utilized the shadows from the desert sun. Young and old alike could enjoy the book purely for the beauty of its illustrations.

Blocks of text appear on each page, giving much interesting information about various plants and animals that live in the desert, including how they are all interrelated. Older children could absorb more, but younger children would certainly glean a good amount of information from the text. Overall, the quality of the text is good but, for this book, the high quality of the illustrations predominates.

□ □ □

Challand, Helen J. *Experiments with Magnets*. Photos; illus. Childrens, 1986. 45 pp. ISBN: LB 0-516-01279-7; pap. 0-516-41279-7. Age levels: 8–11. Grade levels: 3–6. Series: New True.

SUMMARY: What is a magnet? What is it used for? These questions and many more are answered in this book about the properties of magnets and the materials they attract. The reader learns that large and small magnets are used daily in household appliances such as televisions, telephones, and doorbells. There are also industrial uses, such as in the cranes with electromagnets that are used for building as well as demolishing.

The book teaches young scientists how to make their own magnets. A temporary magnet can be made by taking an iron nail or darning needle and striking it in one direction several times with a magnet. The atoms will line up in one direction and make a magnet that can pick up small pins and tacks, but it will not last long. An electromagnet can be made by wrapping wire around a nail several times and connecting the ends to the two terminals of a dry cell. This produces a strong magnet that can pick up many metal objects until the power is shut off.

At the end of the book, the reader learns that the center of the earth has a magnetic field and that all the planets, the sun, and stars have magnetism too.

EVALUATION: This book is well organized and factual, but the style appears too dry and scientific. A table of contents organizes the subject matter into chapters, which may make the material easier for the middle-grade child to digest. Photographs of children aged 8 to 12 performing experiments are included. Diagrams sometimes illustrate properties that cannot be photographed. For example, atoms are drawn in a random pattern inside an ordinary nail. In the magnetic nail, the atoms are all lined up facing north, demonstrating how the nail changed to a magnet.

The author's style may prompt a child to perform the experiments and find the answers to the numerous questions asked at the end of each lesson. However, the answers are not included. Children may therefore have to guess the answers. These should have been included, perhaps at the end of the book. A glossary and an index are welcome additions to the book.

Charosh, Mannis. *Straight Lines, Parallel Lines, Perpendicular Lines*. Illus. by Enrico Arno. Crowell, 1970. 33 pp. ISBN: LB 0-690-77993-3. Age levels: 5–8. Grade levels: K–3. Series: Young Math.

SUMMARY: In contrast to many books that involve experimentation to assist comprehension, this straightforward book allows readers to become participants through simple techniques involving paper, pencil, and string to determine linear measurement. No special materials are required and the tasks can be performed at individual desks in the classroom, if necessary.

EVALUATION: Charosh clearly outlines three basic concepts about lines and presents them in an orderly fashion. He expounds on each before progressing to the next, since the concepts are developmental and require previous understanding in order to build the new terms. The illustrations are large, colorful drawings showing various viewpoints of an observer watching children perform the experiments.

The experiments are illustrated in the text for further clarification. The author's style is simplified for understanding by young readers. His presentations should result in successful comprehension of basic linear measure by the intended age group.

Clark, Margaret Goff. *Benjamin Banneker, Astronomer and Scientist*. Illus. by Russell Hoover. Garrard, 1971. 96 pp. ISBN: LB 0-8116-4564-9. Age levels: 7–11. Grade levels: 2–6. Series: Americans All.

SUMMARY: This biography of Benjamin Banneker, a black astronomer and mathematician, begins when Banneker was five years old and covers his life until his death in 1806. After being lent books on astronomy, he dove into the field immediately and made it a lifelong career. His first efforts were finding errors made by famous astronomers; he also correctly predicted an eclipse of the sun. Soon thereafter he wrote an almanac to help farmers and sailors: he calculated the phases of the moon, the times for the rise and fall of the tides, and the positions of the planets. Unfortunately this book was never published, but Banneker wrote a similar one that was published in 1791. This was the first of several almanacs he published.

In 1791, a friend asked Banneker to assist the U.S. president in surveying and laying out the city of Washington, D.C. Later, he continued his writings, stressing freedom for all Americans. He died at the age of 75. This book has a table of contents and an index.

EVALUATION: This book can be classified as a personalized biography. It contains many personal asides, feelings, and conversations of the characters. These are not necessary but add interest and may help the reader to understand the characteristics that help people attain success. The background leading to Banneker's contributions as an astronomer and scientist is explained, thereby making the reading more interesting. There are a few illustrations done in black, brown, and white, but the text encompasses most of the page.

□ □ □

Cobb, Vicki. *The Long and Short of Measurement*. Illus. by Carol Nicklaus. Parents Magazine Press, 1973, o.p. 64 pp. Age levels: 8–11. Grade levels: 3–6. Series: Stepping-Stone.

SUMMARY: The tallest mountain is Mt. Everest. The longest river is the Nile. The elephant is the largest land animal; the whale is the largest sea animal. The hummingbird is the smallest bird. Although a grain of salt and a speck of dust are small, there are things that are smaller that we need a magnifying glass or microscope to see.

The science of measurement helps us determine size, weight, length, height, depth, temperature, and other important properties. The reader of this book discovers that there are some people whose jobs depend on measurement, such as grocers, butchers, and construction workers. Historically, people estimated size by using parts of their bodies: a foot was the length from heel to toe, a yard was the distance between the tip of a finger and the tip of the nose. An inch was the width of a thumb, and a hand the length of a palm. Since this way of measuring proved inaccurate because of differences in people's sizes, it was necessary to develop a standard measuring unit to be used by everyone.

The reader learns how to measure length, weight, temperature, and time in this book. Included in the text are how-to's for constructing simple measuring devices from household items; specific experiments follow each chapter.

EVALUATION: The writing style of *The Long and Short of Measurement* is simple, straightforward, and easy to read. The reader learns that measurements were once all estimates. We estimate things every day, sticking toes in water to see how cold it is, or trying two bags of groceries to see which is heavier.

The illustrations are appropriately simple. The children in the sketches have a cartoon quality, which adds to the levity of the style. Gauges and measuring devices used by scientists and engineers are drawn to aid the child's understanding.

The intermediate grade child is sure to be stimulated by the breezy style and pace of the book. The author encourages the child to think like a scientist, and to invent measuring devices and perform experiments for fun. The book closes with a few experiments as well as methods of recording the results on a chart. A table of contents and an index enhance the book's use in the library or classroom.

□ □ □

Couper, Heather, and Nigel Henbest. *The Sun*. Illus. Watts, 1986. 32 pp. ISBN: LB 0-531-10055-3. Age levels: 10–13. Grade levels: 5–8. Series: Space Scientist.

SUMMARY: Amazing details about the characteristics of the sun constitute the text of this book for middle grade children. When the sun is overhead, the air in our atmosphere scatters some of its light sideways, giving us blue sky. When the sun appears to be setting, more blue light is removed and the sun appears red. The sun also appears larger because our eyes compare it with other distant objects in the horizon. A flash of green light appears just as the sun disappears over the horizon. The sun appears to be up farther than it really is when going over the horizon because the light is bent upward by the air. Because the earth is tilted on its axis, we have seasons. Sunspots are dark because they are cooler than the other parts of the sun. The number of spots on the sun varies from year to year, although there is somewhat of a pattern when every 11 years a maximum number is discernible.

A total eclipse is an unforgettable experience lasting only a few minutes. When an eclipse occurs, the entire sky goes dark. The stars appear and animals go to sleep, thinking it is night. The sun appears to be able to hold very distant objects in its orbit by gravity and scientists believe the sun's gravitational pull may reach as far as a quarter of the way to the nearest star.

EVALUATION: Unlike Seymour Simon's *The Sun* (Morrow, 1986), which contains a close-up view of the sun, this book takes both a more distant and a close-up view. Included are sunspots, flares, eclipses, seasons, and day and night. While this book is thorough in its information, it is overloaded. The photographs and illustrations add to this impression by providing information beyond what's included in the already full text. If students seek a reference book, this title may meet their needs. However, if they are looking for a book to satisfy their curiosity, this may well be too much. Because of the imbalance of technical data and apparent higher readability level, this book may not be entirely understood by middle grade children.

A curiosity-arousing illustration on page 17 shows the path of the moon over the next decade and the places and dates of total eclipses.

□ □ □

Darling, David J. *The New Astronomy: An Ever-Changing Universe.* Illus. by Jeanette Swofford. Dillon, 1985. 72 pp. ISBN: LB 0-87518-288-7. Age levels: 10–12. Grade levels: 5–7. Series: Discovering Our Universe.

SUMMARY: This book begins with a section on modern astronomy, including information on telescopes and their size, and location. Questions and answers about modern astronomy follow. Boldfaced words are defined in the glossary.

Facts on light waves, telescopes and how they work, and other instruments used to study the stars are covered. Thermal and nonthermal electromagnetic radiation is explained: it is through this that we can see far into the universe. Images can be formed for observation by light and radio waves; the universe is explored through many types of instruments that are briefly explained. The book ends with lists of observatories and amateur astronomy groups, and a glossary.

EVALUATION: This book presents very difficult concepts, and a lot of words have to be looked up in the glossary. These frequent interruptions may make the book frustrating to read. Some parts of the book seem to be out of order. The Question and Answer section in the beginning includes items better explained at the end of the book; this section alone may discourage some readers due to the difficult nature of the subject matter. Some knowledge of astronomy and electromagnetic radiation is required in order to fully appreciate this book. It is too difficult or confusing for the average reader to enjoy. Even as a reference book, it is better suited for older readers.

□ □ □

Darling, David J. *The Stars: From Birth to Black Hole.* Illus. by Jeanette Swofford. Dillon, 1985. 64 pp. ISBN: 0-87518-284-4. Age levels: 10–17. Grade levels: 5–12. Series: Discovering Our Universe.

SUMMARY: Star facts and a question-and-answer format are used in the beginning of the book to establish some baseline information before readers proceed to the narrative section. Chapters 1 through 5 make readers feel they are on a trip to the stars where they learn about the life of stars, star doom, twin suns, binary stars, and stars in swarms among other things. Star types, composition, size, weight, age, heat, and distance are all explained. Star changes and explosions and star "death" are discussed, as is the phenomenon known as supernova. Three appendixes follow the narrative core. Appendix A suggests experiments in constellation location, and tells how to keep a log of celestial observations. Appendix B provides statistics about known stars—distance, color, temperature, and so on. Appendix C lists amateur astronomy groups in the United States, Canada, and Great Britain. A glossary and suggested reading list follow.

EVALUATION: Darling provides an enormous amount of information in relatively few pages. The appealing design of the book includes a precontent question-and-answer section that sets the stage for the facts to come. In most other books of this type, questions come *after* the body of information has been introduced, suggesting a test of what has been read. Furthermore, Darling uses *we* and *you* frequently, making the reader feel part of the discovery process. Drawings and diagrams are interspersed with photographs. Large print and magnificent color enhance the text, which is exceptionally well written. Budding astronomers of every age will surely like this book, although novices may find technical terms such as *speckle interferometry* somewhat difficult.

□ □ □

Dempsey, Michael W., and Angela Sheehan, eds. *Water.* Illus. World, 1970, o.p. 30 pp. Age levels: 6–10. Grade levels: 1–4. Series: Starting Point Library.

SUMMARY: A parent, teacher, or librarian explaining water to children may want this book by Dempsey and Sheehan. It presents facts about water that young children can grasp. Readers are told that water covers most of our planet and is needed for all living things to exist, and that water can be found in wells, springs, rivers, and lakes. Most people use water pumped from rivers, which is then passed through several layers of gravel and sand for purification; this water is stored in reservoirs. Chlorine is often added to kill germs before the water is piped to homes and other buildings. Few children will know that lake or river water held by dams can create electricity through the use of turbines and generators. The transformation of water into vapor and ice is explained, and the many uses for water are mentioned.

EVALUATION: *Water* is in the Starting Point Library series, which is designed to give the school-age child a basic understanding of particular subjects. The carefully controlled vocabulary and detailed illustrations use a scientific approach to present the properties and uses of this important natural resource.

No experiments are in the text, but there are easy-to-understand diagrams of a sewage disposal plant, a generator, and the clean water and dirty water pipes that can be found in a home.

This book is recommended for the classroom and the school and public libraries because of its clear, informative approach to the important subject of water.

□ □ □

dePaola, Tomie. *The Cloud Book.* Illus. by author. Holiday House, 1984. 32 pp. ISBN: 0-8234-0259-2; pap. 0-8234-0531-1. Age levels: 4–8. Grade levels: PreK–3.

SUMMARY: Clouds are the focus of this picture book. Ten specific clouds are identified by their technical names, some of which are compared with familiar objects. For example, a cumulus cloud is shown to resemble a cauliflower, and a cirrostratus cloud a white sheet. Beneath stratocumulus clouds, there is a winter scene. Four smiling snowmen wear top hats. Adding a light touch under the ominous clouds, a jovial snowwoman wears a colorful, flowery hat.

Weather patterns are related to the color, shape, position in sky, and other factors about clouds. Many sayings and rhymes about clouds foretelling the weather are discussed. For instance, if there are clouds in the morning, there will be *mountains* of rain, whereas clouds at night predict only *fountains*. The author-illustrator shows the role of clouds in the mythology of various cultures. For example, the ancient Greeks believed that clouds were cattle belonging to the sun god. They were stolen from him by Hermes, the god of the wind, who was also the messenger god. At the end of the book, there is a nonsensical cloud story.

EVALUATION: This informative and humorous "Reading Rainbow" book can be enjoyed by both adults and children. Accurate terminology about ten different cloud formations is covered; for quick reference, a cloud index is included. The scientific vocabulary in this book allows for use with middle-grade students as well as the intended audience. A classroom unit on weather in middle grades might benefit by using this interesting book.

DePaola employs playful and humorous illustrations throughout the book to convey scientific information. He injects wit even in the scientific explanations that appear at the bottom margins of some pages. There is a sense of anticipation as scientific knowledge continues to be presented page by page.

□ □ □

Dixon, Dougal. *Geology*. Illus. by Chris Forsey, et al. Watts, 1982. 38 pp. ISBN: LB 0-531-04582-X. Age levels: 9–12. Grade levels: 4–7. Series: Science World.

SUMMARY: As part of the Franklin Watts Science World series, *Geology* provides an overview of earth science, focusing on the Earth's history and its physical properties. Colorful and detailed illustrations accompany the text on each page of the book. The beginning chapters explain how the Earth formed, the theories of plate tectonics, and how volcanoes and earthquakes have shaped the Earth's surface. Following these concepts are descriptions of metaphoric, igneous, and sedimentary rocks. Additional information discusses fossils, minerals, time scales, and geological products. A glossary and an index are included.

EVALUATION: Clear and detailed illustrations, coupled with understandable terminology, allow *Geology* to be informative and pleasurable to read. The author also makes good use of analogies to explore geological concepts. For example, in a description of plate tectonics, the Earth's surface is compared to the panels of a football with each panel, or plate, growing along a seam. However, the book tends to oversimplify scientific concepts by giving limited information, making it more suitable for the lower end of its intended age range.

□ □ □

Farr, Naunerle C. *Madame Curie; Albert Einstein*. Illus. by Nestor Leonidez. Pendulum Press, 1979. 63 pp. ISBN: 0-88301-368-1; pap. 0-88301-356-8. Age levels: 8–12. Grade levels: 3–7. Series: Illustrated Biography.

SUMMARY: Complete biographies of two famous scientists are included in this book. Marie Curie's follows her life from birth to death; Albert Einstein's begins at the age of five and ends with his death. Each biography begins with a cover page highlighting the contributions of its subject to the world and the importance of these contributions.

The text is in black-and-white comic strip format; the size of the comics varies. Up to three asterisks per page are used to define vocabulary words at the bottom of that page. There is a checkup at the end of each biography, which includes questions on vocabulary and comprehension, as well as questions for discussion. A table of contents, divided into five areas of the scientists' lives, begins each biography.

EVALUATION: The comic strip format is an appealing technique to entice young readers to read a biography. The book itself is rather small so there are not many frames per page. Though there is a good amount of text, it is interesting reading, and highlights both the lives of the scientists and the historical and political factors that affected their lives. The illustrations are also well done, and accurately reflect the costuming, homes, and life during those periods. The illustrator has utilized the corners, tops, or bottoms of frames to relate "aside" information about the subjects, their family members, an incident relating to the text, or political or historical background information. The biographies are rather short, but quite complete and accurate.

It would have been helpful to students if the answers for the question pages were included on a separate page at the back of the book, but in no way do these pages dominate or detract from the biographies themselves.

□ □ □

Field, Dr. Frank. *Dr. Frank Field's Weather Book.* Photos; illus. Putnam, 1981, o.p. 208 pp. Age levels: 10–15. Grade levels: 5–10.

SUMMARY: The first four chapters of this 12-chapter book deal with career aspects of weather forecasting. Field describes a day in the life of a television meteorologist, including the concerns of the viewing public. Field cites weather disturbances as high on the list, especially as holiday weekends approach. Apparently weather forecasters do get blamed when mistakes occur. Several chapters explain how weather data are analyzed, enabling the meteorologist to formulate reasonably accurate reports. The role of weather satellites and other geosynchronous satellites in forecasting is explained. A chapter on scientific projections of climate change is offered, followed by one explaining how climate changes have been observed to affect humans physically and psychologically. A glossary of weather terminology is included, as is a section on the questions asked most often by television viewers together with Field's answers. Three of Dr. Field's children are television reporters and all of them have served as weather forecasters. Readers will be interested to know that Dr. Field's son's name is Storm Field.

EVALUATION: Field writes in an easy, pleasant manner, making the role of television weather forecaster sound appealing; he has obviously convinced three of his children that this is the case, since they have more or less followed in his footsteps.

While one cannot deny his scientific knowledge, there could and perhaps *should* have been more scientific information included. After coasting through this pleasant volume, the reader cannot help but feel that the title suggests something deeper in content than is delivered.

One interesting section mentions *meteoralopathologie,* a science introduced in France to study human responses to weather. Improvement in satellite feedback involving weather data constantly improves, dating some of the scientific information contained herein. The strongest feature of the book, therefore, is the career information Field shares about being a meteorologist, a job he obviously enjoys.

□ □ □

Fodor, Ronald V. *Meteorites: Stones from the Sky.* Photos. Dodd, Mead, 1976, o.p. 47 pp. Age levels: 9–12. Grade levels: 4–7.

SUMMARY: The origin, composition, and characteristics of meteorites are detailed in this book. Meteorites have been falling to Earth over a long period of time. Scientists have concluded that meteorites come from within our solar system, probably from asteroids. At first glance meteorites may be mistaken for rocks; one may have stepped over meteorites out in a field or in the mountains without recognizing them. However, scientific investigation reveals that meteorites are substantively different: while Earth rocks rarely contain metal, almost all meteorites have some metal in them.

Meteorites are discovered as falls or finds. A fall is a meteorite that was seen falling to Earth and is then recovered. A find is a meteorite that was discovered but not seen actually falling from the sky. This distinction is important to scientists because exposure to different weather conditions over long periods of time affects the chemical composition of a meteorite. It has been estimated that about 500 meteorites reach the Earth annually, but that only 4 or 5 are found. Sizes and shapes of meteorites vary: the largest weigh 60 tons while others are the size of dust.

The terms *meteoroid, meteor,* and *meteorite* are explained as follows: a meteoroid is a meteorite before it has entered Earth's atmosphere; a meteor is a shooting star that glows as it passes through the Earth's atmosphere; and a meteorite is the solid material from space that is found on Earth. The sound accompanying a falling meteor has been described as a moving train, cannon fire, or thunder. There has been no authenticated record of a meteorite causing a fire or a human death. At the end of the book, the reader is told what type of information to gather and report if he or she witnesses a fall or discovers a find. Five addresses are given to which the observations and details should be sent. An index is included.

EVALUATION: Although this book does not have a table of contents or chapters, headings divide the text into convenient units. This proves helpful in finding pertinent information quickly. The black-and-white photographs provide ample examples, which correspond to the text. Most of the photographs show details of actual meteorites such as shapes and sizes. A ruler is often included to indicate size. However, in some photos it is not clear whether the ruler is measuring inches or centimeters. The text is informative and is written in a style suitable for young scientists.

□ □ □

Forte, Imogene. *Beginning Science.* Illus. by Gayle Harvey. Incentive Publications, 1986, o.p. Unp. Age levels: 3–5. Grade levels: PreK–K. Series: I'm Ready to Learn.

SUMMARY: This book is designed to be read aloud to children. After listening to the book all the way through, the children are invited to follow the directions on each page. Content and activities follow the concepts of living versus nonliving plants and animals, five basic bones of the body (head, shoulder, hand, knee, foot), health foods, mammals born alive versus animals that hatch from eggs, where animals live, animals that make good pets, what food we get from animals, the four basic parts of a plant (flower, leaves, stem, roots), fruits and vegetables, keeping the air clean, daytime and nightime, seasons, vehicles and other machines with wheels.

EVALUATION: In an easy and pleasurable way, children can develop basic skills and concepts in beginning science as they learn facts about plants, animals, the Earth, weather, and the human body. The pictures are simply drawn and easily recognizable. The text provides basic information and the directions are very specific. The scientific information is mostly in a comparison format: when items are grouped in a set, only these items are dealt with at one time. A list of experiments is included at the end and can serve as idea starters for further experiments. This is a very practical book for involving young children in science.

□ □ □

Foster, Leslie. *Rand McNally Mathematics Encyclopedia.* Illus. by author; photos. Rand McNally, 1986. 141 pp. ISBN: 0-02-689202-2. Age levels: 9–13. Grade levels: 4–8.

SUMMARY: As an encyclopedia, this book covers the gamut of the mathematical world. The table of contents includes 50 categories with titles like "Hogsheads and Liters" and "Numbers in a Sieve." The beginning chapters discuss the reasons for using math, what math really is, and that it can be fun and a type of magic to stimulate thinking. This is followed by a history of math, beginning with the cave dwellers and their drawings, and ending with the Greeks and Romans. This leads into our present-day Arabic numeral system and its history.

Throughout the book, there are problems or equations for the student to work. The answers, many with illustrations, are found at the back of the book; an index is provided.

EVALUATION: The book is large and contains many illustrations including photos, drawings, museum prints, diagrams, and charts. This format makes this one-volume encyclopedia on mathematics exciting and appealing. At a glance, the reader can see from current photographs how math is necessary in architecture, in everyday life (such as using calculations when crossing a street), and in inventions, particularly those used in space. The illustrations are lively, colorful, and sometimes humorous. In some cases the reader can grasp information instantly from the illustrations; other illustrations entice the reader to find out more. Even the endpapers have interesting illustrations.

Each new topic is begun as if the book is speaking directly to and involving the reader personally. Subtitles then divide the section to aid understanding or finding pertinent information. Many puzzles, comparisons of world records, and other activities invite the reader to enjoy math. The text contains a good amount of mathematics information, problems, and concepts. The reading difficulty varies because of the jargon particular to the discipline.

This book should not be overlooked by teachers, parents, and others to use in teaching or clarifying a mathematical problem or concept. Many of the ideas are directly applicable to classroom use.

□ □ □

Fradin, Dennis B. *Astronomy.* Photos. Childrens, 1983. 48 pp. ISBN: LB 0-516-00533-2; pap. 0-516-41673-1. Age levels: 6–9. Grade levels: 1–4. Series: New True.

SUMMARY: Scientific concepts about the stars, constellations, galaxies, the solar system, planet Earth, and the universe are presented here with accompanying full-page photographs from the U.S. Naval Observatory and NASA, among other sources. Copernicus, Galileo, Sir Isaac Newton, and Edwin Hubble are mentioned and their contributions to the science of astronomy capsulized in a sentence or two. Questions still to be resolved by astronomers are cited, and a glossary of astronomical terms is included at the end of the book. Diagrams are used to explain day and night, and the revolution of Earth around the sun. Two pages use drawings to help children understand how the constellations were identified in ancient times.

EVALUATION: This is a very clear, concise "first" book for children about the science of astronomy. It might be argued that it is *too* concise: in several places a concept is left dangling where some elaboration would have been useful. For example, on page 39 Fradin says that there are many stars besides the sun, and that some of these stars might have planets like Earth. Attentive children will want to know why there is presently no knowledge of these other planets, and how they might be discovered.

Marvelous photographs share the page with very large print and substantial margin space. In the section on constellations, the author says that the constellations are pretend star pictures, named by ancient people who played "connect-the-dot" in their minds.

□ □ □

Froman, Robert. ***Bigger and Smaller.*** **Illus. by Gioia Fiammenghi. Crowell, 1971. 33 pp. ISBN: pap. 0-690-14197-1. Age levels: 5–8. Grade levels: K–3. Series: Young Math.**

SUMMARY: The author indicates that whether something is big or small is relative: there are different kinds of big and small. For example, an object may be bigger because it is taller, or it may be short, but bigger because it is fatter. A full grown man may be taller than a baby elephant, but not be taller than a full grown elephant. Whales are the biggest animals known, but some trees are bigger than the blue whale. There are nonliving things that are bigger than the trees, and there are many things smaller than the smallest insects.

EVALUATION: The pictures in this book are multicultural and present a universal appeal to readers. Also, it is interesting to have visual facts about the smallest man and the largest man, both of which reinforce acceptance of different types of people. The author does a thorough job of exploring and defining the concepts of big and small, and the simple sketches help clarify the concepts. Children gain an understanding that bigness and smallness are relative to the objects with which they are compared. In addition, the author does a wonderful job of incorporating a variety of individual cultural perspectives.

□ □ □

Froman, Robert. ***Hot and Cold and In Between.*** **Illus. by Richard Cuffari. Grosset & Dunlap, 1971, o.p. 45 pp. Age levels: 4–8. Grade levels: PreK–3.**

SUMMARY: This book deals with the concepts of hot and cold. An experiment helps the reader discover what makes hot and cold things feel the way they do. Using three bowls, one is filled with comfortably hot water, another with cold water, and a third with both hot and cold water. The reader is told to put one hand in the hot bowl and the other in the cold bowl and to hold them there for about a minute. Next, the reader is told to put his hands in the third bowl.

The hand that was in the hot water will feel cold, and the hand in the cold water will feel warm. The reason for this is that heat flows from hot to cold. For one hand, heat was flowing from warmer flesh to the cooler water. For the other hand, heat was flowing from the warmer water to the cooler flesh.

Whenever two things touch, there is usually a flow of heat between them. The author explains that when an individual feels cold, heat loss is occurring.

Heat has a flow pattern but cold does not. Since heat is always flowing, it is difficult to locate an object that can be described neither hot nor cold.

EVALUATION: Written for primary children, this book does an excellent job of presenting the properties of hot and cold. Through a simple experiment, the contrasts are discovered. The author follows up with a detailed but simple explanation of why things feel hot or cold. The pictures help clarify the experiment; illustrations of other experiments regarding the flow of heat are included. With clarity and ease of readability, the book can provide the reader with a scientific explanation of hot and cold without using experimental techniques that are unsafe.

□ □ □

Froman, Robert. ***Less Than Nothing Is Really Something.*** **Illus. by Don Madden. Crowell, 1973, o.p. 33 pp. Age levels: 7–10. Grade levels: 2–5. Series: Young Math.**

SUMMARY: This book is about negative numbers and provides excellent examples of how we use them every day. For example, the reader is asked to imagine what would happen if he had two pennies to buy a piece of candy that costs 2 cents (after buying the candy he has no money left). But if he only had one penny and needed to borrow another from a friend, after buying the 2-cent piece of candy he would owe 1 cent to his friend and therefore have negative 1 cent or −1.

Negative numbers are also described by using a thermometer. 0° on the thermometer is very cold but there are places in the world where it is colder: once a thermometer in Antartica measured −125°. A game is illustrated to facilitate the addition and subtraction of negative numbers. The simple rules are explained, and methods of adapting the game for outdoor use are included.

EVALUATION: Many young children have the notion that numbers start at 1 and go all the way up to whatever needs to be counted. Indeed we teach the three-year-old to count by putting a 1 to 1 correspondence between numeral and object. Thinking negatively is an abstract notion most children are unable to comprehend until they reach school age. *Less Than Nothing Is Really Something* explains through the use of the number line as well as other examples how we use negative numbers in our daily lives.

The game P.A.M. (plus and minus) is a valuable addition to the book; the reader is asked to divide a large piece of paper into eight boxes. A positive or negative number is printed in each and the paper is taped to the table. The child is instructed to flick a counter (such as a small disc of paper) onto the numbers and either add or subtract his score each time he receives a turn. The numbers are kept small at the beginning but as the children become more adept, they are encouraged to raise the value and increase the size of the board.

Madden's illustrations help explain the text. A white boy and a black girl are pictured throughout, with the frequent appearance of delightful animals and a curious parrot. Books like this with a breezy style and numerous concrete examples make learning about mathematics fun and easy.

□ □ □

Froman, Robert. *Rubber Bands, Baseballs and Doughnuts.* Illus. by Harvey Weiss. Crowell, 1972, o.p. 33 pp. Age levels: 8–12. Grade levels: 3–7.

SUMMARY: Froman helps the reader understand the principles of topology. Topologists are mathematicians who work with items that when stretched, crumpled, or sliced retain some of their original properties. Examples of these property changes are given throughout the book; for example, when a piece of paper with a straight line on it is crumbled the line is distorted, but it can be observed that the points along the line stay in sequence. A rubber band with knots may be stretched, but the knots stay closest to the end near which they were tied. The author explains that things that don't change are called *invariants,* and readers learn that invariants can also be found, among other places, on simple closed curves. One's face reflected on the side of a toaster may be distorted, but some things are observed to be invariant, for example, one's nose is still between one's eyes.

EVALUATION: At a glance, the book appears to be written for younger children. However, the abstract concepts and the methods of presentation generally require a more mature mind. Two birds accompany the reader through the learning of topology and, through their conversations, add welcome humor and illuminate some concepts presented in the text. The illustrations are an essential component for understanding the book. Because the concept of invariants requires some abstract thought, the questions posed by the author should all be answered, but sometimes they are not. Relatively difficult to define and infrequently taught at the elementary level, topology can prove a precursor for many skills and for logic in later grades. The book is not new, but the subject matter cannot become dated.

□ □ □

Gans, Roma. *Danger—Icebergs!* Illus. by Richard Rosenblum. Crowell, 1987. 32 pp. ISBN: LB 0-690-04629-4; pap. 0-06-445066-X. Age levels: 4–10. Grade levels: PreK–5. Series: Let's-Read-and-Find-Out Science.

SUMMARY: A map and text locate icebergs throughout the world. Several pages are devoted to how icebergs are formed, beginning with the many years it takes the snow to solidify. The text continues with an explanation that a glacier becomes an iceberg when chunks break away from it. Some glaciers are so enormous that towns could be built on them. The speed of icebergs and how they form an ice pack are addressed as well as the fact that some icebergs can turn upside down. Holes in icebergs can provide shelter for seals and polar bears. Finally, the book ends with the tragic story of the *Titanic,* and a description of the modern-day technological instruments that now alert navigators to icebergs.

EVALUATION: The map at the front of the book clearly designates where icebergs form in the world; this is done in white on a green and blue map. The colors in the illustrations are striking and informative, and convey the size and shape of these enormous, natural formations well. Particularly dramatic is the illustration showing approximately one-eighth of an iceberg above water and seven-eighths of it submerged.

The author does not portray icebergs as silent, but rather as mountains of ice that boom when they break apart. Sounds of the newborn seal pups or polar bears are emitted from the hollows within the icebergs.

The author relates explanations to a child's way of thinking: the speed of an iceberg's movement is slower than the pace of a child walking. In today's world, scientific instrumentation makes navigation less hazardous than ever before. This book is a revised edition of Gans's *Icebergs* (Crowell, 1964). Both books could be used as a worthwhile classroom scientific exercise in comparing new and former knowledge about icebergs.

□ □ □

Gans, Roma. *Icebergs.* Illus. by Vladimir Bobri. Crowell, 1964, o.p. Unp. Age levels: 7–10. Grade levels: 2–5. Series: Let's-Read-and-Find-Out Science.

SUMMARY: Facts about icebergs are presented with minimal detail in this controlled vocabulary book. Illustrations are also characterized by a minimum of detail. Alternating blue-green and black-and-white illustrations are highlighted with yellow on some pages. Occasionally humans, ships, animals, and birds appear, although the focus remains on the formation, size, shape, and changes of icebergs. Certain pertinent facts about icebergs are included, such as information on the detection and monitoring of them.

EVALUATION: Experts in the fields of science and education were consulted in the preparation of this book. While the book contains accurate information, it does not answer questions children may ask about iceberg phenomena, such as why icebergs change in color from blue to green or white. Also, it does not explain why most of the iceberg is below the water line. The book reveals that glaciers work their way toward the ocean, but does not explain why they move slowly. Even within the constraints of the vocabulary used, concepts like these could have been explored. A strength of the book is the discussion of the relationship that icebergs have to all of nature.

□ □ □

Gans, Roma. *Rock Collecting*. Illus. by Holly Keller. Crowell, 1984. 28 pp. ISBN: LB 0-690-04266-3; pap. 0-06-445063-5. Age levels: 5–8. Grade levels: K–3.

SUMMARY: Young children are sure to like this picture book, which gives a great deal of simplified geological information in very few pages. Facts are cast into elementary vocabulary: some rocks are easy to find and some are rare—that is what makes them valuable. Thousands of years ago, Romans used rocks to build roads, and Egyptians built pyramids with them. Talc is the softest rock, and diamond is the hardest substance, sometimes found in rock. Rocks cover the entire earth and are located under the ocean. The top layer of the earth's crust is made mostly of igneous rock such as granite, quartz, and basalt. Magma (melted rock inside the earth) comes to the surface as lava from a volcano. Sedimentary rocks are made of layers pressed together, resulting in sandstone, shale, and limestone. Another kind of rock is called metamorphic (meaning change) and includes slate, gneiss, marble, and quartzite formed from shale, granite, limestone, and sandstone, respectively.

EVALUATION: Labeled for ages 5 through 8, this book could be expanded to include older readers when introducing them to rock collecting. It is very simple in its descriptions of the three types of rock, but detailed enough to appeal to advanced students. Rock collecting is made appealing because one can collect rocks anywhere and still have a unique collection. There is a combination of pictures, photographs, and drawings that are enticing but the actual photographs provide the best examples. These will be important for students beginning to categorize their collections since they require as accurate a reference as possible.

□ □ □

Gibbons, Gail. *Dinosaurs*. Illus. by author. Holiday House, 1987. Unp. ISBN: 0-8234-0657-1; pap. 0-8234-0708-X. Age levels: 4–8. Grade levels: PreK–3.

SUMMARY: This book for young children reads like a junior dictionary of information about 14 specific dinosaurs. Included are respellings for pronunciation help and information about how dinosaurs looked, how they lived, and what dinosaurs ate. The work that paleontologists do and theories about the disappearance of dinosaurs are included. Paleontologists continue the study of dinosaurs, even though these creatures became extinct about 70 million years ago. The last page, "Dinosaur Footprints," gives information derived from tracks left by dinosaurs.

EVALUATION: The illustrations' bright, bold colors are a dominant feature of this informative book. They are vital—the dinosaurs appear alive, but nonthreatening. The colorful dinosaurs will attract attention; however, artistic license may have been taken because there is probably no way scientists can know about dinosaur colors from fossil remains. The dictionary features of the text are helpful, especially the respellings of the dinosaurs' names. However, the dictionary format is not followed through because the dinosaurs are not presented in alphabetical order. Children will love the beauty and the excitement of this book as they learn about dinosaurs, but will not be overburdened by facts.

□ □ □

Gibbons, Gail. *Weather Forecasting*. Illus. by author. Four Winds, 1987. Unp. ISBN: 0-02-737250-2. Age levels: 5–8. Grade levels: K–3.

SUMMARY: The activities of two weather forecasters and a meteorologist at a weather station are featured. The work of an immediate weather forecaster and a long-range forecaster is explained. These forecasters are seen reading and checking instruments that provide pertinent information that will be transmitted to a central weather ofice. Their modern weather station has computers, an elaborate observation console, a broadcasting system, a radar monitor, and other technological equipment.

This book is divided into four sections featuring the seasons. Some typical seasonal weather is included in the drawings and text. On a chilly spring day, there are heavy dark clouds and rain as the busy forecasters gather data. A tornado is reported in the Midwest. During a very hot, sunny day in the summer, thunderstorms are predicted. On a cloudy, windy day in the fall the course of a hurricane is followed. Due to the tireless work of competent forecasters who keep track of changing weather, people are prepared for a heavy snowstorm in the winter.

EVALUATION: This colorful book contains a great deal of information about weather forecasting, and could serve as a reference book for young children. The role of the two forecasters is vividly and prominently described; however, the meteorologist's function is underplayed. The text is direct and mainly explanatory. Some of the vocabulary is defined within the illustrations. This pictorial dictionary arrangement adds to clarifying the scientific terminology that is used. However, there is so much technical vocabulary in the text that adult help might also be necessary. Illustrations dominate each page and follow the cartoon format. As the significance of weather predictions on people's lives unfolds through seasonal minidramas, readers' attention will surely be captured.

□ □ □

Goldreich, Gloria, and Ester Goldreich. *What Can She Be? A Geologist.* Photos by Robert Ipcar. Lothrop, 1976, o.p. 48 pp. Age levels: 6–10. Grade levels: 1–5. Series: What Can She Be?

SUMMARY: This nonfiction book is an account of the life of a female geologist who teaches at a university, conducts research in field geology, and also does consultation work for companies. Her interest in teaching extends beyond college students: in her busy life, she finds time to help very young friends examine rocks so that they can learn more about their planet Earth. With them, she inspects and identifies the interesting rocks the children have brought back from a camping trip. She even takes them on an outdoor geologic exploration. Her full life includes time to relax with her family, too.

This scientist keeps current in her profession because she enjoys reading books and journals about geology. A medical geologist, a mineralogist, and a lunar geologist are some of her friends in related occupations.

Her professional field work is varied. She scientifically investigates an area in Maryland in order to write its geologic history. A site in Pennsylvania is studied for the best and safest place to build an electric power plant. In California, the planned construction of a highway is evaluated. The expertise of a competent geologist is shown to be needed by many people.

EVALUATION: This factual portrayal of a female geologist is part of the Lothrop What Can She Be? series that includes an architect, a musician, a veterinarian, and other professionals. The authors introduce this dedicated scientist by name, and readers feel that they have met a knowledgeable person who has an interesting occupation.

The photographs are in black and white, and present the factual information in a straightforward fashion. The reader finds that when this geologist was young she was interested in her environment and asked many questions about it. She wondered why some places were flat and others had mountains. Her studies included many courses in chemistry, math, and physics. She entered graduate school and earned a doctorate. Enthusiasm for her work is evident in the easily read text and plentiful photographs. The details in this simple text might foster children's interest in preparing for a career in this area.

□ □ □

Goor, Ron, and Nancy Goor. *Shadows: Here, There and Everywhere.* Photos. Crowell, 1981. 47 pp. ISBN: LB 0-690-04133-0. Age levels: 4–8. Grade levels: PreK–3.

SUMMARY: This science book explains in language that younger children will grasp what causes shadow formations. Readers are told about different positions the sun must have in order for shadows to appear. The book reveals that multiple lights shining on an object will produce a number of shadows equal to the number of lights. Children will be interested to know that changing the surface, position, or shape of a given object will result in a change in the shadow the object creates. The authors further explain how shadows have been used to tell time on a sun dial, and point out that shadows cannot be seen in the dark, or on a cloudy day.

EVALUATION: This is a strikingly beautiful book. The photographs have captured unique shadows and they are appropriately synchronized with the text. For example, the number of lights shining on a pair of ice skaters is discovered by counting their seven shadows. Changing the surface the shadow falls on gives a zigzag up steps or a bend on the curve of a water tank. A page of black sketches on white backgrounds with a challenge to guess what objects made the pictures is included. Some are quite difficult and will present a challenge to the reader. The surprising answers are provided by actual photographs on the following pages. Very well written and uniquely illustrated, *Shadows* will be an enjoyable reading experience for any reader.

□ □ □

Gutnik, Martin J. *The Science of Classification: Finding Order among Living and Nonliving Objects*. Photos; illus. Watts, 1980, o.p . 66 pp. Age levels: 9–14. Grade levels: 4–9. Series: First Book.

SUMMARY: This book is divided into nine chapters and includes a glossary and index. The first chapter introduces the reader to man's need for order, patterns, and relationships and the idea that the search for order in one's environment is called classification. It is explained how all matter is made up of elements, which are listed in the periodic table. Also discussed are classifying nonliving objects, distinguishing between living and nonliving things, the biosphere and its classifications, classifying living objects through the food chain, evolution and classification, cells and classification, and modern taxonomy. Included are black-and-white diagrams, drawings, photos, and references to people who had an impact on this system.

EVALUATION: The author uses examples that relate to a young reader's way of thinking, for instance, in classifying a cage full of puppies, each having some similar and some dissimilar characteristics, such as size and coloring. There are a variety of ways to classify and the author does not state one particular way to be correct. He suggests the criteria for classification is that the system should be consistent and useful. The drawings and diagrams are simple and many are labeled with specific technical terms. For instance, the amoeba's parts are labeled; for older students this might be quite useful, particularly for research purposes.

□ □ □

Hansen, Rosanna, and Robert A. Bell. *My First Book of Space: Developed in Conjunction with NASA*. Photos; illus. Simon & Schuster, 1985. Unp. ISBN: LB 0-671-60621-2; pap. 0-671-60262-4. Age levels: 7–10. Grade levels: 2–5.

SUMMARY: Photographs highlight this large book about space, which shows what the Earth looks like from space. Discussion about the planet Earth and the solar system follows, including diagrams. What comprises the solar system and information about the force of gravity are contained here as well. A section about the sun includes very large photos and an insert about how the sun began. Each planet is then discussed, with photos supporting the text. Interspersed is coverage of asteroids, comets, meteoroids, and stars.

Each of the photos, mostly from NASA, is explained in a caption; when any enhancements are made, such as the adding of color, the photo caption clearly states it. There is a considerable amount of text for a first book; however, the sentences are generally short and, with the exception of proper nouns, they read fairly well.

EVALUATION: The photos are extremely colorful; in some the color was added by a computer. They are also quite large; much information may be gained even if the text is not read. The cover and inside endpapers are black, aiding the space theme; the cover has a photo of Saturn.

The text flows smoothly but gives much factual information. Considerable space is devoted to Earth; a drawing shows the layers of the atmosphere and a cross-section shows the three parts of the Earth—the core, mantle, and crust. Rotation and revolution are also illustrated and explained. There is also a section about Earth's moon.

□ □ □

Hatch, Shirley Cook. *Wind Is to Feel*. Illus. by Marilyn Miller. Coward, McCann, 1973, o.p. Unp. Age levels: 0–7. Grade levels: PreK–2.

SUMMARY: Almost every aspect of wind is discussed in this book for young children: how wind feels in a light breeze or a strong gust; whether it feels warm or cool, dry or moist; the effects of wind blowing objects; and how wind carries scents and odors. The idea that wind has worked for man is conveyed through a windmill and sailboats. Wind instruments such as the clarinet, flute, and saxophone show that wind can make music and therefore has a fun aspect. The wind as it affects our living environment, such as wind moving over large land and water masses, weather, and clouds or wind carrying seeds in the air, is also discussed. Many simple experiments and activities for children to do are included. The last page is devoted to a short summary of the book.

EVALUATION: The author relates well to a young child's knowledge and experience with wind, making the book one a child can understand and enjoy. There are at least 15 easy activities for children to do, such as blowing a piece of tissue and a piece of cardboard off a table to find out which is easier to move, or watching a leaf float through the air. Blowing across the top of a bottle or letting air out of a balloon slow or fast also help a child learn that wind makes different sounds. In addition, there is a page describing the book and suggesting many additional activities for children.

That wind is approached on a global scale—from blowing on a curtain to its effects on weather and environment—helps to broaden a child's spectrum on a particular subject. In these descriptions, the author uses vivid imagery and colorful adjectives such as *creak, whistle,* and *groan* to make the reader aware of the effects of wind. The illustrations are interspersed throughout the text and the charcoallike drawings depict the effects of the wind. The summary of what wind is and how it affects our lives is a welcome reinforcement to a concept that can be difficult for young children to comprehend.

□ □ □

Hatchett, Clint. *The Glow-in-the-Dark Night Sky Book.* Illus. by Stephen Marchesi. Random House, 1988. Unp. ISBN: 0-394-89113-9. Age levels: 5–17. Grade levels: K–12.

SUMMARY: *The Glow-in-the-Dark Night Sky Book* actually does what the title indicates. Directions for making the book glow in the dark are given in the beginning of the book. More than 30 constellations are pictured on the maps, and the tiny dots that glow show what the constellations look like. The reader is then encouraged to find these constellations in the night sky. The maps are also divided by season: by looking at a chart the reader can find out at approximately what time of the month the constellations in the sky will match those on the map. One of three star groupings—the Big Dipper, the Summer Triangle, or Orion the Hunter—is included on each map to help find the direction south, which is necessary in order to use the star maps. A few simple directions for doing sightings outside are also given.

Each two-page spread has two maps: one shows the constellations as animals, gods, humans, or other creatures and the other shows diagrams of these constellations; each of these maps glows. The time of year that the constellation can be seen and a compass complete the spread.

The book ends with specific information about the constellations, including background on what people used to believe and how the stars were used to identify the seasons for planting purposes and navigation. The two end pages depict the summer and the winter sky.

EVALUATION: This book, written by an astronomer, is an intriguing one for both the child and the adult, as well as the astronomy novice or expert. By exposing the constellations (which were printed using a nontoxic ink) to a bright light indoors, the reader can make the stars in the book glow in the dark (if necessary, the constellations can be recharged with a flashlight). The book is rather large and the constellations are simple yet well defined for easy use. Great care and detail have been given to the illustrations of the map of the "ancients." The colors are outstanding and the lines on the fur of the animals are extremely detailed. A glowing star dots the eye of a whale, a horse, and a fish. Plants or moons occupy a small place in the outside corners of the double pages and the moons are identified with the planet.

The last page is a quick and easy source of reference about the constellations: size, tips for recognizing shapes, or information on the gods.

□ □ □

Herbert, Don. *Mr. Wizard's Supermarket Science.* Illus. by Roy McKie. Random House, 1980. 96 pp. ISBN: LB 0-394-93800-3; pap. 0-394-83800-9. Age levels: 8–12. Grade levels: 3–7.

SUMMARY: The author presents more than 100 experiments using items that can be found around the house or purchased in a neighborhood supermarket. These ideas are a result of the author's preparation for the "Mr. Wizard" television program. The table of contents is ordered according to supermarket classifications: cleaning supplies, dairy, juices, condiments, and the like. In all, there are 19 categories from which to choose, and titles of the actual experiments are listed under the chapter titles.

In many instances, the author gives background information about an ingredient or item that will be used in an experiment. These facts help explain why the experiment works. At times additional uses for an ingredient, such as using baking soda for stopping grease and oil fires, is included. The experiments are then outlined and explained.

Green, yellow, orange, and blue are the only colors used in the illustrations. A humorous illustration may introduce the experiment, but other drawings serve to explain.

In addition to a contents page, there is an index and information about the author—including his photograph—and the illustrator's.

EVALUATION: This book should appeal to young readers because the experiments can be conducted easily and use common household items. Some of these include a cereal box, granulated sugar, a jar, a straw, a potato, and the like. Helpful hints are often given, such as adding a tablespoon of sugar to a vase of marigolds to eliminate a disagreeable odor, or adding vanilla to paint to mask its odor. The author also uses words like *trick, puzzle,* or *secret code,* which can lure the reader to find out more or to do the experiment. The titles of the experiments are also appealing: "Paste-ry," "Spaghetti: Mothball Substitute," and "Glue from Milk." It would have been helpful to list the materials needed for each activity; the reader must read the entire experiment and at times the introduction to the experiment to make a list of the necessary equipment. The illustrations help at times by showing what is needed; in other cases, they illustrate only the stirring of the ingredients, for example. Sometimes the drawings show exactly how to do the experiment.

The background of the author—an author and producer of many science books, films, and television programs—would seem to indicate that the material is tested and sure to be extremely appealing to children.

□ □ □

Hoban, Tana. *Circles, Triangles and Squares.* Photos by author. Macmillan, 1974. Unp. ISBN: 0-02-744830-4. Age levels: 4–7. Grade levels: PreK–2. Awards: The New York Academy of Sciences Children's Science Book Award—Younger Honor, 1975.

SUMMARY: This wordless picture book is a series of black-and-white photographs that capture circles, triangles, and squares in ordinary and familiar objects. Bubbles blown by a child, shoelace eyelets on sneakers, and the wheels of roller skates are some of the round shapes that can be found. The silhouette of a flying crane reveals innumerable triangles. A child wears a triangular sailboat hat, and the spokes on the wheels of a baby carriage form triangles as they radiate from their hubs to the rims. Squares are seen in the strings of a tennis racket, the fabric of a dress, and the wire screening of a rabbit hutch.

In many photographs, all three shapes appear more than once. On a building, the triangular shape of the pediment, the square window panes, and a large circular decoration within the pediment are easily discernible. Upon more careful scrutiny, the observer can pick out these shapes again in the round pull of a window shade, square bricks, and triangular rafters. Scenes with children, cityscapes, and a harbor view constitute opportunities for geometric shapes to be observed in everyday situations.

EVALUATION: No text is needed with Hoban's imaginative and self-explanatory photographs, which encourage children to use their eyes in creative ways. Circles, triangles, and squares become obvious in places where they were not seen before. After experiencing this incredible book, children will surely be motivated to look for geometric shapes in their immediate environment. This exceptional book is a beginning geometry lesson that is also a lot of fun.

□ □ □

Hunter, Nigel. *Einstein.* Illus. by Richard Hook; photos. Bookwright, 1987. 32 pp. ISBN: LB 0-531-18092-1. Age levels: 8–11. Grade levels: 3–6. Series: Great Lives.

SUMMARY: This book relates the life and achievements of Albert Einstein, a man considered by many to be this century's greatest scientist. The book follows Einstein from his birth in Germany to his death in the United States. It chronicles his scientific interests and accomplishments with anecdotes about his early years, schooling, and adult life.

The reader discovers that, as a young boy, Einstein was intrigued with algebra, which his Uncle Jakob said was a science where one chased an unknown animal called "x." Before he was a teenager he was reading books about science and math. Considered a rebel and disruptive, Einstein was expelled from school at age 15. Later, while studying for a teacher's degree, he became preoccupied with physics. Through the language of higher mathematics, he worked on basic problems in physics and developed theories that changed the world of science. His special theory of relativity became the foundation of modern physics. With his general theory, the place of gravity in the space-time universe was explained. Through experiments, other scientists confirmed Einstein's theories.

He received the Nobel prize for physics in 1922. His professional as well as personal life was marked by many changes, especially by Germany during World War I and later Nazi Germany. His scientific research and search for peace in the world continued when he settled in the United States in 1933. Until his death in 1955, he urged that nuclear weapons be abolished. Important dates in Einstein's life are listed at the end of the book. A table of contents, glossary, index, and short bibliography are supplied.

EVALUATION: In this biography the human side of Einstein is traced along with his scientific breakthroughs. The book seems chatty with anecdotes such as how Einstein didn't wear socks because it was a waste of time, that his second wife was also his cousin, and that, after taking a walk, he had to telephone the university to find out where he lived. While divulging this eminent scientist's personal life, the text also strives to explain his complex theories in simple language. The black-and-white photographs provide a visual look into his world. Colorful illustrations supplement the text.

□ □ □

Jefferis, David. *Satellites*. Illus. by Robert Burns, et al. Watts, 1987. 32 pp. ISBN: LB 0-531-10348-X. Age levels: 8–11. Grade levels: 3–6. Series: Easy-Read Fact.

SUMMARY: This book begins with a brief history and background of satellites. It describes how they are constructed and how they operate to orbit the Earth. How satellites track weather and provide information on the Earth in the areas of animals, minerals, mapping, fishing, earthquakes, volcanoes, farming, forests, ice, and snow are also explained. Communications satellites and spy satellites are covered. The use of satellites in science, space exploration, and astronomy, as well as the projected use of satellites for the Strategic Defense Initiative, is also included. Space stations are briefly explained. A short diary of satellite history and events is followed by a glossary and an index.

EVALUATION: This book is part of an easy-to-read series, but has many long, technical words that may exclude it for younger readers. The concepts are presented in a clear, concise manner with many detailed color photographs or drawings of the subject matter. Each illustration is explained by a caption. The glossary contains only twelve words. An extensive glossary would be more appropriate due to the technical nature of the subject matter.

This book is a good introduction to the various types of satellites. The sophisticated subject matter of this book may be interesting to older elementary students, but they may not read it because of its resemblance to a younger child's book.

□ □ □

Jobb, Jamie. *The Night Sky Book: An Everyday Guide to Every Night*. Illus. by Linda Bennett; photos. Little, Brown, 1977. 128 pp. ISBN: 0-316-46551-8; pap. 0-316-46552-6. Age levels: 8–13. Grade levels: 3–8. Series: Brown Paper School.

SUMMARY: This book begins with information on how the ancients used the sky and constellations to determine time and seasons. Games and activities are interspersed throughout, such as the directions for making a solar stone. The night sky and navigation, the zodiac, latitude, longitude, and sundials are mentioned. Text and photographs provide information about the planets and their orbits, light, stars, meteors, and so on. Some illustrations are cartoonlike, while others are scientific drawings.

EVALUATION: Jobb has consulted researchers in the field and obtained photographs from NASA and the Lowell and Yerkes Observatories for this book. Most of the drawings, maps, games, diagrams, and activities are simplified, but the general information contained in the book is rather sophisticated and probably intended for older audiences. However, the historical aspects could be shared by children in literature or social studies classes. The two sections on the solar system and stars may also be appropriate for classroom use. A section on the zodiac with a game and activities could provide hours of fun.

The entire book is done in sepia tone. The large illustrations, drawings, maps, and diagrams are easy to follow and read, and many of the facts are presented in easy-to-read measurements or drawings.

□ □ □

Kightley, Rosalinda. *Shapes*. Illus. by author. Little, Brown, 1986. Unp. ISBN: 0-316-54005-6. Age levels: 3–6. Grade levels: PreK–1. Series: Primers.

SUMMARY: Ten different geometric shapes are introduced in this picture book. Each shape is presented, with its name, on a left-hand page. On the adjacent right-hand page, a large illustration contains the geometric shape, and the reader is instructed to find as many instances of that shape as possible. For example, after the circle is shown, a stylized caterpillar made up of 32 circles is pictured.

As each geometric shape is depicted, the illustrations progress to include all the other shapes that have been introduced to that point. For example, the illustration containing semicircles—a black cat peering over a wall—also includes circles, squares, rectangles, triangles, and diamonds. For reinforcement, examples of the shapes are included below the illustrations. The tenth illustration—a couple of robots interacting with each other—is composed of all the geometric shapes. The reader is given a sequential and final look at all the shapes as the last two pages of the book reproduce all the shapes, their names, and the illustrations containing these shapes that were seen in the previous pages.

EVALUATION: This very colorful book not only provides a very basic lesson in geometry, but also offers entertaining exercises for counting and visual discrimination. The stylized illustrations are bold and striking and could easily be reproduced as collage designs by teachers, parents, and even children. The vibrant color of the geometric forms repeated throughout the endpapers adds to the book's appeal. The large print text uses labels to inform and questions to involve the reader-listener as an active participant. Readers will experience the wonder of geometric forms and the excitement of discovery in this exciting concept book.

□ □ □

Knowlton, Jack. *Maps and Globes.* Illus. by Harriett Barton. Crowell, 1985. 42 pp. ISBN: LB 0-690-04459-3; pap. 0-06-446049-5. Age levels: 7–10. Grade levels: 2–5.

SUMMARY: The history of mapmaking is briefly described at the beginning of this book. The use and reading of globes and maps are explored, with an explanation given as to why globes are more realistic than world maps. Some basic language used in cartography, such as *equator, scale, latitude,* and *longitude,* is defined.

Although world maps give some distorted geographical views, the importance of maps in imparting information is discussed. By using different techniques, mapmakers produce all kinds of maps. Natural features, such as terrain and vegetation, are shown on physical maps. The boundaries between countries are seen on political maps. Local maps can provide much detail about a community, including the location of important structures.

Four different ways of looking at the state of Arizona are highlighted in large illustrations. The first map shows the counties and county seats of Arizona; the second map presents the state's national parks and monuments. Locations of gold, silver, and copper deposits are featured in the third map. The routes of nineteenth-century stagecoach lines run across the fourth map, which tells a little history about Arizona. All kinds of helpful information about maps are introduced in this book, even the Dewey decimal classification number for finding atlases in the library.

EVALUATION: This "Reading Rainbow" selection is a wonderful teaching tool to aid beginning map readers to understand globes and the intricacies of maps. Mainly devoted to map instruction, this book relates technical information with simplicity and clarity. Large print adds to the ease of reading. The bright, childlike illustrations are large and appealing. Some young geographers might be disappointed that while interesting facts are related about the highest place on earth and the deepest canyon under the ocean floor, they are not indicated on any map in the book. The variations in maps and map reading are presented in an interesting manner. This information book would be a welcome addition to primary social studies curricula.

□ □ □

Kraske, Robert. *Riddles of the Stars: White Dwarfs, Red Giants, and Black Holes.* Photos; illus. Harcourt, 1979, o.p. 95 pp. Age levels: 9–14. Grade levels: 4–9.

SUMMARY: This book about stars is not a riddle book, as the title may imply, but contains numerous facts about the birth and death of stars. Approximately one-fourth of the contents discusses the sun, it characteristics, and its relation to other stars, including its similarities to and dissimilarities from them. Scientists' and astronomers' contributions to our knowledge about stars, the importance of gravity and how it affects stars, black holes, and modern scientific instruments used today—and those of tomorrow—are also covered. The last chapter contains questions the author poses on the possibilities for humankind for the future. This book is illustrated by drawings and photographs, and contains an index.

EVALUATION: Although the concepts are presented easily and clearly, the abundance of facts may make this book appealing to a student with some prior knowledge of the subject.

Many interesting questions and speculations about the future in regard to energy, pollution, gravity, space travel, life on other planets, and whether the universe can expand or not are explored in the final chapter. Kraske ends by telling us where humankind plays a part in the immense universe.

There are many black-and-white photographs throughout the book that help to clarify the text.

□ □ □

Krupp, Edwin C. *The Comet and You.* Illus. by Robin Rector Krupp. Macmillan, 1985. 48 pp. ISBN: 0-02-751250-9. Age levels: 8–11. Grade levels: 3–6.

SUMMARY: Beginning with information about comets in general, the book quickly moves to include Halley's Comet, what it is composed of, and an analogy between the nucleus of the comet and a huge dirty ice cube. The orbits of comets, particularly Halley's Comet, and the speed at which they travel are discussed. When and how comets can be seen, the study of comets by astronomers through telescopes, and what people used to think comets were and looked like are also included. The author tells what was happening in the United States in 1910 when Halley's Comet appeared. Information is given about Edmund Halley and Isaac Newton, who were contemporaries, and how Halley used Newton's law of gravity to predict when the comet could be seen. The reader also discovers that a comet can be named after whomever is the first one to find it, but that such a discovery would probably require the use of a telescope.

EVALUATION: The reader is lured into the book, especially through the artwork. It is humorous and simple and explains concepts from the viewpoint of a young reader. Analogies are used throughout the book for better comprehension of concepts presented. For instance, the orbit of Halley's Comet is shown using hot dogs and an analogy is made to how many hot dogs each person on Earth could eat for 22 years to complete the string of orbit.

Little asides such as a child's notes to Mom, balloons showing the sun or moon talking, and children asking questions in a classroom add interest and humor to the book. In brief text and many illustrations, the book covers the history of Halley's Comet and its orbit very well, as well as the composition of comets, how they are named, and how they fit into our solar system.

□ □ □

Laithwaite, Eric. *Force: The Power behind Movement.* Photos; illus. Watts, 1986. 32 pp. ISBN: LB 0-531-10181-9. Age levels: 9–12. Grade levels: 4–7. Series: Science at Work.

SUMMARY: Laithwaite discusses force by associating it with familiar tasks. People are conscious of force because they feel it. It can be measured just as length and time are measured. Examples of how force is used are given: the force of gravity can be used to find the vertical plane for building houses; a skier can use it to move down a hill; by the use of dams, the force of water can be channeled to pass through a turbine, thus generating electricity. Force is needed to start and stop an object's movement: a heavier weight, a longer handle, and extra pressure all create a greater force.

Friction is used to slow down or to stop a force. The author elaborates by showing that friction can be reduced by lubrication, and indicates that water and oil can serve as lubricants. Gravity, inertia, and friction often work together as when a plane lands or a person walks. Twisting forces are common, for example, turning a door knob, faucet, screwdriver, or lid on a jar. Levers, gears, and inclined planes can help reduce the amount of force necessary to move an object, whether the object has to be lifted or only pushed or pulled. A particularly interesting section reveals that heating gases causes them to expand, which may cause huge forces. Rockets use the greatest forces that man has ever made to date.

EVALUATION: The pictures and drawings in this book are superb. Each reinforces the text, without giving more information. The illustrations effectively reinforce what the student just read on the corresponding text pages and the captions are worded simply while explaining some complex concepts of force. This book is an excellent introduction on force for the intermediate grade reader.

□ □ □

Lambert, David. *The Solar System.* Photos. Bookwright, 1984. 48 pp. ISBN: LB 0-531-03803-3. Age levels: 9–13. Grade levels: 4–8. Series: Planet Earth.

SUMMARY: In this book the solar system is explained in a manner designed to engage the interest of children in grades 4 through 8. The planets are discussed, and the size, surface, temperature, rotation pattern, and degree of brightness of each is revealed. Gravitational pull and its effects on tides is explained. Examples make phenomena such as atmospheric pressure understandable; for example, the atmospheric pressure on Saturn is such that even though its core is solid, the planet itself is so light that it would float on water. Throughout the book the author uses terms that are simplified yet accurate and within the reader's grasp.

EVALUATION: A superb presentation of our solar system, this book provides recent factual information in a manner that will generate intense interest by the reader. The photographs are fabulous as they clearly depict the planets and some of their moons. The photography corresponds directly with the text, and the captions provide additional information. The simple yet detailed text is exciting without being ponderous or boring. At the end of the book a facts and figure page contains information about the relative size of the sun and planets. A glossary gives definitions of bold terms used in the text. An index and a list of further readings complete this book.

□ □ □

Lambert, David. *Volcanoes.* Illus. by Michael Roffe, et al. Watts, 1985. 32 pp. ISBN: LB 0-531-10009-X. Age levels: 8–11. Grade levels: 3–6. Series: Easy-Read Fact.

SUMMARY: Lambert holds the reader's attention with a wealth of basic information about volcanoes. Hot molten rock from inside the earth, *magma,* bursts to the earth's surface forming a volcano. The earth's crust, through which the magma escapes, is made of plates, which are slowly moving. Most of the eruptions occur along the edges of these plates because they may collide or separate. There are thousands of volcanoes at the bottom of oceans. If they are hidden underwater, they are called seamounts. Magma flowing from a volcano is called lava. Lava may flow fairly quietly from volcanoes when there is no trapped gas. Cooled lava forms basalt rock. Hot springs and geysers are found where volcanoes have been active because of the heated rock underground. Terrible damage has been done by volcanoes such as Vesuvius, Krakatoa, and Mt. St. Helens. Fortunately, scientists often can predict a volcanic eruption by using special instruments, although they cannot prevent them from occurring. Volcanoes can be useful by creating new land and providing good soil for crops. Valuable minerals come from rock deep below the earth's surface. Volcanoes are even found on moons and other planets in our solar system.

EVALUATION: This book presents a variety of information about volcanoes, such as their formation and destructive and beneficial effects. Three major volcanoes in history are cited, and their captivating stories are told. The illustrations and photographs highlight the topic well. Captions supply the reader with additional knowledge. At the end of the book, there are a glossary, an index, and a "Fascinating Facts" page, which truly is interesting. For instance, one learns that *volcano* comes from Vulcano Island, off the coast of Italy, which was named by the Romans after their god of fire, Vulcanus. This book is enjoyable to read and should attract and maintain the interest of readers.

□ □ □

Lambert, David. *Weather.* Illus. by Christopher Forsey, David Mallot and Michael Roffe. Watts, 1983. 32 pp. ISBN: LB 0-531-04621-4. Age levels: 7–11. Grade levels: 2–6. Series: Easy-Read Fact.

SUMMARY: Lambert distills many facts about weather in this science book with controlled vocabulary to accommodate unsophisticated readers. He relates that the Earth's atmosphere is always changing because the heat from the sun stirs up the air, causing warm air to rise and cooler air to move in and take its place. When water evaporates from a liquid into water vapor, it cools down and then condenses around specks of dust to form water droplets, which collect into a cloud and fall as rain when they become big and heavy. Snow is formed from freezing water vapor, and approximately ten inches of snow is about equal to one inch of water. Hailstones are formed when the droplets get tossed up and down by the wind, continuing to be layered with ice until they become too heavy and fall to the ground.

Readers will learn that there are ten basic types of clouds that form at different heights; different weather conditions result in the formation of fog, dew, and frost. Anticyclones are air masses that slowly sink over the land and press hard against the Earth's surface resulting in thunderstorms, hurricanes, and tornadoes. During the summer one may see what appears to be a pool of water on a flat, dry surface because the sky is reflected from a layer of warm air above the ground. At times circles of light appear around the sun or moon before rain because the light bends as it passes through ice crystals in cirrostratus clouds; rainbows appear after a rain, reflecting sunlight. Lambert then discusses the fact that man has made changes in the weather using planes to drop special crystals in the air, around which moisture can condense and fall as rain. Ecological concerns about chemicals released into the air from factories, which cause acid rain to fall back and poison plants, are expressed. Lambert tells children that by observing weather on other planets—for example, Mars in its deep freeze; fiery, hot Venus; or icy, windy Saturn—scientists can learn more about weather on Earth.

EVALUATION: *Weather* is an exceptionally well-written book. A broad overview of types of precipitation and weather is given; each section contains some interesting fact that makes the information meaningful. For example, on Mt. Waialeale, Hawaii, it rains up to 350 days a year; the biggest hailstone ever recorded was 7.5 inches long. The illustrations are colorful and attractive and aid in the explanation of concepts. At the end of the book there is a glossary of technical words used in the book and an index. Also included is a page explaining weather lore, such as "Red sky at night . . . ," "Clear moon, frost soon," "Rain before seven, fine before eleven," and "Ants run fast when the mercury rises." This book includes an appropriate and interesting amount of scientific information.

□ □ □

Lauber, Patricia. *Volcano: The Eruption and Healing of Mount St. Helens.* Photos. Bradbury, 1986. 60 pp. ISBN: 0-02-754500-8. Age levels: 8–15. Grade levels: 3–10. Awards: Newbery Honor Book, 1987.

SUMMARY: *Volcano* tells the story of Mt. St. Helens. Beginning in March 1980, anticipation builds—scientists are seen making observations, trying to predict when the volcano will erupt. In May, a tragedy occurs that seems incredibly devastating. The reader sympathizes with the loss of plant, animal, and human life caused by the volcanic eruption. Still the story continues, to the amazement of all, including most scientists. Life renews itself there. Some animals and insects survive and continue to live in the area; others move in from surrounding areas making Mt. St. Helens their home. From a distance, it looked as if all life had been extinguished, but a close examination proved that life would once again flourish here in time. Though volcanoes can be disastrous, they build land and soil; they add gases to the atmosphere and water to the ocean. There is promise that Mt. St. Helens will once again be a pleasant home for many plants and animals.

EVALUATION: This is an outstanding book. Its story form is extremely captivating, making it difficult to put aside. Emotionally the reader is taken through the great tragedy and the triumph as life returns to the area. *Volcano* can easily capture and maintain the attention of a group. It is quite readable and can be enjoyed by individuals of almost any age. The illustrations greatly enhance the text. The contrasts of the before and after seemed devastating, yet the pictures of new plant life growing are hopeful. This book is outstanding in its research and report of Mt. St. Helens.

□ □ □

Liem, Tik L. *Invitations to Science Inquiry.* Illus. by author. Ginn, 1987. 470 pp. o.p. Age levels: 9–17. Grade levels: 4–12.

SUMMARY: Although this book appears to be intended for adults, younger students with an interest or curiosity in science experiments may find it valuable. There are approximately 450 science experiments outlined (one per page). A list of materials needed, the procedures to follow, questions that can be asked about the experiment, and an explanation of why or how the experiment works are given. A drawing is also included for each experiment. There are four sections: "Environment," "Energy," "Forces and Motion on Earth and in Space," and "Living Things." Within each section the author has further divided the subject into subcategories and these are explained in an introduction for each section.

The author feels that adults can help instill a love of science in the younger generation, and much of his introduction discusses this topic. He highlights particular characteristics, such as presenting a certain attitude about science to foster a sense of eagerness in young people toward this discipline. Many techniques for achieving his goal are included, often with research to support his ideas. A "Further Readings" list for reference material and an index complete the book.

EVALUATION: The experiments are easy to follow even for the uninitiated, and the explanations are clear enough that it is easy for the experimenter to understand what happened and convey it to a class or group of friends.

The sketches of the experiments are equally easy to follow and the explanations often contain drawings, too. The entire book is done in black and white.

The experiments have catchy and interesting titles: "The Leaping Egg," "The Two Bottles in Love," and "The Confused Flashlight" are just a few.

A page of caution—in large print at the beginning of the book—gives precautions on six of the included experiments. Many of them can be done with easily obtainable materials, but others involve chemicals or materials not usually found in the home.

□ □ □

Linn, Charles F. *Estimation*. Illus. by Don Madden. Crowell, 1970. 34 pp. ISBN: LB 0-690-27028-3; pap. 0-690-27033-X. Age levels: 5–11. Grade levels: K–6. Series: Young Math.

SUMMARY: In this book in the Young Math series, Linn explains what estimation is and when scientists employ it. The fact that estimation as a practice is endorsed at all may come as a surprise to some young readers, who probably think that scientists measure everything with precision. Examples show that estimation is an important function for groups of people with which children are familiar: grocers, builders, cafeteria cooks, and airline pilots.

EVALUATION: Children perusing this book will see that there are countless familiar situations in which estimation is practiced every day. Theory translates into practice as the author offers readers the opportunity to perform simple "experiments" in estimation: estimate how many grains of rice are in a small portion, then estimate how many grains are in a box of rice. This experiment can be changed by a grownup to use a commodity taking less time, for example, estimating how many beans are in a cup, then how many are in three cups.

□ □ □

Linn, Charles F. *Probability.* Illus. by Wendy Watson. Crowell, 1972, o.p. 33 pp. Age levels: 5–9. Grade levels: K–4. Series: Young Math.

SUMMARY: Probability is the focus of this science book in the Young Math series. One example of the uses of probability involves weather forecasters, who use probability to predict the weather. By using records of what has happened in the past, weather forecasters are better able to make predictions. If, for example, a weather forecaster discloses that there are seven out of ten chances it will rain tomorrow, it means that past records where conditions have been the same as today's weather have been examined, and the following day it rained seven out of ten times. Readers are invited to toss tacks and note how many landed point up to use to predict further throws. From a discussion of the uses of the charts, the author explains what graphs are, and shows that graphs are easier to use for comparisons than trying to study individual numbers. A basic understanding of sampling techniques is presented.

EVALUATION: This introduction to probability shows the practicality of basing predictions on past outcomes, even though a prediction will not always come true. It involves the reader in scientific experiments to see why this is the case. The cartoon pictures of two mice characters who are involved in the experiments add humor with their separate conversations in balloons. The charts and graphs are clear, and the information good. The cuboctahedron should have been illustrated or possibly omitted. This concept could probably best be included in a slightly more advanced book.

□ □ □

McGowen, Tom. *Album of Rocks and Minerals.* Illus. by Rod Ruth. Rand McNally, 1981. 61 pp. ISBN: 0-528-82400-7; pap. 0-02-688504-2. Age levels: 9–12. Grade levels: 4–7. Series: Rand McNally Albums.

SUMMARY: A history of rocks and minerals is given in this book. Featured are 22 of these objects with details on how they were discovered, their uses in ancient times, and how people depend on them today. Small black-and-white sketches appear with the text, while full-page color illustrations are included on the facing page for most entries. Rocks and minerals not featured in this part of the book are covered in a separate section with color illustrations and one-paragraph descriptions. A pronunciation guide and an index are also included.

EVALUATION: Although a wide variety of information is included in *Rocks and Minerals,* some descriptions lack clarity and may even be misleading to the reader. On the verso page a statement says that the text and illustrations are authenticated by Dr. Edward J. Olsen, Curator of Mineralogy, Field Museum of Natural History in Chicago. While the illustrations and information on the rocks and minerals appear accurate, some of the anecdotes about their use in history seem questionable. The section on jade, for example, includes references to cannibalism and states that knives of jade were "used for cutting up human flesh to be cooked and eaten." Also, the introduction discusses igneous, metamorphic, and sedimentary rocks in a perfunctory manner, without explaining why rocks are given these terms. The illustrations themselves match the context of the text well, but a few seem unrealistic: do diamond miners actually work without wearing shirts, or protective goggles when drilling? The pronunciation guide is very helpful, but again some of the entries are awkward, such as *NYS* for the rock gneiss, which is pronounced "nice." *Album of Rocks and Minerals* has potential as an information source, but there may be difficulty with the technical terminology.

□ □ □

McGrath, Susan. *Fun with Physics.* Photos; illus. National Geographic, 1986. 104 pp. ISBN: 0-87044-576-6. Age levels: 9–12. Grade levels: 4–7. Series: World Explorers.

SUMMARY: The principles of physics are explained within contexts that might occur in a child's life. The physics involved in a roller coaster ride, in making ice cream, in running an electric train, and in playing tennis are described. The book is divided into four chapters: physics of fun, physics of nature, physics at home, and physics of sports. Sliding down a fiberglass slide, kite-flying, and riding on a swing are a few of the fun things presented to explain friction, Bernoulli's principle, and the energy of motion, respectively. A rainbow, the aurora borealis, lightning, and a peacock's colors are some of nature's wonders examined from a physicist's point of view. The screwdriver that becomes a lever when a lid is pried off a paint can, the invisible radio waves in the electromagnetic spectrum transmitted by radio stations, and the atmospheric pressure that allows drinking through a straw are some of the examples of physics in the home. In sports, an athlete's center of gravity plays an important role in high jumping and gymnastics, centripetal force affects skiers and bobsledders, and Newton's third law of motion is demonstrated in the pushoff of sprinters in a race.

Important terms stand out in bold type and are also found in the glossary.

Throughout the book, many very simple experiments are suggested. Activities at the end of each chapter tell the reader the materials that are needed, the instructions to be followed, and the physics involved.

EVALUATION: This informational book is well organized and presents the difficult topic of physics in a pleasant and straightforward manner. Understanding is aided by defining terms that have been highlighted in bold type, and showing them in usage. For instance, friction is defined and then demonstrated at an amusement park's giant slide, on the brakes of a bicycle, and simply by rubbing one's palms together. The photographs are colorful and attractive. Simple activities and experiments explain sophisticated concepts with clear illustrations and directions. Young physicists will feel as if they are magicians as they make eggs float, keep tennis balls in the air without touching them, and make water stay in an overturned glass. The explanatory text has a conversational tone and the captions are full of interesting information. The study of energy and matter is made easy and fun in this book, which can serve as an excellent reference source and activity book.

□ □ □

Markle, Sandra. *Digging Deeper: Investigations into Rocks, Shocks, Quakes, and Other Earthly Matters.* Illus. by author; photos. Lothrop, 1987. 120 pp. ISBN: 0-688-05986-4. Age levels: 9–14. Grade levels: 4–9.

SUMMARY: As its title indicates, *Digging Deeper* encourages readers to develop an understanding of various geological phenomena by conducting their own experiments. The author introduces each topic using the correct scientific terminology in an understandable way, enabling readers to easily comprehend the concept presented. Black-and-white illustrations and photographs are also included for many of the topics. The investigations include discussions and experiments about various forces that shape, build, break down, and move the Earth's natural resources. The experiments can easily be done by upper elementary school children, and use materials typically found in the home. The subject of rocks and minerals is also covered. Glossary and index are included.

EVALUATION: The author, a science teacher and science writer for children, is impressive in her ability to "translate" difficult concepts into easier terminology. The detailed experiments are the highlight of the book, as they correspond quite well to the various geological theories. To illustrate how earthquake vibrations spread through the Earth's crust, the reader is instructed to put a small amount of water in a clear glass. Then, clear plastic wrap is stretched tightly over the top and the edges are pressed together to seal them. Working over a sink, the reader is instructed to turn the cup upside down so the water is resting on the clear wrap and gently tap the plastic with a finger. The focus of the earthquake—the point in the Earth's crust where something happens to cause the quake—is like the spot that was poked. By tapping the clear wrap again, the reader can see the effect of the shock waves moving outward to the sides of the glass. These waves imitate how an earthquake's shock waves travel, even across great distances. Some of the other discussions and experiments concern plate tectonics, volcanoes, fossils, crystal growing, and glacial movement. This is an extremely well-written book that should interest everyone, especially children and adults who have an aversion to science.

□ □ □

Martin, Paul D. *Science: It's Changing Your World.* Photos; illus. National Geographic, 1985. 104 pp. ISBN: LB 0-87044-521-9. Age levels: 9–17. Grade levels: 4–12.

SUMMARY: The futuristic focus of this volume explores developments in computers, lasers, industry, fuel, food, transportation, medicine, and space. Art ranges from amusing cartoon-style drawings to sophisticated photography. The format selectively examines topics of particular interest. For example, lasers are used to husk peanuts, slice through steel, drill holes through baby bottle nipples, and accurately measure a railroad tunnel. Holograms made by lasers may one day appear on paper money, thereby catching counterfeiters.

Information about topics such as robots, ultrasound, electronic pollution, computers, CAT scanners, solar energy, space stations, and nuclear power plants is included. There is an index and an additional reading list accompanying each section.

The Careers in Science section provides sources of information on science-oriented occupations.

EVALUATION: A variety of topics are presented here in a dynamic manner. The graphics add to the dramatic quality of the book. The exciting and beautiful format makes this a superb book in the fields of science and technology. The uses of high technology in today's world lead to tomorrow's developments and inventions. Sheer excitement is generated by the book itself, which in turn may foster a desire to become a part of the scientific world of tomorrow.

□ □ □

Milgrom, Harry. *First Experiments with Gravity.* Illus. by Lewis Zacks. Dutton, 1966, o.p. 56 pp. Age levels: 9–12. Grade levels: 4–7.

SUMMARY: The force of gravity is explained through 20 experiments. The experiments are described and illustrated, and a list of easy-to-obtain materials is included.

Since any explanation of gravity must begin with the definition of *force,* the experimenter is asked to drop a ball and observe what happens: the ball falls to the ground through the force of gravity.

Next, the experimenter is asked to hold a lead sinker; the sinker is held up by the force the student puts on it. Then the sinker is put into a glass of water. The sinker drops to the bottom of the glass because the water cannot balance the pull of the gravity; the lead weight sinks. The experimenter learns that the pull of gravity on an object is balanced by the opposing force. If the pull of gravity is stronger than the force (the water), the object will sink.

The book concludes with experiments that explain how the earth, sun, and moon are kept in orbit through the force of gravity. Scientists use their knowledge of gravity to keep spaceships in orbit as well as regulate their speed. A glossary is included, defining words used in the text.

EVALUATION: The author motivates the reader from the very beginning with a description of how our knowledge of gravity has helped the space program. A few of the experiments deal with launching and orbiting; directions for making a simple parachute are included.

The text is kept interesting and relatively exciting through the frequent use of experiments. The insertion of the list of needed materials at the beginning of the book, rather than with each experiment, cuts down on unnecessary text. Each experiment is well illustrated, but older readers may not need to perform each experiment to understand these concepts.

The illustrations are simple drawings and graphs in pen and ink. Unfortunately, most of the children depicted are male but the 1966 publication date could explain that limitation. The text is written with clarity, and each scientific term is well described. This book is sure to motivate the older reader to find out more about the subject. The younger reader should have little trouble understanding the rather sophisticated concepts, as the author has outlined the laws of physics by using elementary objects and explanations.

□ □ □

Nozaki, Akihiro. *Anno's Hat Tricks.* Illus. by Mitsumasa Anno. Philomel, 1985. 44 pp. ISBN: 0-399-21212-4. Age levels: 6–13. Grade levels: 1–8.

SUMMARY: This book introduces young readers to the concept of binary logic. Addressed as Shadowchild, the reader is asked to guess the color of his own hat. A tricky hatter provides red and white hats to the Shadowchild and two friends. Beginning quite simply, the trickster poses questions and later explains the reasons for the correct choice. As it progresses, the book becomes increasingly complex. If one does not understand the previous problems, the correct solutions will be difficult to determine with confidence. The final situation does not even include the answer, assuming that the problem has been solved with logic. At the end of the book, "A Note to Parents and Other Older Readers" more fully explains the puzzles of binary logic.

EVALUATION: The book is simply written and designed, yet it is very challenging. The more difficult logic requires an older reader. The author systematically explains why Shadowchild's hat is a particular color. The pictures are simple, colorful, and necessary to the text because they provide a visual image of the hatter's proposed trick. The reader is required to turn the page in order to find some answers; this is an extra boost to challenge children to draw their own logical conclusions. The questions posed by the hatter in this book will stretch the minds of readers of all ages.

□ □ □

Oleksy, Walter. *Experiments with Heat.* Photos; illus. Childrens, 1986. 47 pp. ISBN: LB 0-516-01277-0; pap. 0-516-41277-9. Age levels: 7–10. Grade levels: 2–5. Series: New True.

SUMMARY: The numerous uses and applications of heat are described in this book. Readers probably already know that heat keeps houses warm and cooks food, but may not know that heat is used in factories to make cars, appliances, and other goods.

The reader learns that heat is a form of energy and most of the heat used comes from the sun. The sun makes living things grow and contributes to the supply of coal, wood, oil, and gas that are used as fuel today. Heat travels in three ways: conduction, convection, and radiation. Each term is explained and appropriate examples are given.

The book includes a chapter on temperature and how a thermometer works: molecules vibrate in warm temperatures and take up more room in the glass of a thermometer. This causes the liquid in the thermometer to go up. The liquid will go down as the temperature cools because the molecules vibrate less and need less room.

EVALUATION: This is another fine book from the New True series. Heat, its practical applications, its sources, and how it affects present and future travel are explained for the middle-grade student. The importance of the sun as the source of heat is explained simply: we get warmth from the sun's rays directly; indirectly, the sun gives us energy from the fossil fuels. Coal, wood, oil, and gas were once living things dependent on the sun for growth. The means by which heat is transferred from one object to another is explained less simply and clearly.

Although the print is as large as one would find in a primary reader, the material is rather sophisticated and difficult for its intended audience to comprehend. Photographs help explain the experiments and colorful paintings describe the role heat will play as the source of energy in the space program.

□ □ □

Olney, Ross, and Patricia Olney. *How Long? To Go, To Grow, To Know.* Illus. by R. W. Alley. Morrow, 1984. 40 pp. ISBN: LB 0-688-02774-1. Age levels: 7–11. Grade levels: 2–6.

SUMMARY: This book of facts has been divided into several sections. The section on growth tells how long it takes things to grow, such as fingernails, hair, trees, and animals. It is followed by a section on how long certain things live. Also covered are facts on speed and how long things last, such as the top layer of skin, a television set, an automobile, and granite. The authors discuss how we discover things through our senses, such as how long it takes to see things close up and at a distance or to feel vibrations of sound waves. The final section deals with random facts such as the longest animal, river, and word, and the smallest bird. Cartoons and drawings in tones of blue, green, and yellow showing motion enhance the facts.

EVALUATION: In some instances, the authors use comparisons to clarify the concepts of growth, speed, duration, and change, which may be difficult for some children to understand. In other cases, the authors simply state unusual or interesting facts such as the flying distance of the Arctic tern or the jumping distance of a flea.

The illustrations are clever and children can easily relate to them. For example, to show that hair grows one-half inch a month, the illustrator shows a knight looking up at Rapunzel and in cartoon ballons asks her how long she's been locked up in the tower (she replies that it's been years). Also in the illustration, her hair is shown flowing down to the ground.

The authors end the book by asking the reader how long it has taken to read the book, thereby leading the reader to make observations of things about and around himself or herself.

□ □ □

Osband, Gillian. *Our Living Earth: An Exploration in Three Dimensions.* Illus. by Richard Clifton-Dey. Putnam, 1987. Unp. ISBN: 0-399-21447-X. Age levels: 6–12. Grade levels: 2–7.

SUMMARY: This three-dimensional manipulative pop-up book begins with two theories about the formation of the Earth as part of the solar system, which is part of the Milky Way galaxy. The structure of the Earth is compared to an onion and its layers around a center core; the composition of the Earth's crust is also discussed. Rotation and revolution are explained as the influence of the sun. A history of the Earth from about 4 or 5 thousand million years ago to 250-or-so million years ago is included. The continental drift and geological changes that cause volcanic eruptions and earthquakes are included. In the last section, the development of life on the Earth is compared to a clock, beginning at midnight with soft-bodied creatures and jellyfish and running to 11:58 A.M. with Ice Age mammals and saber-toothed tigers. The reader is left with the thought that many changes will occur as the clock advances to 1:00 P.M. in another 50 million years.

EVALUATION: This dynamic book presents its subject through illustrations in both pop-up and manipulative form. The solar system pops up in the center of the book and the Earth and its moon and sun move to show relative distances and sizes. Lift-up cutaway views and arrows to push and pull help explain the Earth's crust and rotation/revolution. A fantastic pop-up volcanic eruption illustrates its corresponding two-page text. The reader can shift the movable plates of the Earth, and pull-outs show life on Earth during different periods of time. All illustrations, diagrams, and manipulatives are in full color. Bold type for titles and subtitles makes the text easy to read. This timeless book is one that will be read over and over and can be handed down from generation to generation.

□ □ □

Papy, Frédérique, and Georges Papy. *Graph Games.* Illus. by Susan Holding. Crowell, 1971. 34 pp. ISBN: LB 0-690-34965-3. Age levels: 6–9. Grade levels: 1–4. Series: Young Math.

SUMMARY: Games that use graphs are found in this book that involves mathematical thinking. With dots that represent children and through a series of games with clues, the reader is drawn in to play a graph game by arriving at logical conclusions about sets and one-to-one correspondences. By looking at a graph game called "Point to Your Sisters," determinations are made about which are girl dots and which are boy dots. Another game involves finding which dot represents the birthday boy by the number of letters he received from the mail carrier, because the letters can be separated into sets that go to different children. Knowledge about math is gained through graph games without using difficult terminology.

EVALUATION: The authors immediately draw the reader into a graph game by the challenge to determine whether a dot is a boy or a girl. Also, the point of view presented is in the first person singular, which makes the book much more personal.

The illustrations are simple but essential to young readers in understanding and solving logical problems. The reader is invited to participate by tracing pictures and using a blue and red crayon to map out answers. The solutions are then illustrated on the following page. The authors are quite adept at presenting complex ideas requiring logic without using scientific jargon that may confuse amateur mathematicians.

□ □ □

Parish, Peggy. *Dinosaur Time*. Illus. by Arnold Lobel. Harper & Row, 1974. 30 pp. ISBN: LB 0-06-024654-5; pap. 0-06-444037-0. Age levels: 4–8. Grade levels: PreK–3. Series: I Can Read.

SUMMARY: Eleven different dinosaurs are briefly described in this beginner reader. Each dinosaur's multisyllabic name is accompanied by a phonetic pronunciation. Distinctive features in appearance, some eating habits, and various habitats are included. Dinosaurs of all sizes are shown: Compsognathus was no larger than a cat and was a speedy runner. Brontosaurus was gigantic but had a very small mouth. Anatosurus' ability to grow a new tooth to replace a broken one and the five horns on Pentaceratops' face are some particulars mentioned about these animals that inhabited the prehistoric world.

At the end of the book the author explains that not much is known about dinosaurs, since they are extinct. Scientists continue to study fossils in order to gather more information about dinosaurs, although there is a possibility that complete knowledge about these animals might never be attained. A table of contents is provided.

EVALUATION: The simple organization of this informational book presents details in a direct and clear manner. Each dinosaur is described with one page of text and illustrated on an adjacent page. Atop the textual page, the name of the dinosaur is headlined in bold print. Within the text and after a phonetic pronunciation is given, the dinosaur's name is repeated again, thus allowing for the mastery of an unusual and difficult word before another one is presented in the same fashion. The large print text is simple and concise. The earth colors used in the illustrations provide a feeling of realism. Lobel's illustrations are dynamic. This reference book on a popular subject is a "Reading Rainbow" selection that will intrigue many young children.

□ □ □

Parramon, J. M., Carme Solé Vendrell, and Maria Rius. *Water*. Illus. by J. M. Parramon. Barron's Educational, 1985. Unp. ISBN: 0-8120-3599-2. Age levels: 0–6. Grade levels: PreK–1. Series: Four Elements.

SUMMARY: In this book a single line of text on the left-hand page suggests the events occurring on the right-hand page. Two children do something with water, although *water* is not used, except once as an adjective and once as a verb, suggesting that the book might be used in a guessing game fashion for very young readers. The characters—accompanied by their dog—drink, wash, water flowers, sail paper boats, dash away from rain clouds, admire a fountain, watch fire fighters use a hose, notice new growth in a marsh, take a sailboat ride, watch cows find a stream in a pasture, and gaze at a rainbow.

On the last two pages the authors proclaim that this wonderful substance that is found everywhere is indeed water. Three other books in this series explain the elements earth, fire, and air. Two pages of information about water are included at the end of the book, to be used by the individual who reads this book aloud to children.

EVALUATION: The gorgeous colors and very simple text convey how important water is to people, plants, and animals. The bold letters are effective on a stark, white background. A question arises as to why the authors used the words *watering can* and took the guesswork out of the game of what is being identified. Since this book is translated from Spanish, it is possible that the give-away does not occur in the same way in the original. Several of the illustrations are questionable: the little boy should not be using his finger to test hot water, children drink from a river, and the "marsh" looks more like a meadow—again, this may be a distortion due to translation.

□ □ □

Pollock, Penny. *Water Is Wet*. Photos. Putnam, 1985. Unp. ISBN: 0-399-21180-2. Age levels: 0–6. Grade levels: PreK–1.

SUMMARY: In black-and-white photographs, water is presented as useful and fun. This book shows what can be done with water. It can be used for drinking and washing. Water helps living things grow. It provides amusement on a city street, at the beach, in the house, or in the country.

Within some pages, the sound of water is suggested by the smack of a child's boot in a puddle, and the splash heard when objects are tossed into the puddle. Sounds of delight can almost be heard from the pages that show children playing in water. Rain causes a sensation when it falls into an open hand, and also causes a pinging sound on an umbrella.

In one scene the lightness of water is seen in bubbles, and in another water is mixed with dirt to make heavy mud pies. At the beach, a child floats in the water, while another youngster is knocked down by the waves.

Seasonal change is revealed when water is turned into snow and ice in winter. Water in this form is also enjoyed in such pastimes as skiing, sledding, skating, and making a snowman.

EVALUATION: A multiethnic assortment of children experiencing the pleasures of water are revealed in these black-and-white photographs. Water sounds and sights are included throughout this action-packed picture book. This book graphically displays the properties of water, both as a liquid and a solid, year round.

Readers selecting this book might be misled by the colorful book jacket, as the inside photographs are all black and white. However, this in no way detracts from the vitality of the photographs or the book itself.

The text is sparse, and the print is large and easy to read. Some sentences are in rhyme, which provides interest and surprise. The use of second person allows young readers to identify with the children in this book. Since children like to play in water, they are sure to enjoy this realistic picture book.

□ □ □

Provensen, Alice, and Martin Provensen. *Leonardo da Vinci: The Artist, Inventor, Scientist in Three-Dimensional Movable Pictures.* Illus. by authors. Viking, 1984. Unp. ISBN: 0-670-42384-X. Age levels: 0–11. Grade levels: PreK–6.

SUMMARY: This short biography of Leonardo da Vinci tells about his many scientific and technological contributions, as well as his artistic masterpieces. Reference is made to the machines he invented, his study of birds in flight, and his own attempts at flying; his five thousand or so pages of notes and drawings on various subjects; and his work as an astronomer. Mention is also made of his giant statue of a man on a horse (the fearful monster he made), his *Last Supper* mural in Milan, Italy, and the *Mona Lisa*. The book begins with a quote attributed to Giorgio Vasari—a contemporary of da Vinci—describing how talented a person da Vinci was and the fact that this was acknowledged in his time.

EVALUATION: This outstanding pop-up and manipulative book was authored and illustrated by Caldecott Medal winners. Brief text highlights significant contributions by one of the most talented and inventive minds in history. Da Vinci's lifetime is stated as paralleling Columbus's, which helps the reader to form a time line of events. The colorful illustrations, pop-ups, and manipulatives enhance the text as the reader is able to move machines and inventions, watch birds in flight, observe the monstrous man on a horse, turn wheels showing drawings, see da Vinci viewing the stars, and, of course, view his paintings, including the *Mona Lisa*. This timeless book will help today's readers become familiar with a classical figure from the Renaissance in an exciting way.

□ □ □

Razzell, Arthur G., and K. G. O. Watts. *A Question of Accuracy.* Illus. by Ellen Raskin. Doubleday, 1969, o.p. 46 pp. Age levels: 10–13. Grade levels: 5–8. Series: Exploring Mathematics.

SUMMARY: Concepts of measurement form the focus of this slim volume. Designed to synchronize with the mathematics curriculum of the middle school, it contains information that can inform and enlighten such areas of study as history and biography. To introduce concepts of measurement, the book covers the concept of accuracy, and its importance through the centuries to the builders of civilization.

From the outset, the narrative includes simple experiments to prove points, for example, on page 6 it is suggested that children line up single-file around the buildings of the school complex and that the "line" they create be measured by ruler, yardstick, and tape rule. Readers are asked not only to guess which response is more correct, but why. Spark plug calibration and the thickness of several stacks of paper of various heights are examined, as are medical thermometers, micrometers, and stop watches, making the point that fine discrepancies occur when "common," everyday measurement devices are used. Once convinced of the importance of standard units of measure, the reader moves on to measures of time, "true north," great linear distances (highways), liquids, and electric current.

EVALUATION: This charming little book does not pretend to be a text on accuracy in measurement. Instead, it utilizes folk humor—how fast is two shakes of a lamb's tail?—and experiments with simple, familiar tools to convince the reader of the need for standardizing measurement. Bold, clear, orange-colored drawings effectively illustrate the fundamental principles revealed throughout. The book will make readers want to learn more about measurement, and, of course, this is exactly the authors' intent.

□ □ □

Reidel, Marlene. *From Ice to Rain.* Illus. by author. Carolrhoda, 1981. Unp. ISBN: LB 0-87614-157-2. Age levels: 6–9. Grade levels: 1–4. Series: Start to Finish.

SUMMARY: This book of facts begins in winter with children skating on ice and moves to spring and the melting of the ice. In summer the evaporation of water is discussed and the process continues until it falls to the ground in the form of rain. The misty clouds of fall give way to the cold air of winter once again, bringing with it snow and ice. A one-page illustration accompanies each page of text.

EVALUATION: A big concept is well presented in a small book. Short sentences and few words along with simple, colorful illustrations say it all. Three scientific words—*water vapor, evaporated,* and *water dust*—are highlighted using bold black ink. The naming of the seasons helps very young children relate to rain, snow, and ice. Most of the illustrations depict children relating to the weather. The author-illustrator gives additional information when possible, such as the fact that under ice there is water (an illustration shows fish swimming under the ice). In spring when the ice is breaking up, the ice floats because it is lighter than water. This artist-illustrator has won many European honors and awards for her work in children's literature, including the German Youth Book Prize and the Most Beautiful German Book of the Year.

□ □ □

Rinkoff, Barbara. *Guess What Rocks Do.* Illus. by Leslie Morrill. Lothrop, 1975, o.p. Unp. Age levels: 8–11. Grade levels: 3–6.

SUMMARY: Rocks give clues to the history of our planet, and minerals that come from rocks are prized for many reasons. Through a discussion of how rocks were used by people and animals long ago as well as a discussion of some beliefs and traditions concerning rocks, the reader learns many fascinating things about rocks and minerals.

An introduction to the characteristics, uses, and scientific and practical value of rocks and minerals is provided for children to begin to understand what rocks do.

The inclusion of folklore concerning rocks from prehistoric time on is very interesting and provides the young readers with a sense of the importance of rocks, which today are generally taken for granted. Fascinating information, as well as suggestions for the beginning rock collector, is presented in this introductory book of rocks and minerals.

EVALUATION: The text is accurate and written in a manner easily understood by children in the early elementary grades. Even though rock names are presented, there is no nomenclature problem since descriptions and examples are given that children can relate to and comprehend. The use of large bold print as well as the vocabulary used tells young children this book is for them.

Rocks and minerals used in everyday things that children see and use—chromium in refrigerators, aluminum in pots and pans, and iron and steel in cars and tools—help children relate to the information presented. Ways in which people and animals used rocks and minerals gives the reader a broader sense of the history of life on our planet.

The illustrations further the information the text provides. The drawings are large and multicolored but do not represent sparkling or glittering rocks, which are the sort that often attract young children to rock collecting. Including a few of these rocks in the illustrations would entice young children to choose this book. Suggestions for beginning rock collectors are presented at the end of the book.

Children will enjoy reading this book for its information and may even be inspired to begin their own collections—if they can first be enticed to pull this book from the library shelf.

□ □ □

Ritter, Rhoda. *Rocks and Fossils.* Illus. by Pamela Mara. Watts, 1977, o.p. 48 pp. Age levels: 8–11. Grade levels: 3–6.

SUMMARY: In this detailed book, Ritter explains many facts about rocks and fossils. Although much of the Earth's rock is hidden from view, a lot of it is visible. At the beach different kinds of rocks are washed up from the ocean floor in the form of sand, which is fine granules of stone produced by waves tossing rocks around for many years. Hardened volcanic material, such as lava, obsidian, and pumice, is discussed and other igneous rocks are briefly described, as are sedimentary and metamorphic rocks. Also mentioned are the caves, stalagmites, and stalactites that develop in limestone rocks from the action of water. Rocks from space occasionally land on Earth; although they may look like other rocks, they are not made of the same substances. Dead plants and animals leave their mark as fossils, imprinted in rocks. Geologists gather and study rocks and fossils that provide information about the history of the Earth.

EVALUATION: This broad overview of rocks and fossils contains a great deal of scientific information and can serve as an elementary reference book. However, the generalizations offer little in the way of appeal. Further, a young reader might be overwhelmed by the number of new terms per page while an older reader might need more detail to make the terms meaningful. The black-and-white illustrations help to clarify the text. The pronunciation of difficult words, such as *stalagmites* and *stalactites,* is given in parentheses in the body of the text. A topical index is included at the end of the book.

□ □ □

Rius, Maria, and J. M. Parramon. *Air.* Illus. by authors. Barrons Educational, 1985. Unp. ISBN: pap. 0-8120-3597-6. Age levels: 3–6. Grade levels: PreK–1. Series: Four Elements.

SUMMARY: According to modern science, air, fire, water, and earth are not technically considered elements today, though they were believed by the Greeks to be the qualities that made up all substances. This book gives scientific information about the effects of the air through the text and illustrations. It begins by stating that air cannot be seen but that it moves. The intensity of the movement of air begins with a gentle breeze, and moves on to a wind and hurricane. What wind can do for us or how we use the wind is described and illustrated—fly a kite or balloon, move a boat or sails on a windmill, allow a plane or bird to fly, and, of course, breathe.

The book has text on the left-hand page and full color illustrations on the right-hand page. A few pages contain a two-page spread illustration.

EVALUATION: Soft but colorful illustrations depict the sparse text very accurately and appropriately; the page on hurricanes shows the effects of one, but in a way that will not be frightening to a very young reader.

The book is an excellent introduction to air and the ways that air can work for us. The authors also address the fact that the wind can be bothersome, as shown in the drawing of the wind blowing an umbrella inside out together with hats blowing away.

Because children ask many questions of a scientific nature, the authors have included a section of scientific information and facts to use in explanations for young peopele. These are divided into sections on atmospheric pressure, the composition of air, why the wind blows, and the fact that plants help cleanse polluted air.

□ □ □

Robin, Gordon de Q. *Glaciers and Ice Sheets.* Photos. Watts, 1984. 48 pp. ISBN: LB 0-531-03801-7. Age levels: 9–12. Grade levels: 4–7. Series: Planet Earth.

SUMMARY: *Glaciers and Ice Sheets* explains that in some cold countries, the summer never becomes hot enough to melt all the snow, including ice and snow trapped in valleys and between large rocks. These valleys are rocky, uneven areas and were formed by the erosion caused by ice and rock, which occurred over millions of years. Readers are told that if the sun is hot during the summer, heat will eventually melt some of the ice and snow, usually from the underside first. This causes the deep accumulation of snow to flow down the hill, and the resulting river of ice and snow is called a glacier.

The study of glaciers is relatively recent, although a crude form of glacier watching must have taken place previously, or farmers and other settlers would not have known where to build their homes. Scientists have been studying glaciers for the past 200 years and now use echo sounding to measure ice thickness and carbon dating to study the age of glacial material. Although Antarctica contains 90 percent of the world's glaciers and ice caps, glaciers can also be found in northern Europe.

Scientists think there have been ten ice ages in the last million years. We presently live in an interglacial period, that is, between ice ages. A "Facts and Figures" section at the end of the book contains interesting information, such as that the huge Ross ice shelf in Antarctica is larger than the country of France and reaches a height of nearly 100 feet while another 360 feet of it may be submerged. A glossary of the scientific terms used in the book, as well as an index and a bibliography, completes this factual book written for the upper elementary grade reader.

EVALUATION: This book provides much factual information on the subject of glaciers. The reader learns of the erosion caused by ice sheets and glaciers; sheets of ice carry rock and debris that shape the land. The movement of the rocks embedded in the flowing ice carves away the piedmont and makes the mountains appear larger.

The breathtaking photographs included in the book help the reader understand the somewhat technical description of the glaciers and ice sheets. A few colored diagrams describe the text further.

□ □ □

Rutland, Jonathan. *The Violent Earth*. Illus. by Francis Phillipps and Charlotte Snook. Random House, 1987. 24 pp. ISBN: LB 0-394-98970-8; pap. 0-394-88970-3. Age levels: 8–11. Grade levels: 3–6. Series: All-About Books.

SUMMARY: The author portrays the Earth as restless and sometimes violent because of the forces of nature like earthquakes, tidal waves (*tsunami*), volcanoes, avalanches, and weather such as tornadoes and hurricanes. The reasons for these occurrences are explored and explained as well as the destruction that can accompany these happenings. Particular attention is given to Vesuvius and the Indonesian islands for volcanic eruptions, China for earthquakes, and The Netherlands for flooding. Whenever appropriate, technological instruments used to detect disasters before they occur are explained and sometimes illustrated; an illustration explaining lightning is also included.

The book contains color drawings and illustrations, an index, and a foldout panorama cross-section of a volcano with a numbered explanation of its sections. An explanation of how the Earth was formed using the big bang theory and the plate movement of the Earth is included. The conclusion is devoted to the harnessing of energy from the same forces that can cause such potential danger.

EVALUATION: Natural disasters are simply yet thoroughly explained and the accompanying illustrations—many with captions and small diagrams—enhance the text. A young reader could absorb much information from the illustrations alone, but there is a great amount of information in the text, as can be seen by the many entries in the index. The drawings are very vibrant in color and show much action. Care has been taken to show the enormity or size of certain elements, such as an avalanche or tsunami, by depicting people next to the occurrence. Small inserts such as maps, historical paintings, photos from a museum, and pictures of technological instruments add interest and information in a unique way. The panoramic insert (which can be easily removed) is also vividly and simply done so that the young reader would find it easy to understand the workings of a volcano. The positive forces of nature are shown as well: the use of wind power for mills to produce food supplies, and for both wind and water power generating electricity, and geothermal power produced from volcanic areas. Possible future developments are also mentioned.

□ □ □

Schwartz, David M. *How Much Is a Million?* Illus. by Steven Kellogg. Lothrop, 1985. Unp. ISBN: LB 0-688-04050-0. Age levels: 7–10. Grade levels: 2–5. Awards: American Library Association Notable Book, 1985.

SUMMARY: The author and illustrator attempt to help a child understand the concepts of three large numbers: a million, a billion, and a trillion. Using each of the three concepts, a human tree is made with kids standing on each other's shoulders, goldfish in a bowl, and stars printed on the pages of the book. The reader is told and shown how high the tree would reach, how large the goldfish bowl would have to be, and how many pages would have to be printed for a million, a billion, and a trillion. The concept is also illustrated by indicating how long it would take one to count to these three numbers. An author's note at the end of the book explains his calculations.

Soft but colorful illustrations dominate this large-sized book and cover most of the page; there is a limited amount of text and the print is very large.

EVALUATION: The concept of large numbers is one that children often find difficult to understand. This book says it all in only a few words and illustrations, and could be easily and quickly used in a mathematics class to clarify the concept.

For years young people have enjoyed the illustrations of Steven Kellogg, and the pictures in this book are equally appealing. They relate to a child's sense of humor, and the details offer more humor than noticed at first glance. Marvelosissimo, the Mathematical Magician who leads the reader through the book, is also sure to please.

The author invites children who want to go on an arithmetic journey to follow him as he explains the calculations for the text. Although this is enticing, most students will not be able to keep up. However, it is the basis for his calculations as many students ask how the author knows how high the human tower would stretch and so on. The inside back cover of the jacket quotes Pulitzer Prize winner and mathematician Douglas R. Hofstadter as saying that there is an alarming amount of mathematical illiteracy or innumeracy in our society.

□ □ □

Seddon, Tony, and Jill Bailey. *The Physical World.* Illus. by David Antsey, et al. Doubleday, 1987. 159 pp. ISBN: 0-385-24179-8. Age levels: 9–13. Grade levels: 4–8.

SUMMARY: *The Physical World* provides an understanding of Earth and all of its properties. Concepts about the origins of mountains and volcanoes, the making of rivers, oceans, and glaciers, plus the many forms of matter and energy, are explained in great detail. Principles such as heat, sound, magnetism, gravity, and time are also presented in this ambitious book. Included are trivia quizzes, suggested reading lists, a glossary, an index, and hundreds of color photographs and illustrations.

EVALUATION: Using an unusual graphic layout, *The Physical World* combines clear and precise scientific information with an incredible variety of color illustrations and photographs. The format consists of short paragraphs of text interspersed with graphics, while related pieces of information are offset in squares or bubbles on the same page. Instructions on the bottom right corner tell the reader to turn to another page for more information on the subject. Written with great clarity, this is an enjoyable book well suited for upper elementary through high school students.

□ □ □

Selsam, Millicent E. *Sea Monsters of Long Ago.* Illus. by John Hamberger. Four Winds, 1977. Unp. ISBN: 0-02-778050-3. Age levels: 3–7. Grade levels: PreK–2.

SUMMARY: This book describes the physical appearance and eating habits of dinosaurs of the sea. These monsters were not fish with gills but reptiles with lungs; they needed to come to the surface to get air. The solution to the mystery of how sea-living reptiles bore their young is revealed: ichthyosaur skeletons were discovered with baby skeletons inside them. The scientists deduced that the eggs were hatched inside the mother's body, and the babies were born alive into the sea. In addition to ichthyosaurs, other prehistoric sea creatures, Plesiosaurs, and masasaurs are featured.

EVALUATION: Young children are simultaneously attracted to and repelled by the huge, ugly creatures that inhabited our planet long ago. One can easily find juvenile books devoted to the subject of land-living dinosaurs, but this title is unique in that it deals exclusively with the subject of the dinosaurs that inhabited the oceans. Selsam exhibits a gift for simplifying and organizing the material so the very young children who are attracted to this subject can understand these awesome creatures. Each "monster" is introduced with an accompanying lllustration. Its "relative" is described with appropriate similarities or differences noted. She often relates its size to an object the children are familiar with, for example, "its eyes are as big as cereal bowls." Each new name is both capitalized and spelled phonetically in parentheses.

The illustrator used watercolor drawings to enhance the text. The monsters are not drawn flat and one dimensional, but rather twisting and jumping through the foam with powerful jaws snapping shut on their prey. Underwater movement and action on one page gives way to peace and tranquility on the next. Plesiosaurs bob among great stone canyons bathed in the orangy glow of smoking volcanoes, giving the Earth a look quite different from its twentieth-century appearance. A detailed pen-and-ink drawing of the first sea monster skeleton found is a fine addition to the book. A time chart in the appendix tells when each class of reptiles lived.

The author and illustrator acknowledge the help of Dr. Eugene S. Gaffney of the American Museum of Natural History, New York, for giving a stamp of authenticity to the material. Children will love the sharp teeth and strong jaws of the huge sea monsters, safe with the knowledge that they exist no more.

□ □ □

Selsam, Millicent E. *Strange Creatures That Really Lived.* Illus. by Jennifer Dewey. Scholastic, 1987. 32 pp. ISBN: pap. 0-590-40493-0. Age levels: 3–7. Grade levels: PreK–2.

SUMMARY: The prolific science writer Millicent Selsam has written a book about the odd creatures that inhabited the Earth millions of years ago. She describes their physical appearance, their extraordinary size, when and where they lived, what some of them ate, and how scientists learned about these extinct animals.

Details about these aquatic, land, and flying creatures include comparisons that promote a mental picture of these prehistoric animals. For instance, the colossal crocodile was longer than a school bus; the giant land sloth, which looked similar to a great hairy bear, was as tall as a telephone pole.

A paragraph of information is supplied for each creature featured. Names that might be unfamiliar, such as uintatherium and archaeopteryx, are italicized and a phonetic spelling is provided in parentheses. Descriptions are given credence by evidence found in fossil remains and insects entrapped in amber.

A world map shows where some animals lived and died. The author suggests that the next animal to become extinct may be the African elephant and urges that its habitat be protected. The final page lists the animals found in the book along with page number, scientific name, approximate date of existence, and habitat.

EVALUATION: The interesting text is complemented by dramatic illustrations. Sufficient details are given in the text and drawings to make these extinct animals come alive. The subject matter will captivate children while the tongue-twisting scientific names of the animals offer a challenge and are likely to produce giggles.

Readers will gain some insight into how scientists untangle the mysteries of the past. While the author mentions that today many kinds of animals are disappearing, she delivers a strong message that the elephant in particular is endangered and needs protection. An illustration of the skeletal remains of an African elephant is very effective in drawing attention to the immediate problem. This book can be coordinated with a visit to a science museum.

□ □ □

Shapp, Martha, and Charles Shapp. *Let's Find Out about the Moon.* Illus. by Brigitte Hartmann. Watts, 1975, o.p. 43 pp. Age levels: 5–9. Grade levels: K–4.

SUMMARY: Basic facts about the moon are presented for young children in this book. Because the information is not new, it may be used to show what was known about the moon in 1975 and to contrast this attention with what is known today.

Children are told that appearance is deceiving: although the moon seems to be larger than stars because it is closer to Earth, it is really much smaller. The orbit of the moon around the Earth and the orbit of the Earth around the sun are discussed. The book highlights discoveries of the *Apollo 11* astronauts in their 1969 journey, for example, the moon has no life, water, and air. Also, days and nights on the moon are each two weeks long, and although the temperature is very hot during the day, it is extremely cold at night. The authors also hypothesize about future developments in exploration of the moon.

EVALUATION: Although not a recent publication, this book does contain some good information and is written in an interesting manner designed to appeal to children. The moon is a fascinating topic for children, including toddlers. The illustrations are colorful, detailed sketches that aid in explaining such concepts as the orbits of moon and sun. Critical thinking could be encouraged by comparing this early book with any one of several newer ones on the topic of moon exploration. Children will no doubt be able to tell some new things that science has discovered about the moon since this book was written.

□ □ □

Shimek, William. *The Celsius Thermometer.* Illus. by George Overlie. Lerner, 1975, o.p. Unp. Age levels: 6–9. Grade levels: 1–4.

SUMMARY: *The Celsius Thermometer* is full of useful facts about different ways of reporting temperature. A thermometer is used to measure how hot or cold something is, and its name comes from two Greek words meaning heat measure. The liquid (mercury or alcohol) in a thermometer expands and takes up more space when something is hot. Anders Celsius's centigrade thermometer, which was named after him, uses a scale from 0 to 100 with 0° used to mark temperature of ice and 100° used to mark the temperature of boiling water. Each mark is called a degree. Degree marks go above 100° and below 0° so measurements above boiling water and below ice can be ascertained. Children will be interested to know that Isaac Newton made a thermometer that ranged from 0 to 12 on which 0° marked the temperature where water freezes and 12° marked the body temperature. Gabriel Fahrenheit used 0° on his thermometer to indicate the temperature of ice and salt mixed together. Water alone freezes at 32° on his scale and boils at 212°. Lord Kelvin made a thermometer in which 0 marks absolute 0°—the coldest temperature that just about anything can be. Water freezes at 273°K and boils at 373°K. The Kelvin thermometer is used most frequently in laboratories.

EVALUATION: This book explains the basics of what a thermometer is and does, and then highlights the Celsius thermometer, with some attention paid to three other types of thermometers. The colorful pictures serve more to break up the text than to clarify any confusing or difficult concepts. The lack of repetition of text and illustrations may leave the reader with only a general notion of the types of thermometers, although the title suggests that the reader will leave the book retaining some information on the Celsius thermometer. Detail is here, but it is not presented in the best way for younger children to encourage recall.

□ □ □

Simon, Seymour. *Galaxies*. Photos; illus. Morrow, 1988. Unp. ISBN: LB 0-688-08004-9. Age levels: 9–13. Grade levels: 4–8.

SUMMARY: Simon gives a very detailed account of galaxies, beginning with our own, the Milky Way. The photographs are the focal point of the book. Different angles of the Milky Way serve as a basis to compare it with other galaxies. Several other galaxies are photographed, for example, the Andromeda galaxy, a galaxy much like our own, as well as two smaller satellite galaxies. The text serves to describe and enhance the pictures. Galaxies can be classified by their shape: spiral (like our own, which has three arms), elliptical, barred spiral, irregular-shaped, and dwarf. A final picture shows several galaxy clusters, each of which contains about 150 galaxies of its own.

EVALUATION: This book seems too technical for the intended audience. A reader would need some background terms such as *galaxy, nucleus, satellite galaxies, spiral,* and *elliptical*. Diagrams depicting comparisons of Earth to our solar system, our solar system to the galaxy, and finally our galaxy in comparison to the unending universe would be helpful. The book also fails to make clear in the beginning that *Milky Way* is the name of our own galaxy. The pictures are breathtaking photographs and serve as the basis for the text. The following books by Seymour Simon and published by Morrow are appropriate for young readers interested in astronomy: *Jupiter* (1985), *Mars* (1987), *Saturn* (1985), *Stars* (1986), *Sun* (1986), and *Uranus* (1987).

☐ ☐ ☐

Simon, Seymour. *The Long View into Space.* Photos. Crown, 1979. Unp. ISBN: 0-517-53659-5. Age levels: 8–11. Grade levels: 3–6.

SUMMARY: The author takes the reader on a journey from Earth to its moon, then to the solar system, the sun and other stars, and finally to the Milky Way and other galaxies. Five sections of the book explore different parts of outer space. The sun, moon, planets, stars, nebulae, and galaxies are discussed as well as other things, such as rocks, dust, and large clouds of gases. The relationship of our solar system to other galaxies is introduced. Each chapter with its many photographs gives interesting facts and analogies to which children can relate.

EVALUATION: Seymour Simon is a well-known author of science books for children. In *The Long View into Space,* he gives readers a sense of where they fit in the universe. On the last page Simon shows how a child's return address could include not only name, street, city, state, and country, but also planet (Earth), system (solar), and galaxy (Milky Way).

Detailed and close-up black-and-white photographs enhance almost every paragraph of text; many observatories and NASA are credited for these photographs. Each chapter is introduced with white text on black background, conveying darkness or depth, which may give the reader the feeling of entering a new phase of the universe. The photographs occupy more space than text, enhancing the sense of a photographic journey into outer space.

☐ ☐ ☐

Simon, Seymour. *Look to the Night Sky: An Introduction to Star Watching.* Illus. by author; photos. Puffin Books, 1987. 88 pp. ISBN: pap. 0-14-049185-6. Age levels: 9–13. Grade levels: 4–8. Series: Science.

SUMMARY: In his introductory chapter, Simon suggests that the reader use this book as a field guide to stars, planets, and constellations. The author would like the reader not to consider this an astronomy book but rather an "observing book." With good eyesight it is possible to see colors, or a band of light, or planets that are bright. Many helpful hints are offered to the observer: problems and considerations covered include where the observer is viewing (from the city, a rooftop, or the suburbs), smog or air pollution, the adjustment time required for the pupils of the eyes, preparing a flashlight, time of night and month of the year, and the Earth's rotation. Subsequent chapters deal with the constellations, time of year, finding the planets, and noticing particular things about the moon, comets, and meteors. Information about purchasing a telescope and various special events or sightings such as eclipses, the northern lights, and star clusters is also given.

The book consists mainly of text with several black-and-white photos and diagrams interspersed throughout. An appendix about astrology and the zodiac is included, although the text explains why astrology is not considered a science. In addition, there is a list of sources for telescopes and a reading list of books and periodicals. An index and a brief sketch about the author follows.

EVALUATION: This well-known science author presents the material in such a way that the young reader will want to try his suggestions and become a novice astronomer. For those already into night sky watching, the book offers many new helpful ideas, although there is much interesting information for readers who do not actually do star watching too. The text is easy to follow and clearly encourages the reader to become familiar with and understand the night sky with the naked eye.

□ □ □

Simon, Seymour. *Mars.* Photos; illus. Morrow, 1987. Unp. ISBN: LB 0-688-06585-6. Age levels: 6–10. Grade levels: 1–5.

SUMMARY: Mars, the fourth planet from the sun, has two moons, is smaller than Earth, and has a year that's about twice as long as Earth's. This book supplies some early speculations that astronomers have made about Mars, a few artists' conceptions, and much photographic and experimental data that recent space probes have provided about our neighboring planet. In the 1970s, four *Mariner* and two *Viking* spacecraft reached Mars and sent back thousands of pictures to Earth for analysis. Mars is a dusty planet with huge volcanic peaks, craters, valleys, plains, and polar ice caps. There is no water on the surface of Mars; however, scientists believe that there are underground reservoirs. No life has been found to exist on this planet.

Because of its reddish appearance, Mars was named by the ancient Romans for the god of war, who was associated with blood and combat. The dusty soil and rocks of Mars have an orange-red color because they contain iron oxide, commonly known as rust. Space probes have provided much information about this harsh planet; however, many areas are unexplored and questions still persist.

EVALUATION: The NASA photographs of Mars, taken at close range by various space probes, are magnificent and incredible. They help clarify the text about the planet's characteristics. Scientific details are clearly given in large print and unencumbered language. The gathering of factual information is subtly made exciting by Seymour Simon, who also wrote *Jupiter* (1985), *Saturn* (1985), *Stars* (1986), *The Sun* (1986), *Uranus* (1987), and *Galaxies* (1988), all published by Morrow. Imaginations are tantalized when the author suggests in closing that a human might set foot on Mars within the readers' lifetime. This book is not only eye-appealing but also thought-provoking.

□ □ □

Simon, Seymour. *Mirror Magic*. Illus. by Lisa Campbell Ernst. Lothrop, 1980, o.p. 48 pp. Age levels: 5–9. Grade levels: K–4.

SUMMARY: Children often express curiosity about mirrors and this book will answer some of their questions. The author explains that a mirror reflects light and presents images in reverse. Familiar experiences with mirrors are related, for example, the function of a rearview mirror on cars and bikes, and the working principle of a kaleidoscope. Children are invited to use mirrors in new ways to prove how they work: for example, the reverse image shown in a single mirror will change when two mirrors are hinged together and the child looks at the middle section where the two mirrors meet. Also, the back of one's head can be seen when two mirrors are used. Children are told how to make a periscope, an activity all will enjoy. Readers will be particularly interested to know how curved mirrors used in a fun house attraction create amusingly distorted images.

EVALUATION: This book presents some intriguing experiments to do with mirrors. Children are fascinated by looking at themselves in mirrors, and this natural interest may be the initial attraction to experimenting further with mirrors. The illustrations are very effective in explaining the projects, concepts, and experiments offered in the book. Children should enjoy this hands-on type of activity book in using mirrors they have around the house.

□ □ □

Simon, Seymour. *Stars*. Photos. Morrow, 1986. Unp. ISBN: LB 0-688-05856-6. Age levels: 9–13. Grade levels: 4–8.

SUMMARY: Seymour Simon, a prolific science writer for all ages, discusses the stars in this detailed book for readers in middle grades. He explains that although stars are huge balls of gases, even through a telescope they appear as small points of light. Thousands of years ago, stars and groups of stars were given names. What astronomers saw back then is not what the sky looks like now because the stars are moving at a speed of about ten miles per second. Because the distances between stars are so great, scientists use the light year (186,000 miles per second) as the basis of measurement. Younger stars are hotter than middle-aged stars, and old stars first collapse on themselves, then balloon out as giant red stars. Sometimes an old star that explodes into fragments appears brighter, causing star watchers to think that new stars have appeared in the sky. These stars are incorrectly named *novas,* which means "new." Massive old stars that collapse become black holes. There are billions of stars in a galaxy, and billions of galaxies. The book leads the reader to believe that the more we learn about the mystery of stars, the more questions we will have.

EVALUATION: *Stars* is a highly informative collection of photographs and data about stars. It seems to be aimed toward intermediate grade readers. While current information about stars can be highly complex, Simon presents the topic in an understandable and readable form. The number concepts used, such as in the distance light travels per second or the distance to the closest star besides the sun, are mind-boggling. Certainly a reader's interest would be piqued by the facts presented, such as that on a clear night one can only see thousands of stars but there are billions more. The photographs aid understanding as well as spark curiosity.

□ □ □

Simon, Seymour. *The Sun.* Illus. by Frank Schwarz; photos. Morrow, 1986. Unp. ISBN: LB 0-688-05858-2. Age levels: 8–12. Grade levels: 3–7.

SUMMARY: Simon informs the reader early on that the sun is a medium-sized star that appears bigger than other stars because it is closer to the Earth. While it is about 93 million miles away, the next closest star is about 25 trillion miles away. Pluto is the ninth and last planet in our solar system and is 4 billion miles from the sun, and on Pluto the sun would appear like a bright star on a dark night.

Interesting phenomena orbit the sun. An asteroid belt made of rocks is between Mars and Jupiter. These rocks range in size from small pebbles to giant rocks approximately 500 miles aross. Comets, which are clumps of ice and dust, also orbit the sun, and when they near the sun they produce a shining tail.

The author explains that the sun could hold up to 1.3 million Earths. It is a ball of fire, but doesn't burn up like a huge bonfire would because it burns what appears to be a cloud of gases called the corona, or the outer atmosphere of the sun. The inner atmosphere, called the chromosphere, can be seen by the naked eye only during a total eclipse. Sunspots are defined as giant storms on the surface of the sun. During a one-hour sun storm enough energy is released to provide electrical power for the United States for a million years. The flaming streams of gases in an electrical storm may fly off into space or they may loop back to the sun. Flares are powerful magnetic explosions that last only briefly, but may black out communication or throw off magnetic compasses on ships and planes. Because flares send out many particles into space, they often get caught in the Earth's magnetic field near the North and South poles. These result in shimmering curtains of light called the northern and southern lights.

EVALUATION: Expertly written for middle-grade children, *The Sun* takes the reader through amazing information about our closest star. The basics about the sun are easily understood but the author pushes beyond these facts to fascinating documentation, all the while relating the details in simplified language. The pictures enhance the text by giving real examples of what scientists are looking at and searching for. This would be a primary resource to consult in any study to be done on the sun.

□ □ □

Simon, Seymour. *Uranus*. Illus. by Frank Schwarz and Todd Radom; photos. Morrow, 1987. 32 pp. ISBN: LB 0-688-06583-X. Age levels: 8–13. Grade levels: 3–8.

SUMMARY: In this book about Uranus, information about its discovery, name, features, and current flyby photographic data is provided. The book begins with the discovery of Uranus in 1781 by William Herschel. Eventually it was named after the Greek god of heaven who ruled the world.

A description of the planet is given. It is made up mostly of gases, and about 50 Earths could fit into it. In 1986, four years after it had obtained data about Saturn, the spacecraft *Voyager 2* flew by Uranus. Thanks to this amazing craft scientists now know a great deal more about Uranus than they had known before. Scientists now know that Uranus lies on its side, possibly due to a collision with another space object. Additional information that scientists gleaned from *Voyager 2* include the following: a full day on the planet lasts 42 Earth years. A strong magnetic field is unusually located at Uranus's equator instead of its poles, as on Earth. There are blue-green clouds over the planet, and they travel at a speed of about 200 miles an hour. The temperature above the clouds is about −350 degrees Fahrenheit; however, below the clouds the temperature is thousands of degrees hotter. There is speculation among some scientists that Uranus does not have a surface but is hot and watery with a rocky molten core. The outer space of the planet, such as the 11 spinning rings of unknown black substance and the 15 moons of Uranus, is also examined in the book.

EVALUATION: Current and educational, *Uranus* provides some top-rate photographs of the seventh planet. It is interesting to read, look at, and think about. Modern technological abilities to combine several photographs provide an incredible view of what it would be like to be living on a moon with Uranus looming above. Based almost entirely on information received from *Voyager 2,* this book is an excellent science book for children.

□ □ □

Sitomer, Mindel, and Harry Sitomer. *Spirals.* Illus. by Pam Makie. Crowell, 1974. 33 pp. ISBN: 0-690-00180-0. Age levels: 5–8. Grade levels: K–3.

SUMMARY: The authors discuss the differences between circles and spirals, and how spirals can be found in many places including nature. They describe how a merry-go-round moves in a circle and the center is called an axis. Children can draw different sized circles using a compass. Or they can roll strips of clay to make circles, or roll one long strip of clay around itself to make a spiral. A circle begins and ends at the same place and so it is called a closed figure; however, a spiral does not start and end at the same point so it is called an open figure. All points on a circle are the same distance from the axis, but not so with a spiral. Many things appear to be circles but in fact they are spirals, such as the grooves on a record, the bottom of a straw basket, or rag rugs. In ancient times, the Greeks utilized spirals to adorn their buildings and clothing.

Detailed descriptions for making round and flat spirals are included in the book. A simple experiment demonstrates how wood screws differ from nails because wood screws have a spiral thread, allowing them to hold better than nails. Children's hopping games are shown to make use of spirals. With careful observation, spirals can be found everywhere, even in nature, for example, tornadoes, hurricanes, and pine cones.

EVALUATION: The book highlights many spirals found in nature: sunflowers, snail shells, coiling snakes, and galaxies. It is amazing how many shapes are thought of as circles when they are actually spirals. Also of great benefit is the initial comparison between a circle and a spiral; the two appear so similar until examined closely. One section of the book seems too complex compared with the rest of the text. The directions for using graph paper to draw spirals are so detailed that they may keep the reader from experimenting, because of the emphasis on precision in using graph paper and counting out points. Although this may be an appropriate experiment with spirals, it seems too complicated for the intended audience.

Overall, the text is of good quality. The reading flows smoothly and is relatively clear and the illustrations aid in clarifying the text. This seems to be a wonderful introduction for students noting distinctions between circles and spirals.

□ □ □

Sitomer, Mindel, and Harry Sitomer. *What Is Symmetry?* Illus. by Ed Emberley. Crowell, 1970. 33 pp. ISBN: pap. 0-690-87618-1. Age levels: 5–9. Grade levels: K–4.

SUMMARY: Line, point, and plane symmetry are explained in this book, using colorful pictures of natural and manmade objects. Line symmetry is covered first and most thoroughly. It is explored in a person, a butterfly, a leaf, and the letters *M* and *E*. The lines of symmetry are drawn vertically or horizontally across the pictures, and the text explains that each side balances the other. The book provides a folding test where the reader can see if a picture has line symmetry; the line of symmetry is the fold. If the reader were to fold a paper along the middle and cut out any shape, the unfolded shape would be symmetrical. Folding a paper twice and punching holes gives three lines of symmetry: one in the middle for the entire paper, and the two outer lines for the sides they separate. A chain of paper dolls shows lines of symmetry; a ball thrown in the air has a line of symmetry. Circles have several lines of symmetry. The orbit of the Earth is elliptical and has two lines of symmetry.

Point symmetry occurs when one object can turn halfway around to match a second object; if it does not match, there is no point symmetry. The third type of symmetry covered is plane symmetry and is seen in mirror reflections or ponds.

EVALUATION: This is a delightful book for young children, stretching their minds to see symmetry in familiar things. It thus provides a new way to look at the world. The most difficult concept to grasp seems to be point symmetry. After some explanation, the summary of imagining the half twist of an object to end up looking the same as the original clears up the confusion. The pictures also provide entertainment with the guidance of an alligator. Especially humorous are the alligator's shy and bold facial expressions in front of the mirror. The well-presented concept of line symmetry prepares the reader for the more abstract concepts of point and plane symmetry.

□ □ □

Smith, Henry. *Amazing Air.* Illus. by Barbara Firth, Rosalinda Kightley, and Elizabeth Falconer. Lothrop, 1982. 45 pp. ISBN: LB 0-688-00973-5; pap. 0-688-00977-8. Age levels: 9–13. Grade levels: 4–8. Series: Science Club.

SUMMARY: This book offers a variety of experiments to discover properties of air. Some of the properties discussed are that plants produce oxygen; coldness slows down oxidation of fruits; fire uses oxygen to burn; water is in the air; water will evaporate from foods; coldness causes the water in the air to appear on the outside of a glass of ice water; plants give off water; dirty air is trapped by warm air; the more energy used, the more air needed; hot air expands and rises; air pressure pushes in all directions; and things fly on streamlines for least resistance.

EVALUATION: This book presents simple concepts of air, to be explored through experiments. Most of the experiments are complex and require materials that might not always be readily available—nor would children be familiar with them. For example, sodium bicarbonate, a skewer, and plastic-covered flex are required. However, a supply list at the beginning of the book describes the supplies and tells where one could purchase them. Some of the terms used are of British origin, as that is where the book was first published. The author uses questions that would intrigue the reader to perform the experiments. Since he does not directly answer these questions, the experiments must be performed to discover the answer. The illustrations are very clearly drawn and help define the experiments, which are written in a step-by-step form. Indeed, air is discovered to be amazing as shown by the variety of experiments presented.

□ □ □

Srivastava, Jane Jonas. *Area*. Illus. by Shelley Freshman. Crowell, 1974. 33 pp. ISBN: LB 0-690-00405-2. Age levels: 5–8. Grade levels: K–3. Series: Young Math.

SUMMARY: Srivastava explains the concept of area by identifying what some familiar areas are. Experiments to measure area are offered through simple activities in which the reader can participate. A rationale for measuring area is presented in terms understood by children approximately five to eight years of age; for example, measurement is necessary to know how much paint is needed to cover a room, or how much paper is required to wrap a present. A few mathematical terms are defined and illustrated, such as standard unit and surface area.

EVALUATION: The style and content of this book will appeal to younger readers. The elements of the text are presented in a logical order, and the illustrations are clear and colorful. By taking time to work the simple problems, the reader can benefit from the understanding that is the result of involvement.

□ □ □

Stone, Lynn M. *The Arctic*. Photos. Childrens, 1985. 45 pp. ISBN: LB 0-516-01935-X. Age levels: 7–11. Grade levels: 2–6. Series: New True.

SUMMARY: This book describes the land of the midnight sun—the name given to the Arctic. This term covers both the land and sea at the northern part of the world that touches seven countries. In the summer the sun never sets in the Arctic, and on some days in the winter the sun never rises. The Arctic is the coldest place on Earth; winter temperatures are −70° Fahrenheit, and many areas are constantly covered by ice, even in summer. The Arctic winter lasts about eight months. During this time most animals leave, although some remain hidden in protective snow tunnels. In the spring, snow melts as the days become longer. Animals return and plants appear by June. Average summer temperatures are 70° Fahrenheit, but the summer season is very brief.

As one travels north to the Arctic one passes great evergreen forests, and gradually the trees appear scattered and somewhat smaller; the ground is covered by caribou moss in the tundra. Further north is the Arctic prairie where there is open grassland; birch and willow trees grow here, although they are only a few inches tall. Moss and lichens grow here in spite of the fact that the Arctic receives about as much annual rain as a desert. Ponds form in the summer where ice underground stops water from draining further. Even though the Arctic Ocean is salt water, the extremely cold temperature causes it to freeze entirely. Many animals live in the Arctic, although people think the land is barren. Eskimos have lived there for thousands of years; they call themselves *Inuit,* which means "human beings." Changes in Inuit life include being able to buy food and clothes at stores instead of having to hunt for them, and occasionally using snowmobiles, although dog sleds are still very much a part of the Arctic scene. Valuable minerals and oil have been discovered in the Arctic, and environmentalists see an intensified need to protect the Arctic from damage and pollution.

EVALUATION: The photographs in the book are excellent in showing the beauty of the Arctic. They coincide with the topics covered: the midnight sun, the varieties of plants and animal life that live there, and the life-style of the Eskimos. This book, simply written in large type, provides a great deal of information in seven topical areas. The use of scientific terms is limited but a great deal of information is still presented. An index and "Words You Should Know" list are included at the end. The book provides a wonderful blend of information and enjoyment.

□ □ □

Tyler, Margaret. *Deserts.* Illus. by David Farris and John Plumb. Day, 1970, o.p. 47 pp. Age levels: 8–11. Grade levels: 3–6. Series: Finding Out about Science.

SUMMARY: Wind and rain as elements that have a direct cause-and-effect relationship to the formation of deserts are introduced in this science book for middle grade children. It tells that in the desert, the wind can easily blow away any loose earth because there are very few trees, and thus few roots to hold the land surface in place. The pieces of loose earth vary from fine dust to coarse stones; heavier sand grains are usually swept around near the ground until the wind drops them or they are blocked by some obstacle. Sand dunes usually have a long windward slope and a steep leeward slope caused by the blowing wind. Barchan dunes display crescent-shaped patterns when the wind is blowing from the same direction, and seif dunes result from cross winds. So-called star dunes are formed when desert winds blow in all directions.

Wind and temperature change the shape of rocks in the desert. When it does rain in the desert, it may come as a downpour. Rain may not fall for months or years but some plants still survive such as cacti, which are well adapted to survive with their spiny shapes and protective resin coats. The author tells readers that camels can go several days without water but will drink large quantities when water is available. The date palm tree is useful for fruit, fuel, and building supplies, and is found in oases, places in the desert where water is found. Although civilizations have survived in deserts and man has tamed parts of deserts, lack of water is still a problem and overall desert populations have remained small.

EVALUATION: This book provides information on deserts, different sand dune patterns, and plant life that survive in deserts. Groups of people who have lived in deserts are also mentioned. The illustrations are colorful and suggest the hot, dry feel of the desert. There are no actual photographs of deserts, which might have been more helpful than the colorful illustrations. There are many specific terms in the text—people groups, plants, cultural objects—which are difficult to pronounce. A list at the back of the book includes 35 new words, but does not include the pronunciation keys that are essential for so many foreign words. The book provides an abundance of information but has a feeling of being written "down" to children. The style of writing seems geared for primary children, while the information given is more appropriate for the intermediate grades.

□ □ □

Updegraff, Imelda, and Robert Updegraff. *Continents and Climates.* Illus. by authors. Penguin, 1982. 24 pp. ISBN: 0-14-049188-0. Age levels: 8–11. Grade levels: 3–6. Series: Turning Points.

SUMMARY: The shape of planet Earth is compared to a spaceship spinning around the sun in this detailed picture book. A discussion of Earth's surface and rotation leads to explanations of climate, temperature zones, growing conditions, and characteristics of life around the planet. The six continents and the features of each, including rain forest, desert, polar region, and temperate zone, are included. On the back page, "Things to Do" gives directions for making Eskimo sunglasses and a miniature tropical rain forest called a jungle jar. The book is one in a series of six, and contains a cross-referenced index so a reader can easily find out which book expands on topics discussed here.

EVALUATION: A great deal of information is distilled in very few pages. Consequently, there is the possibility that scientific concepts are oversimplified. This does not mean the book lacks value, as it appears to present facts accurately and is well written. An added advantage is the use of large print, offsetting the rather dense pages. The illustrations are pleasant but don't do much to elaborate on the scientific information. First published in the United Kingdom, this book appears to be a low-budget production; there are no photographs, and the colors used in drawings are somewhat limited. Since the United Kingdom publication date is 1980, the authors can perhaps be excused for using the word *Eskimo* instead of the currently more acceptable term *Inuit*.

□ □ □

Updegraff, Imelda, and Robert Updegraff. *Mountains and Valleys*. Illus. by authors. Penguin, 1980. 24 pp. ISBN: 0-14-049189-9. Age levels: 9–11. Grade levels: 4–6. Series: Turning Points.

SUMMARY: The Updegraffs use a cutaway page technique to show the many changes that occur in the Earth's surface. The book begins with a discussion of how mountains are formed and explains that fossils buried in mountains prove that even the highest mountain ranges were formed from rocks once covered by the sea. Distinctions are made between block mountains and fold mountains, and these differences are highlighted by cutaway pages. The causes of erosion—ice, glaciers, plants, rain, wind, and even people—are discussed in some detail. The formation of "cones," or surfaces that rise to a peak, and valleys are explained. The illustrations' subtle changes in color, which are earth tones, call attention to the many layers of the Earth's surface. The last page includes some delightful activities designed to help children understand the scientific concepts in the text. One activity invites children to make sedimentary fudge from instant vanilla pudding and cocoa powder. A one-page index is located inside the back cover.

EVALUATION: *Mountains and Valleys* cleverly teaches basic geological information in an appealing manner. The illustrations are somewhat primitive but appropriate because the amount of information, while not exactly skeletal, lacks detail that could be burdensome for young readers. The book was first published in the United Kingdom and has not been altered for an American audience. For example, on the "Things to Do" activity page, the term *cotton reel* is used. If a child does not grasp the meaning from context, this may require explanation that even an adult may be hard pressed to provide.

□ □ □

Walter, Marion. *Make a Bigger Puddle, Make a Smaller Worm*. Illus. by author. Evans, 1971. Unp. ISBN: 0-87131-073-2. Age levels: 4–8. Grade levels: PreK–3.

SUMMARY: *Make a Bigger Puddle, Make a Smaller Worm* is an activity-oriented picture book that involves readers in seeing, doing, and thinking creatively. This is accomplished through the use of drawings, the visual perception of which can be changed by using the small steel mirror included in a pocket attached to the front cover of the book. The reader can make a whole moon out of a half-moon; a shorter block of houses out of a longer block. Short stripes can appear longer, and straight lines can be made to look bent. Dots will seem to multiply, and a puddle can look bigger or smaller. The text is spare and simple and is directly related to the illustrations on the opposite page. Simple mathematical concepts, such as more or less, are introduced and reinforced, and the terms used to discuss these concepts may be learned in an incidental way.

EVALUATION: This is a manipulative book with a specific purpose, not merely a toy, as is often the case with manipulatives featuring moving parts. The book is an extremely clever learning device. The only problem that might arise with the use of the mirror is not associated with safety—it appears to be safe enough. If an adult or older child does not help the young reader use the mirror at first, the math operations might be confusing and a degree of frustration could occur.

□ □ □

Watson, Clyde. *Binary Numbers*. Illus. by Wendy Watson. Crowell, 1977. 33 pp. ISBN: LB 0-690-00993-3. Age levels: 7–10. Grade levels: 2–5. Series: Young Math.

SUMMARY: Binary numbers are presented by use of examples that younger children can grasp. Watson tells how binary numbers, although they are represented by familiar numerals, are used differently from the way in which readers are accustomed to using numerals in arithmetic computation. The use of binary numbers in computers and secret codes is explained.

EVALUATION: Very well written, *Binary Numbers* offers a detailed introduction to the use of binary numbers that are commonly used in computers. This book is designed to be shared with an adult or someone familiar with binary numbers because questions are frequently raised but left unanswered. For readers desiring a self-check, this may become frustrating: they may not feel secure about their comprehension of the concepts. Thus, a knowledgeable adult could affirm their answers or redirect their thinking. The illustrations are essential to aid understanding the concepts; simply designed, they afford a good condition for learning by not distracting with excessive detail. Overall, this book is a good introduction to the system of binary numbers.

□ □ □

Watson, Philip. *Light Fantastic.* Illus. by Clive Scruton and Ronald Fenton. Lothrop, 1982. 46 pp. ISBN: LB 0-688-00969-7; pap. 0-688-00975-1. Age levels: 9–13. Grade levels: 4–8. Series: Science Club.

SUMMARY: The properties of light are explored through experiments that will lead the young scientist to conclude that light *is* fantastic. Properties explored include the attraction plants have for light, the ability to begin a fire through concentrated light on a magnifying glass, reflection of light off paper or planets or mirrors, the principle used in making kaleidoscopes, the seven colors of white light, and the refraction of light in water.

EVALUATION: Though the concept of light may be commonplace and readily acceptable, the experiments bring a fresh curiosity and interest to the qualities of light. The experiments are simple to do and yet interesting in their results. The illustrations serve to clear up instructions by way of example. The level of difficulty of experiments varies greatly, as indicated by the span of ages for which the book would be appropriate. Although the concepts are understandable by intermediate age children, they provide proven answers to the "why" questions younger children ask. A "Supplies and Skills" list at the outset prepares the young scientist for equipment and skills needed for the experiments. A glossary and an index at the end aid quick referencing. The book contains a large amount of scientific information, but it is presented in a highly interesting hands-on activity approach.

□ □ □

Watts, Lisa, and Jenny Tyler. *The Children's Book of the Earth.* Illus. by Bob Hersey. EMC, 1978, o.p. 32 pp. Age levels: 8–11. Grade levels: 3–6.

SUMMARY: Most of this book's 19 "chapters" are two-page spreads depicting characteristics of the Earth: rocks, fossils, seasons, weather, caves, Earth's natural resources, earthquakes, volcanoes, and glaciers. The authors also include a theory about how the Earth was formed, information on how towns grew, and data on life near the equator and the North Pole. Each page is brightly colored and usually several illustrations are depicted on a page. These may be scenes, diagrams, or pictures in small squares, rectangles, or circles. In the chapter "The Journey of a River," the route of rainwater coming down a mountain in streams leading into a river, and eventually ending up at the mouth, is shown in illustrations and text. Comic strip balloons describe these illustrations.

A glossary called "Earth Words" is included at the end, and each word has an accompanying small black-and-white illustration. There is also a brief index.

EVALUATION: This book has instant appeal because the illustrations are bright and use many colors. Cartoonlike rectangular balloons are also used to convey facts. The presentation of material tempts the reader to want to know what is in the circle, square, or scene. Only one page or a two-page spread is allotted to each chapter. The facts are very concise and interesting to readers, such as the stegosaurus having three brains: in its head, its tail, and its back. The illustrations often convey more detail than the text. In the chapter "How a Town Grows," the reader can go from an old castle with its fortification, to houses and modern buildings, and even a sports stadium; not all of these things are mentioned in the text.

□ □ □

Whitfield, Philip, and Joyce Pope. *Why Do the Seasons Change?* Illus. by Richard Orr, et al. Viking, 1987. 96 pp. ISBN: 0-317-62526-8. Age levels: 7–10. Grade levels: 2–5.

SUMMARY: The answer to the title rests in the repeating patterns of events occurring in the world, called rhythms and cycles by the authors. These rhythms and cycles give a better understanding of why and how the world of nature works. The animals and plants on Earth are all governed by these forces. Answered are 113 questions ranging from the birth of the Earth, to weather and its effects on the life cycles of plants and animals. Readers learn why fireflies flash, why one's tummy rumbles, and what factors resulted in the development of the ice age.

Most questions are accompanied by photos, drawings, or diagrams in color or black and white, giving clear definition to the text and visual appeal. The information included herein comes from the Natural History Museum in London.

EVALUATION: Entries are short, easy to read, and easily understood. Scientific terms are explained. In major categories questions are linked together, and there is a logical flow from one category to another. A list of abbreviations used in the book is given in the introduction. Measurements are given in both metric and English conversions. Many cross-references are found within the text. An index with illustrations is also included. This book is very interesting, informative, and fun to read for anyone who has scientific questions about the world.

□ □ □

Whyman, Kathryn. *Chemical Changes.* Illus. by Louise Nevett; photos. Gloucester, 1986. 32 pp. ISBN: LB 0-531-17032-2. Age levels: 8–11. Grade levels: 3–6. Series: Science Today.

SUMMARY: The many ways in which chemicals and chemical compounds can improve the living conditions of human beings form the basis for this book in the Science Today series. Readers will be interested to learn about chemical elements and the combinations thereof, and the familiar products formed by chemical combination, filtration, and distillation, such as the fact that a combination of iron, water, and air form rust. Boiling and subsequent evaporation also cause dramatic chemical changes. Science buffs will readily absorb the basic information found here, partly because chemistry and chemical change fascinate and unfailingly hold attention.

EVALUATION: *Chemical Changes* is a reasonably well-written book appropriate for the middle- and upper-elementary grades. Explanations are simple but too much detail is included. The illustrations flow with the text, but only occasionally does the text refer one to the illustrations. Also included is an experiment that can be performed with chemicals found around the home. The smooth writing style makes this a recommended introductory book in the field, but if children are too young to understand the examples, their interest may waiver.

□ □ □

Whyman, Kathryn. *Heat and Energy.* Illus. by Louise Nevett. Gloucester, 1986. 32 pp. ISBN: LB 0-531-17022-5. Age levels: 8–11. Grade levels: 3–6. Series: Science Today.

SUMMARY: One of the books in the Science Today series, *Heat and Energy* demonstrates convincingly that every action taken involves energy, and explains that ultimately all energy comes from the sun. Readers learn that energy flows in a cycle and the result is that energy, though changed, is never lost or used up. Potential energy and kinetic energy are discussed, as is convection, conduction, and radiation.

EVALUATION: The illustrations successfully complement the text in this slim volume, one of the better books in this series. Whyman's style of writing is easy to read, as she presents information without getting bogged down in excessive technological jargon. Because energy is commonly taken for granted, what is learned herein is sure to surprise readers as well as interest them. Whyman knows how much detail is enough; she seldom overloads her books with facts and avoids tiring her young readers.

□ □ □

Whyman, Kathryn. *Light and Lasers.* Illus. by Louise Nevett. Gloucester, 1986. 32 pp. ISBN: LB 0-531-17033-0. Age levels: 8–11. Grade levels: 3–6. Series: Science Today.

SUMMARY: The importance of light, both natural and artificial, is emphasized in this book. It reveals how light travels and how it is reflected and refracted. Information is provided on how colors are produced from light and how they are perceived by the human eye. An explanation of lenses and how we see is also included.

Information is given about the kinds of light and wave length produced by lasers. Laser technology is used in the making of tools, surgery and surgical instruments, and holograms.

The book contains a glossary and an index.

EVALUATION: Complex scientific and technological concepts are presented in a simplified manner. The headings and text utilize large clear print. A concise presentation of text helps to make concepts understandable, which is further enhanced by the color photos, diagrams, and experiments.

Readers will find explicit directions for making a periscope. They could also make a sundial and a pinhole camera.

At the end of the book a final page is devoted to more information about light waves and measuring with light.

□ □ □

Wilson, Francis. *The Weather Pop-Up Book.* Illus. by Philip Jacobs. Simon & Schuster, 1987. Unp. ISBN: pap. 0-671-63699-5. Age levels: 9–14. Grade levels: 4–9.

SUMMARY: The book contains five two-page spreads, each covering a different aspect of weather, telling what weather is and why it changes. In the middle of the book, part of the globe pops up to show portions of both the eastern and western hemispheres surrounded by other illustrations and a manipulative. This is followed by a spread about highs, lows, and frontal systems supported by several illustrations and manipulatives. The next covers tropical storms, and as the book is opened a partial hurricane pops up; there are two other manipulatives on these pages. Clouds and rain are discussed on the next spread, taking into consideration how clouds form, rain, snow, hail, and thunderstorms; this is supported by seven manipulatives. Finally, in the section on weather forecasting, a satellite pops up as the pages are opened; three manipulatives can be found on these pages as well.

The illustrations show detail—as is necessary in cloud formation pictures—and are colorful and easy to understand.

EVALUATION: This book is highly recommended for a unit on weather or for anyone interested in the subject. The illustrations are excellent. Though done on a black background, they are very bright in color and the colors were chosen wisely. Arrows indicate directions for clarification. The pop-ups are outstanding and both they and the manipulatives appear durable. Also, there are many manipulatives in comparison with other books having this feature. Depending on the purpose for which the book is used, it could span many age levels. Because there is much detailed text, this book could be enjoyed by adults as well as very young children.

□ □ □

Wyler, Rose. *Real Science Riddles.* Illus. by Talivaldis Stubis. Hastings House, 1972, o.p. 48 pp. Age levels: 5–10. Grade levels: K–5.

SUMMARY: Riddles and questions (with answers) on the subjects of the sun, solar system, weight and gravity, and light refraction are in this science book. Plants and animals and how they affect our environment are also interspersed throughout. "Silly Science Time" riddles provide background and amusement; some simple science experiments are also included. Questions are generally in a larger type size than answers. The text and illustrations are in three different colors: yellow, blue, and orange. The illustrations are appealing (the sun is shown wearing sunglasses, and a carrot has a face). Illustrations provide labels where necessary such as to the parts of a root vegetable and the planets in the solar system.

EVALUATION: The author has a natural flair for teaching children science in a way that is humorous and fun. The varied subjects and the jokes entitled "Silly Science Time" should also be appealing to children. Simple experiments also provide interest: how smell affects taste (using an apple and an onion) is explained, as well as how to make a rainbow and grow a vine. The answers to riddles sometimes circumvent the illustrations. The answers give sufficient information to answer the question, and often additional interesting facts are included. The language is easy to read, for example, *bellies* (of penguins) is used instead of *stomachs*. Children will want to share this riddle book with friends by asking the questions and giving answers, a sure way to spread science knowledge and facts.

□ □ □

Wyler, Rose. *Science Fun with Mud and Dirt.* Illus. by Pat Stewart. Simon & Schuster, 1986. 48 pp. ISBN: LB 0-671-55569-3; pap. 0-317-56794-2. Age levels: 7–10. Grade levels: 2–5. Series: Science Fun With . . . .

SUMMARY: In this book the upper-grade child is provided information about the composition of mud. He or she is asked to shake mud and water together in a large jar. As the mud settles, it will do so in three layers: a bottom layer of pebbles, a middle layer of course grains, and a top layer of fine grains. In the murky water, the child will find tiny bits of leaves and twigs—the vegetation. Even after being strained, the water is cloudy with clay.

All of this matter is found in dirt. Dirt varies in different locations; some dirt contains a lot of clay, and some a lot of sand. The best type of soil for growing plants contains humus. An experiment is explained and illustrated: the reader is asked to place four kinds of dirt —the sandy soil, the loam, the clay, and the silt—into four paper cups, then to plant grass seeds, water the soil, and put the cups near a sunny window. After a few weeks of care and observation, the reader is asked which kind of soil has the better crop.

The book also includes a chapter on animals and insects who live underground. Their tunnels, dens, and the food they eat are described and illustrated. In the past, men used mud to make homes and cooking pots but today man uses clay for pottery and bricks.

EVALUATION: It is refreshing to find a book that explains a most basic form of matter: dirt. Children will make mud anyway but by validating and explaining the results, the child's play is elevated to a science. The reader is certain to become more observant of changes in his own backyard after reading *Science Fun with Mud and Dirt*.

The illustrations are simple, three-color drawings. Brown, beige, and orange earth tones enhance the subject of the text. The animals who live underground are well labeled and drawn in their natural habitat. Their footprints are also illustrated and the reader is asked to be a detective and match the prints in the books with tracks found while exploring.

Possibly the author should have included a few pages on the importance of rich soil to our nation's farmland. The production of rich humus is a slow process and much organic matter is needed, especially in areas of our country where soil is naturally sandy or contains too much clay.

□ □ □

Wyler, Rose. *Science Fun with Toy Boats and Planes.* Illus. by Pat Stewart. Simon & Schuster, 1986. 48 pp. ISBN: 0-671-55573-1; pap. 0-317-56816-7. Age levels: 8–11. Grade levels: 3–6. Series: Science Fun With . . . .

SUMMARY: *Science Fun with Toy Boats and Planes* gives directions for making these objects out of materials available around the house, such as milk cartons, aluminum foil, spools, coffee can lids, paper clips, scissors, and various kinds of paper. Basic principles of flotation, displacement, weight, gravity, action and reaction, and aerodynamics are introduced. Children are told how paddles, propellers, and gliders work, and are given instructions about making and using toys that illustrate each principle.

EVALUATION: Bright yellow highlights the toy objects in this simple book, which will be fun for children. The toys are easy to make and, for the most part, can be done by children aged eight and above independent of adult supervision. On a few pages, the reader *is* told to enlist the aid of a grown-up, such as page 20 when a milk carton has to be cut lengthwise. Some parents may object to the illustration of a child launching a toy plane while standing on a bed, but for the most part the handmade toys are fun to make and their manipulation will be harmless, though the scientific principles may not always "take." One cute touch in the book is that the reader is referred to thoughout as "captain."

□ □ □

Zim, Herbert S. *The Universe*. Illus. by Gustav Shrotter and Rene Martin; photos. Morrow, 1973, o.p. 63 pp. Age levels: 9–13. Grade levels: 4–8.

SUMMARY: This book informs the reader that the universe as we know it is not only immense but is growing. Scientists use distance and time to calculate measurement of the universe, and the author contrasts this with methods employed by ancient civilizations. Readers may be astonished to learn that at one time in history scientists could be put to death for advancing radical theories. Zim's compendium of facts also includes the information that Uranus and Neptune were discovered more or less by accident, and that Pluto was discovered in 1930 only after scientists had mastered the measurement of distances using the speed of light.

EVALUATION: Zim gives an abundance of technological information on the universe, but his style flows so smoothly that children will be able to read sections touching on difficult concepts without confusion. The illustrations are somewhat disappointing: they add information without clarifying the text, and the photographs are of rather poor quality. Indeed, one wonders why some blurry shots were included at all. Probably the most fascinating characteristic of the book has to do with the historical background of individual scientists and the development of scientific language about the universe.

□ □ □

Ziner, Feenie, and Elizabeth Thompson. *Time*. Photos. Childrens, 1982. 45 pp. ISBN: LB 0-516-01651-2; pap. 0-516-41651-0. Age levels: 4–8. Grade levels: PreK–3. Series: New True.

SUMMARY: *Time* provides a historic overview of many types of clocks used in ancient civilizations through modern times. The function of shadow sticks and sundials, candle clocks, rope clocks, water clocks, and hourglasses is explained. Universal military time, using a 24-hour clock, is included even though this represents the way time is reported rather than the way it is calculated.

EVALUATION: The book is certainly useful. The manner in which each device for telling time is explained and then criticized presents a benefit beyond the information itself: it gives young children the opportunity to exercise fledgling critical thinking skills. However, until children are in grades one or two, comparisons of ancient clocks will not mean much: children four and five years old are still learning to tell time via conventional modern clocks.

□ □ □

Zubrowski, Bernie. *Bubbles*. Illus. by Joan Drescher. Little, Brown, 1979, o.p. 64 pp. Age levels: 5–9. Grade levels: K–4.

SUMMARY: *Bubbles* encourages the reader to experiment with bubbles to find out what they can do. Readers are told that higher-priced dishwashing soaps work best, and that glycerine or gelatine powder can be added to make stronger bubbles, although they will not necessarily make the bubbles last longer. All sizes of bubbles can be made, and bubbles can be made inside of bubbles. A bubble cannot be touched with a dry hand or other dry object without breaking it. Bubble sculpture is discussed, and children learn how to use a wet ruler to measure the dimensions of bubbles. The author suggests that children may want to enlist the help of an adult to record the many observations arising from these experiments with bubbles.

EVALUATION: The book sparks a new curiosity about bubbles in its departure from the fabricated soap solutions and blowers from stores. It presents a challenge to explore bubble-making using new techniques and encourages the experimenter to work with different bubble-making tools such as cans, lids, or wires bent into various shapes. Also, several questions concerning the properties of bubbles discernible through activities such as adding food coloring and looking at bubbles through sunglasses are asked and answered. The underlying philosophy of this book—that science and experiments should be done for the fun of it—is not prevalent in all science books. Because of its simple reading style, the book is probably intended for older readers but older children and even adults will have fun with this one.

Technology: Fiction

The modern wonders of technology—spaceships, robots, video games, and computers—are integrated within plots in these books of fiction. Hot-air balloons, ferryboats, motorcycles, unicycles, time machines, and other means of transport convey readers into new experiences. Woven into fictional backgrounds, information is presented about machinery on farms and at construction sites. Technological methods involved in protecting food and water supplies are also presented in imaginative settings.

□ □ □

Alexander, Martha. *Marty McGee's Space Lab, No Girls Allowed.* Illus. by author. Dial, 1981. Unp. ISBN: 0-8037-5156-7; pap. 0-8037-0018-0. Age levels: 3–7. Grade levels: PreK–2.

SUMMARY: In the privacy of his room, Marty attempts to perfect his invention, a space helmet that can fly. A sign on the door of his space lab emphatically bars his two sisters from entering. However, his youngest sister, who can't read yet, enters his secret domain. She dons the space helmet, shakes her musical rattle, and succeeds in making the space helmet work. Out the window she flies and soars over the house.

Marty is astounded to learn that his little sister, by wearing his space helmet, has been able to fly. However, instead of complimenting her, he scolds her because she didn't heed his warning sign about entering his lab.

Frustrated that he doesn't know how his own space helmet works, he finally realizes that he needs the help of both his sisters in order to discover how to make his invention function. Marty assigns his sisters as space pilots and gains new insights.

EVALUATION: In this amusing little book, an inventive young boy excludes his sisters from his space project. The antagonism that often exists among siblings is dramatically presented through illustrations that show hostility and language that is usually used by angry children. The text is brisk and brief. An important message is relayed in a nondidactic manner. With cleverness and wit, the author-illustrator takes a very young girl, who is still in diapers, on an unusual flight and ultimately shows that high-technology inventions know no discrimination and can use the input of both girls and boys.

□ □ □

Allen, Jeffrey. *Mary Alice Returns.* Illus. by James Marshall. Little, Brown, 1986. Unp. ISBN: 0-316-03429-0. Age levels: 5–8. Grade levels: K–3.

SUMMARY: Mary Alice, a duck, has a seemingly ordinary occupation, but it leads her to a series of unusual events. She's Operator Number Nine, a reliable switchboard operator, who announces the correct time to the exact second. Everyone in town depends on her precision. One day when she hears a cry for help over her headphones, she consults authorities to find out where the problem might be. Neither her immediate supervisor nor the police and fire departments are of any assistance.

She persists in her quest and places an advertisement in the personal column of the newspaper. She also hires a biplane to carry a message asking that the needy person get in touch with her. Compassionate Mary Alice even moves her bed near the switchboard in order to receive every call; some strange requests come over the telephone lines before the call for help is again heard. Mary Alice hastily dons her hard hat, climbs a telephone pole, and intercepts the call through a special line. She rushes to the home where the call was made and finds that Karen Squirrel has put her talking doll on the telephone. Mary Alice convinces the apologetic Karen Squirrel that telephones are not toys. Later Karen Squirrel is given the opportunity to wear headphones and see how Mary Alice's switchboard functions.

EVALUATION: The main character is energetic and able, and perseveres in her work. Through her persistence, a problem is solved and an important message is delivered to young readers. The text is animated and humorous. The repetitive nature of Mary Alice's announcement of the exact time every five seconds allows children to add their voices to the next notification. The illustrations are zany, colorful, and bold. The animal characters are given comical human qualities by the illustrator. Text and illustrations combine to make this an amusing book for youngsters.

□ □ □

Barton, Byron. *I Want to Be an Astronaut.* Illus. by author. Crowell, 1988. Unp. ISBN: LB 0-694-04744-4. Age levels: 3–6. Grade levels: PreK–1.

SUMMARY: A mission into outer space results from a young girl's desire to become an astronaut. She imagines living conditions in the space shuttle and possible tasks to be performed. As part of a crew, she experiences the effects of zero gravity, repairs a satellite, and helps to build an orbiting factory. Within the shuttle, she dons the top of her space suit by jumping into it upside down, and her prepackaged meal floats when she releases it. On a space walk, she wears a propulsion unit on her back and helps to fix a satellite and to construct a factory in space. A cutaway of the space shuttle on a two-page spread reveals the crew at work in front of control consoles, the sleeping quarters, and a horizontal connecting passageway between them.

EVALUATION: This excellent picture book about a young child's yearning for a flight into space is replete with visual references to developments in space technology. The complexities of space travel are kept in check by the simplicity of the colorful and bold descriptive illustrations. The black outlines on the childlike illustrations seem to give a certain amount of stability to this outer space voyage. Weightlessness, spacewalking, and the astronauts' teamwork are portrayed well. As seen from space, the curvature of Earth and its landforms are pleasing and add an element of comfort to this faraway imaginary trip. The multiracial crew comprises both genders. Space terminology, such as *orbit, zero gravity, satellite,* and *shuttle,* is contained in the sparse text. The uniqueness of space is superbly captured in this exceptional mental voyage.

□ □ □

Barton, Byron. *Machines at Work.* Illus. by author. Crowell, 1987. Unp. ISBN: LB 0-690-04573-5. Age levels: 0–6. Grade levels: PreK–1.

SUMMARY: A crew of multiracial workers, both male and female, arrives with tools and lunch boxes to begin a day's work at a construction site. Some construction workers operate machines, such as a dump truck, a steamroller, and a cement mixer. To clear the site, one machine operator knocks down a building with a wrecker's ball. Another bulldozes a tree. Other workers use their tools—jackhammers, sledgehammers, and pickaxes—to dig up a road. Before new construction can begin, there is much tearing down and tearing apart.

The crew takes time out for lunch; then the work quickly resumes. The busy men and women use a backhoe to dig a hole, a crane to lift beams for a new building, and a steamroller to build a road. At the end of the working day, all activity ceases as the machines are stopped and the crew heads for home. Tomorrow is another day.

EVALUATION: This lap book about machines is very appealing with its bright and bold colors and simple text. Except for the last page, the illustrations are two-page spreads. The childlike illustrations have heavy black outlines that give them a serious and stable quality. In a subtle manner, the illustrations convey sexual and racial equality in the workplace. The text is sparse and direct as it tells what work is done by people and machines during a day at a construction site. There is only a single three- or a four-word sentence per two-page spread. Adults who share this book with children might take the opportunity to identify the machines and tools by name because the text does not do so. This book for very young children, with its distinctive illustrations, is a wonderful introduction to the world of work and machines.

□ □ □

Berenstain, Stan, and Jan Berenstain. *The Berenstain Bears and the Big Road Race.* Illus. by authors. Random House, 1987. Unp. ISBN: LB 0-394-99134-6. Age levels: 2–6. Grade levels: PreK–1. Series: First Time.

SUMMARY: This time the Berenstain Bears are in an automobile race involving four large cars—named Orange, Green, Yellow, and Blue—whose drivers all think that they will reach the finish line first. Little Red—a much smaller and slower automobile—also enters the race. The cars are modern, up-to-date racing cars, except for Little Red. As one driver after another tries to trick or outsmart the others, the race finally boils down to two cars, Green and Little Red. In this two-car race, Green is certain he will win, but one mistake on his part allows Little Red to take the trophy. The racetrack has miles of Grand Prix-type curves and obstacles.

EVALUATION: The illustrations depict the automobiles as large, sometimes taking up almost a quarter of the page, and similar to ones used on the racetrack today. Two of the four drivers are wearing moustaches, leaving it to the reader to decide if the other two are female. The drawings show much action and are very colorful.

Some of the words may be difficult for young children: *putt-putt, cough,* and *boastful* are examples. The book can be used for teaching the concepts over, under, around, through, up, and down because they describe the racetrack as the cars move along through the town and the country. Children may all pick up the theme that if one is determined and persistent, and doesn't give up, things may work out well.

□ □ □

Berenstain, Stan, and Jan Berenstain. *Berenstain Bears and Too Much TV.* Illus. by authors. Random House, 1984. Unp. ISBN: LB 0-394-96570-1; pap. 0-394-86570-7. Age levels: 5–8. Grade levels: K–3. Series: First Time.

SUMMARY: Mama Bear decides that the Bear family is watching too much TV and therefore issues a decree that there will be no more TV for a week. Papa Bear is not amused to discover that TV is off limits for him as well as Brother Bear and Sister Bear. The family immediately begins to complain about having nothing to do. Then Mama Bear announces that the family will sit and watch the stars come out; Brother Bear whispers that the excitement may simply be too much for him. However, bats in the night sky garner everyone's attention until the stars come out. It is noted that the night sky is "sharper" and has a much bigger "picture" than the TV screen. As the week goes by, the family shops and takes nature walks; only Papa Bear sneaks a peek at a TV in a store window. When the week is over, Sister Bear is knitting a rug and Brother Bear works on a puzzle. Only Papa Bear goes back to routine TV viewing, but even he decides to go fishing now and then.

EVALUATION: The story conveys a useful message, but is saved from an unrealistic ending by Papa Bear's backsliding. The Berenstains' books all have a charming undercurrent of humor, often more readily discernible by adults than children. Children love them because the bears are cuddly, funny, and resemble real kids (as in Brother Bear's sarcastic remark about stargazing).

Adults and older readers will love the sly looks and hilariously revealing facial expressions as Mama Bear puts her foot down and Papa Bear sheepishly goes along with the plan.

□ □ □

Berenstain, Stan, and Jan Berenstain. *The Berenstain Bears on the Job*. Illus. by authors. Random House, 1987. Unp. ISBN: LB 0-394-99131-1; pap. 0-394-89131-7. Age levels: 2–6. Grade levels: PreK–1. Series: First Time.

SUMMARY: The Berenstain Bears tell about the many kinds of work done in Bear Country. As the bears make their tour through Bear Country, they pass bears who work as firefighters, repair persons, construction workers, astronauts, computer programmers, bankers, doctors, engineers, and scientists. The Berenstain Bears are usually onlookers but sometimes become involved with the job. The nature of the work is very briefly and simply explained; for instance, a programmer can work with computers, and an engineer tries to find better ways to do things.

Illustrations cover all the pages, with text written on the drawings. Soft colors are used and the pictures show activity.

EVALUATION: No full uniforms or specialized clothing is shown for most jobs; accessories frequently suggest the profession; a bear is a doctor because of the stethescope, or a construction worker because a hard hat is worn. It is also good to see that the "fixer" or repair person wears safety glasses while using power tools.

Bears and unisex clothing suggest that either male or female could perform any of the work described in the book. The authors make clear two points: if the reader hasn't found an interesting job by reading the book, it may not have been invented yet; and work is fun if the right job is chosen.

□ □ □

Berenstain, Stan, and Jan Berenstain. *The Berenstain Bears on the Moon*. Illus. by authors. Random House, 1985. Unp. ISBN: LB 0-394-97180-9. Age levels: 5–8. Grade levels: K–3. Series: Bright and Early.

SUMMARY: From their backyard, with an enthusiastic crowd cheering them on, two Berenstain bears and their pup blast off in a rocket ship. Their destination is the moon. They view Earth from space, float around in their rocket ship, and fly through a meteor shower.

Upon landing safely on the moon, they look at the moon's craters, mountains, and valleys. Wearing space suits and space helmets, they proceed to plant their flag on the moon's surface, record observations, and explore the surrounding terrain in a moonmobile. Their pup is helpful in digging up moon rocks.

After their information-gathering work is finished, they wonder if their rocket ship will function correctly and bring them back to Earth. The return trip is uneventful, and a joyful crowd awaits and welcomes them back home. At night, the Berenstain bears and their pup look at the sky and ponder the possibility of traveling to a star in the future.

EVALUATION: This fictional easy-to-read book about moon exploration is enjoyable. The characters are likable, adventurous, and skillful. They jestfully perform some activities that former NASA astronauts actually undertook, such as collecting rock samples, planting a flag, and riding around in a mooncar. The distinctive and bold illustrations portray the liveliness of the characters, the excitement of space travel, and the interesting features of the moonscape. The use of rhyme keeps the text flowing briskly in this action-packed book for youngsters.

□ □ □

Berenstain, Stan, and Jan Berenstain. *The Bike Lesson.* Illus. by authors. Random House, 1964. 62 pp. ISBN: LB 0-394-90036-7. Age levels: 5–7. Grade levels: K–2. Series: Beginner.

SUMMARY: Before he is allowed to get on his new bike, Small Bear is told that he must have lessons; however, his first lesson is anything but a success. Father Bear tries to demonstrate how to get on and off a two-wheeler, but loses his balance; he tells Small Bear that he was simply showing him what not to do. The next lesson involves stopping, but Father Bear forgets how to use the brakes and is stopped by a tree branch instead. The lessons continue: how to turn, how to steer, how to go down a hill, knowing where the road is going, and riding on the right-hand side of the road. In every case, Father Bear's performance teaches Small Bear what not to do. On the last page, an illustration tells it all: Small Bear rides his bike home, carrying a bruised and rumpled Father Bear on the handlebars. Mother Bear's expression informs readers that she is not at all surprised.

EVALUATION: Learning to ride a bike is something most children will learn how to do. Even the youngest reader will not miss the point of this humorous story: Father Bear's lessons turn into minor disasters because he is showing off rather than paying attention to what he is doing. Riding a bike requires knowledge, skill, and coordination; readers realize this, and furthermore they guess that Small Bear could learn better *without* Father Bear's lessons. The simple drawings are humorous and the facial expressions reveal much about human nature.

□ □ □

Brewster, Patience. *Ellsworth and the Cats from Mars*. Illus. by author. Houghton Mifflin, 1981. 32 pp. ISBN: 0-395-29612-9. Age levels: 6–10. Grade levels: 1–5.

SUMMARY: Ellsworth the cat has dreams that launch his space adventure. At first his dream about Cat-Martians worries him, and at breakfast his clever mother sets his mind at ease by creating a space adventure using a plate of eggs with knife and fork antennae. However, when she leaves to go shopping, Ellsworth dreams of communicating with Margaret, a friendly Cat-Martian, after donning a space hat with antennae. Ellsworth tells Margaret about Earth-Cat friends. Later, when Margaret hears about his desire to ride in a spaceship, she zaps a one-cat spaceship to him for his use. She tells him how to operate it, but warns him about its dangers. Adventurous Ellsworth can't resist pushing forbidden buttons, and propels himself into deep space and deep trouble. Luckily, he is wearing his space hat, and calls Margaret for help. She sends a rescue spaceship that tows Ellsworth back to Earth. He is happy to be reunited with his mother, even though she thinks he is a daydreamer.

EVALUATION: The Cat-Martians that Ellsworth encounters are gigantic and colorful with their green skin and white and yellow eyes. They are also friendly and technologically knowledgeable; they know about communicating with Earth creatures, and about spaceflight. Contact is made with Cat-Martians in a fashion children will find plausible. Ellsworth is adventurous, so it is no surprise that he gets into a dangerous situation. Throughout, the action is depicted in separate frames of different sizes, suggesting an animated filmstrip. The text advances the action from home to outer space to home again in a swift and logical manner. Ellsworth's mother understands and provides security as well as a verbal wink when she says that he has his head in the clouds.

□ □ □

Brown, Laurene Krasny, and Marc Brown. *Dinosaurs Travel: A Guide for Families on the Go*. Illus. by authors. Little, Brown, 1988. 32 pp. ISBN: 0-316-11076-0. Age levels: 4–10. Grade levels: PreK–5.

SUMMARY: Dinosaurs take the reader through various ways to travel, how to prepare, things to think about or do while riding in a particular vehicle, and points to remember while traveling. Methods of transportation include bicycle, automobile, subway, bus, train, boat, and plane. In traveling in vehicles it is often necessary to pay the fare, sit forward or backward, go through security checks, and operate special buttons or equipment.

Illustrations are divided into frames—usually three or four to a page—and are quite colorful. Vehicles are often named after classifications of dinosaurs. A table of contents is included.

EVALUATION: The young reader learns about fast and slow methods of transportation, moving about, and using certain buttons and switches. The book gives a good overview of current passenger vehicles, covering many interesting aspects. This is further enhanced by the humorous illustrations; older students may also enjoy and relate to the funny situations depicted. The text is to the point and printed within the frame of the illustration. Many children travel alone today, and the helpful preparation and planning provided here by the authors could alleviate the apprehension that may accompany this experience.

□ □ □

Brown, Marc, and Laurene Krasny Brown. *The Bionic Bunny Show.* Illus. by Marc Brown. Little, Brown, 1984. 32 pp. ISBN: 0-316-11120-1; pap. 0-316-10992-4. Age levels: 4–8. Grade levels: PreK–3. Awards: School Library Journal Best Book of the Year, 1984; American Library Association Notable Book, 1984; Parents' Choice Remarkable Book, 1984.

SUMMARY: An actor rabbit plays the lead role in a television series, and a behind-the-scenes look reveals some of the staff and electronics involved in the production. With the aid of the Wardrobe Department and the wonders of technology Wilbur, a weak and scrawny rabbit, becomes the Bionic Bunny, a superhero with superstrength. Camera angles allow him to look taller, and the teleprompter helps him with his lines. The editor's assemblage of posed shots produces a bionic leap that seems to take Wilbur from the ground to the roof of a three-story building. A wind machine and the efforts of sound engineers in the control room create a storm. As he views the taping of an episode from within the television studio and also sees it framed on the monitor, the reader realizes that Wilbur's heroic deeds need the assistance of special effects. When his day of work at the television station ends, Wilbur is shown at home unable to use his bionic strength to open a jar; ironically, his infant child succeeds where he fails. A glossary of television terminology completes this picture book, with the words that are actually used in the story highlighted with asterisks.

EVALUATION: This is a very funny tale about what's real and what's created in the make-believe world of television. The comical and colorful illustrations are very effective and cleverly arranged to take readers subtly from the activities at a television studio to the story's sequence of events on the television screen. The main character, with all of his ineptitude and wimpishness, arouses sympathy along with laughter. The text is wry and has a quality of spontaneity. In the midst of TV cables, video and audio equipment, and studio lights, the illusionary effects of television are skillfully and amusingly presented in this "Reading Rainbow" selection.

□ □ □

Buchanan, Heather S. *George Mouse Learns to Fly.* Illus. by author. Dutton, 1985. Unp. ISBN: 0-8037-0172-1. Age levels: 0–6. Grade levels: PreK–1.

SUMMARY: George Mouse lives in the hollow at the base of a tree with his dreams of flight. He reads in the newspaper about a plane that has made a transcontinental trip, and decides to make a plane of his own. He fashions his plane from odds and ends of wood from his workshop. He has cushioned seats in the cockpit, a horn to warn birds to get out of his way, a stop and start lever, and two hairpins for a propeller. His sisters find a rubber band around the rim of a pickle jar at a picnic site, and George uses it to activate his propeller.

His family sees him off on his maiden flight. His first trip comes to a sudden end when the rubber band completely unwinds. Clever George rewinds the propeller, and zooms off again. George's family awaits him at an airstrip lit with candles. George safely returns and is happy to have made his dream come true.

EVALUATION: George is an imaginative, innovative mouse who sets out to accomplish his dreams of flying with the help of a supportive family. Even when George is impatient about making his airplane work, he is considerate of others' needs. For example, he doesn't make noise in the morning because he doesn't want to wake the others. He also installs a horn in his flying machine so he will not harm birds. The reader will recognize that George understands some principles of flight. The book is filled with delightful, colorful drawings containing an abundance of exquisite detail. There is a nice flow, and perfect synchronization of text and illustration. The size of the book is small for comfortable use by little hands, but George's story is enormously appealing.

□ □ □

Buchanan, Heather S. *George Mouse's First Summer.* Illus. by author. Dutton, 1985. Unp. ISBN: 0-8037-0173-X. Age levels: 0–6. Grade levels: PreK–1.

SUMMARY: The mouse family that lives in Tree Stump House delights in the birth of a baby boy. He joins a family that already has five girls who are named for flowers: Bryony, Clover, Cowslip, Campanula, and Daisy. The baby receives his name when the father accidentally steps on a thistle and exclaims, "By George!" When the baby smiles during this incident, he is named George.

As George grows older, he learns to read and enjoys using his hands to construct things. His active sisters enjoy climbing trees and swimming, but he does not. He likes to spend most of his time in his secret workshop.

One day at the end of summer, the family begins to collect food for the winter months. The father is angry because George does not help them at this important time. A heavy downpour of rain distracts the family, and they later discover that their food pile, so carefully stacked, has disappeared. They fear that the squirrels have taken advantage of all their hard work. Tired and discouraged, they arrive home and find, to their surprise, all the food safely stored there.

George shows them the wagon he has built from twigs in his secret workshop. He tells them that he created it because he knew that he was too little to gather as much food as the others. So he constructed this wagon in order to help on an equal basis. In addition, he saves the day because the squirrels had indeed planned to rob the family. George's family now knows that the hours he has spent in isolation have not been wasted ones. They look forward to practical products emerging from his workshop. However, George dreams of fashioning a machine that flies.

EVALUATION: A strong message of praise for inventiveness is conveyed in this book intended for little tots. The loving family communicates security, tenderness, and understanding. They come to realize that they have an ingenious member in their midst. There is a special exquisiteness in this book that is small in size. The artwork displays lovely scenes; especially beautiful are the detailed illustrations of flowers. The text clearly and simply states that a creative mind not only is full of ideas but can also produce practical results. One anticipates that the wheeled creation that the clever and industrious main character constructs is only the beginning of many inventions to come.

□ □ □

Burton, Virginia Lee. *Mike Mulligan and His Steam Shovel.* Illus. by author. Houghton Mifflin, 1939. 44 pp. ISBN: LB 0-395-06681-6; pap. 0-395-25939-8. Age levels: 3–7. Grade levels: PreK–2.

SUMMARY: Mike Mulligan has a red steam shovel named Mary Anne. Mike thinks Mary Anne can dig better and faster than a hundred men. Together they dig canals, help flatten the land for highways, and dig cellars for many of the skyscrapers in the city. And they always dig their fastest and best when they have an audience.

However, gasoline-, electric-, and diesel-powered shovels make Mary Anne obsolete. They are out of work until Mike decides to take Mary Anne to

Popperville to dig the cellar for the new town hall. Mike brags that they can dig the cellar in one day or they will do the work for free. Mike and Mary Anne dig and dig; as more people come to watch them, they dig deeper and faster. Finally the cellar is dug on schedule, but Mike forgot to leave a way to get out. A little boy suggests that Mary Anne stay in the cellar and become the furnace. The town hall is built around it and Mike stays on as the janitor.

EVALUATION: This delightful story explores the relationship between man and machine. We see that Mike takes good care of Mary Anne and together they are able to help grade airstrips, train beds, and highways needed for the technological advancement and improvement of humanity. But just as Mike and Mary Anne take an active role in man's improved life-style, ironically, the steam shovel becomes obsolete. The steam engine uses coal for fuel and therefore is replaced by shovels powered by gasoline, electricity, and diesel fuel.

Since Mike had taken such good care of Mary Anne she was still as good as new. He could not sell her for scrap; besides, it is evident that he loves her. Here the reader learns that if you treat an engine well it will serve you well.

Mike is portrayed as a cheerful Irishman—hardworking, lovable, and loyal. The solution to the problem of Mary Anne being enclosed in the cellar, as well as the illustration of the junked steam shovels, can lead to discussions about recycling. The book is illustrated by the author in an unpretentious pen-and-ink style with a watercolor wash. In the sketch where Mary Anne digs the canal, we see her red body in stark contrast to the brown earth as it cuts through the land from one body of water to another. At the beginning of the book, a diagram of the steam shovel and its parts adds to the knowledge of how she works. The words flow freely around the page rather than the typical style with words on the bottom and picture on the top.

Unfortunately, the steam shovel and steam engine are unknown to today's young reader. Few parents and teachers can remember the use of coal to heat homes or power machines. A simple verbal explanation could satisfy the confused child.

□ □ □

Calhoun, Mary. *Hot-Air Henry.* Illus. by Erick Ingraham. Morrow, 1981. Unp. ISBN: LB 0-688-00502-0; pap. 0-688-04068-3. Age levels: 4–8. Grade levels: PreK–3.

SUMMARY: Henry, a Siamese cat, has always been excluded from hot-air ballooning, a sport that his human family enjoys. During an unguarded moment, Henry jumps into the basket of a hot-air balloon that's ready for flight. He accidently claws apart the cord that fires the burner, and, to the consternation of all, soars away.

Aloft, he is overwhelmed by the scenery and exhilarated by his unique aerial experience. However, he soon realizes that he doesn't know how to return to earth. By experimenting, he learns to control his craft by pulling the air-spilling cord and the burner cord. He puts his knowledge to use as he meets a flock of playful blackbirds, a threatening eagle, and a bold, hissing goose. Henry finally lands the hot-air balloon safely and purrs his way into forgiveness for his lofty escapade.

EVALUATION: This unexpected solo flight in a hot-air balloon is beautifully portrayed with exquisite illustrations that capture a breathtaking trip. The text is spirited, and reveals the adventurous feline's point of view.

As Henry attempts to gain control of his wild flight, information about hot-air ballooning is provided to the reader. The graphic details of the construction of the basket and the balloon, seen from many different angles, are extraordinary. A balloon being prepared for flight is also shown. In this fantasy, the specifics of hot-air ballooning are presented in an understandable manner. The clear text and the exceptional illustrations make this an absorbing "Reading Rainbow" picture book.

□ □ □

Carrick, Donald. *Milk.* Illus. by author. Greenwillow, 1985. Unp. ISBN: LB 0-688-04823-4. Age levels: 2–5. Grade levels: PreK–K.

SUMMARY: In this story, the preschooler learns how milk is processed. The cows eat grass in a field in the summer; in the winter they are kept and fed in the barn. Twice a day, the cows' udders are washed and then hooked up to milking machines. The warm milk is poured into a transfer machine and pumped into a cool tank located in the barn. The milk is transferred to a tank truck and taken to the dairy; here it is pasteurized to kill germs and homogenized to keep it from separating. Next the milk is poured into cartons, sealed and stamped with a date. The cartons are taken by refrigerated truck to the store, where milk is kept in the cool dairy case with other dairy products.

EVALUATION: At some point every child inquires, "Where does milk come from?" "Cows give us milk" is an appropriate answer, but by reading *Milk,* both adults and children discover the near around-the-clock attention paid to getting milk from the cow to the consumer.

Hand milking has been replaced by large gleaming machines that milk each cow at dawn and at dusk. The milk is never exposed to air, being pumped from one machine to another for processing.

The technology of the modern dairy farm is explained in a simple style to help preschoolers comprehend. The author's breathtaking illustrations are in watercolor. We see lazy cows eating grass in a pasture shaded by a large tree. On another page, the rosy fingers of dawn serve as a backdrop for the busy dairy farmers who milk the cows and then take the milk to the dairy.

The pipes, dials, and valves are evident on the pasteurizing machine. The observant child can see the cartons stacked flat in the loader, ready to be opened and filled with milk. In each picture cleaning buckets and hoses indicate how sanitary the machines and work area must be. This book will introduce the child to the importance of technology and transportation. *Milk* should help make the young child more aware of the world around him.

□ □ □

Cleary, Beverly. *Lucky Chuck.* Illus. by J. Winslow Higginbottom. Morrow, 1984. Unp. ISBN: LB 0-688-02738-5. Age levels: 8–11. Grade levels: 3–6.

SUMMARY: Chuck is the industrious, proud owner of a used motorcycle. After school he pumps gas but still finds time to enjoy his beloved motorcycle. He knows the motorcycle driving laws and has earned a license. Knowledgeable about the operation of his bike, Chuck feels secure. However, his worried mother appeals to him to be cautious.

One day he puts on his safety helmet and protective clothing, and after checking and starting his bike in a careful step-by-step procedure, he heads for the open road. Shortly into his ride, Chuck manages to escape safely from a growling, ferocious dog that chases him by upshifting into second gear. Gaining courage, he is soon speeding in third gear, weaving in and out of heavy traffic, and riding down the center of a main road.

The laws Chuck has learned are ignored in his enjoyment of daredevil motorcycling. After dangerously passing a truck, he barrels along in fourth gear until he sights a highway patrolman in his rearview mirror. Nervous, Chuck attempts to slow down but makes errors that result in skidding and falling. The stern officer gives Chuck a ticket and a lecture about recklessness. On his return home, Chuck abides by all the laws. Even with a hurting body and a fine to be paid, Chuck realizes that he's lucky to be alive.

EVALUATION: Although this story covers one daring but negligent motorcycle ride, it contains a great deal of information about the mechanics of motorcycles. For instance, the functions of the choke, clutch, throttle, and tachometer are some of the details that are explained in an interesting fashion as the main character operates his bike. The use of "This is" at the beginning of most sentences gives the book a predictable cadence that's enjoyable. The text is brisk and informative, and teaches lessons about safety and respect for laws. The mother's warning is made in bikers' language, and thus she is portrayed not only as caring but also as understanding and sensitive to her son's love for his sport. The black-and-white illustrations contain abundant detail and extend the text. The endpapers display large, labeled diagrams of the right and left sides of a motorcycle. This realistic story, with its informative text and drawings, is a delight.

□ □ □

Coerr, Eleanor. *The Big Balloon Race.* Illus. by Carolyn Croll. Harper & Row, 1981. 62 pp. ISBN: LB 0-06-021353-1; pap. 0-06-444053-2. Age levels: 4–8. Grade levels: PreK–3. Series: I Can Read.

SUMMARY: In the year 1882 a young girl's famous aeronaut mother, Carlotta the Great, is a participant in a hot-air balloon race. Ariel wants to accompany her mother, but is told she's much too young. However, she's allowed to sit in the basket of the balloon until the start of the race. Tired Ariel falls asleep in the odds and ends box, and doesn't awaken until after the balloon is airborne.

During the flight, she sees her competent mother in action as they have to clear treetops, are sucked into a rain cloud, and almost hit a church steeple. Ariel learns about ballast, updraft, airstream, and what a ripcord does. But Ariel shows her own mettle and quick-wittedness when they land in the shallow part of a lake: she pulls the floating balloon to the finish line, which happens to be on that very shore, and is rewarded with a gold medal and her mother's assurance that she is indeed old enough to fly. On the last page, an author's note tells about the real family upon which this story is patterned. There is a table of contents, and three chapters divide this book.

EVALUATION: Based on stories about a true family, this adventure about a fearless balloonist and her young daughter is entertaining and informative. Within the experiences of an exciting balloon race, some fundamentals about ballooning are related, such as the fact that hydrogen, a lighter-than-air gas, is used to fill some balloons, and cold air outside a balloon causes the gas to cool and the balloon to lose altitude. The text in this easy reader is lively, and the color illustrations effectively capture a thrilling voyage through the air. The spirit and daring of both mother and daughter are well portrayed. This story of the past, a "Reading Rainbow" selection, will give pleasure to children of today.

□ □ □

Cole, Joanna. *The Magic School Bus at the Waterworks.* Illus. by Bruce Degen. Scholastic, 1988. 40 pp. ISBN: pap. 0-590-40360-5. Age levels: 7–10. Grade levels: 2–5.

SUMMARY: When Ms. Frizzle announces a trip to the waterworks, her class anticipates boredom because that is a rather ordinary idea from their unconventional teacher. In preparation for the trip, the children are sent to the library in search of ten interesting facts about water and how it is supplied to the city.

Then the students and teacher embark on an unusual trip through the waterworks—dressed in scuba diving gear! Their route takes them through the reservoir, the mixing and settling basins, and above the sand and gravel filters. The water pressure pushes them through the city pipes and finally through the school's basin faucets. With the knowledge gained on this fantasy journey, the students produce a class mural and chart about the waterworks.

EVALUATION: This clever book, embellished with humorous and informative art, clarifies concepts and processes of waterworks systems. This type of information is often presented to children in an uninteresting manner. However, in this case the author and illustrator have combined original devices, such as a magic bus, some cartoon format such as balloons for conversational text, and composition notebook fact sheets, to explain the topic. The creativity of the author and illustrator effectively engages the interest of children in what could have been a boring science lesson. All readers will want to be in Ms. Fizzle's class, regardless of what she teaches.

In this fantasy field trip, many facts about water purification and distribution are presented within a delightful framework. At the end of the book, the author is careful to include a page-by-page distinction between science facts, fantasy, and jokes. The very last page contains a surprise that is bound to create interest in more of Cole's books.

□ □ □

Crews, Donald. *Harbor.* Illus. by author. Greenwillow, 1982. Unp. ISBN: LB 0-688-00862-3; pap. 0-688-07332-8. Age levels: 0–8. Grade levels: PreK–3.

SUMMARY: A busy harbor is the focal point for this examination of the many types of ships and boats that are part of the water transportation system. The view of the harbor includes a port for ships, warehouses for storage, wharves, piers, and docks. Activity abounds as the ships and ferryboats move through the waters. Larger vessels include tankers, barges, liners, and freighters. Some of the relatively smaller boats shown are tug and police boats. Pleasure craft, such as sail and power boats, are also seen in this active harbor. A display of a fireboat in operation forms the finale. For identification purposes, a full page of silhouettes is included to distinguish more clearly the size and shape of boats and ships.

EVALUATION: The modern-day vessels that move through the waters of a harbor are simply and beautifully illustrated. Interesting details capture the eye and invite the reader to inspect more closely for further information. The colors are striking and varied. The text is simple and almost reads like a reference to the illustrations. Often the sentences are incomplete, but this in no way detracts from the message of this book.

The setting appears to be New York harbor, and, as such, it portrays a bustling place with a great variety of vessels going from one destination to another. Young readers will enjoy the ships and boats as they move through the book. They can refer to the last page for further descriptions of the shape and size of the vessels they have seen. No people are visible in the harbor scenes: the boats and ships themselves, prominently named, are the industrious main characters.

□ □ □

Douglass, Barbara. *The Great Town and Country Bicycle Balloon Chase.* Illus. by Carol Newsom. Lothrop, 1984. Unp. ISBN: LB 0-688-02232-4. Age levels: 5–9. Grade levels: K–4.

SUMMARY: Two bicyclists, a grandfather and his granddaughter, prepare to take part in a balloon chase. They know that they must be the first to touch the hot-air balloon when it lands. Since the wind will determine where the balloon lands, they bike all over town in order to be familiar with all the possible shortcuts to landing sites. They want the free balloon ride that is being awarded to two lucky winners.

On the day of the chase, they find many contestants and all types of bicycles at the starting place in the park. Soon they spot the woman aeronaut, with her pet parrot on her shoulder, being helped to assemble the balloon and gondola. They watch as a fan and a burner inflate the balloon to an incredible height. Once the ropes that keep the balloon stationary are released, the chase begins.

The grandfather and granddaughter find their shortcuts useful and twice come close to winning. In one instance, the aeronaut decides not to land because of beehives. In the other instance she decides not to land in a field with an angry bull. However, frightened by the roar of the bull, her pet parrot flies away. Granddaughter and grandfather abandon the contest in order to chase the parrot.

They succeed in finding the parrot on a sunflower. After capturing it, they are the last to arrive at the balloon's landing site. The grateful aeronaut invites the grandfather and his granddaughter to join her, her parrot, and the winners on a balloon ride. Finally, they all soar up into the sky, with the parrot again perched on the aeronaut's shoulder.

EVALUATION: This picture book is beautifully illustrated with fantastic scenes of hot-air ballooning. Both ground and aerial views are works of art. The important part that wind plays in ballooning is emphasized in a subtle and enjoyable manner within the plot of a balloon chase.

Details of ballooning and bicycling are given in both text and illustrations.

Clarifications of ballooning terms are found in the informative and interesting text. Readers are provided with the opportunity to identify three different kinds of cycles—tandem, penny-farthing, and unicycle—in a clever integration of illustration and text.

The main characters demonstrate caring and skill. This outstanding book ends with an illustration from an unusual perspective, and readers are welcomed to join an incredible flight as the balloon floats away.

□ □ □

Drescher, Henrik. *The Yellow Umbrella.* Illus. by author. Bradbury, 1987. Unp. ISBN: 0-02-733240-3. Age levels: 0–8. Grade levels: PreK–3.

SUMMARY: A father and daughter visit a city zoo. At the monkey cage, the father's yellow umbrella accidently falls into the moat. A mother monkey and her child retrieve the umbrella, and open it. Excited spectators watch as the wind carries the umbrella away with its two surprised passengers hanging on. They are carried far above city skyscrapers, over mountain peaks, against a flock of flying geese, and over choppy seas.

A sudden lightning storm forces them to land in perilous waters where the umbrella becomes their sailing craft. Next, the umbrella-ship sails into a river where exotic tropical life abounds. They successfully navigate a waterfall, and encounter wild animals on their winding river journey. At nightfall, they beach their umbrella-ship and discover monkey inhabitants just like themselves. The final page shows a peaceful scene in a palm tree where the monkeys are part of a family trio resting under an umbrella-roof.

EVALUATION: The use of the bright yellow umbrella against a pen-and-ink background provides an effective thread of continuity. The umbrella itself is a symbol of security since it is not only a flying machine and sailing vessel as needed, but also a home. This small, wordless picture book can generate oral language as children make up a story line to go along with the illustrations, and share their travel experiences. Children will like the action-filled pictures and amusing situations.

□ □ □

Florian, Douglas. *Airplane Ride.* Illus. by author. Crowell, 1984. Unp. ISBN: LB 0-690-04365-1. Age levels: 3–7. Grade levels: PreK–2.

SUMMARY: In his biplane a pilot flies over diverse areas at different speeds and various altitudes. His departure from the airport begins with his setting the engine in motion by flipping the propeller of the plane by hand. In the sky, he soars over a coastline that's dotted with many boats and ships, and even a whale. Soon, he's making loops over a roller coaster at the beach and flying upside down. His daredevil antics are halted as he encounters heavy air traffic over a city that's crisscrossed with heavily traveled highways. Climbing high into the sky, he finds snow at that altitude.

His aerial adventure continues when he descends and has a close call flying between mountain peaks. He skywrites a greeting to the Indians in the desert. At low altitudes, the shadow of the plane accompanies him as he flies solo. He cruises over a canyon and starts his descent near a waterfall. Upon landing, a warm welcome from other pilots, with small planes of unusual design, awaits him.

EVALUATION: With minimal text, the flight of a daring and able pilot is told in colorful and sweeping illustrations. The aerial scenes contain geographical clues that can be useful in a social studies curriculum. For example, the Grand Canyon can be identified by the distinctive feature of its mesas and the gorge formed by the Colorado River. In addition, the plotting of the pilot's course can lead to some map work.

From takeoff to landing, the pilot is shown to be in control of his aircraft, despite some daredeviltry. Similar actions in the air and on the ground are sometimes juxtaposed effectively, as when the pilot does a loop in the sky above people on a roller coaster who are about to hurl down the steepest part of their ride. The author-illustrator lifts the reader into and above the clouds and conveys the thrill of flying. This enjoyment is reinforced on the last page where pilots (who seem to be members of a flying club) greet the arrival of one of their own, the main character. This book will make children feel they have also taken part in a very special experience.

□ □ □

Gibbons, Gail. *Fill It Up! All about Service Stations.* Illus. by author. Crowell, 1985. Unp. ISBN: LB 0-690-04440-2; pap. 0-06-446051-7. Age levels: 5–9. Grade levels: K–4.

SUMMARY: In *Fill It Up!,* Gibbons describes the activities that take place during a busy day at a gas station. Even before the title page is reached, a tow truck is seen coming to the rescue of a disabled car. Soon the same tow truck pulls up at John and Peggy's service station, and a mechanic discovers that the disabled car needs a new fan belt. Meanwhile some of the day crew attend to customers at the gas pumps; others are in the garage making repairs. Where gasoline is stored and how it is pumped are graphically shown. Also pictured are the operation of a hydraulic lift and the repair of a flat tire.

Directions are given to travelers, tires are balanced, wheels are aligned, and many other services are performed during this busy day. Office work, such as ordering supplies, is also part of the chores. An evening crew, which doesn't do repair work, replaces the day attendants and mechanics. After midnight, when there are few customers, only one attendant is present. When the next morning arrives, the day crew reappears and the cycle of activity begins again. On the last page, some of the tools that are used at a service station are diagrammed.

EVALUATION: The author-illustrator uses bright illustrations, diagrams, a comic strip format, and simple and clear text to inform young readers about the work at a service station. The operation of a hydraulic lift and the storage and pumping of gas are demonstrated in clear and simple diagrams. The day crew is shown in a labeled photograph that allows each person to be identified by name and the work that he or she does. The owners and crew comprise a gender and racial mix. Some readers might question why this station and its employees are so clean. Nevertheless, this colorful book transmits a great deal of information in a pleasing manner.

□ □ □

Gibbons, Gail. *Trucks*. Illus. by author. Crowell, 1981. Unp. ISBN: LB: 0-690-04119-5; pap. 0-06-443069-3. Age levels: 0–8. Grade levels: PreK–3.

SUMMARY: Many kinds of trucks occupy all the pages of this picture book. Crisscrossing a city street an armored truck, a tank truck, and a cement truck, to name a few, can be seen. Along a country road a dump truck, a cattle truck, and a logging truck move their loads. At a construction site many trucks help with the heavy jobs that need to be accomplished.

Trucks are involved in seasonal jobs, too. A pickup truck equipped with a snow plow can be used in the clearing of streets. In warmer months ice cream and other frozen treats can be purchased from a vendor who drives a refrigerated food service truck. At harvesttime, fruit can be conveniently reached from a truck that is outfitted with a cherry picker.

Trucks help the community in many ways. They pick up garbage, clean the streets, put out fires, deliver mail, supply meals to the needy, and transport the sick. Disabled cars are towed by trucks, and lines on roads are painted by special trucks. Specially furnished vehicles provide rolling conveniences: a miniature library is available through a bookmobile. Blood can be donated and X rays taken in some mobile units. Vehicles are seen everywhere, in motion and helping people. On the last page of the book, four parts of a truck—the cab, engine, body, and chassis—are shown as they appear in a standard truck and on a tractor-trailer.

EVALUATION: This brightly colored picture book is packed with a variety of trucks that will certainly catch the attention of readers from the title page and hold it until the last page is turned. Even though a large number of trucks is presented, the artwork is uncluttered. In the illustrations, trucks are labeled in red print for identification purposes. After a group of trucks is featured with their designated labels, the same trucks, without labels, are illustrated as they are engaged in a variety of activities. Through color coding, the author-illustrator cleverly provides readers with an opportunity to play a game of "name that truck."

The text is very sparse and directs the reader's concentration to becoming familiar with trucks. Cranes, a bulldozer, and tractors are included; since these machines are not conventionally considered trucks, some confusion might arise. However, all the machinery in this book will surely capture the attention of young people.

□ □ □

Hanrahan, Mariellen. *My Little Book of Boats.* Illus. by Tom Dunnington. Western, 1974, o.p. Unp. Age levels: 4–7. Grade levels: PreK–2. Series: Tell-a-Tale.

SUMMARY: No space is wasted in this small book, which provides a generous amount of information about different kinds of boats from the simplest canoe to the hydrofoil. The purposes of different kinds of boats are explained in uncomplicated vocabulary intended for young children. The accompanying drawings show ocean liners, a tug boat, sailboats, a rowboat, a Coast Guard boat, a houseboat, a fireboat, a submarine, a kayak, and a toy boat with which a child plays in a bathtub, to name a few.

EVALUATION: What various kinds of boats look like and how they are powered is the focus of this book in the Tell-a-Tale series. Colorful drawings represent each kind of boat on every page, including this board book's endpapers. Also, Dunnington was careful to anticipate reader interest by showing an inside view of a ferryboat. It is unfortunate that boats powered by human energy—the rowboat, kayak, and so on—are not grouped together, rather than interspersed with mechanically powered boats and ships. The one- to three-sentence descriptions are concise and informative.

□ □ □

Harris, Mark. *The Doctor Who Technical Manual.* Illus. Random House, 1983. 64 pp. ISBN: LB 0-394-96214-1; pap. 0-394-86214-7. Age levels: 9–17. Grade levels: 4–12.

SUMMARY: This is a profusely illustrated technical manual showing devices used in the British television series "Dr. Who." Characters, robots, props, gadgets, and machines are described in great detail, with cutaway views used in the many illustrations and diagrams. Also included are instructions for making a "tardis" or model police box such as those seen on the TV program.

The center illustrations utilize photography, while the illustrative technique for the rest of the guidebook is comprised of line drawings.

EVALUATION: Fans of futurism and science fantasy will find the highly diagrammatic and detailed content appealing. Familiarity with the "Dr. Who" series will cause many readers to select this book, but one need not have seen the program to understand the text. Although it appears as much a game book as a guidebook, the manual can serve as an excellent vehicle for stimulating creative and futuristic thinking.

□ □ □

Hoban, Lillian, and Phoebe Hoban. *Ready—Set—Robot!* Illus. by Lillian Hoban. Harper & Row, 1982. 64 pp. ISBN: LB 0-06-022346-4. Age levels: 5–8. Grade levels: K–3. Series: I Can Read.

SUMMARY: Sol-1 lives in a robot colony on Zone One. He is preparing for the Digi-Maze race. As he recharges his solar cap using the sun's energy, he feels confident that he will be the winner. He knows that he will also need his power pack to supply energy when he enters the Outer Zone during the race. His sister, Sola, helps him by finding his power pack in his messy room. She encourages him, telling him that he is very smart and that she will be at the finish line to see him win. If she can find the lost power pack of their dog, Big Rover, she promises to bring him too.

At the race's starting line, Sol-1 finds seven other eager robots. Fax, who is thin and big, is there; Rocko has a jukebox that's built in his controls. Super Scan has X-ray vision; Micromax is a new robot. Arla, Ming, and Ping complete the roster of contestants. They must successfully maneuver through the Spectra Zone, the Laser Link Zone, and the Outer Zone. In the Spectra Zone, floating 3-D shapes offer a challenge to the racers. Light beams present difficulties in the Laser Link Zone. There are sharp rocks and deep craters in the dark Outer Zone.

Sol-1 manages to get through the Spectra and Laser Link zones. However, his pack doesn't work properly in the Outer Zone, and Sol-1 finds himself precariously hanging near the edge of a crater. Sola and Big Rover come to his rescue. Sola discovered that there was a mixup of power packs due to the mess in Sol-1's room: Sol-1 is wearing Big Rover's power pack rather than his own. By using his intelligence, Sol-1 crosses the finish line before the others. Big Rover proves that he's indeed a robot's best friend, and Sol-1's mother plans to give a big Digi-Party, after Sol-1 cleans up his room.

EVALUATION: This easy reader is fast-paced and entertaining; although mechanical, the robots have human characteristics and are appealing. The main character possesses intelligence, but his untidiness almost leads to disaster. Children can easily relate to the main character's problem. The large-print text is simple and contains a few songs in rhyme that are rather catchy. The illustrations are interesting and enhance the text. This space age adventure will delight beginning readers.

□ □ □

Keats, Ezra Jack. *Regards to the Man in the Moon.* Illus. by author. Four Winds, 1985. Unp. ISBN: pap. 0-02-044130-4. Age levels: 5–8. Grade levels: K–3.

SUMMARY: Louie is sensitive to the remarks that other children make about his father, who is a junkman. His father feels that people who see no value in junk are lacking in imagination. He convinces his son that junk is material that can take one to another world.

With his parents' help and parts found in the junkyard, Louie builds a spacecraft, *Imagination I.* Susie, a friend who also possesses much imagination, wants to join Louie on his trip to outer space. Some playmates scoff at them, making sarcastic comments.

Despite criticism, Louie and Susie successfully travel into outer space, passing wondrous planets and going through a rock storm. However, their journey is brought to a standstill when they are tethered by the spacecraft of two other adventurous friends who have suddenly run out of imagination. It takes a great deal of imagination to set all of them into motion again and head safely home.

After their arrival home, they share their adventures with others. Even their skeptical and critical playmates are now ready to blast into new worlds using the junk they had thought worthless.

EVALUATION: Using an urban setting as a stepping stone for extraterrestrial experiences, Keats has produced a marvelous story about the limitless distances that one can travel by using imagination. His powerful illustrations of spaceflight and celestial bodies are awe-inspiring. The distinctive features of planets make them identifiable, and some planetary illustrations give the impression of color photographs.

Keats's light touch is seen in the whimsical spacecrafts that are invented from a discarded washtub and a bathtub. A pot and sieve become space helmets.

The text contains everyday children's language. Striking illustrations accompany this imaginative "Reading Rainbow" story about space travel.

□ □ □

Keats, Ezra Jack. *The Trip.* Illus. by author. Greenwillow, 1978. Unp. ISBN: LB 0-688-84123-6; pap. 0-688-07328-X. Age levels: 0–7. Grade levels: PreK–2.

SUMMARY: Louie has moved to a new neighborhood where he has no friends; in fact, he has not even seen any pets, and there are no doorsteps to sit on. Lonely Louie creates a city inside a shoe box by using scissors, paint, colored paper, crayons, and glue. He adds a paper airplane that he suspends from the top of the box. Then Louie enters a fantasy world by pretending to fly his plane over the moon, finally landing in his old neighborhood. He is frightened by some costumed creatures who turn out to be his old friends. He invites them to take a plane ride with him. After Louie returns from his fantasy trip to his new home, his mother suggests that he put on his Halloween costume so he can join the neighbor children. He sees a robot, a butterfly, a hobo, a witch, and a bunny. Louie joins the group as a walking ice-cream cone.

EVALUATION: The unusual peephole diorama with an airplane is an effective device to transport the main character into a fantasy world. The three-dimensional features of the diorama create a dramatic urban landscape. The intensity of the collage technique is striking because a dismal urban setting is transformed into a vibrant locale. Parts of colored photographs interspersed within the collage on a two-page spread reinforce the ethnic mix of the city neighborhood. The spectacular color overshadows the thin story line, although the Halloween parade provides a believable situation in which Louie can mingle with new children on the block. The author may be suggesting that children can create fantastic and comforting worlds by using simple and inexpensive art materials.

□ □ □

Kellogg, Steven. *Aster Aardvark's Alphabet Adventures*. Illus. by author. Morrow, 1987. Unp. ISBN: LB 0-688-07257-7. Age levels: 5–8. Grade levels: K–3.

SUMMARY: This book uses alliteration in sentences to introduce the alphabet, and many technological devices are found in the adventures of Aster Aardvark. At the Acorn Acres Academy, she finds it quite difficult to learn the alphabet. However, when her ingenious aunt gives her an airplane as an incentive, Aster Aardvark excels in school and is soon skywriting letters.

After receiving an alphabet award, adventure-bound Aster flies to Brooklyn, Japan, Louisiana, Paris, and other alphabetically sequenced locales. From her aerial viewpoint, she sees interesting and incredible happenings. A group of dogs are served tasty food that ducks deliver via a dirigible. Cameras click to record the doomed flight of some foxes; they had fashioned winged outfits with feathers, à la Icarus, that were inadequate for flying.

A young koala bear maneuvers a kayak. A mailman wrapped like a mummy delivers messages on his motorbike. There is a mishap at sea when a skiff and a sloop collide. A turkey, annoyed over antics in a tournament, throws a television outside her window in disgust. On unicycles, unicorns perform acrobatic tricks. Visitors from Venus hijack the sailing vessel that vultures and voles would have taken to view Vesuvius. A yak on a huge yacht ignores the message from the skipper on a small yawl that's nearby. Aster, a daring aardvark, soars above a great number and variety of busy animals and observes the intriguing activities in which they are engaged.

EVALUATION: This alphabet book provides fast-paced action and tongue-twisting fun with its alliterative descriptions of unusual settings and goings-on. One can imagine the listener delighting in the reader's stumbling over many of the passages. Adult readers will appreciate the clever manner in which the author weaves mythology, geography, modern life, and a futuristic touch to tell a fantastic tale.

Even though this story is farfetched, readers might question why the television set that is discarded out of a window—with its disconnected cord prominently displayed—still has a colorful picture on its screen. Most important, the illustrations are full of wonderful scenes and extraordinary animals that demand to be viewed repeatedly. Many means of transportation carry the story line and readers' imaginations to places that are likely to become memorable.

□ □ □

Lapham, Sarah, reteller. *Max Dreams of Flying*. Illus. by S. Rainaud. Derrydale, 1985, o.p. Unp. Age levels: 0–6. Grade levels: PreK–1.

SUMMARY: Rather than practicing his violin, Max the mouse would like to be flying, just like the butterflies and birds that he sees around him. He consults knowledgeable Mole to find out how to accomplish this feat. Mole informs him that, although he won't be able to fly as the birds he admires, he can perhaps find an answer in a book, which he gives Max.

Seeing a picture in the book of a flying balloon, Max thinks that the only materials needed are a balloon and a basket. Eager Max sets out to lasso the biggest round object he's ever seen. Soon he realizes that this yellow thing is not a balloon, but the sun. After some experimenting, he comes to the conclusion that clouds or butterflies attached to a basket won't work either. For a final test, he climbs a tree and hopes that the wind and his umbrella will help him fly away. His wish is realized for only a moment. He lands right back in the tree. Max decides to complete practicing his violin and attract the birds to come hear his music. Thus he can study them closely; perhaps his dream of flying as they do will come true some day.

EVALUATION: The main character is a creature who uses observation, available recorded knowledge, and experimentation to attempt to accomplish a scientific endeavor. Although he is not successful, his gargantuan effort is captivating and humorous. The pastoral setting is colorful and whimsical. The text is simple and gives the impression that an interesting conversation is taking place between the reader and the main character. This enjoyable book touches on the need for scientific knowledge to achieve technological success.

□ □ □

Lloyd, David. *The Stopwatch.* Illus. by Penny Dale. Lippincott, 1986. Unp. ISBN: LB 0-397-32193-7; pap. 0-06-443107-X. Age levels: 0–7. Grade levels: PreK–2.

SUMMARY: The stopwatch that Tom receives from his grandmother is a most engrossing gift. He times himself—to the exact second—while he eats, holds his breath, stares at his sister, takes a bath, and stands on his head.

After Tom loses his stopwatch, he frantically begins a search; he realizes how much he has come to depend on this timepiece. When his sister Jan announces that she knows exactly how long it has taken her to ride her bike to the store, eat a Popsicle, go to the park, meet her friend, and climb a tree, Tom is convinced that she has taken his watch. A spirited fight ensues. Grandma arrives on the scene and successfully ends the confrontation. Tom can now tell his grandmother how many minutes the fight lasted by looking at his stopwatch.

EVALUATION: This book reveals that a young child can operate and utilize a mechanical device. The sequence of some events is timed in minutes and seconds, providing an opportunity for meaningful use of a stopwatch for the first time. The action is realistic, the characters are believable, and the text is sparse and appears to be readable by beginning readers. The realistic scenes are illustrated with appealing artwork. A secure family feeling is established by the three characters involved. One feels that the time that this family spends together will ultimately be warm and happy.

□ □ □

Lyon, David. *The Brave Little Computer.* Illus. by R. W. Alley. Simon & Schuster, 1984. Unp. ISBN: 0-671-52455-0. Age levels: 5–8. Grade levels: K–3.

SUMMARY: Betsy, the personal computer owned by the Roberts family, comes to the rescue when Mongo, the very large and important electric company computer, breaks down. Mr. Roberts, who works for the power company, receives a call late one night from his coworker Frank Del Gado telling him that half the city of Cowabunga, including Mercy General Hospital, has no electricity. Little Betsy has to take over. She wonders if she can handle the information load and is scared that she will overheat, causing all her circuits to burn up like toast. However, Betsy saves the day, and is praised by everybody, including Mongo, who apologizes for calling her a pest. Best of all, Betsy is rewarded with a new disc drive by her grateful owners.

EVALUATION: Personification of equipment is a little unusual in books for children, but in this case the device appears to work well. Betsy is shown to be a comfortable fixture in the Roberts household. When she is not paying the family bills, she amuses herself by playing tic-tac-toe with the cat. A crisis reveals that small personal computers can do many things. This concept may stretch the imaginations of personal computer users and encourage creative as well as routine uses of household computers. The colorful art has a cartoon flavor. The people are happy, funny-looking folks. Children will like the story, although it might appear silly to some adults.

□ □ □

McNaughton, Colin, and Elizabeth Attenborough. *Walk Rabbit Walk.* Illus. by Colin McNaughton. Viking, 1977, o.p. Unp. Age levels: 3–6. Grade levels: PreK–1.

SUMMARY: *Walk Rabbit Walk* tells about various types of transportation. One day, Rabbit receives a letter from Eagle. Eagle has invited all his friends for tea and would like Rabbit to join them. Rabbit sets out on foot; although Rabbit is aware that it is a long journey, he is not troubled because it is a beautiful day. On the way, he spies Fox drifting high overhead in his hot-air balloon. He offers Rabbit a ride but Rabbit refuses. He replies that he likes "to look at the flowers." Bear drives past in his new sports car followed by Cat on a motorbike, Pig in a helicopter, and Donkey on skates. Each graciously offers Rabbit a ride to Eagle's house. Rabbit politely refuses but assures them that he will be at Eagle's house in time for tea.

Fox creates a terrible accident by not watching where he is going—he crashes into Pig in the helicopter. The helicopter hits the ground and Bear, in an attempt to avoid hitting the helicopter, hits a tree. The tree falls across the road and Cat can't avoid hitting it. Donkey can't stop and skates into the tree as well.

The animals are disappointed; they fear they will never make it to Eagle's house in time for tea. Rabbit happens upon the scene, and tells them that since no one is injured they can surely make it if they walk. He goes ahead to tell Eagle that they are on their way and adds, "sometimes it's quicker to walk."

EVALUATION: The toddler is sure to find this book, originally published in Britain, enchanting. Only a few short sentences are found on each page and the illustrations are a feast for the eyes. Any bright three-year-old could point to and name the animals in the story as well as the modes of transportation each uses. The crash scenes are well drawn with twisted metal, flying bolts, and stars circulating around Bear's head. McNaughton uses soft pastel watercolors to portray the English countryside where the story takes place.

Eagle looks appropriately regal and his home is a castle high atop a mountain. Rabbit shuns all the modern forms of transportation not because he is afraid, but because it would not afford him the opportunity to look at the flowers. His choice proves beneficial since he gets to Eagle's house faster after all—a modern twist to "The Hare and the Tortoise."

The British prize walking as a form of transportation as well as exercise, although most Americans would prefer a car for short trips and an airplane for long trips. The balloon, motorbike, helicopter, and skates are well chosen: young children are fascinated by these wheeled and airborne vehicles and will love the crash that ensues. Logical questions to ask the youngster at the end of the story are: Which vehicles are used for pleasure? Which for industry? Which for both? How would you get to Eagle's house?

□ □ □

Maestro, Betsy, and Ellen DelVecchio. *Big City Port.* Illus. by Giulio Maestro. Macmillan, 1983. Unp. ISBN: LB 0-02-762110-3; pap. Scholastic 0-590-41577-8. Age levels: 3–6. Grade levels: PreK–1.

SUMMARY: All kinds of ships and boats of different sizes are featured at a busy port of a major city. Through the harbor, small tugboats help large ships reach the docks safely. Workers, cranes, and machines unload and move cargo from freighters to trains, trucks, and storage sheds. Not only oil and gasoline but also fruit juice are pumped from tankers.

Ocean liners and ferryboats provide convenient transportation for people traveling long or short distances. At daybreak, fishermen head out of the port for work. The safety of the port is protected by harbor police and fire fighters in their special vessels. In addition, the Captain of the port and his staff are essential in controlling the port's traffic. There is a buzz of around-the-clock activity at this port where both passengers and freight are transported.

EVALUATION: In this view of a bustling port, large and small craft are both given importance in the functions they perform. The realistic setting pictures both men and women at the port. Specific freight, such as the gasoline and fruit juice cargo, is likely to raise questions about storage mishaps. The text is brief and clear while the illustrations are colorful and detailed, and provide different perspectives. The vastness of the port and ships' decks, heights from different angles, and panoramic views engage the eye. This "Reading Rainbow" selection captures busy nautical scenes at a city port where children would enjoy watching the events taking place.

□ □ □

Maestro, Betsy, and Giulio Maestro. *Ferryboat.* Illus. by authors. Crowell, 1986. Unp. ISBN: LB 0-690-04520-4. Age levels: 5–7. Grade levels: K–2.

SUMMARY: A family drives onto a busy ferryboat to cross a river; the Chester-Hadlyme Ferry is identified by a sign near the loading ramp. A ferryman directs them into a tight parking position. Cars, a truck, a van, and people on foot quickly fill the boat to capacity. The family watches the operation of a ferry crossing, as the captain expertly navigates across a heavily traveled river from up in the pilothouse. They also experience sensations from the humming and vibrating engine, and the water's cool spray.

Reaching the other side, the captain sounds the whistle and steers the ferry into the narrow slip between pilings. The ferryman works the pulleys, and the ramp is lowered. After the gates are raised, all the vehicles and passengers leave. Preparations then begin for another crossing; since the ferry is designed the same in the front and the back, it need not be turned around for the return trip.

The family plans to take the last ferry run that evening. However, they know that if they miss it, they will have to drive to a bridge that's quite a distance up the river. The bridge is used in the winter when ice makes the river impassable. During the winter months when the ferry is not in operation, it is repaired, painted, and made ready for another busy year. Thus there is year-round activity on a ferryboat.

The last page of the book contains a map of the Connecticut River, the ferry crossing between Chester and Hadlyme, and the bridge near the town of East Haddam. Also included is historical information about the ferry's operation in 1769, as Warner's Ferry, up to the Selden III service in 1950.

EVALUATION: The text and illustrations allow the reader to experience a pleasant ferryboat ride as well as gain information; simple terminology produces clarity. The illustrations are appealing and invite the eye to inspect particular details at the loading and unloading ramps and on the deck of the ferry. An aerial view of the loaded ferryboat sweeps the docking area and provides visual explanation of the text. The river, sky, and scenery along the shore are beautifully portrayed. Again and again, readers will be drawn to inspect many of the pages in this book.

The factual page at the end of the book about the history of the Chester-Hadlyme Ferry in Connecticut adds to the realism of this narrative. During the Revolutionary War the ferry was in operation and carried supplies. As technology advanced, different types of ferries serviced this area. This book may heighten the reader's interest not only in ferryboats, but also in historical and geographical aspects of river transportation.

□ □ □

Marshall, Edward. *Space Case*. Illus. by James Marshall. Dial, 1982. Unp. ISBN: LB 0-8037-8007-9; pap. 0-8037-8431-7. Age levels: 4–8. Grade levels: PreK–3.

SUMMARY: It is Halloween night when the thing from outer space lands, so no one pays any attention. In fact, a group of trick-or-treaters think the thing is the new kid on the block dressed up as a UFO with lights and antennae.

Invited by Buddy to spend the night at his house, the thing learns to communicate by studying a dictionary. At breakfast, the thing atomizes a glass of orange juice. Buddy's parents simply think that children have imaginative toys nowadays; they ignore the morning headlines announcing the sighting of a UFO.

The thing accompanies Buddy to school in order to observe life on this planet. This is the day that the students' space projects are due. When a homemade rocket and a model of Mars are demonstrated, the thing knows they're too simplified and inaccurate. Poor Buddy has forgotten to do his project; however, the thing rescues him by doing some addition on the blackboard. The teacher believes that the thing is a robot project and rewards Buddy with an A minus.

The thing confesses that it doesn't like school and would rather trick-or-treat. Finding out that this occurs only once a year, the thing suddenly remembers an appointment on Jupiter. Not wanting his newly found friend to leave, Buddy tells him about Christmas. Even though the thing speeds away, one is quite sure that it'll return soon for more holiday fun.

EVALUATION: This warm "Reading Rainbow" story about a friendly young boy and an inquisitive thing from outer space sparks with imagination. It seems appropriate that the thing arrives on Earth on Halloween night and is easily integrated with trick-or-treaters in costume; it is a night when the unexpected could occur. An innovative word, *zyglot,* is readily deciphered when it is explained that Christmas is two zyglots (months) from Halloween.

The thing wins a friend through its intelligence and propensity for fun. The team of Edward Marshall and James Marshall wins young readers with spirited text and zany, humorous illustrations. In a sequel, *Merry Christmas, Space Case,* by James Marshall (Dial, 1986), the thing from outer space returns for another holiday visit with his friend.

□ □ □

Marzollo, Jean, and Claudio Marzollo. *Jed and the Space Bandits.* Illus. by Peter Sis. Dial, 1987. 48 pp. ISBN: LB 0-8037-0136-5. Age levels: 4–8. Grade levels: PreK–3. Series: Easy-to-Read.

SUMMARY: The sudden arrival of a strange spaceship gives young Jed and his Junior Space Patrol the chance to make a new friend and combat criminal activity in space. The drama unfolds when an unknown spaceship is caught in a gigantic magnetic net while it dangerously heads toward the spacecraft under the command of Jed's parents. A young girl, the sole occupant of the mysterious spaceship, explains that her parents are scientists who have been kidnaped; space bandits want her parents' formula that can make people invisible.

Jed and his Junior Space Patrol, which is made up of a teddy bear robot and two cogs (part cat and part dog), prove helpful. The robot supplies computer information that locates the bandits on the cold, damp Planet X32. The two cogs transmit thought waves that are crucial in the rescue mission. The young girl, who has the ability to become invisible, plays an important role in confusing the evil bandits and saving her parents. In addition, Jed's courage and his parents' assistance are instrumental in the successful release of the young girl's parents from captivity. An air-cushion taxi, a space houseboat, and egg-shaped prisons are some of the factors leading to the defeat of the enemy. At the end of this space adventure, Jed discovers that he enjoys the young girl's companionship and invites her to become a member of his Junior Space Patrol.

EVALUATION: This space fantasy is entertaining and full of action. Beginning readers will share the exhiliration of fighting crime in an ethereal region beyond the confines of Earth's atmosphere. The imaginative and colorful illustrations dramatize the futuristic setting while the sprightly and simple text transmits the plot and the characters' thoughts and actions. The credibility of this adventure is augmented by the advances made in space age technology. The division of this book into five chapters provides convenient stopping points if this outer space story is read aloud.

□ □ □

Marzollo, Jean, and Claudio Marzollo. *Ruthie's Rude Friends.* Illus. by Susan Meddaugh. Dial, 1984. 48 pp. ISBN: LB 0-8037-0116-0; pap. 0-8037-0378-3. Age levels: 4–8. Grade levels: PreK–3. Series: Easy-to-Read.

SUMMARY: Ruthie has recently arrived at the space station on Planet X10 and is lonely. Her parents are busy scientists who work at the laboratory. Ruthie and her parents are the only Earth people at the space station. To investigate this new planet, other scientists have also come from different parts of the galaxy. Their families have joined them, but bashful Ruthie has been unable to make friends with the strange children: scales cover one child, sixteen eyes protrude from another, and one has three horns.

Ruthie feels not only alone but restricted. Her mother has told her to stay within the walls of the space station. Finally, she meets a fish with butterfly wings and a boy with the ears, hooves, and tail of a pig. Using her word belt, she is able to decipher their grunting, puffing, and clicking sounds. They think she's strange and laugh at her appearance. All three exchange cutting remarks. Angry that they have treated her rudely, Ruthie becomes determined to gain their respect by showing that she is courageous. She decides to explore the forbidden territory that lies beyond the walls of the space station.

Aided by low gravity, she easily jumps over the high wall. When she enters a thick pine forest, she begins to feel afraid and lost. Soon she discovers that Pig and Fish have followed her. All three become terrified by a large cannibalistic monster with three flaming heads. Since Pig and Fish can fly, they lift Ruthie into the air to escape the monster's mouths. However, they are unable to hold onto her for very long. Ruthie falls to the ground near the monster and manages to kick one of its heads with all her force. To her surprise, the head stretches out high into the air and safely away from her. She does the same thing to the other heads, and the monster flees in consternation. Apparently, the difference in gravity on Planet X10 has made Ruthie exceptionally strong. Pig and Fish are quite impressed. All finally admit that they have been rude, apologize, and become friends.

EVALUATION: This easy-reading space adventure is lively and imaginative. The large-print text is spirited and conveys typical children's thoughts and emotions, even though some of the young characters are extraterrestrial beings. The main character shows spunk and daring; however, it is disappointing that her disobedience in not heeding her mother's warning to stay within the space station's walls is not addressed.

The unusual creatures at the futuristic space station and the planet's vicious resident, a three-headed monster, offer animated entertainment. The strange world of an unfamiliar planet is portrayed in simple, colorful illustrations. Young readers will be transported into a fantastic future where friendship and being sensitive to other's feelings are still important.

□ □ □

Mayer, Mercer. *Astronaut Critter.* Illus. by author. Simon & Schuster, 1986. Unp. ISBN: 0-671-61142-9. Age levels: 0–6. Grade levels: PreK–1.

SUMMARY: In this board book for young readers, Astronaut Critter, an imaginative creature, enters his carton-rocket and prepares for his flight into space. He uses his camera to record the sights outside his spacecraft. His space snack consists of crescent moon- and star-shaped cookies. A paper towel roll serves as a telescope for Astronaut Critter. With the moon—which is really a little girl's yellow balloon—viewed close at hand, the eager astronaut is ready for his imaginary moon landing.

He hops onto his lunar rover—his tricycle—and drags along a vacuum cleaner. The terrain is bumpy—he runs over a hill of flowers and into a sandbox. There are moon rocks to collect—from a stone wall that is being constructed—so that he can give scientists specimens to examine. To verify that he has accomplished his goal, he plants a flag—in the wet cement of a neatly troweled walk. A hasty retreat is made back to his spacecraft when he is suddenly attacked by an odd moon creature—a frisky puppy. He finds a committee ready to welcome him upon his return home. The president delivers a speech that certainly must refer to the rambunctious activity that occurred during this space escapade that never leaves the ground.

EVALUATION: Space terminology and the excitement of space travel fill this board book. The main character exudes enthusiasm and zest for space age play. His excesses of imagination provide humorous situations that a young child can fully appreciate. The young astronaut sees his space voyage from an adventurous and pleasurable angle, while the others around him, who are disturbed and inconvenienced, view it in a totally different way. The text is brief, simple, and spirited. The illustrations are zany, colorful, and full of action. This picture book transports the reader to an extraordinary exploration of the moon without leaving Earth.

□ □ □

Moché, Dinah. *We're Taking an Airplane Trip*. Illus. by Carolyn Bracken. Golden, 1982. Unp. ISBN: pap. 0-307-11869-X. Age levels: 4–7. Grade levels: PreK–2. Series: Golden Look-Look Book.

SUMMARY: Jimmy and Elizabeth journey alone on an airplane to see their grandparents. Every phase of the trip—from arrival at the terminal with their mother to the retrieval of baggage at their destination point—is shown, including check in, security clearance, a visit to the cabin to meet the pilot, takeoff, meals on board, use of bathrooms, and movie viewing. The children are given close supervision by a flight attendant. Such details as seat belt buckling, stopped-up ears, and the bump experienced at touch down are mentioned. At the journey's end, Grandma and Grandpa are waiting at the arrival gate to greet Elizabeth and Jimmy with open arms.

EVALUATION: Nothing about a first experience on an airplane has been overlooked by this delightful book in the Golden Look-Look Book series. Bright, cheerful colors and lots of detail, including cutaway cabin views, are informative and as authentic as they need to be for this age group. Some vocabulary will need explanation, for example, security, terminal, conveyor, taxied (children will probably know taxi only as a noun), and galley. A nice touch has Jimmy, the younger of the two children, holding on tight to his toy, Mr. Bear, because he wants to make sure Mr. Bear is not afraid of anything.

□ □ □

Myers, Bernice. *The Extraordinary Invention.* Illus. by author. Macmillan, 1984, o.p. Unp. Age levels: 5–8. Grade levels: K–3.

SUMMARY: Sally and her father are inventors. Their inventions are found throughout the house. The sneaker finder, the door closer, and the clothes sucker-upper are just a few of their ideas put to practical use. They make their inventions from spare parts and junk. One day they decide to make something special for Mother. After days of hard work, their new invention, a time machine, is completed.

Mother puts on her coat and hat and is coaxed into the machine. The machine rattles and shakes, and Mother comes out unchanged. Father goes into the contraption to fix it and comes out as a chicken.

Father sits around the house each day, growing more and more depressed at his appearance. Mother makes him sleep in the barn, and Sally scatters corn on the ground for him to eat each day. Finally he succumbs to Sally's pleas and gets back into the machine to be returned to his former self. However, each time the machine is activated, Father is transformed into a different animal. Finally the machine sputters and falls apart. There in the middle of the heap of rubble sits Father.

Sally and Father again try to improve their invention. As soon as it is finished, they invite Mother to try the time machine again. However, the machine fails and instead of sending Mother to Florida, she is sent into orbit in outer space.

EVALUATION: This humorous book features a girl as one of the main characters. The father and daughter team seems to share equally in the planning and creation of the inventions. More girls should be included as active participants in science books, as each year, more women enter fields of science, such as medicine and space, fields once labeled "Men Only."

The term *time machine* is a misnomer here. A time machine takes one to the past or future; it does not turn one into a chicken or launch one into outer space. This can be confusing.

The illustrations show a curious array of inventions the family uses to improve their life-style. The young reader is sure to realize that the appliances we use today, such as the vacuum cleaner, remote control TV, and electric mixer, are simply the inventions of the past. The book is a delightful way to encourage children to invent machines or devices they would like to see marketed.

The passages relating to Father's habits as a chicken are too long and silly; the preschooler will become bored and the school-aged child will find them predictable. At the end Mother is orbited into outer space and apparently left there, until Sally and her father can invent a machine to bring her back.

The book contains the raw materials for a successful introduction to technology. However, the technology is overshadowed by the humor.

□ □ □

Neumeier, Marty, and Byron Glaser. *Action Alphabet.* Illus. by authors. Greenwillow, 1984. Unp. ISBN: LB 0-688-05704-7. Age levels: 0–6. Grade levels: PreK–1.

SUMMARY: From *A* to *Z,* these letters of the alphabet are kinetic. The letter *A,* as in acrobat, moves across the page on a tightrope. A drip falls from a faucet and forms the letter *D*. An *I* skates across the ice. *J* rhythmically jumps up and down. Floating in the air, a kite is the letter *K*.

Entering the space age, the letter *O* is the path formed by a spacecraft that's orbiting Earth. An *X* is displayed within a child's chest to indicate how medical technology allows an X ray to penetrate a body and image the interior. The top part of the letter *Y* is an open mouth that's emitting a yell. Finally, a group of *Z*'s zigzag diagonally across the last page.

EVALUATION: Although illustrated in black and white, this alphabet book is vibrant in its vigorous motion. In addition to the letters of the alphabet, the words that represent the action of the letters are included as text. Familiar words are most often used in alphabet books for children. An X ray can indeed be part of many children's experiences. To see *O* for orbit is an indication that space terminology has become appropriate in a lexicon for the very young. This clever alphabet book is dynamic.

□ □ □

Newton, Laura P. *William the Vehicle King.* Illus. by Jacqueline Rogers. Bradbury, 1987. Unp. ISBN: 0-02-768230-7. Age levels: 0–7. Grade levels: PreK–2.

SUMMARY: William's bedroom rug becomes the background for a transportation network. William needs a car to play with; however, he soon realizes that only one car is not sufficient for the game he has in mind. From his toy box, he selects many different vehicles including a van, a racing car, and a sedan. Random activity with his moving vehicles leads to a pileup. Then William decides that it is also necessary to have his ambulance, police cruiser, and fire engine on the scene.

Thinking things over, William comes to the conclusion that he needs more roads to avoid further mishaps. He brings out his construction vehicles—a grader, a steamroller, and a dump truck—in order to complete his road-building project. Blocks and socks are spread out to become roads, and sneakers become parts of bridges. Even his toy robot is integrated into the highway landscape on his bedroom rug. A contented William surveys his domain and knows that he is truly a Vehicle King.

EVALUATION: Mechanically minded William uses his imagination and toys to create a transportation system in his bedroom. His robot, blocks, motor vehicles, clothing, and assorted toys all become part of his construction panorama. Intense watercolor illustrations heighten the visual pleasure of this construction fantasy.

Primary and some secondary colors are identified by name in the text as the vehicles are introduced. This indicates that color discrimination can be taught incidentally as the book is enjoyed. Almost imperceptibly, the vocabulary becomes more sophisticated as the text progresses; however, this doesn't impede understanding, since the text is brief and the visual context is firm. This is a pleasing book about a little boy who succeeds in creating an exciting superhighway scene within the confines of his bedroom.

□ □ □

Oppel, Kenneth. *Colin's Fantastic Video Adventure.* Illus. by Kathleen C. Howell. Dutton, 1985. 104 pp. ISBN: 0-525-44151-4. Age levels: 9–12. Grade levels: 4–7.

SUMMARY: Colin Filmore meets the two tiny men who pilot the spaceship of his favorite video game, Meteoroids. They relate amazing goings-on that occur in the secret world that they inhabit. Colin realizes that, with their cooperation, he can use their skills to win video contests.

After gaining national recognition and prize money for his unbelievable scores, Colin concludes that he is indeed cheating by using his little friends. In order to become a truly expert player, he agrees to enter their world via the Meteoroids video screen to receive special training from his miniature friends.

EVALUATION: This fast-moving story effectively focuses on the intensity of the protagonist's involvement with a video game. Electronic game lovers will surely be attracted to this book and might easily identify with the main character. This teenage author has a great deal of experience with video games; his youth and knowledge are likely to appeal to young readers. The author takes a fanciful approach to the highly technical aspects of video games, rendering the text imaginative and entertaining. The illustrator lends credibility to the story with her controlled and seemingly realistic black-and-white drawings.

□ □ □

Phelan, Terry Wolfe. *The S.S. Valentine*. Illus. by Judy Glasser. Four Winds, 1979. 42 pp. ISBN: 0-02-774570-8. Age levels: 8–11. Grade levels: 3–6.

SUMMARY: Through the eyes of Andy, the reader learns about his middle-grade classroom. The whole class will be performing in a play on Valentine's Day. One day an unexpected element enters the picture when Connie Robinson arrives in the class in a wheelchair. Mrs. Jacobs, the teacher, tells the class that they will get her cooperation for any play they prepare but insists that they include *everyone*. The challenge of how to include Connie and her wheelchair is met by Andy, who suggests that the wheelchair be a spaceship called the "S.S. Valentine." Building Connie's spaceship and designing the play around her takes over the plot.

EVALUATION: While the literary stature of the story isn't exactly impressive, it *is* pleasant, coherent, humorous, believable, and sympathetic. The shaded pencil illustrations in black and white are well done and add to the flow of the plot. The story conveys charm, warmth, and humor. One of the best things about the book is the honest and convincing way the author shows the first reactions of Connie's classmates to her arrival, and how effectively everyone adjusts to the situation. However, one does not finish the book with the sense that its central focus is the inclusion of a handicapped individual in a class activity, and this is a plus. Nothing is forced; events unfold logically as a satisfactory solution to a challenge is met. The idea of turning Connie's wheelchair into a space vehicle is certainly creative; thus the book promotes positive behavior while telling a cute story.

□ □ □

Pieńkowski, Jan. *Robot*. Illus. by author. Delacorte, 1981. Unp. ISBN: 0-440-07459-2. Age levels: 6–13. Grade levels: 1–8.

SUMMARY: A robot mother and father, somewhere in outer space, receive a letter from their son on Earth. His letter becomes the text of this story. The son misses his family and home cooking. As he asks about every member of the family, the reader meets some very unusual robots doing some ordinary tasks in extraordinary surroundings.

Multihanded Mom does her dusting, vacuuming, washing, and reading all at the same time. Her arms move and the wash tumbles in the washer. Her utility room is full of mechanical gadgetry. Earth can be seen outside the porthole of their home. A very metallic Dad lifts weights. Sis, with four arms that look like pneumatic tubes, busily blow-dries her hair, combs it out, puts on lipstick, and at the same time uses an aerosol spray as she prepares for a date. Without restraint, the young twins are superactive and get into all sorts of mischief.

Grandpa takes care of an Earth creature pet that curls and jumps out of the pages of the book, while Grandma, with her two sets of hands, busily knits a coat for the creature. The precocious new baby operates a computer whose screen displays kinetic art. Continuously getting into mischief, the twins manage to ignite a rocket that dramatically blasts off in three-dimensional wonder.

EVALUATION: This is a spectacular pop-up book with many manipulative devices. By using text that is a brief computer printout letter, Jan Pieńkowski accents the highly technological life of a robot family. The illustrations are boldly rendered in strong colors that capture the reader's attention.

These mechanical robots are appealing because their everyday life, although outlandish, is somewhat familiar. Mother is involved with household chores, Dad engages in body-building exercises, Sis primps as she gets ready for a date with her boyfriend, and the young children create mayhem. This is a most attractive and exceptionally creative book.

□ □ □

Provensen, Alice, and Martin Provensen. *Play on Words*. Illus. by authors. Random House, 1972, o.p. Unp. Age levels: 7–10. Grade levels: 2–5.

SUMMARY: The idea behind this book is the enjoyment of words for their own sake: the way they look and the way they sound. Time words, space words, noisy words, "watch" words, words-within-words, and words totally unfamiliar to most children are included. Words are playthings throughout, shown sideways, upside down, in cursive and elaborate calligraphy, and as part of the illustrations. For example, machines, vehicles, nose cones, and buildings are built out of word formations. Words are occasionally used to convey specific concepts but the whole idea is basically supposed to be fun. For example, an astronaut is shown bent into a conical shape because he is pushed by a string of letter *g*s, hinting at gravitational pull.

EVALUATION: This book is best categorized as a work of artistic technology. The play on words throughout, beginning with the foreword, is clever but unfortunately could also be misleading. The deliberately nonsensical way in which words are manipulated reveals that enforcement of learning principles was not the authors' intent. Words offered in total abstraction may confuse young children, as well as amuse them. Furthermore, words are misused: *lettuce* when *let us* is intended, and *no cents* for *no sense*. Also, the authors select examples of concepts unfamiliar to all but a very few. For example, *roe* is included so the authors can use the sound represented by letter *O* in the middle position in a collection of one-syllable words; the average young child will not know that *roe* means fish eggs. Some of the illustrations are fun to look at, especially a full-page marvelous machine at work—but a line using the word *din* may cause children to think a *din* is what the machine is rather than the sound it makes. Slightly older children who already have a basic reading vocabulary may think the book is fun.

□ □ □

Rey, H. A. *Curious George Gets a Medal.* Illus. by author. Houghton Mifflin, 1957. 47 pp. ISBN: 0-395-16973-9; pap. 0-395-18559-9. Age levels: 5–8. Grade levels: K–3.

SUMMARY: George, a curious monkey, gets into all sorts of misadventures after receiving a letter from the Museum of Science that he can't read. While waiting for his friend to arrive and read the letter to him, George manages to flood a room in the house. In order to clean it up, he takes a heavy portable pump—without permission—from a neighboring farmer's shed. He creates further mayhem when he attempts to borrow farm animals to drag this needed equipment. With angry farmers chasing him, George escapes by hopping onto a truck, which is on route to a delivery to the Museum of Science.

At the museum, he gets into more trouble at the dinosaur exhibit. He also discovers that his letter came from a professor at the museum who wants George's help for an experiment in space. Assured that his naughtiness will be overlooked if he participates, George agrees to take a spaceship into flight. His voyage into space is monitored on a special television screen. With his space suit and helmet, oxygen tank, emergency rockets, parachute, and courage, George performs well and succeeds in becoming the first living being to return to Earth from a flight in space. George's mischievousness is forgiven by everyone, and he receives a well-deserved medal.

EVALUATION: Colorful and lively illustrations and spirited text blend to make this action-filled story hilarious. Rey produces slapstick humor that is still funny after three decades and continues to be popular with children. Curious George's trip into space may raise questions and discussions about the role of animals in space exploration. The main character captures a spot in the history of space and also captures readers' hearts.

□ □ □

Rockwell, Anne. *Planes.* Illus. by author. Dutton, 1985. Unp. ISBN: 0-525-44159-X. Age levels: 0–6. Grade levels: PreK–1.

SUMMARY: This book with rabbit characters introduces very young children to a variety of airplanes, as a mini-overview is provided. Information about what makes planes fly, where they take off and land, and where they fly to is given. Flying is achieved by means of wind, jet engines, or propellers. Planes take off from an airport; a sea plane lands on a river, while helicopters touch down atop a skyscraper. Going around the world, from pole to pole, pilots fly over oceans, mountains, and urban and rural areas. From the ground, youngsters fly model airplanes.

EVALUATION: This delightful lap book about planes contains large and bright illustrations. That planes go around the world is cleverly illustrated by the book itself, since it begins and ends with the same illustration and text, giving a circular effect. Different aircraft are portrayed in interesting settings. For instance, the sea plane lands on water in an unidentified lush, tropical location; a propeller plane flies over the North Pole populated with ice floes and polar bears. At the South Pole, penguins look up toward another type of propeller plane. In the illustrations, rabbit characters are charming and actively involved; this may stimulate the adult reader or the young listener to compose extemporaneous stories, since there is no story line—the text relates directly to the planes in the illustrations. The simple text is appropriate for a read-aloud book.

□ □ □

Rockwell, Anne. *Things That Go.* Illus. by author. Dutton, 1986. 24 pp. ISBN: 0-525-44266-9. Age levels: 0–6. Grade levels: PreK–1.

SUMMARY: All kinds of things that move on land, water, and in the air are pictured and labeled in this book with animal characters. Under different headings, things that go and are seen in familiar places, such as in the country, the city, the house, and the park, are divided into two-page sections. There are tractors, combines, and hay wagons in the country; subways, garbage trucks, and delivery bikes in the city; vacuum cleaners, bathtub boats, and kiddie cars in the house; tricycles, scooters, and ice-cream wagons in the park.

Cement mixers, bulldozers, and robots are busy at work sites. In the air, satellites, rockets, and space shuttles are illustrated along with blimps, helicopters, and hang gliders. On the water, tugboats, ocean liners, and canoes sail by. Winter scenes include sleighs, snowplows, and sleds. Motorcycles, wheelchairs, pogo sticks, school buses, and tractor-trailers are some of the many moving things that are demonstrated in 10 mini-chapters. There is also a table of contents.

EVALUATION: This is an excellent lap book for helping children identify words by using pictures. The pictorial dictionary format with its colorful artwork is appealing. The moving objects are organized into groupings that are likely to be meaningful to children. There is no story in the text; however, the pleasing animal characters can be used by both adult and child to create story situations. A touch of ethnicity is seen in the hay wagon illustration through characters dressed in Amish-type clothing. The satellite, space shuttle, and rocket bring the space age into these pages. The bold illustrations are full of action as the animal characters enjoy using and making things go.

□ □ □

Rockwell, Anne. *Trucks*. Illus. by author. Dutton, 1984. Unp. ISBN: pap. 0-525-44432-7. Age levels: 0–6. Grade levels: PreK–1.

SUMMARY: Many kinds of trucks and how they are used are discussed in this book about special types of transportation vehicles. The smallest are toy trucks that are fun to play with in imaginary ways. In the workaday world, much larger trucks can actually carry away garbage or plow snow. Some are able to dump gravel or tow away disabled cars. Certain long, flat trucks haul steel girders for building purposes.

Other trucks help deliver flowers, move furniture, or bring vegetables and meat to the stores. Trucks that are used to clean streets, repair power lines, or put out fires are of great service to a community. There are also trucks that bring particular enjoyment to youngsters, such as ice-cream trucks, bookmobiles, and campers. A wide variety of trucks perform important functions not only in heavy work situations but also in pleasurable circumstances.

EVALUATION: All sorts of trucks occupy every page of this book and offer children a close look at these marvelous conveyances from an interesting angle. The trucks are illustrated from the side only. This provides a clear and wide view of some specific detail or details that make each type of truck unique. The colors are bright and appealing. Two-page spreads engage the eye and invite longer inspections for specifics.

The characters are cats that have human characteristics and are delightful as they go through their everyday activities. The text is sparse and simple. The identification of trucks is paramount in this picture book. Both text and illustrations seem to make this a pictorial dictionary of trucks that children are likely to see on the road or in their neighborhood. The author-illustrator cleverly begins and ends the book with toy trucks and successfully keeps her topic within the boundaries of a child's interest.

□ □ □

Round, Graham. *What If . . . I Were an Astronaut?* Illus. by author. Barron's Educational, 1987. Unp. ISBN: 0-8120-5773-2. Age levels: 3–6. Grade levels: PreK–1.

SUMMARY: A youngster wonders what it would be like to be an astronaut. In his private spaceship, he would take a trip into outer space. He'd successfully land on a faraway planet that would be inhabited. Since there would probably be a meeting with a weird creature, the young astronaut would then find out about its home and what food it eats.

In addition, he imagines that he would explore and discover a futuristic city. There would be unusual skyscrapers, the moon overhead would be pink, and there would be a lake that is purple in color. In a space garden, he would gather spectacular and rare flowers and some stars from a tree to give his mother when he returns home.

EVALUATION: In this small book, an imaginary interplanetary trip is presented in bold colors and blazing action. The story unfolds immediately, right from the inside cover of the book and continuing to the inside of the back cover. All available space is enveloped with attention-grabbing illustrations. The main character, wearing a space suit, is engaging with his vivd imagination, as he anticipates the possibility of taking a trip to outer space, suspects the existence of friendly extraterrestrial life, and welcomes the thought of a warm reunion with his mother upon his return to Earth. The text is simple, and quickly and directly answers the question posed by the title of the book in a whimsical fashion.

□ □ □

Sadler, Marilyn. *Alistair in Outer Space.* Illus. by Roger Bollen. Prentice-Hall, 1984. Unp. ISBN: 0-13-022369-7. Age levels: 4–8. Grade levels: PreK–3.

SUMMARY: Although he is seized by blue creatures—the Goots—and carried away in a spaceship, young Alistair seems only concerned with his books that are due at the library. Not only is he punctual, organized, and neat, but he is also a great reader. Alistair is accustomed to an orderly life and reacts to this unusual turn of events in a predictable fashion. He wants to read an exciting book in order to experience adventure. However, instead of reading, Alistair finds himself hurtled out of the spaceship and meeting new space creatures, the Trollabobbles.

When he later returns to the Goots' spaceship, he is still merely interested in getting his books to the library before he has to pay a fine. He convinces the Goots, who are not very good at directions, to let him fly the spaceship to Earth.

Sensible Alistair takes a shortcut and is sure that now he'll get to the library in time. However, for the final part of the voyage, the not-so-sensible Goots take over, and Alistair is about to embark on another unexpected adventure. The final illustration reveals that Alistair's books will certainly be overdue.

EVALUATION: The bookish and bespectacled main character, who is portrayed as methodical, prompt, and unimaginative, is both amusing and strange. Although Alistair finds himself in fantastic circumstances, he can only enjoy himself vicariously through books. His adventures in space are illustrated in colorful and dramatic scenes, while he is shown in sedate suit and tie, hair neatly combed and parted, and wearing his cap on top of his bubble space helmet. The contrast is effective and humorous. The human qualities of the space creatures make them nonthreatening and appealing. The brief text is accompanied by intriguing illustrations of a volcanic planet and exterior and interior views of the spaceship. Young readers are likely to be entertained by this funny out-of-this-world "Reading Rainbow" tale.

□ □ □

Sadler, Marilyn. *Alistair's Time Machine.* Illus. by Roger Bollen. Prentice-Hall, 1986. Unp. ISBN: 0-13-022351-4. Age levels: 4–8. Grade levels: PreK–3.

SUMMARY: Brainy and orderly Alistair is deeply interested and involved in scientific experiments and inquiry. He confidently enters a science competition, using his microscope to examine cells, mixing chemicals, collecting specimens in jars, and studying the night sky with his telescope. He finally decides to build a time machine for his project. With this invention, he's certain that he'll win first prize.

When he tests his machine, however, he discovers it doesn't perform as expected. Pushing the day-before-yesterday button, he finds himself with knights at a round table. After lunching with these men, who have very hearty appetites, Alistair attempts to return home. However, the time machine sequentially takes him to a royal ball at a French palace, to a pirate ship, to a Roman arena, and finally to a prehistoric campsite.

The cave people are suddenly attacked by woolly mammoths, and quick-thinking Alistair uses his time machine to save them: by twisting the time machine toward the menacing beasts, he makes them vanish. The grateful prehistoric people record this event in cave paintings. Eventually Alistair succeeds in operating his machine correctly and is transported home safely.

Once at the science competition, the time machine is inoperative. A disappointed Alistair watches another contestant win with a model of the solar system. Although he has no proof, Alistair is convinced that his machine is functional. As he leaves the building where the science competition was held, Alistair is oblivious to mammoths looming behind him.

EVALUATION: This zany fictional story propels the reader backward into time via a machine that seems to have a mind of its own. The illustrations are bright, large, colorful, bold, and comical. Readers will be amused by the inventive and methodical main character, dressed in suit and tie, as he travels through time unperturbed by unexpected circumstances. Scenes from the different time periods lend themselves to discussions in primary social studies. The model of the solar system, Alistair's refracted face as seen through the glass of a specimen jar, the appearance of the mammoths in the twentieth century, and the time machine itself are topics appropriate for scientific discussions. The witty text combines with the humorous and interesting illustrations to keep readers entertained and challenged throughout this book.

□ □ □

Scarry, Richard. *Richard Scarry's Things That Go.* Illus. by author. Western, 1987. Unp. ISBN: LB 0-307-61817-X; pap. 0-307-11817-7. Age levels: 3–6. Grade levels: PreK–1. Series: Golden Look-Look Book.

SUMMARY: Things that move provide the action throughout the pages of this little book. In the beginning pages, animal characters enjoy toys that rock, spin, and fly. This is followed by things on wheels, such as bicycles and wagons. A new road is constructed with the help of an asphalt mixer, a roller, and other heavy machinery.

Racing cars speed down a mountain road, and trains transport animal characters and their goods. On the farm, machines pick corn, cut hay, and gather wheat. Planes and balloons rise above the ground. As a waterwheel turns the machinery in a mill, pleasure boats are rowed and paddled in the nearby stream. In the home, some objects, such as a fan and a vacuum cleaner, whirl and whirr. The last page of the book shows a father cat reading a book with his daughter at bedtime about things that go.

EVALUATION: This clever book is busy with activity and packed with scenes that will keep youngsters going back to look at the pages again and again. Many humorous illustrations depict the theme of things in motion. For instance, to show that water is able to make things go, an unsuspecting moose in a canoe—with a mandolin for a paddle—is carried to the edge of a waterfall whose falling waters are used to power a mill. At a railroad station, airport, farm, racecourse, and other locales, machines are operated by endearing animal characters. The delightful illustrations are colorful and packed with action.

The text is simple and sometimes engages the reader/listener by asking questions relating to the illustrations, such as determining the problem of a race driver who has not maneuvered a curve. The observant child might notice that the term "people-powered" is used, when only animal characters are featured. The ending, with animal characters reading a bedtime book just like the one that the child is enjoying, comes as a surprise. As the tired kitten falls asleep, the author succeeds in slowing the action of the prior pages and adding a soothing touch to a lively book.

□ □ □

Schoberle, Cecile. *Beyond the Milky Way.* Illus. by author. Crown, 1986. Unp. ISBN: 0-517-55716-9. Age levels: 5–7. Grade levels: K–2.

SUMMARY: On an early summer evening, a young girl peers out of her bedroom window. Between the tall city buildings she notices the special color of the sky at twilight. Up high, an airplane with its blinking lights passes by. There are hazy clouds beyond the plane, and stars and the Milky Way can be seen in the distance. Her mind wanders beyond the Milky Way to an imaginary planet that is populated.

She wonders if another young girl is looking out of her window. Can she see a colorful summer sky beyond the tall futuristic buildings of her ultramodern city with its high technology transportation systems? Is the girl able to see the flashing lights of a spaceship? Can she also see above the clouds and notice the stars and the Milky Way? Is she capable of catching sight of the planet Earth? Does she wonder if an Earth-girl is doing the same thing? All of these thoughts go through her mind before she falls asleep.

EVALUATION: The main character is effectively portrayed in text and illustrations as having an active mind that allows her to travel in her imagination to outer space and to wonder about the existence of life there. The fascinating illustrations on glossy pages are strong in color. The futuristic world on a distant planet is displayed as a dynamic place. The clouds, stars, planets, and Milky Way are beautiful to behold. The planet Earth with its circling clouds and visible continents commands a two-page spread that is most dramatic. One is reminded of the photos that the NASA astronauts took in space of our planet.

The text is simple and clear, and has part of a sentence or one sentence on a page. The continuity and effectiveness of text is maintained as a key word in one sentence is repeated in the sentence on the following page. This repetitive writing style appeals to young children. The central idea of another living being existing on a faraway planet and thinking similar thoughts is executed in an impressive manner. An extraordinary voyage is taken in this book and includes cosmic wonders that will delight readers.

□ □ □

Sloan, Carolyn. *The Friendly Robot.* Illus. by Jonathan Langley. Derrydale, 1987. 28 pp. ISBN: 0-517-65134-3. Age levels: 5–8. Grade levels: K–3. Series: Children's Library.

SUMMARY: Paul's father brings a robot home because it doesn't work properly in the factory. After the robot is reassembled, young Paul takes it to his bedroom in order to get it out of his mother's way. There he shines it with his T-shirt, and suddenly it begins to work without interruption. Toys are put in order, clothes are picked up, and socks are matched. Trying to slow the robot, Paul pulls a lever that causes the robot to speak. They become friends and have a lot of fun together. Paul personalizes the robot by giving it a name: Bob.

When Paul's father sees that the robot is now operating, he returns it to the factory. However, Bob will not work there and is disassembled. Finding Bob in his father's car, Paul fears that his friend will be thrown away. Paul puts Bob together again and takes him to school. Bob becomes a great hit in the classroom. Not only is he a wonderful organizer, but he is also quite creative. The teacher convinces Paul's father that the right place for Bob is with Paul.

EVALUATION: This book is delightful and shows the importance of consideration and friendship. A young boy befriends and humanizes a robot; their special relationship is illustrated in soft and colorful illustrations. The setting of home and classroom is contemporary, realistic, and believable.

First published in the United Kingdom, this book is intended to be read to children by adults. The text reads easily, although some words, such as *dustbin* and *muddle,* may have to be explained to very young American children. The endpapers of this book are full of illustrations of nuts, bolts, resistors, and transitors; however, the story conveys the idea that a robot can be more than the sum of its mechanical parts.

□ □ □

Stevenson, James. *No Need for Monty.* Illus. by author. Greenwillow, 1987. Unp. ISBN: LB 0-688-07084-1. Age levels: 4–7. Grade levels: PreK–2.

SUMMARY: A ferry service is provided by Monty, an alligator, to three young animal friends who have to cross a river to go to school. Dependable Monty carries his passengers through wind, snow, and rain. Because he is slow, impatient and concerned grown-ups look for a faster means of transportation for the children.

One idea that they test is a rowboat. Due to lack of planning, this idea fails when the boat gets mired in mud. Next, they devise an overhead tram, which also meets disaster. When an assortment of animal friends crowd on for this popular ride, they cause an overload, and the cable snaps; they all fall into the water. The third idea involves a balloon trip. Aloft, the balloon is carried far afield by an unpredictable wind. Then the grown-ups come up with a fourth alternative. They provide the children with stilts. When the children meet another watery downfall, all concerned hail Monty's return as the most efficient ferry service.

EVALUATION: Stevenson's cartoons and comic strip format provide a visual treat. There is abundant humor for reader and listener through illustrations and clever use of language. The adults' lack of insight into transportation technology will amuse children.

The art is filled with action. There is also a special beauty in the treatment of the natural elements of rain, wind, and snow. Underlying the story is a positive image involving the value of tried and true friends and proven means of doing things. The alligator makes an unlikely transport vehicle, adding irony to a delightful tale.

□ □ □

Testa, Fulvio. *The Paper Airplane.* Illus. by author. Henry Holt, 1988. Unp. ISBN: 0-8050-0743-1; pap. 0-8050-0744-X. Age levels: 5–8. Grade levels: K–3.

SUMMARY: A child flies a paper airplane from a window, feeling that he's flying also. He imagines himself flying over the ocean, watching ships below. Then, on dry land, he proceeds to fly a paper kite in the shape of a bird. He wonders about all the things that birds can do, such as soaring, perching, and singing. He then questions the state of birds and animals in captivity, and thinks about them in their natural environment.

Studying the globe helps him to understand that a continent is only part of one's world. He then launches his imagination into outer space. Traveling in a spaceship, he realizes that the world looks small and round.

Then his imagination takes him up in a large balloon from which he releases many sheets of paper. He suggests that one can make an airplane from a piece of paper, fly it through a window, and travel through the world and beyond.

EVALUATION: A paper airplane is introduced in the title, and on the first page; however, the concept of a paper airplane is quickly lost and not found again until the end of the book. Visual instructions for making a paper airplane are shown through nine diagrams on the outside of the book's back cover. Because of their location and since there are no verbal instructions given, the diagrams seem to be an afterthought.

The central focus often strays and leads to confusion. Possibly through translation into English (from the original German), grammatical errors occur such as the use of the word *through,* and an inaccurate referent. In one case, an illustration does not match the text. Adding to the confusion, multiple characters appear in the illustrations even as the protagonist uses first person singular. The illustrations themselves, however, are clever and vivid, and reveal subtle humor.

□ □ □

Titus, Eve. *Anatole and the Robot.* Illus. by Paul Galdone. McGraw-Hill, 1960, o.p. 32 pp. Age levels: 5–10. Grade levels: K–5.

SUMMARY: The classic character Anatole, the French mouse, is a cheese taster in a French cheese factory in this story. Using a flashback technique, the author invites the reader to find out what happens when the president of the company contracts measles and puts his fifth cousin La Rue, an inventor, in charge. La Rue tries out his latest invention—a robot that is a cheese taster. Anatole is crushed when a memo informs him that he must find another job. At the same time, the mouse looks upon the mechanical cheese taster as a challenge and checks up on the impressive, metal, man-sized robot with many moving parts. After some time Anatole realizes the problems the programmed robot is causing (it works too quickly for the men to keep up with it and requires supervision). Anatole knows that the factory workers want him back. Eventually, Anatole goes happily back to work as the cheese taster, with the factory workers overjoyed as well.

EVALUATION: One of the most outstanding features of the story is the author's use of language. Titus writes with the knack of a storyteller so that the reader feels that the story is actually being told. The prose is almost poetic in the way the lines of type are set, and in the use of alliteration. In the most appropriate places, Titus draws from nursery rhymes or folktales and adds a different twist. Going along with the folktale theme, Anatole is a hero who is very clever and knows how to protect his interests. The use of the number seven and little moral themes throughout also provide a folktale quality to the story. Several French words are used throughout, such as *tout de suite, oui,* and *magnifique.*

The author devotes a full page to the description of the robot, how it operates, and what its duties are. The primary reason for the robot not being able to replace Anatole is that it works too quickly for the factory workers, illustrating how one piece of technology can have a snowball effect on a total operation. In this case, it shows the limitations of the effects of the robot.

Paul Galdone's illustrations—many in the French tri-colors of red, white, and blue—add amusing appeal to the book. The characters show much expression and the illustrations capture the text in an amusing way.

□ □ □

Wildsmith, Brian. *Bear's Adventure.* Illus. by author. Pantheon, 1981. Unp. ISBN: LB 0-394-95295-2. Age levels: 5–8. Grade levels: K–3.

SUMMARY: Two balloonists descend on a mountainside to enjoy a picnic. As they look for a perfect site, a brown bear climbs into the balloon's basket, which he mistakenly thinks is an unusual den. In no time, he is sound asleep, and the wind carries him away. His adventure starts when he awakes to find himself adrift over skyscrapers. He thinks these are tall leafless trees in a strange forest.

Without warning, a bird bumps into the balloon and punctures it. The bear safely lands in front of a parade whose human participants think that the bear is wearing a costume and has arrived in a unique way to begin the proceedings. The confused bear is frightened by all the to-do and noise. He flees the scene, only to end up at a TV station. More misinterpretations occur, and the bear runs away during an unusual TV interview. A motorcyclist, who has a portable television set on his bike and has seen this "interview," gives the bear a lift.

The bear is left at a sports stadium where he wins a race, but escapes before the mayor awards him a medal. A helicopter (which the bear thinks is an odd, five-winged bird) provides transportation to another adventure. After alighting at a rock concert, the bear wins the admiration of the audience with his dancing. However, the crowd's enthusiasm scares him and forces him to flee.

A fireman comes to his rescue and hoists him high on a fire truck's ladder. At that moment, the two balloonists float by. The bear joins them and is soon fast asleep in the basket. When they approach the mountains, the balloonists decide to eat again. They bring their craft down, and the men enjoy their picnic by a stream. The bear wakes up and departs, confused by the day's events in a modern world that he didn't know existed.

EVALUATION: The author effectively presents a bear's experiences that take place in a human environment. Technological advancements, such as skyscrapers, a flying balloon, a TV station, a motorcycle, a fire truck, and a helicopter, befuddle the main character. The several modes of transportation allow the adventures to occur in a plausible and humorous manner.

In this oversized large-print book, a fascinating tale with its lively text progresses quickly. The artwork commands readers' attention as bright, colorful illustrations convey a playful tone.

□ □ □

Wildsmith, Brian. *Professor Noah's Spaceship*. Illus. by author. Oxford University Press, 1980. Unp. ISBN: 0-19-279741-7; pap. 0-19-272149-6. Age levels: 6–10. Grade levels: 1–5.

SUMMARY: The animals and birds of the forest are threatened by air pollution. Realizing that their lives are in danger, they consult Owl, who directs them to Professor Noah. They learn that the professor is building a spaceship that can fly to a planet in the future that has unspoiled forests. Enthusiastic about finding a healthy environment, the animals help robots complete the spaceship. Food for forty days and forty nights is stored in the spacecraft.

Before the forest is totally destroyed, Professor Noah and all the animals blast off into space. During their journey, an exterior time zone guidance fin needs repair. Elephant dons a special space suit and tries to fix the damaged fin containing the system that will propel the spacecraft into the future and to a new planet. Accidently, Elephant twists the time guidance system backward, and the spaceship returns to an Earth that existed centuries ago. The animals disembark and are thankful to find that the land is unpolluted and a wonderful place to live.

EVALUATION: This oversized book carries an important message in large print and eye-catching illustrations. The seriousness of pollution is intermingled with a modern version of the biblical story of Noah and the Ark. Brian Wildsmith uses space technology to solve the animals' problem in a delightful manner. The animal kingdom is saved from doom by being thrust backward in time rather than projected forward into the future, in an interesting setting with elements of the space age. The illustrations are colorful and include a detailed plan of the spaceship, an amusing portrait of an elephant in a space suit, and intricate mechanical contraptions, some of which help in the assembling of the spaceship, and others that are installed in the spaceship. The characters encounter a contemporary problem and travel through a time warp in order to avoid catastrophe. A warning for everyone is sounded in this "Reading Rainbow" book.

□ □ □

Yolen, Jane. *Commander Toad and the Big Black Hole*. Illus. by Bruce Degen. Coward, McCann, 1983. 64 pp. ISBN: LB 0-698-30741-0; pap. 0-698-20594-0. Age levels: 4–8. Grade levels: PreK–3. Series: Break-of-Day.

SUMMARY: Intelligent and courageous Commander Toad is in charge of the spaceship *Star Warts,* which is the fastest in the star fleet and has flown many successful missions. His crew consists of Lieutenant Lily, the bright engineer, Jake Skyjumper, the capable navigator, Doc Pepper, the caring physician, and Mr. Hop, the wise philosopher. Although they appreciate their commander's leadership, they do not enjoy his singing or his meals, which are only pills.

During a spaceflight across the galaxy, they sight a black hole that mystifies them. Commander Toad is certain that the hole is someone's home. When their smooth flight comes to a jarring halt they again are bewildered, this time by the long, pink, sticky thing that has stopped their spaceship. Neither Commander Toad nor the computer can provide answers. It's Doc Pepper who knows that they are stuck on a tongue. Soon all realize that the black hole is the mouth of an Extra Terrestrial Toad, E.T.T. for short, that is about to swallow them.

Conventional maneuvers are attempted in order to release the spaceship from the sticky tongue. When these fail, Commander Toad orders the crew into the sky skimmer. From this vantage point, further means of making the monstrous space creature free the spaceship are unsuccessful. Remembering something his mother had told him, Commander Toad decides to try his secret weapon. He starts to sing, and soon E.T.T. joins him. *Star Warts* comes hurtling off his tongue. Mother's advice not to sing with a full mouth was remembered at the right moment and saves the day for Commander Toad and his crew. Away from the black hole, they rendezvous with their spaceship.

Relieved that another space adventure has ended well, the crew enjoys a meal of pills and singing with Commander Toad.

EVALUATION: This zany space story gives a humorous explanation of a black hole. A parody of *Star Wars,* this tale results in a clever portrayal of a space fantasy that contains puns, silly adages, and revised familiar songs. The illustrations are appealing and blend with this funny, ridiculous story. Beginning readers are likely to find the antics of the winsome toad characters hilarious. The text is simple and clear in this easy-to-read book that is a "Reading Rainbow" selection. Readers who wonder about the existence of black holes might find Franklyn M. Branley's nonfiction book *Journey into a Black Hole* (Crowell, 1986) informative.

□ □ □

Yolen, Jane. *Commander Toad and the Dis-Asteroid.* Illus. by Bruce Degen. Coward, McCann, 1985. Unp. ISBN: LB 0-698-30744-5; pap. 0-698-20620-7. Age levels: 4–8. Grade levels: PreK–3. Series: Break-of-Day.

SUMMARY: Intrepid Commander Toad and his crew embark on a mission to help an inhabited asteroid that has suffered a disaster. Having received a confusing SOS message about swell beans and bad beans, they have filled the hold of their spaceship, *Star Warts,* with all sorts of beans because they think that the asteroid's bean crop has failed. On their voyage they encounter a shooting star with bad aim and the cow who jumped over the moon and produced the Milky Way.

A star map is consulted when an unknown watery world appears on the screen of the control console. To their dismay they discover that this is their destination and that the asteroid is completely flooded. While one member stays with the spaceship, Commander Toad and the rest of the crew take a shuttle craft to investigate the disaster that has befallen this place.

When they first meet the mayor they find it difficult to understand his pigeon talk, but soon Commander Toad's copilot deciphers the mayor's sentences and unravels the problem. This particular asteroid has lots of holes for the drainage of rain water; however, a recent downpour caused the asteroid's beans to swell and plug up the drainage system. Heroic Commander Toad swims underwater and deflates the swollen beans with his pocket knife. This adventure ends as the Commander earns a medal for saving the asteroid from total disaster and the spaceship heads to its base still carrying a cargo of beans.

EVALUATION: The toad space travelers in this humorous adventure are a zany and entertaining group. Readers will enjoy how the problems they encounter are resolved in unusual ways. The text is full of puns that will delight readers—for example, the communication between the pigeon mayor and his rescuers is called pigeon-toad—and even the title contains a pun. In one instance, the use of a pun leads to vocabulary enrichment. After amphibean is used in jest, the word *amphibian* appears in the text and is defined for young readers. Within this fictional space story, the fact that asteroids are found between Jupiter and Mars is disclosed. The lively illustrations are in black and white as well as color. This "Reading Rainbow" selection is sure to make readers search for the other Commander Toad books that Jane Yolen has created.

□ □ □

Yolen, Jane. *Commander Toad in Space.* Illus. by Bruce Degen. Coward, McCann, 1980. 64 pp. ISBN: LB 0-698-30724-0; pap. 0-698-20522-7. Age levels: 4–8. Grade levels: PreK–3. Series: Break-of-Day.

SUMMARY: During a space mission, Commander Toad and his crew of four discover a new and unusual planet. The intrepid leader decides to use the sky skimmer, a shuttle craft, to land on this new planet. However, the computer aboard their spaceship, *Star Warts,* reports that this planet is covered with water. Thinking quickly, the commander realizes that his large rubber lily pad and hand pump will now come in handy. The sky skimmer, with brave Commander Toad and three crew members, sets off toward the watery planet. The mother ship is left with the navigator in control.

Tethered to the sky skimmer, the flight engineer uses her dexterity and know-how to pump up the lily pad and declare it ready for the sky skimmer's landing. Shortly after they feel safe and secure on their floating pad, a gigantic and chameleonlike monster, Deep Wader, makes its sudden appearance and topples the sky skimmer. The frightened and helpless space explorers feel doomed.

Commander Toad's ingenuity and good planning provide the means of escaping the threatening jaws of Deep Wader. For birthdays and other emergencies, the captain always carries matches and special candles that don't blow out. As the crew distracts the monster with riddles and songs, Commander Toad transforms their raft into a hot-air balloon. Successfully aloft, they rendezvous with *Star Warts* and look forward to more galactic adventures.

EVALUATION: This "Reading Rainbow" selection is an easy-to-read space fantasy that's fast-paced and funny. Youngsters will enjoy meeting unusual space explorers and an incredible extraterrestrial sea monster. The text is simple and reveals an exciting adventure with lovable and gifted toads who display special knowledge about space travel. Color and black-and-white illustrations delightfully punctuate the text. There are puns, a riddle, and exciting reading in this spoof of *Star Wars*. Beginning readers will be tickled by this amusing space age story.

□ □ □

Yorinks, Arthur. *Company's Coming*. Illus. by David Small. Crown, 1988. Unp. ISBN: 0-517-56751-2. Age levels: 8–11. Grade levels: 3–6.

SUMMARY: A flying saucer carrying visitors from outer space lands on Earth and life at the home of Shirley and Moe changes. The story begins when Moe is mowing his lawn and a flying saucer arrives. His wife thinks it's a large barbecue. However, two spacemen, looking like large insects with elongated noses, emerge from the spacecraft. Unbeknownst to Moe, the hospitable Shirley allows them to use the facilities and also invites them to stay for dinner. The men from outer space have some unfinished business to attend to, but agree to return for dinner.

Shirley prepares dinner for her family and guests. After Moe finds out what happened in his absence, he hysterically calls the FBI. The armed forces are informed. Before the space visitors return, tanks have surrounded the house, and helicopters are hovering above the neighborhood.

Although Shirley is jittery, she warmly welcomes the spacemen when they arrive for dinner with a hostess gift. They explain that their planet is overpopulated and their mission is to scout for new places to live. The space visitors ask Shirley to open her gift. All the relatives, Moe, and the armed forces are extremely nervous when polite Shirley unwraps the present. Expecting a bomb, deadly gas, or a laser, everyone is relieved to see that Shirley has received a blender. In a happy ending, the outer space visitors enjoy a spaghetti and meatball dinner with their hosts and the military people they have met.

EVALUATION: This book is a satire on the range of human reactions to an unexpected and unfamiliar situation. The illustrations and text manifest the fear of the unknown and also the natural inclination to be friendly and hospitable to strangers. The cartoonlike illustrations are clever and humorous. The text is simple yet meaningful in its wry humor and remarks on the space age. Although all the characters are adults, children will be entertained by the grown-ups' coping mechanisms. Sophisticated readers will find social and political aspects in this story.

□ □ □

Young, Miriam. *If I Flew a Plane*. Illus. by Robert Quackenbush. Lothrop, 1970, o.p. Unp. Age levels: 4–8. Grade levels: PreK–3.

SUMMARY: In this book, a young boy has dreams of becoming a pilot. His thoughts are alive with the sights and sounds of flight. Through his fantasies, he has great adventures and arrives at interesting destinations via a variety of aircraft.

He stunt-flies, skywrites, directs traffic, rounds up cattle, and rescues lost pets. He also picks up astronauts at sea and manages to take a space trip to the moon. He imagines himself a hero, and dreams of recognition from the mayor and the President. The book ends with the young boy confident that he will invent an original and unusual aircraft. The finale is a rhymed verse about a spacecraft featuring pontoons and helicopter rotors.

EVALUATION: The setting is a boy's room, replete with airplane models, pictures, and building materials. Immediately, the reader is taken into the boy's imaginary world, which is complete with loud aeronautical sounds and descriptions of faraway places. A variety of aircraft is shown, and believable childlike activities are described. The color scheme of the illustrations is bold and bright, and the boy's thoughts include inventive scenarios. The space age is affirmed in this 1970 book that provides much information about aircraft.

□ □ □

Zokeisha. *Chubby Bear Goes to the Moon.* Illus. by Yasuko Ito. Simon & Schuster, 1984. Unp. ISBN: pap. 0-671-50953-5. Age levels: 0–5. Grade levels: PreK–K. Series: Chubby Board Books.

SUMMARY: Chubby Bear tells Mama Bear that he is going to the moon; Mama Bear tells him to be home in time for supper. Chubby Bear says, "Roger, over and out," and kisses her good-bye as he puts on his space helmet. Chubby Bear climbs through a door in a tree, which he uses as a rocket ship, and blasts off. Suddenly Chubby Bear's makeshift tree rocket looks like a real rocket, and he is in the sky looking at dreamlike animals suggested by star formations. He passes a smiling sun, feeds Pegasus an apple, and stops by the Milky Way for some milk to go with his cookies. He thinks aloud that he will bring home one of Saturn's rings for Mama Bear. Next Chubby Bear lands his rocket ship on the moon, where he is greeted by moon children. He tells his new friends he has to go home because Mama Bear has baked honey pie for supper. On the last page, we see Mama Bear enjoying her ring from Saturn, which has turned into a ring of blossoms.

EVALUATION: This adorable board book is one of a series first published in Japan. The plot, which makes a smooth transition from realism to fantasy and back again, has a plausible beginning, middle, and end. Family values are obvious as Mama Bear plans Chubby Bear's homecoming supper, and Chubby Bear thinks of Mama in space when he remembers to bring her a ring from Saturn. The colors are bright, and the binding appears sturdy. The only element that might confuse little children is on the page that shows Chubby

Bear feeding an apple to Pegasus. Questions may be asked about the mythical winged horse, and an adult sharing the book with a child may be unable to answer. Mythology seems out of place here, and is an unnecessary intrusion.

□ □ □

Zokeisha. *Let's Go for a Ride.* Illus. by author. Simon & Schuster, 1982. Unp. ISBN: 0-671-44896-X. Age levels: 0–5. Grade levels: PreK–K. Series: Chubby Board Books.

SUMMARY: *Let's Go for a Ride* was first published in Japan for very young readers. On every page a vehicle is pictured, with captions identifying it. These captions constitute the only text. Included are a bicycle, motorcycle, car, bus, horse and wagon, train, jeep, helicopter, balloon, airship, raft, sailboat, ship, submarine, airplane, and rocket. A merry-go-round is on the front cover.

EVALUATION: This little book is pretty to look at with its pastel and primary colors, and appears to be durable like all the books in this series. However, the items pictured look more like toys than real vehicles: there is a pink submarine; a pink, blue, and yellow rocket; and so on. Also, the vehicles are not presented in a logical order; for example, a horse and wagon are shown *after* a car and a bus. A raft—probably the most primitive form of transportation—is found in the middle of the book. It can be argued that for young children, organization is not important. However, one particular illustration may confuse preschoolers: the hot-air balloon is labeled simply *balloon*. Younger children have a simpler concept of what a balloon is, so the word *hot-air* probably should have been included as part of the vehicle label.

□ □ □

Zokeisha. *Surprise Robot!* Illus. by Yasuko Ito. Simon & Schuster, 1987. Unp. ISBN: 0-671-63635-9. Age levels: 0–5. Grade levels: PreK–K.

SUMMARY: This board book is really a toy that looks like a robot. Both front and back covers tell the reader that the robot opens up to be both a robot *and* a book. The parts of this toy/book move to reveal simple robot functions. There is practically no text, but on each page, a few words are used to show and tell what robots can do. Several word and number concepts appear but they are introduced in an incidental way, and there does not appear to be a conscious effort to use this book as a teaching vehicle. Board "pages" fold back and forth, and the robot parts swivel on revolving rivets.

EVALUATION: As manipulative books go, this one seems to be reasonably durable. The surfaces are hard and smooth, making them easy to clean. The swiveling rivets work, and would appear to withstand use—even abuse—by little children. The robot itself is very cute; he smiles, his eye winks, his arms turn, and his legs move. The bright colors are cheerful and add to the appeal. Well-made manipulative books such as this one serve a useful purpose because they show that books are objects that can offer hours of fun.

Technology: Nonfiction

Some of the books examined in this section provide information about what technological equipment is used in weather forecasting, how crayons are made, why lasers can cut steel, and what computers can do. Other books contain facts concerning holograms, spacecraft, locomotives, and radar. Incorporating attention to detail, there are descriptive accounts about the building of a skyscraper, the moving of a bridge, and the production of a movie. Different kinds of cranes, trucks, snow removal equipment, trailbikes, and customized cars are also featured. Readers can learn about the operation of a hydraulic lift in a service station, how carbonated beverages are bottled, and how patents can be obtained for inventions. Activity books include instructions for using a computer and also directions for making paper airplanes and spaceships that fly.

□ □ □

Adler, David A. *Calculator Fun.* Illus. by Arline Oberman and Marvin Oberman. Watts, 1981, o.p. 32 pp. Age levels: 5–10. Grade levels: K–5.

SUMMARY: Books introducing the uses of calculators are common, probably because directions that accompany the calculators are often incomplete and difficult to read. This short book presents rudimentary information to young users. Although all calculators may look somewhat different, they function in basically the same way. Some practice problems are given using the plus (+), minus (−), multiplication (×), and division (÷) symbols followed by "string" problems that include a mixture of operations that all equal 0. Also included are some interesting facts about odd and even numbers. The highlight of the book is a series of riddles and games. The riddles can be solved by doing a mathematical problem and turning the calculator upside down to read the answer spelled out. Also, there are some story problem puzzles whose answers prove interesting in their visual patterns. The games are for two or more players and are varied in their use of the calculator.

EVALUATION: This book provides a delightful introduction for students familiarizing themselves with the calculator. Because calculators will probably play a large part in their upper schooling and adult life, this knowledge is crucial. It is particularly appropriate to acquaint them with the calculator through fun and games, which will spark curiosity. Adler's presentation is clever and his writing style is simple so the reader may take a step-by-step approach to solving the problems, riddles, puzzles, and games. The pictures add to the sense of fun one can have in experimenting with the device. This is a good book for all younger children.

□ □ □

Apfel, Necia H. *Space Station.* Photos; illus. Watts, 1987. 72 pp. ISBN: LB 0-531-10394-3. Age levels: 9–12. Grade levels: 4–7. Series: First Book.

SUMMARY: Beginning with a description of the sensations an astronaut experiences during a rocket launch into orbit and weightlessness in space, the author provides information on the development and history of rockets. Newton's three laws of motion are described and explained through examples. The third law is presented and explained to help readers understand rockets and how they operate.

The purposes of the space shuttle are explained, and life aboard the shuttle during a routine flight is described. Living conditions in zero gravity are given particular attention presumably because children are intrigued about gravity as a phenomenon. The final chapter predicts what space colony life may be like.

EVALUATION: This book provides a lot of basic facts about space stations. The black-and-white photographs and drawings support the text and provide a better understanding of what is being presented. There is no glossary to define the many technical words, and this would have been helpful. This book would be of special interest to students who are particularly interested in space.

□ □ □

Ardley, Neil. *Using the Computer.* Illus. by Janos Marffy. Watts, 1983. 32 pp. ISBN: LB 0-531-04518-8. Age levels: 6–11. Grade levels: 1–6. Series: Action Science.

SUMMARY: Ardley is both a scientist and the author of many science books. In this book he details the basics for using a computer, including on/off procedures, commands, and line-by-line descriptive language using the BASIC program. The student is shown how to make a display; build a program; use already-made programs, graphics, and charts; and is provided line-by-line descriptions of some games. Also included are contents, glossary, and index.

EVALUATION: The large print and exact line-by-line description of the BASIC program make this an easily understood introductory book to computers. The two-page spread for most headings allows the user to follow directions quite easily and computer language is shown exactly as the program appears on the computer, making for a clear and concise presentation of the material. The author compares the computer to a robot parrot. The last section, "Tips and Troubles," assists the user in understanding how a computer responds to certain inappropriate commands and provides aids for getting out of and preventing trouble.

□ □ □

Ardley, Neil, et al. *How Things Work: A Guide to How Human-Made and Living Things Function.* Illus. by Bob Bampton, et al. Simon & Schuster, 1988. 128 pp. ISBN: pap. 0-671-67032-8. Age levels: 9–14. Grade levels: 4–9. Series: Young Readers.

SUMMARY: *How Things Work* is a question-and-answer encyclopedia that informs on the operation of scientific instruments, household equipment, appliances, and even parts of the human body. Among the scientific and technological entries, the reader will find facts about how refrigerators keep things cold and vacuum bottles keep things warm; how different items such as aerosol sprays and lightning rods work; what antennas do; how plastic is manufactured; what solar power is; and what robots are used for. Young people may find the sections on electric guitars and synthesizers, citizen's band radios, television, computers, and holograms particularly appealing. Whatever one's interests or hobbies may be, this encyclopedia-style volume probably touches on them.

EVALUATION: The design of this book is impressive because topics are easy to locate and the facts are well illustrated. Only essential facts are presented. Diagrams, photos, and cutaway illustrations in bright colors add to the information. The table of contents is very detailed, as specific questions are listed, such as "How Is Oil Obtained?," followed by "How Does an Oil Rig Work?" The print is easy to read and comparatively large. Questions are printed in blue, and boldface print is used for the first paragraph of every answer. It is a book that can and probably will be used for reference, but it is really fun to read from cover to cover.

□ □ □

Ault, Roz. *Basic Programming for Kids*. Houghton Mifflin, 1983. 184 pp. ISBN: pap. 0-395-34920-6. Age levels: 9–14. Grade levels: 4–9.

SUMMARY: As computers become an indisputable fact of life, and their uses in the future have only been imagined, the user of computers is becoming younger and younger. This book is geared for those inexperienced with basic computer use, and functions as a hands-on guide in the familiarization with and use of six major home computers on the market. The text helps the reader understand computers, explains the many ways they can be helpful, and tells how much fun they can be. The reader learns by actually doing. Each of the 12 chapters explains a program, gives a practice exercise on what was learned, and presents the answers to particular problems. By the end of the book, the reader should have a good foundation on how a computer works.

EVALUATION: The text imparts a great amount of useful information in simple terms, and each chapter builds on the previous one. Key terms and symbols are presented in bold type with definitions, whys, and how-tos. A cartoon drawing illustrating the appropriate context begins each chapter. There are numerous definitions of terms and examples of how these terms are used. The text is followed by four appendixes, the last of which explains exactly how a computer works. There is a concise index. This is a good book for beginning computer programmers.

□ □ □

Barrett, Norman. *Custom Cars.* Illus. by Rhoda Burns and Robert Burns; photos. Watts, 1987. 32 pp. ISBN: LB 0-531-10273-4. Age levels: 8–14. Grade levels: 3–9. Series: Picture Library.

SUMMARY: Automobiles that have been altered, completely rebuilt, or constructed from a kit are featured in this visual reference book. Color illustrations of two custom cars and their features are graphically explained on a two-page spread. Different types of customized cars are described and shown in photographs, for instance, a lowrider has had its body lowered and seems to hug the road. Vehicles based on models built before 1949 are called street rods. Show cars often reflect a special topic, such as the space shuttle, and are usually only exhibited and not driven.

Attention to bodywork, paint schemes, interiors, and engine detail are some of the concerns of the designers and builders of custom cars. Kits provide the opportunity for all kinds of cars to be reproduced by custom car enthusiasts; the customizing of vans and trucks is also included.

There is a brief history about the hobby of customizing cars. A facts and records page is devoted to explaining the terms *chopping* and *cut 'n' shut*. The former relates to the lowering of a car's roof and the latter to the lengthening or shortening of a car. A table of contents, a glossary, and an index are also included for reference purposes.

EVALUATION: *Custom Cars* shows what imagination and mechanical know-how can produce. Automobiles with engines exposed, a van that looks like a stagecoach, and a vehicle with eight wheels and eight headlights are some of the machines that will intrigue young car lovers. Bright and clear color photographs dominate this book. The captions supplement the text with additional information about custom cars. The text is informative and easily understood; the high interest factor of this book will also attract older readers. Children will be stimulated to create their own designs and perhaps think about building their own dream car some day.

□ □ □

Barrett, Norman. *Trailbikes.* Illus. by Rhoda Burns and Robert Burns; photos. Watts, 1987. 32 pp. ISBN: LB 0-531-10277-7. Age levels: 8–14. Grade levels: 3–9. Series: Picture Library.

SUMMARY: In *Trailbikes,* motorcycle sports are defined as motocross, supercross, and enduro. These sports take place away from main highways in a variety of settings. Motocross, also called scrambling, requires wearing special protective clothing because it occurs on rough, hilly natural terrain. The course is characterized by sharp bends, steep slopes, and many land surfaces, and is usually about ¼ mile long. There are international meets and world championships where prize money can be won. The average speed of a motocross bike is 30 mph, but they can reach a top speed of 70 mph. Supercross takes place in a stadium on a man-made obstacle course so spectators do not have to follow the bikers to watch races, as in motocross. Enduros are long distance runs that take place over deserts and along coastlines. The stages of the enduro events are timed at checkpoints. Riders must repair their own bikes and cannot make major changes during the race. One famous event is the international Six-Day Enduro, a world championship team effort. Other trailbike sports include sand or beach races with all-terrain vehicles (ATVs), grass-track racing, arena trails, and sidecar competitions. In some arena trails driving over an automobile is part of the event. The final chapter of the book contains a short history of trailbikes. A glossary and index are at the end of the book.

EVALUATION: After a short introduction, a two-page colored illustration of an off-road bike is shown, pointing out many technical details. The handlebars are featured separately at the top corner of the page to draw attention to this special part. The diagram allows the reader to learn technical vocabulary associated with biking such as tachometer, throttle, brake cable, and drive chain. The photographs taken at the event site are colorful and full of action. The captions clearly explain what is happening in the photographs and give additional information about trailbiking. The table of contents and index enable readers to locate the portions of the book that interest them. There is also a glossary of trailbiking jargon. An interesting facts and records page discusses world championships in motocross racing and the distance in enduro events. This book could be used with older readers because of its relatively basic vocabulary and the high interest factor.

□ □ □

Becker-Mayer Associates. *Tough Trucks.* Illus. by Nina Barbaresi. Grosset & Dunlap, 1986. Unp. ISBN: pap. 0-448-09884-9. Age levels: 3–5. Grade levels: PreK–K. Series: Fast Rolling Books.

SUMMARY: Part of the Fast Rolling Books series, *Tough Trucks* is a board book toy on wheels. Every "page" shows a gigantic piece of machinery maneuvered by an appropriately helmeted operator, one of whom happens to be a woman. Detailed drawings of a cement mixer, bulldozer, scraper, grader, front-end loader, crane, concrete pump, and cherry picker are all shown. A double snap keeps the book closed and the plastic wheels in place.

EVALUATION: Although this board book appears to be durable, it carries a warning on the back that says it is recommended for children three and over, because it contains small parts. For this reason, it is perhaps unfortunate that the plastic wheels and snaps are attached because little children probably would like the book almost as much without them.

□ □ □

Berger, Melvin. *Computers.* Illus. by Arthur Schaffert. Coward, McCann, 1972, o.p. 46 pp. Age levels: 8–12. Grade levels: 3–7.

SUMMARY: Today's computers are capable of processing information more quickly and accurately than most human beings. The challenging thing to remember is that computers will not operate without being provided information by people.

This book of basic facts about computers explains what they are and how they work. It is one of the first books about computers for children and, because it reveals primary operations, will not be outdated even though the equipment used to perform the operations will constantly be updated.

EVALUATION: This author-illustrator team usually manages to produce a quality book no matter what the topic, but their topics are almost always scientific in nature. Although *Computers* is dated in terms of state of the art technology, the basic, stark illustrations by Schaffert (credited to the use of UNIVAC, the Data Processing Division of Sperry Rand) ensure that the book will probably continue to be a standard beginner resource. However, there is a problem with the text. In an effort to make sophisticated technological concepts perfectly clear, Berger has made the mistake of using only primary grade vocabulary and sentence structure. Therefore, a ten-year-old reader who needs basic information will be turned off by this book because of the overly simple vocabulary. A young child, however, will probably like the book for the same reason a ten-year-old won't: the young reader can actually read the five- and six-word sentences without the aid of an adult intermediary to provide explanations. This can and probably will foster the erroneous idea that the actual operations are as simple as the words. In any event, the book will motivate young children as well as satisfy many of their questions about computer technology.

□ □ □

Blocksma, Mary, and Dewey Blocksma. *Easy-to-Make Spaceships That Really Fly.* Illus. by Marisabina Russo. Prentice-Hall, 1985. 40 pp. ISBN: pap. 0-13-223199-9. Age levels: 7–11. Grade levels: 2–6.

SUMMARY: Instructions for assembling 15 spaceships are included in this book. Each spaceship "recipe" lists the materials that are necessary to be gathered, enumerates instructions, and guides the crafter with diagrams and illustrations. Even a hanger can be built for the flying craft. However, since hooks have to be screwed into walls, the authors state that parents' permission is needed before starting on this particular project. Since these spaceships can fly in one of two ways, direction in launching is also provided.

The spaceships have names such as Zoom-a-Rang, Floating Saucer, Galactic Glider, and Flying Flapper. The Zoom-a-Rang is formed from five paper plates and tape or staples. The addition of a paper cup modifies it into another spacecraft. With two paper cups, the Zoom-a-Rang is transformed again. Attaching recommended landing gear allows this spacecraft to avoid belly landings. Readers are urged to attempt to make their own creations. A table of contents is included.

EVALUATION: Test pilots will surely have hours of fun reading, making, and flying the spaceships found in this "Reading Rainbow" book. The authors provide support with touches of humor, and encourage creativity by pointing out opportunities where it might be expressed. Some knowledge of aerodynamics might be intuitively gained while constructing and flying these spaceships.

The text is clear and exudes enthusiasm. The diagrams are detailed, and illustrate the directions step-by-step. This clever how-to book has pizzazz and will help readers' imaginations soar along with the spaceships that they make or plan to make. More spacecraft construction can be found in the Blocksmas' *Space-Crafting: Invent Your Own Flying Spaceships* (Prentice-Hall, 1986).

□ □ □

Blocksma, Mary, and Dewey Blocksma. *Space-Crafting: Invent Your Own Flying Spaceships.* Illus. by Art Seiden. Prentice-Hall, 1986. 47 pp. ISBN: 0-13-823998-3. Age levels: 7–11. Grade levels: 2–6.

SUMMARY: Instructions are given for creating 24 flying spaceships. Most of the materials needed are generally found around the house, such as plastic lids, cardboard cartons, picnic ware, tape, glue, and scissors. A trip to the hardware store is suggested for the vinyl tubing that is used in some of the space age flyers.

Lists of tools, materials, and decorations are given, followed by instructions and diagrams for making a kit of the OR-BIT parts utilized in many of the models. Simple designs—the Mach-Mite, Tri-Mite, and Petal Ship—are introduced at the beginning.

Builders are given suggestions on how to modify completed spaceships. The Cupnik flyer, for instance, can be converted into a satellite by adding a bottle cap and two plastic straws. The Rim Spinner is heralded as the steadiest spacecraft in the book; licorice or modeling clay can be substituted for the required tubing if tubing is unavailable. Using a large plastic trash bag, tape, and a Hula Hoop, a giant space station—the Hurla Hoop—can be assembled. The authors urge space-crafters to incorporate their own ideas into these designs.

EVALUATION: Many hours of pleasure in constructing and flying spaceships are promised in this how-to book. The step-by-step directions are easy to understand and brief, but touches of humor alleviate the formality of instructions. The names of the spacecraft are clever and appropriate for out-of-this-world aircraft. The diagrams are clear and helpful in providing the proper perspective for assembling. The text effectively communicates the authors' enthusiasm for the highly imaginative spaceships that young readers will ultimately produce on their own. The Blocksmas' *Easy-to-Make Spaceships That Really Fly* (Prentice-Hall, 1985) is a "Reading Rainbow" book that might motivate further assembling of futuristic aircraft and foster more ideas.

□ □ □

Buehr, Walter. *The Story of the Wheel.* Illus. by author. Putnam, 1960, o.p. 48 pp. Age levels: 8–13. Grade levels: 3–8.

SUMMARY: What would life be like if the wheel had not been invented? This question is addressed in the book as the author zeroes in on some areas in which the wheel has been invaluable. Examples of what life would be like without the wheel and how it has changed the ways in which we do many things, such as in transportation, communication, and clothing, are provided. The author also suggests what early wheels may have looked like, what they were made from, and how they were constructed. Much of the book is from a historical perspective.

EVALUATION: This book is timeless because of its subject—the invention of the wheel and the way in which it continues to advance and change all areas of technology. Many bits of information are presented that provide interesting reading and enjoyment. The speculation on how the wheel was actually invented could be turned into a classroom activity, with students writing their own ideas of how this happened. The many illustrations are done in tones of blue, yellow, and green. However, even with the illustrations, the text is long and some two-page spreads have only one small drawing. The print is in blue ink on off-white paper.

□ □ □

Charles, Oz. *How Does Soda Get into the Bottle?* Photos by author. Simon & Schuster, 1988. Unp. ISBN: pap. 0-671-63755-X. Age levels: 8–11. Grade levels: 3–6.

SUMMARY: The author toured a Pepsi-Cola bottling facility to learn how soda gets into a bottle and photograph the process. After learning that soda is the second most popular beverage in America (water is the first), the reader can vicariously enter the syrup room of the factory as the journey begins.

In the syrup room are tanks holding 15,000 gallons of liquid sugar and barrels containing syrup concentrate in such flavors as vanilla extract. After water is mixed with the liquid sugar, it is tested on a refractometer for strength. The syrup concentrate is then added and the finished syrup is tested as above and is also tasted by a technician. This finished syrup mixture then passes through pipes to a flow-mix machine where it is blended (five to seven cups of water per one cup syrup). The carbo-cooler puts the effervescent into the soda and cools it. As soon as the empty bottles are rinsed inside and out and air-dried, they are filled and capped by special machines. They are then warmed so that condensation does not form on the outside of the bottles after they are packed into cardboard cartons to be shipped away.

The photos dominate this factual book.

EVALUATION: The subject matter is one that would be of interest to students and the photos showing employees performing their tasks are current. The names of different kinds of machines may be difficult to pronounce but their uses are apparent. The reader should undoubtedly be impressed by the enormity and complexity of this process; no mention is made of the number of bottles processed a day, however.

□ □ □

Charles, Oz. *How Is a Crayon Made?* Photos by author. Simon & Schuster, 1988. Unp. ISBN: pap. 0-671-63756-8. Age levels: 6–11. Grade levels: 1–6.

SUMMARY: *How Is a Crayon Made?* is the result of the author's tour of the Crayola crayons facility. Using concise word-by-word descriptions, the book begins at the color mill where chemicals are mixed to create different pigments. After the extraction of excess water, the "cakes" are placed on plastic trays, kiln dried, blended for special formulas, and then pulverized. The powder is then sent to the manufacturing facility, where it is mixed with liquid paraffin wax. The wax is poured into mold tables and cooled. Each crayon is then inspected in a special frame; damaged ones are remelted and remolded. Labeling machines wrap and glue the labels. After a quality control inspection, the crayons are crated and moved to the packing machines. A packing machine that can sort colors then boxes them from 6 to 64 in a box. At this point they are shipped to a warehouse where they are stored until delivered to stores.

Brightly colored up-to-date photographs comprise most of the book, many covering an entire page. Due to the nature of the subject matter, the text must necessarily contain more mature vocabulary; however, the photos parallel the text.

EVALUATION: Few items are more ubiquitous in schools than crayons, and this matter-of-fact book explains the process of how crayons are made in as few words as possible. In one year in the United States alone, the number of crayons made, if laid end to end, would circle the globe at the equator four and a half times. Many of the facts presented, particularly about sizes, gallons, or numbers of molds, would be very impressive to young readers. These facts, the tremendous use of crayons by children, and the excellent color photographs would interest both young people and adults.

The process can be understood even if the text is not read, as the photographs correspond closely to the text. The photos are very intense in color: the crayon colors are reproduced as true to those in a crayon box. The photos document workers performing their tasks with phases outside the plant, such as the storage tanks, pipelines, and the quality control lab, included as part of the manufacturing processs.

Although very young children may not be able to read the text, the subject matter and photos would surely be of interest to them.

□ □ □

Crump, Donald J., ed. *How Things Work*. Illus. National Geographic, 1983. 104 pp. ISBN: LB 0-87044-430-1. Age levels: 9–17. Grade levels: 4–12.

SUMMARY: This book distills scientific and technological information about familiar objects and processes, such as the operation of video games, vending machines, radar, lasers, hot-air balloons, and the space shuttle. Through the means of photographs, diagrams, and drawings—including cutaways—much information is given about things that specifically interest children. The book includes everyday devices, such as clocks, toasters, cameras, and neon lights. The inner workings of some musical instruments are explored in addition to highly electronic inventions. Safety factors for emergency situations, including smoke detectors, traffic signals, and laser applications for the blind, are featured. Progress in engineering also applies to recreational vehicles such as bicycles and sailplanes. A glossary, an index, and a list of additional readings are to be found at the end of the book. All of the scientific and technological items are highlighted in the table of contents.

EVALUATION: The illustrations dominate the content of this book. The photographs, diagrams, cutaways, and enlargements ensure reader interest and provide a great deal of information. National Geographic has been effectively selective in its choice of scientific accomplishments because these are certain to appeal to young readers. Multicolored illustrations and bright photography add much to the glossy pages. The glossary appears brief considering the concepts and inventions introduced; for example, *xenon* is listed in the index but is not referred to in the glossary or specifically defined in the text. This dynamic book is exciting to look at and concentrates essential information into a very compact volume.

□ □ □

Darling, David J. *Computers at Home: Today and Tomorrow*. Illus. by Tom Lund and Billy Fugate. Dillon, 1986. 77 pp. ISBN: LB 0-

87518-314-X. Age levels: 9–12. Grade levels: 4–7. Series: World of Computers.

SUMMARY: This book reviews computers and their current household uses. An introduction to computers and how they operate, from the central processing unit to chips, begins the book. The reader is then projected into the year 2010 for a view of what life may be like when computers are used in most facets of everyday life. More detailed descriptions of computer operations, laser disks, databases, and the future of robotics follow.

Current home uses of computers include games, word processing, creating and playing music, and drawing. While computers are making our lives easier, the book states that there is a danger of people becoming less socially and physically active because of them. The book concludes with suggestions for future use of computers in teaching and business and contains a section of questions for discussion. It also has a glossary; words in the text that appear in the glossary are in bold print.

EVALUATION: This book offers general background information about how computers work. The more technical information is followed by speculation on how computers may affect our lives in the home, which is quite interesting.

The glossary is helpful, as are the photos and illustrations. But this book will quickly become dated due to the technological advances in computers. Additionally, this book is average in coverage and appeal.

□ □ □

Darling, David J. *Fast, Faster, Fastest Supercomputers*. Illus. by Tom Lund. Dillon, 1986. 75 pp. ISBN: LB 0-87518-316-6. Age levels: 9–14. Grade levels: 4–9. Series: World of Computers.

SUMMARY: Information about what computers are and do launches this exploration of supercomputers—the inevitable next step in computer technology. Supercomputers can solve complex problems very quickly, and already are being used for designing aircraft, making movies, performing experiments with molecules and atoms, developing spacecraft, and making extensive studies of the stars.

The author asserts that at the time this book was published (1986), over 200 supercomputers were in use, primarily in scientific and military research. At that time only a few companies manufactured supercomputers, which ranged in cost from $4 to $20 million. Scientists are constantly looking for ways to make the complex array of switches on a supercomputer simpler to increase its speed, safety, and efficiency. Already scientists are working on supercomputer technology that will produce one trillion calculations per second compared with 500 million calculations per second performed by current models.

EVALUATION: This book is complex due to its subject matter of supercomputers and their current and future use. A thorough glossary and index are included, as well as a list of suggested reading materials. The photographs clarify concepts but overall the text is difficult for the designated age level to comprehend because of technological detail.

□ □ □

D'Ignazio, Fred. *The Star Wars Question and Answer Book about Computers*. Illus. by Ken Barr. Random House, 1983. 61 pp. ISBN: LB 0-394-95686-9; pap. 0-394-85686-4. Age levels: 9–15. Grade levels: 4–10.

SUMMARY: The question-and-answer format of this book presents many useful facts about computers such as how they were invented, who invented them, and how they work. The information is presented through examples with which children will be familiar: computers are good at arithmetic, remembering information, making decisions, and controlling other machines, but they are not good at thinking on their own. Computer chips today are the same size as the old ones, but scientists are finding ways to put more pathways and transistors onto a single chip, which increases their power. Computers used to be the size of a school gym, but newer ones can fit on the tip of your tongue.

Computer games are wired to computer chips that cause the computer to act as referee, scorekeeper, timekeeper, director (providing the pictures, music, voices, and sound effects), and opponent. The first computer games were played secretly because computers were popularly thought of as serious, expensive machines, so playful uses were taboo. In 1962, Steve Russell secretly invented "Spacewar" on his boss's computer and in 1971, Nolan Bushnell invented "Pong," a game played like table tennis, which made him a millionaire. Bushnell later founded the Atari Company. Because computers follow a limited number of programmed orders, people can become game experts by watching for patterns. The work of Charles Babbage inspired computer inventions, but his projects were too big, too expensive, and too progressive to be successful during his lifetime.

Interesting facts readers will learn from this book include the fact that today's computers use transistors, whereas they once utilized vacuum tubes. People have long been interested in building their own home computers, which in the past cost thousands of dollars; now it can be done for a few hundred dollars. One of the earliest computers, the Mark 1, once gave an incorrect answer to a question submitted by an operator; a dead moth was stuck in one of the switches, and ever since then a mistake in a program has been termed "a bug."

Robots—machines that utilize computer "brains"—are discussed and some of their functions mentioned. Interesting facts about the relative time and various uses of supercomputers are also included.

EVALUATION: This book is divided into a unique question-and-answer format. The questions flow logically from one section to another, and include questions that youngsters are often curious about. This layout also allows the reader to easily select areas of interest. Technological information is found throughout the book, but can be skimmed without losing the focus of the text. Also, the book is tuned in to children's popular interests: the pictures of R2-D2 and C-3PO throughout the book were a wise addition. Technological information is presented in a clear, meaningful, and interesting manner. A short glossary of basic terms is located at the end of the book. Overall, the text makes for very enjoyable reading and provides a wealth of information about computers.

□ □ □

Donner, Michael. *Calculator Games*. Illus. by Lynn Matus. Golden, 1977. 48 pp. ISBN: pap. 0-307-12350-2. Age levels: 9–14. Grade levels: 4–9.

SUMMARY: *Calculator Games* is a collection of games, puzzles, and riddles that can be played on a calculator. In a simple introduction to the calculator the book explains how to operate it, and then moves on to more complicated games. Some of the games can be played alone, but many of them are designed for two or more players. One game walks the player through steps to calculate how old the player is in terms of seconds. Another series of tricks shows how special the number 9 is. There are brainteasers and quizzes for nimble fingers; the back of the book includes the answers to the puzzle and riddle sections.

EVALUATION: Although the book begins simply with a basic introduction to calculator operations, it progresses to much higher and increasingly challenging games. The directions must be read slowly and carefully to catch the purpose of the riddle or trick. The variety of challenges offered would be appealing to a person interested in math facts and calculations. The colorful pictures add a flavor of fun to the book. Some of the pictures are integral to the games being played, such as the calculator maze where problems must be solved correctly in order to move the pig through the garden to the corn. Another example of an essential illustration is the number tree that has 162 numbered apples: the object is to find out why certain apples are on particular branches. The topic is highly appropriate for upper-elementary age and older students as well, since they are exposed to computations requiring the use of the calculator in the classroom. This book provides practice and stimulates an interest in this modern tool.

□ □ □

Encyclopedia of Computers and Electronics. **Photos; illus. Rand McNally, 1985. 140 pp. ISBN: 0-02-689260-6. Age levels: 8–14. Grade levels: 3–9.**

SUMMARY: This comprehensive compilation on electronics and computers for young people includes information about present-day knowledge and technology as well as possible future developments.

Approximately 50 chapters, most about two pages long, give information on the practical uses of electronics and computers: "Light That Cuts Steel," "Electronic Telephones," "Computers and the Police," and so on. Historical information about individuals who made significant contributions to the field is briefly covered. An illustrated glossary and an index are included at the end of the book.

The illustrations and large captions generally dominate, but some chapters contain mostly text. Photos comprise the primary illustrations; there are fewer drawings, charts, or other pictorial information.

EVALUATION: Through the illustrations, it is possible to see at a glance the many uses for computers and electronics. Both photos and drawings are informative, appealing, colorful, and show activity. Sometimes background information is provided in the captions explaining, for instance, why a machine is necessary.

Many topics will appeal to the young reader: videos, transistors, robots, lasers, sound recordings, computer games, drawing on the computer, satellites, three-dimensional pictures, and the like. Even though the book is entitled an encyclopedia, it need not be used specifically as a reference book; it can be a source of enjoyable reading.

Also included are concepts and theories about the way in which electronics may change the way we live. The illustrated glossary, which includes both black-and-white photos and drawings, is most appealing and helpful to the reader. Two photos are two-page spreads; the remainder of the illustrations vary in size from one-half page to approximately four by three inches.

□ □ □

Fisher, Leonard Everett. *The Statue of Liberty.* Photos. Holiday House, 1985. 64 pp. ISBN: LB 0-8234-0586-9. Age levels: 9–15. Grade levels: 4–10.

SUMMARY: This is a retrospective look at the Statue of Liberty, or "Liberty Enlightening the World." Beginning with the dedication of the statue, this book explores the inception of the idea to honor liberty, the history of Bedloe (Liberty) Island, and the background of the artist Bartholdi. Fisher describes techniques used to construct the statue, the groups involved, the various lighting systems used in the torch, fund-raising problems, and the dedication in 1886. The events leading to the statue's rededication in 1986 are also reported. The text concludes with a brief statement about Emma Lazarus, and her heartfelt poem describing the full meaning of the statue.

EVALUATION: This is an easy-to-read text that imparts much factual information about the statue as well as background about the social and political climate of the time. Many interesting bits of trivia attached to the story of Liberty are revealed. The "Liberty Cantata," written for and sung at a fund-raising gala in 1876, was performed for the second time in 1985. Photographs from the archives of the National Park Service, the Statue of Liberty, and the American Museum of Immigration, along with black-and-white photos of the principal people involved in the statue's erection, illustrate this book. This is an interesting book for quick reference.

□ □ □

Fradin, Dennis B. *Movies*. Photos; illus. Childrens, 1983. 48 pp. ISBN: LB 0-516-01699-7. Age levels: 8–11. Grade levels: 3–6. Series: New True.

SUMMARY: People can express themselves through many art forms including painting, dancing, music, and storytelling, to name a few. In this introduction to the movie industry, a comparison is made between the earliest paintings of cavemen and movies. This fairly new invention is less than one hundred years old, during which time movies have been called flickers, films, moving pictures, and motion pictures.

The author suggests that the reader make a movie by drawing a series of pictures showing a pitcher throwing a ball to a catcher, with each drawing showing the progressive movement of the ball. When finished, the pictures are put in sequential order; if they are flipped through quickly, the viewer gets an idea of how movies work.

Beginning with the early 1800s, the history of movie viewing is described. The work of inventors, such as Frenchman Georges Melies, is cited through a chronology of movies from *Cinderella* up to the present day.

The people and the special equipment necessary to make a movie are discussed at length. Another chapter covers the different kinds of movies made such as cartoons and documentaries.

The use of home video recorders to tell stories or document family events leads into a discussion of the film industry as a career. A glossary, an index, and information about the author complete this book.

EVALUATION: The author appears to have done a thorough review in a very concise manner. The historical aspect should be especially interesting to students, with its amusing photographs. Black-and-white photographs naturally predominate, but color is used whenever possible. Photographs or diagrams appear on every two-page spread. The text print is large and the vocabulary is easy to read. A glossary includes both pronunciations and definitions.

The science section, which covers what film is made of and the workings of the camera and sound, has classroom applications. In particular, a diagram of a movie projector and its workings could be shown and explained on a school movie projector. Since many students attend movies and know a great deal about the subject, they could add to the discussion.

□ □ □

Friskey, Margaret. *Space Shuttles*. Photos; illus. Childrens, 1982. 47 pp. ISBN: LB 0-516-01655-5; pap. 0-516-41655-3. Age levels: 6–9. Grade levels: 1–4. Series: New True.

SUMMARY: This book describes the operation of a space shuttle and its advantages for space travel and exploration. The reader discovers that a space shuttle uses its three rocket engines only during lift-off, and that two rocket boosters also help to take the shuttle past the pull of Earth's gravity. A space shuttle can be controlled on board by the commander or pilot or by computers on Earth.

This reusable spacecraft carries payloads, which include communications and weather satellites to be put into Earth's orbit. Spacelab, built by the cooperative efforts of ten European countries, is described. The space program has introduced many new products into our daily lives such as microwave ovens, heat-resistant materials used in the clothing of fire fighters, and pocket calculators, to name a few. Space colonies are foreseen in the future, and space travel may become common someday. Brief descriptions of the living quarters and cargo bay of a space shuttle are given. Also included are a table of contents, a glossary, and an index.

EVALUATION: In this book, facts about space shuttles are clearly written and presented with superior photographs and colorful illustrations, which invite close inspection. This is a wonderful introduction to space shuttles and satellites, and is easy to read with its large print and concise text. This 1982 edition is likely to stimulate questions and further reading about space. Observant children might want to know why the *Challenger, Discovery,* and *Atlantis* are not included in this book about shuttles. Nevertheless, this is a good "first" book for children to learn about aerospace vehicles.

□ □ □

Gaffney, Timothy R. *Kennedy Space Center.* Photos. Childrens, 1985, o.p. 48 pp. Age levels: 7–11. Grade levels: 2–6. Series: New True.

SUMMARY: This book covers space flights at the John F. Kennedy Space Center on Merritt Island, Florida, from a historical point of view. A description of the location and information about the island itself such as the flora and fauna, and the first Native American settlers, is given. The space center's former name, Canaveral, came from Spanish explorers who named it after the fields of sugarcane that grew there. The text moves chronologically from some of the early satellite and rocket launches and the U.S.-U.S.S.R. space race in the 1950s to the early days of NASA and President Kennedy's involvement with it. Three chapters are devoted to the Apollo project, the first moon flight, the Skylab space station, and the meeting of the U.S. and the Soviet Union in space. The last chapter includes some projections about future NASA work. In addition to a table of contents, the book also contains a short glossary and an index.

EVALUATION: This small book with large print covers much historical and present-day information about the John F. Kennedy Space Center and its work. Chapters are appropriately titled and sequenced, beginning with a blast-off to capture the reader's interest. Bright, current color photographs were selected from the collections of NASA, the Kennedy Space Center, and other organizations; these photos will entice the reader. "Words You Should Know" contains explanations of technical terms that the young reader will be able to understand.

□ □ □

Gibbons, Gail. *Flying*. Illus. by author. Holiday House, 1986. Unp. ISBN: LB 0-8234-0599-0. Age levels: 3–6. Grade levels: PreK–1.

SUMMARY: In this lap book a young child is told the history of flight. The book explains that hundreds of years ago man studied flying creatures in hopes of duplicating their ability to soar through the air. In France in 1783 two men went up in a balloon; this was followed by dirigibles, balloons that could be steered. In 1903, the Wright brothers invented a plane that flew for 30 minutes. A little more than 20 years later, Charles Lindbergh flew solo across the Atlantic Ocean. The history of these developments in aircraft constitute the focus of this book.

Flying shows how airplanes, jets, rockets, and balloons are used for scientific and recreational purposes. Crop dusting, weather and traffic reporting, rescue missions, as well as transportation of people and cargo, are some of the functions of aircraft today. At the cutting edge of aviation, rockets and space shuttles carry astronauts into space so scientists can learn more about our solar system.

EVALUATION: *Flying* is clear and well organized. The listener is taken from a time when man merely dreamed of flying to his present accomplishments. Different types of flying machines—balloons, gliders, propeller planes, sea planes, ski planes, as well as jets and helicopters—are illustrated and described in the text.

Each page wisely has no more than one or two sentences. Instead, boldly colored drawings of each flying machine dominate and enhance the appropriateness of this text for its age group.

A wonderful cross-section diagram of a passenger plane covers two pages. The cargo section, cockpit, and passenger areas are detailed. The reader learns that powerful engines propel a jet forward, and that jets smaller than commercial jet airliners not only carry passengers and cargo, but are also used in time of war.

This book should be read to children prior to their first trip on an airplane. A chart in the back of the book, which chronicles important dates in aviation history, benefits the adult reader more than the child. Even though *Flying* was published as recently as 1986, the transoceanic flight of the Concorde is not included in either the text or the chronological chart.

□ □ □

Gibbons, Gail. *Sunken Treasure*. Illus. by author. Crowell, 1988. 32 pp. ISBN: LB 0-690-04736-3. Age levels: 7–11. Grade levels: 2–6.

SUMMARY: Here Gail Gibbons, well known for nonfiction books, tells the story of the finding of treasure on the Spanish galleon *Nuestra Senora de Atocha*, referred to as the *Atocha*. A hurricane sank this treasure-laden ship along with several of its sister ships sailing from South America to Spain in 1622.

The book is divided into several areas. "The Sinking" relates background information. "The Search" gives the chronology of events that finally lead to "The Find" of the sunken treasure. Sonar and metal detectors and new evidence on the location of the wreck helped lead to the actual finding of the ship—laden with gold, silver, and jewelry—in the 1970s and 1980s. "The Recording" discusses how marine archaeologists on the ocean floor work to preserve information for future use. In "The Salvage," boats take on numerous baskets of recovered treasure. "Restoration and Preservation" and "Cataloging" tell about the work done on land to clean, preserve, and record every item; "Distribution" explains to whom the treasure is given. Also included are four more treasure hunts, including the more recent search for the *Titanic*. A final page is devoted to diving and its history.

The book utilizes single-page illustrations with text printed on them or illustrations in frames, ranging from two to several on a page. These detail specifics of the text as well as interesting objects like the marine plant and animal life.

EVALUATION: Most people are curious about sunken treasures, and would be interested in the actual recovery of such a large treasure (worth hundreds of millions of dollars). The inclusion of work done by archaeologists shows children that information about the past will remain forever; this is a very strong point in the book. Through this book the reader can develop a better appreciation for the amount of effort, ingenuity, determination, and patience that goes into such an undertaking.

The text focuses on key events occurring over hundreds of years and the illustrations show details, such as the markings on silver bars or the archaeologists' grid of plastic pipes pinpointing the treasure. An outline describing other sunken ships and the history of diving can act as a beginning reference for further research on these subjects.

□ □ □

Gibbons, Gail. *Up Goes the Skyscraper!* Illus. by author. Four Winds, 1986. Unp. ISBN: 0-02-736780-0. Age levels: 5–9. Grade levels: K–4.

SUMMARY: A skyscraper is erected from the ground up to its finished state. The reader, along with city people who stop by to look through holes in the wooden wall at the site, sees construction machinery such as a backhoe and a payloader clear the site. Before the foundation is dug, a site survey is carried out to study the ground, and architects and other personnel plan and process necessary paperwork. After the digging of the hole for the foundation, steel piles, concrete work, steel columns, floorbeams, and other building materials start to shape the framework of the skyscraper. When the substructure is finished at ground level, the strongest part of the building—the core—is begun with beams in the very center of the skyscraper.

Tower cranes lift materials and assist in the completion of two floors before the cranes are raised to the next level to continue this high-rise construction. Lifts and hoists also help the male and female construction workers, and soon more workers arrive and begin fireproofing the beams and installing the exterior walls and windows. Interior work is performed by finish workers, electricians, plumbers, and carpenters, and a good luck charm—an evergreen—is put on top of the structure. The reader is told that it took 300 people and two years to construct this skyscraper, which will be used for businesses and residences.

EVALUATION: This wonderful book gives a detailed look at a skyscraper's rise on a small city block. Readers will feel as if they are at the site and know exactly what's happening. A full-page diagram of a skyscraper acquaints readers with construction terminology. Large and colorful illustrations—most of them with labels—are informative and pleasing. Workers and passersby reflect a multiracial population. The text is clear and coordinates well with the labeled illustrations. The tall and narrow shape of this oversized book suggests a skyscraper and indeed contains a comprehensive view of the construction of a skyscraper for beginners.

□ □ □

Gunston, Bill. *Aircraft*. Illus. by Gerald Browne; photos. Watts, 1986, o.p. 32 pp. Age levels: 10–13. Grade levels: 5–8. Series: Modern Technology.

SUMMARY: This book reveals many details about aircraft. Aircraft must be equipped with the ability to overcome gravity for lift-off, an engine to fly through the air, and systems to navigate and provide communication with people on the ground. Computer technology has provided the greatest advancement in aircraft over the last ten years. Because of its speed and wing shape, an aircraft has the ability to stay in the air. Changing the airflow over the wings can change the airplane's direction. A good engine must first of all be safe, and then it must be fuel-efficient. Although kerosene is costly, military aircraft are fueled with kerosene, allowing for rapid acceleration, which is crucial in a war situation. Propellers can be found at the front or back of the plane to pull or push it, although today most propellers are in the front.

While pilots now primarily program and monitor the plane's instruments, they still go through extensive training using sophisticated hydraulic flight simulators. Before an aircraft is used for normal flights, it is tested by putting greater stresses on it than it would ever encounter: aircraft are smashed downward, their wings are vibrated and contorted, and frozen chickens are hurled into their windshields at 620 miles per hour. Airlines also carry "the black box," which records all the details of the flight so that the information can be reviewed to see what caused the failure in case of an accident. A discussion of the design goals of warplanes is included. For example, because of their shape and engine types, stealth bombers can fly undetected by electronic detection systems.

Of the 20,000 aircraft built every year, most are smaller models used for business, farming, or sport. Improvements and new designs are always being made, and one day we may fly in aircraft that are controlled by ground personnel. A table of contents, glossary, index, and a "Datasheet" with important firsts are included.

EVALUATION: The writing style seems appropriate for upper-elementary students; it is moderately technical but all the concepts can be understood by the student. The pictures and diagrams are somewhat correlated to the text, but usually go a step beyond, adding much technological information.

□ □ □

Hamer, Martyn. *Comets*. Illus. Watts, 1984. 32 pp. ISBN: LB 0-531-03779-7. Age levels: 7–10. Grade levels: 2–5. Series: Easy-Read Fact.

SUMMARY: Many basic facts are presented in this book for children in elementary grades. Children will learn that because comets are a mixture of snow, dust, and rock, they are similar to dirty snowballs. As a comet passes the sun, it begins to melt. The comet's double tail, stretching for millions of miles, is made of gas and dust. Scientists assume that comets come from left-over matter in space, and classify them as long or short, depending on how much time it takes for them to complete an orbit. Halley's comet is the best known of all comets, and also serves to illustrate the fact that comets are named after the scientist who discovers them.

New comets appear each year, and we now realize that many may have been sighted in the past but were not recorded. Children will be interested to know that although comets are ice cold, they are still warmer than space. Hamer mentions that comets that pass too close to the sun may break up and fall apart, but does not elaborate on why this happens.

EVALUATION: Although this book is part of the Easy-Read Fact series, there are several terms in *Comets* that young readers may not understand, such as *cosmic* and *nucleus*. The book is interesting, but might be more suitable for middle-grade readers who would gain general knowledge with simple vocabulary. The pictures are helpful, but they seem to provide more information than is discussed. A glossary at the end clarifies some terms used in the book. For the more advanced young scientist, there are some additional interesting facts about comets; also included is an index. Because the information is quite dense, this should not be used as an introduction to comets but rather be reserved for readers who already have basic facts on this topic.

□ □ □

Jacobsen, Karen. *Computers*. Photos; illus. Childrens, 1982. 45 pp. ISBN: LB 0-516-01617-2; pap. 0-516-41617-0. Age levels: 6–9. Grade levels: 1–4. Series: New True.

SUMMARY: Computer functions are explained in detail in this early book on the subject. The author begins by asserting that the brain is a computer, and that computers can do things that human brains cannot; however, computers do not think for themselves. Different kinds of computers are introduced; for example, an abacus is a simple computer used for mathematical computations. In 1886 Hollerith invented a machine that could read information off of a card with holes punched in it. These cards were used for taking a census in the United States. Today's computers read similar input from information cards. The computer changes the information into a binary code. Computers send output information to their memory and/or the computer terminal. Microchips are durable and inexpensive, and are found in microcomputers, computer games, and microwave ovens. Although scientists could do all of the mathematics a computer does, computers are much faster, and they never get bored.

EVALUATION: This book gives a good overview of computers. It tells about early forms of computers that led to the development of the modern computer, and also gives some information on how they work. At the end of the book is a list of words one should know about computers, as well as an index. The illustrations and photographs help clarify ideas of older inventions. They also provide concrete ideas of modern computers and their places of use. This book can be appreciated as a supportive text in gathering information on computers.

□ □ □

Jefferis, David. *Lasers*. Illus. by Robert Burns, et al.; photos. Watts, 1986. 32 pp. ISBN: LB 0-531-10164-9. Age levels: 8–11. Grade levels: 3–6. Series: Easy-Read Fact.

SUMMARY: This book provides detailed information about the characteristics and varied uses of lasers. The differences between laser and ordinary light are explained. The light rays of a laser are packed tightly together, are often the same color, and move in a stepwise fashion. Energy, an "active medium" such as a ruby crystal, and a mirror system are needed to make a laser work. Surgeons as well as military personnel utilize this modern tool. Lasers of various types are used to cut through steel, for long-distance communication, in scanning bar codes at the supermarket checkout, and in light shows at discos and rock concerts. Also holograms, which are three-dimensional pictures, are produced via lasers. Besides the experimental lasers mentioned in the text, a page at the end of the book briefly discusses future possibilities for lasers. A table of contents, glossary, and index are included.

EVALUATION: The organization into 13 short sections, each two pages in length, serves to clarify this complex subject. Color photographs and illustrations are also helpful, with captions extending the text. Technological terms are explained in the text, and readers are provided with both common and unfamiliar examples of laser use. The inclusion of a section on the potential uses of lasers adds anticipation of more technological wonders to come.

□ □ □

Kaufman, Joe. *What Makes It Go? What Makes It Work? What Makes It Fly? What Makes It Float?* Illus. by author. Western, 1971. 94 pp. ISBN: 0-307-15767-9. Age levels: 6–12. Grade levels: 1–7.

SUMMARY: As inventions become more complex, it also becomes more difficult to understand how they operate. This author explains how approximately 88 inventions work, which should help answer children's questions about the mechanics and operation of such machines as vehicles and appliances. Many everyday things such as the bicycle, automobile, toaster, ballpoint pen, and television are included. Also explained are some items such as a skating rink, musical instruments, the electric toothbrush, radio, and telephone. Space technology and possible future developments are mentioned as well. The principles of science are included to help explain the technological aspects of the inventions. The book is large in size and contains color illustrations for all items discussed.

EVALUATION: This book, with its many diagrams and explanations of useful inventions, would be appealing to readers of many ages. The asides, such as people asking questions about an invention, complement the text and give additional information. Two-page spreads are used for some significant contributions, for example, the automobile, airplane, bicycle, and ocean liner. Other pages have four inventions per page and generally the layout is varied to make it more interesting. The reader is enticed to find the answers to the questions posed on the endpapers by looking for the answers on the pages indicated.

Although the note to parents recommends this book for children ages six to ten, some of the technical vocabulary may be difficult for younger children, who may need assistance with it.

□ □ □

Larsen, Sally Greenwood. *Computers for Kids: Apple II Plus Edition.* Illus. Creative Computing, 1981, o.p. 73 pp. Age levels: 8–13. Grade levels: 3–8.

SUMMARY: This book explains what a computer can and cannot do: a computer can extend the use of man's brain, but just as a hammer needs a hand to hold it, a computer needs a programmer to tell it what to do. But computers cannot do everything a brain can do; for example, a computer cannot feel emotion. Computers can work faster with data than a brain can.

The Apple II Plus™ is called a microcomputer because of its small size, but size does not necessarily denote function. The Apple II Plus™ uses a computer language called BASIC, which uses words with which most individuals are familiar. A flowchart keeps track of the steps involved in a computer job, but flowcharts have no dead ends, that is, the next step in the program must always be indicated. Do-loops are directions on a flowchart indicating that a certain step has to occur repeatedly. A description of the hardware and practical safety rules for using the computer are given. Directions for saving a program on a diskette or cassette, directions for programming, and directions for printing programs are specified.

EVALUATION: Practically used, this book could benefit young computer programmers. But without an Apple II Plus™ to work directly on while reading through the book it is not very valuable. It is designed to be used as a resource or instruction book rather than for pleasure or information. It is best worked on slowly so the reader does not take too much information at one sitting. There are only a few pictures in the book; it is mostly text, with an occasional computer printout included. The book seems to be quite technical and would best be used with adult guidance. Simple programs, a glossary of statements and commands, and some hints for teachers and parents who would be guiding students through a study of computer programming are most helpful in giving background and for understanding the author's ideas on the most effective use of this book.

□ □ □

Lasson, Kenneth. ***Mousetraps and Muffling Cups: One Hundred Brilliant and Bizarre United States Patents.*** **Illus. Arbor, 1986. 219 pp. ISBN: pap. 0-87795-786-X. Age levels: 8–17. Grade levels: 3–12.**

SUMMARY: Lasson has compiled over 100 patents, dating from 1836 to the present day, in this book. The illustrations are as close to the actual drawings as possible, and the text is excerpted and in some instances edited from the original. Included is an introduction on the historical aspects of inventions and the innovativeness of Americans and their ideas; the procedures for applying for patents is also explained.

The book is divided into nine categories including weapons, patents to aid in cleanliness, farm tools, and things having to do with motion. Each of these topics is preceded by an introduction and a quotation from a famous person.

EVALUATION: The selection of patents from the approximately 5 million issued to date shows thought on the part of Lasson. Some are very humorous, such as a bird diaper and a raincoat with a drain, while others were chosen for their impact on humanity, such as the light bulb and the cotton gin. Still others have historical significance, such as weapons.

The book can be inspirational to both children and adults. The American dream of having an idea patented is clearly obvious, and quotations by Ralph Waldo Emerson, Rudyard Kipling, and others are very appropriate. This book can be used and enjoyed in almost any academic area or as an entity unto itself.

□ □ □

Lipson, Shelley. *It's BASIC: The ABCs of Computer Programming.* Illus. by Janice Stapleton. Holt, Rinehart, 1982, o.p. 46 pp. Age levels: 10–14. Grade levels: 5–10.

SUMMARY: One of the earlier books on computer program creation, *It's BASIC* presents the step-by-step sequence a programmer would follow to develop an original program, including codes used in BASIC, essential commands, and directions to input, output, and print information. Also included are directions for essential arithmetic computations. Readers learn that BASIC stands for Beginner's All-Purpose Symbolic Instruction Code.

EVALUATION: This is not a book for a student completely unfamiliar with computer programming. However, anyone who has seen a computer programmer at work or who has some experience with a computer will find this book easy to follow. Lipson has done a superb job of breaking down the sequential information into step-by-step procedures. Instructions, examples, and illustrations are integrated with the text. Some illustrations show characters who ask questions and make use of the given information. The book is very enjoyable to read, even without a computer handy with which to experiment, although that would be the ideal situation for students reading this particular text.

□ □ □

Markle, Sandra. *Kids' Computer Capers: Investigations for Beginners.* Illus. by Stella Ormani. Lothrop, 1983, o.p. 128 pp. Age levels: 9–15. Grade levels: 4–10. Series: Computer.

SUMMARY: This book imprisons the reader, pictured as a boy of indeterminate age, in a storage room where he tries to escape pursuers named Peterson and McThug. The text explains that although escape isn't a computer game, it is similar in that escape tactics and computer functions require that choices be made. The book then evolves into a history of computers, basic computer operations, and the writing of programs. Clues to help the boy escape through a series of tunnels crop up all along the way. (On page 88 the reader escapes through one of them, laden with treasure.) Also included are jokes, games, and occasional biographical passages about individuals who have had an impact on the invention and use of computers. To check on the grasp of the information presented, mini-mystery tasks are posed throughout, and the answers are revealed on the pages that follow. Robots are mentioned in the final chapter; a glossary of computer terminology is found at the end.

EVALUATION: Probably the only unfortunate thing about this delightfully informative book is its title; a book that uses "kids" in the title will probably be shelved in such a way that only individuals below high school age will find it. The cover illustration, which shows a boy who looks about ten years of age, may also serve to limit the book's audience. While young children will love the book, practically anyone will enjoy it.

Some black-and-white photographs are interspersed with clever and enlightening black-and-white line drawings and diagrams. The book has absolutely no color, but it is not missed. What makes *Kids' Computer Capers* so much fun to read is the amusing way in which a wealth of information is dispensed. To begin with, a mystery story framework is established, which serves as a "grabber." Young people will want to find out what happens to the captured boy, and the author has ingeniously put clear though very technical ideas in their path.

□ □ □

Marston, Hope Irvin. *Snowplows*. Photos. Dodd, Mead, 1986. 48 pp. ISBN: 0-396-08818-X. Age levels: 7–10. Grade levels: 2–5.

SUMMARY: Both light and heavy equipment to remove snow and ice are described in this nonfiction book. Although children enjoy many winter activities, snow and ice can cause many problems. These problems and possible solutions are described. For example, people and cars can become stranded on roads, traffic accidents can occur, and homes can become buried. Special equipment is needed to clear sidewalks, railroad tracks, airport runways, and shipping lanes for reasons of safety.

Some simple equipment is shown, such as shovels, a backpack power blower, snow throwers, and a small plow attached to a motorbike. Snowblowers—a hand-guided one, a rotary blade mounted in front of a locomotive, and massive machines with blade plows, V-plows, and wings that help to push the snow—are displayed. Sanders and snow melters also help keep transportation routes passable in winter. Icebreakers enable ships to travel on waters that are clogged by ice and snow. Also mentioned is a snow-fighting school in New York State that holds snowplowing contests.

EVALUATION: Black-and-white photographs clearly show the removal of light and heavy snow and ice; the text contains technical facts and reads easily. Interesting lightweight and heavy equipment are viewed in operation, illustrating how technology facilitates the removal of snow and ice from walkways, roadways, and waterways. These machines will intrigue many readers.

□ □ □

Maurer, Richard. *The NOVA Space Explorer's Guide*. Photos. Crown, 1985, o.p. 118 pp. Age levels: 9–13. Grade levels: 4–8.

SUMMARY: This book is in the form of a guide to astronauts about to embark on a spaceflight. The imaginary trip begins with a lift-off and orbit around the Earth, then to the moon, toward Venus, up to Mercury, a turn around Mars, past the asteroid belt, to Jupiter, Saturn, Neptune, and Pluto. The characteristics of the planets are detailed and compared to familiar qualities of Earth and the moon.

The principle of a rocket is described and comparison made to several past launches. Life aboard the spaceship is discussed including food and drink, sleeping, and spacesuits. After the travelers reach Pluto, the edge of the frontier of other solar systems and galaxies within the universe are briefly explored.

EVALUATION: A highly informative book, *The NOVA Space Explorer's Guide* leaves the reader in awe of exploration techniques and the magnitude of the solar system. The photographs are breathtakingly beautiful and give an inside perspective to an astronaut's views. The pictures are well coordinated with the text and often clarify size comparisons. It is distracting, however, to have to pause to look at a picture and then resume the text, especially when the final sentence sometimes continues over to a new left page: it interrupts the flow of reading.

The reading is simplified but contains a massive amount of information. The main text itself would be readable to upper-elementary students; however, many of the captions beside the pictures seem very technical.

□ □ □

Moché, Dinah. *If You Were an Astronaut.* Photos. Western, 1985. 24 pp. ISBN: pap. 0-307-11896-7. Age levels: 5–8. Grade levels: K–3. Series: Golden Look-Look Book.

SUMMARY: The opening page of *If You Were an Astronaut* reveals two astronauts floating weightless in "zero g"; gravity as a concept is not mentioned until two pages later. The assumption is that the astronauts are already in a spaceship and in outer space; however, lift-off is shown on the following page. The astronauts are photographed in and out of their space suits in activities such as washing, playing a saxophone, cleaning up, working out, sleeping, and exploring the space environment. Food storage areas, a book of compartments used to store experimental data, a toilet, and medical supplies are also shown through photographs. Interspersed with the photographs are diagrams that show the trajectory of the ship and cutaways that reveal the complex compartmentalization of a spaceship according to the functions carried on therein.

EVALUATION: A book with a tantalizing title such as *If You Were an Astronaut* promises much, but this one actually delivers very little. This is particularly unfortunate because acknowledgements on the title page indicate that the author received professional consultation from experts in the field of space exploration throughout the United States. There is no continuity of events in the book, yet all of the information presented is unquestionably accurate. The short captions accompanying the photographs and diagrams (ranging from one descriptive sentence to a paragraph) are somewhat frustrating because the reader is sure to want to know much more about the topic than is provided here. "Fracturing" of the subject is probably a consequence of the Golden Look-Look series format, which consists of twenty-four pages. Nevertheless, the book will whet the appetite of children who want to know what goes on in space, and will therefore send them to other sources.

□ □ □

Murphy, Jim. *Guess Again: More Weird and Wacky Inventions.* Illus. Bradbury, 1986. 91 pp. ISBN: 0-02-767720-6. Age levels: 7–12. Grade levels: 2–7. Series: Patents.

SUMMARY: This book is divided into chapters on 45 actual patented inventions, including successful and famous ones and others that failed. The reader is invited to guess what the invention was made for through the drawings and multiple-choice answers. Upon turning the page to find the answer, the reader can read the background and history of the invention. The chapters are divided into inventions regarding the animal world, apparel, transport and transportation, pleasure and games, personal hygiene, and odds and ends that do not fall into any particular category. The last chapter consists of five stories about the inventors (Benjamin Franklin, John Fitch, Catherine Beecher, Johann Gutenberg, and Arthur Rayment), the creative processes, and the historical frameworks that surrounded the inventions. An introduction and afterword are included as well as information on how to obtain a patent, and pamphlets and books to write for about inventors and inventions.

EVALUATION: The guessing game approach is inviting and challenging to the reader, and the illustrations accompanying the explanations are often very humorous. In fact, the author suggests that the reader laugh and have a good time using the book. The selection of patents appears to be varied. How inventions came about in American history is discussed in the Introduction through the example of the Kentucky long rifle. This title follows the author's first book, *Weird and Wacky Inventions* (Crown, 1978), which won the American Library Association's Notable Book award, 1979, and was chosen by *School Library Journal* for Best Books of the Year, 1979.

□ □ □

Paige, David. *On the Move . . . Moving a Rocket, a Sub, and London Bridge.* Photos. Childrens, 1981, o.p. 48 pp. Age levels: 8–11. Grade levels: 3–6.

SUMMARY: Taking three special moves—the moving of London Bridge, a submarine, and a rocket—the author explains how these feats were undertaken and some of the complexities involved. Special equipment, such as a hydraulic excavator, a winch, and mechanical jacks, was employed. One piece of equipment—a Marion 8-Caterpillar Crawler—weighed six million pounds and was specially constructed to move eighteen million pounds. The author also explains the snowball effect of such a monstrosity, such as how a special roadway needed to be built to support it.

Historical background and information on the necessities of each of the three moves, as well as some of the planning required, are included. Many black-and-white photographs accompany the fairly large print in the text. A glossary (with pronunciation guides) and an index are included at the end of the book.

EVALUATION: The history and background provided for the three events make for very interesting reading—the reader may even feel armed for the popular trivia games. A footnote explains that any underscored words may be found in the glossary, which is comparatively long; the definitions are relatively short and easy to read, however. Even though some two-page spreads do not contain photographs, there is a limited amount of text on these pages and the margins are wide and varied, so that it is not visually monotonous. The photos are predominantly taken at the actual sites, lending credibility to the retelling of the events.

□ □ □

Provensen, Alice, and Martin Provensen. *The Glorious Flight Across the Channel with Louis Blériot, July 25, 1909*. Illus. by authors. Viking, 1983. 39 pp. ISBN: pap. 0-14-050729-9. Age levels: 4–7. Grade levels: PreK–2. Awards: Caldecott Medal, 1984.

SUMMARY: In the year 1901 in Cambrai, France, automaker Louis Blériot became fascinated by the sight of a dirigible floating high overhead. Blériot thought he could improve the technology of flight. His first experimental vehicle, *Blériot I,* was so small that one could not sit in it. The wings flapped like a bird's, but it managed to become airborne nevertheless. *Blériot II* was a glider plane that was airborne for a moment before crashing into the water. *Blériot III* and *IV* had motors and propellers as well as a double set of wings; Blériot was able to pilot his own plane, fly in a circle, and make a safe landing. *Blériot V* and *VI* had a single set of wings and wheels with which to taxi before taking off.

Finally, after numerous models, *Blériot VII* was a success with its motor, propeller, single wings, and wheels. Blériot became renowned and accepted a challenge to fly across the English channel. Blériot successfully flew the 20 miles from Calais, France, to Dover, England, in spite of the dense fog. The glorious flight of his monoplane lasted 37 minutes.

EVALUATION: The book demonstrates the tenacity and patience that were needed to invent a flying machine. At the beginning of *The Glorious Flight,* we see that Blériot's family—his wife and five young children—attend each attempt at flight. The passage of time is marked by the growth of the children as well as by the improvements in the airplane models. Blériot endures the pain of broken bones to keep his dream alive. His flying machine undergoes several modifications, and eventually Blériot proves himself a first-rate inventor in his successful flight across the English channel in 1909.

A husband and wife team researched, wrote, and illustrated this interesting lap book. The drawings of the people are rigid and one-dimensional, but the flying machines defy land and gravity as they soar and dip far over the heads of the spectators. Land, air, and sea blend into the same dull-hued green as Blériot crouches low in the cockpit and lands successfully in England.

The young reader is sure to enjoy this exciting prizewinning storybook. The technology of the airplane has advanced tremendously in the past century due to the genius and invention of men such as Louis Blériot.

□ □ □

Relf, Patricia. *Big Work Machines*. Illus. by Tom LaPadula. Western, 1984. 24 pp. ISBN: pap. 0-307-11897-5. Age levels: 4–10. Grade levels: PreK–5. Series: Golden Look-Look Book.

SUMMARY: A picture book dominated by illustrations, *Big Work Machines* shows all kinds of equipment used in construction, agriculture, shipping, and mining. Views from every angle and cutaway drawings convey information about monster-sized machines' functions in great detail. Perspective is observed by the illustrator through the use of cars, trees, people, and other elements to show the comparative size of each piece of large equipment. Occasionally the sparse text, which has been carefully designed to support the illustrations, contains specific vocabulary or concepts that will need explanation (e.g., *overburden* is used as a noun). Also, the reader is told that a tunneling machine called a mole is directed by a laser, but the laser beam is not shown.

EVALUATION: *Big Work Machines* will be a big hit with anyone who has a natural curiosity about this kind of technology (the crowds at demolition sites indicate that people are as interested in seeing how structures come down as in how they are constructed). Unfortunately, many older readers may not reach for this picture book because it is in the Golden Look-Look series. While this suggests it is appropriate for younger children, they may have a little difficulty understanding some of the illustrations. In some cases the top of machinery is cut off by the top of the page (e.g.,the height of the sonic hammer and gantry crane can only be imagined because of page size limitations). Nevertheless, the art is good—bright, exciting, and probably as realistic as possible this side of actual photography.

□ □ □

Ride, Sally, and Susan Okie. *To Space and Back*. Photos; illus. Lothrop, 1986. 96 pp. ISBN: 0-688-06159-1. Age levels: 8–13. Grade levels: 3–8.

SUMMARY: Photographs and text take the reader from the crew's quarters to the launch pad, up the launch tower, and then into the 30-story-high space shuttle that is ready for flight into space. Astronaut Sally Ride, the first American woman in space, relates personal experiences. The noise during blast-off, the beauty of Earth as seen from the space shuttle, and the adaptation to space flight are mentioned. Weightlessness in orbit is a challenge, and many daily routines such as preparing meals, eating, drinking, using the toilet, brushing teeth, and sleeping are discussed. Some details of satellite launching, scientific experimentation, and spacewalk preparations are given. This book also includes information about the retrieval of a broken satellite by spacewalkers. Finally, the astronauts' preparation for reentry and their touchdown on Earth at a speed of two hundred miles an hour are described.

The living and working quarters of the crew are shown in a two-page split-view drawing of the space shuttle nose. A glossary and index are included.

EVALUATION: This fascinating book makes the reader feel as though he or she has received a personal invitation to join the astronauts on a rare opportunity to experience space flight. Much interesting information is provided, such as the fact that astronauts are about an inch taller in space because spines are not compressed in weightlessness, drinking glasses and salt shakers aren't used during meals due to lack of gravity, and spacewalking astronauts have some food and drink in their helmets. Questions that children ask astronauts are answered in a way that suggests a comfortable conversation is taking place. Extraordinary color photographs accompany the clear and large text.

□ □ □

Robbins, Ken. *Trucks of Every Sort*. Photos by author. Crown, 1981. 45 pp. ISBN: LB 0-517-54164-5; pap. 0-517-56640-0. Age levels: 5–9. Grade levels: K–4.

SUMMARY: Eighteen different trucks and their features are described in this oversized nonfiction book. A special feature of each truck is highlighted. For instance, the funnellike body of a potato truck, with its wide top and narrow bottom, has a trapdoor and conveyor belt at the back to help unload potatoes with minimal damage. A tanker truck can carry a total of 8,000 gallons of three different grades of gasoline because its tank has separate compartments inside.

Some trucks transport only one type of cargo and have unique characteristics. For example, a glass truck—with its narrow ledges and slanted frames—carries large sheets of glass that are held securely by rubber bumpers. The ice cream truck depends on its refrigeration unit, which is located under the body of the truck and behind the front wheels. A pickup truck and a dump truck are some of the general purpose trucks included. Service trucks such as garbage trucks, tow trucks, and fire engines are also presented. The function of levers, valves, winces, and other mechanical parts is explained.

EVALUATION: With a keen eye and insight into the thoughts of children, the author-photographer emphasizes intriguing aspects about his topic. General and detailed visual descriptions are provided in the large and clear black-and-white photographs. Further clarification is provided by the direct and explanatory text.

□ □ □

Rosenblum, Richard. *Wings: The Early Years of Aviation.* Illus. by author. Four Winds, 1980. 63 pp. ISBN: 0-02-777380-9. Age levels: 7–10. Grade levels: 2–5.

SUMMARY: The book begins with the "flying apparatus" designs of Leonardo da Vinci and ends with the transatlantic flight of Charles Lindbergh. Many milestones in aviation history are recounted: the 59-second flight of Wilbur Wright in 1903, the flight of Louis Blériot across the English channel in 1909, and the successful landing of Eugene Ely on the deck of the U.S.S. *Pennsylvania* in 1911. All of these events improved the technology of aviation, but none spurred its growth as did the outbreak of World War I.

At the beginning of the war, airplanes had been invented only ten years prior. Initially they were used mainly for reconnaissance, although occasionally a pistol was fired by one enemy pilot at another. The war spurred the growth of aviation; the engineering and manufacturing of bigger and faster aircraft would normally have taken years to develop. But due to pressing need, technology advanced in a matter of months.

Several heroes are noted with brief descriptions of each plane and the number of enemy planes downed by the pilot. The United States did not enter the war until 1917, but American pilots were flying for the French and British before that. After the war, pilots had difficulty convincing the government and private sectors of the need for airplanes. Building and technology slowed; commercial flights were unheard of because it was considered too hazardous and too uncomfortable.

In 1925, the U.S. Postal Service bought old fighter planes, built runways, and successfully flew the mail from coast to coast. Then, in 1927 Charles

Lindbergh flew from Roosevelt Field, New York, to Le Bourget Air Field in Paris, France, in 33½ hours. This event piqued the public's imagination around the world and led to a new era of commercial flight.

EVALUATION: *Wings* is a richly detailed book that should not be limited to the young reader for whom it was supposedly written. Each page is full of events and details on the history of aviation that would also stimulate the curiosity of and inform the older reader.

The pen and ink illustrations are well detailed so that each new technological advancement is visually discernible. The positioning of the guns on the cockpit would be of special interest to the young reader, and some illustrations show dogfights. America's role in the airplane's history is clearly noted in the book. Brave, adventurous men such as the Wright brothers, Eddie Rickenbacker, and Charles Lindbergh are covered here, as are the technological achievements of the builders of the Curtiss and Douglas aircraft companies. An index citing key events and people covered in the book is included.

□ □ □

Sarnoff, Jane. *A Great Bicycle Book.* Illus. by Reynold Ruffins. Scribner, 1973, o.p. 31 pp. Age levels: 8–13. Grade levels: 3–8.

SUMMARY: Many diagrams illustrate the components of a bicycle, purchasing a bicycle, its repair and maintenance, and bicycle racing. Particular attention is given to the fact that there are many styles of bicycles but all have the same basic parts. Two diagrams compare a high-rise with a touring bicycle; subsequently some of their parts are shown in close-up and explained more thoroughly. An entire chapter is devoted to the importance of fitting the bike to the rider, and includes how-to information. Maintenance procedures are given step-by-step, and tools needed for maintenance and repair are listed. Instructions for gear and brake adjustment and chain repair are also included. Much space is devoted to removing and replacing wheels and tires; it is cautioned that a flat tire is inevitable. The book ends with some general rules about racing in several types of races, and safety tips. Also included are a bibliography about bicycles and a list of organizations to contact for information about bicycle riding in the United States and abroad.

EVALUATION: Much has been done to organize the book and make the text as easy to read as possible: some explanations are numbered, and the numbers are shadowed for clarification; red and blue stars highlight certain subtitles; blue cursive is interspersed between columns so that it won't be missed; and different type styles are used. Some print looks like waves and follows the lines of the illustrations. In addition to an illustration of tools needed, there is a small illustration of each tool next to the paragraph that discusses it. Some illustrations and text are sideways on the page, and the drawings tend to be very simple, rendering them very understandable.

In general, the book is very appealing and should keep the reader's interest. Humor is present, as in the beginning of the book when the author cautions that the best way to read the book is on the ground.

□ □ □

Scarry, Huck. *Aboard a Steam Locomotive: A Sketchbook.* Illus. by author. Prentice-Hall, 1987. Unp. ISBN: 0-13-000373-5. Age levels: 8–13. Grade levels: 3–8.

SUMMARY: An abundance of information about steam locomotives and how they function is contained in this oversized book. Sketches, captions, and text supply details about past and present locomotives. Mostly European and American engines are featured. Diagrams demonstrate the workings of specific components of a locomotive, such as the throttle, brakes, and pistons. The tasks performed by the fireman and engineer in preparing a locomotive for a run and its operation are shown. The skills and dedication of these men appear throughout the pages of this book. The book covers a day in the life of a locomotive, from the engine being fired up for a run through the hooking on to the train, the actual run, the necessary cleaning at the end of the run, and finally the refilling with water and coal for the next day's trip. The numerical classification of a locomotive's wheels and signaling systems are explained. Also the history of the steam locomotive and some facts about eight famous locomotives are included.

EVALUATION: Readers with a fascination for locomotives will be amply rewarded by this book. In fact, any reader is likely to find something that is engrossing in this informative book that is full of interesting details. The soft sketches effectively capture the billowing smoke, the metal of tracks and locomotive, and the people who work on the railroad. The details in the illustrations are simply astonishing and reflect careful observation by the artist. The text is clear, and the italicized captions convey important additional facts. This is an outstanding book about steam locomotives.

□ □ □

Scarry, Huck. *Things That Fly.* Illus. by author. Derrydale, 1986. Unp. ISBN: 0-517-61657-2. Age levels: 9–14. Grade levels: 4–9.

SUMMARY: This book attempts to use comparisons, descriptions, and analogies to present historical and scientific facts about flight. Technical vocabulary, sometimes combined with poetic narrative, explains the aerodynamics of bird flight, planes, balloons, dirigibles, and helicopters. Historical as well as scientific information is imparted through text and illustrations. Numerous drawings of flying machines provide information through captions and labels. Both propeller and jet-powered civil and military aircraft from the early twentieth century are drawn to scale. Foreign contributions, such as the hot-air balloon in France in 1783, are mentioned. There is also a discussion about the properties of air.

EVALUATION: This book encompasses the history of mankind's knowledge about flight as well as the current technology of air transportation and airports. Some vocabulary words—such as *myriad*—and analogies used may be too sophisticated for some readers, but many drawings and diagrams can be interpreted without the text. However, one two-page diagram is ambiguous because it appears that aircraft are flying below the mountaintops when indeed they would be flying above them. Also, the organization of the material appears to be haphazard: propeller aircraft is discussed, and then helicopters inserted before continuing to jet aircraft. Some material appears to be chronological, but then comparisons are inserted, such as earlier flights, and this breaks up the chronology of events.

Young readers will enjoy the humor in both the illustrations and language. Varisized print helps to hold the reader's interest, and some activities are scattered throughout.

□ □ □

Scarry, Huck. *Things That Sail.* Illus. by author. Derrydale, 1984. Unp. ISBN: 0-517-61656-4. Age levels: 8–11. Grade levels: 3–6.

SUMMARY: Scarry begins this book by showing how much of the Earth's surface is covered by water. Man's dependence on water for transportation, the principle of why things float, and how Archimedes discovered this principle are discussed. From this point on, a chronological pattern is followed beginning with very primitive boats, to boats of ancient times, historical sailing ships such as the one Columbus sailed, the *Mayflower,* and clipper ships. With the invention of the steam engine, it was possible to develop paddleboats, and the propeller was first used on boats in the early 1800s. Following these were the larger steamships and passenger vessels. The historical emphasis ends with the invention of the submarine and hydrofoil.

Interspersed within this chronology of inventions is information about "water sleds," or boats that were towed; oars; sails, tacking, and trimming; warships; locks; navigation; harbors; and docking. All pages are illustrated, with the text printed around the drawings.

EVALUATION: Much information can be gained from the easy-to-understand illustrations, many with captions. Some of the drawings are humorous, which also lends appeal to the book. Some explanations of illustrations are listed using numbers that make the illustrations easy to follow. Diagrams help to explain an invention or a concept, or compare, as in the illustration featuring a duck's webbed feet and a paddle for a canoe.

The text is usually limited to a paragraph per page. Therefore, most of the information is in the captions and diagrams. The drawings are colorful and hold the reader's interest. It may be difficult to read the text in some instances because it is printed on a dark portion of the illustration, but this should in no way preclude an adult from recommending this book to a child.

□ □ □

Simon, Seymour. *The BASIC Book.* Illus. by Barbara Emberley and Ed Emberley. Harper & Row, 1985. 32 pp. ISBN: LB 0-690-04473-9; pap. 0-06-445015-5. Age levels: 7–10. Grade levels: 2–5. Series: Let's-Read-and-Find-Out Science.

SUMMARY: Within the context of a birthday party for a child, Adam, the fundamentals of programming in the BASIC computer language are explained. Almost each page of the book is divided in half: on one side Adam's friends learn how to use the computer to write him a birthday message while the other side explains each part of the BASIC program as the friends use it. What can be done with the program and simple exercises throughout, showing exactly what happens on the computer screen, are reflected in the illustrations in which the children's words are in comic strip balloons, and the computer screen print appears exactly as it would on a real computer monitor. Some asides help the reader to further understand computers; for instance, the reader is told that the computer may appear to be talking to the user but that it really is not that smart. A "Dictionary of BASIC Words" with 16 entries and definitions appears at the end; however, no pronunciation key is provided.

The print is large and explanations are brief; illustrations help show how the computer operates.

EVALUATION: The illustrators are winners of a Caldecott Medal for *Drummer Hoff* (Prentice-Hall, 1976). Upon opening the book it is clear that this book would also appeal to children. There is variety, color, action, and humor in the pictures. The text has been carefully selected to make the subject as easily and quickly understood as possible. The fact that the subject matter is presented within a human interest framework—Adam's birthday party—makes it even more appealing to the young reader.

The author is a highly acclaimed writer of books about science; together with the illustrators he presents a most delightful book about computer language for young children.

□ □ □

Simon, Seymour. *The Paper Airplane Book.* Illus. by Byron Barton. Viking, 1988. 48 pp. ISBN: LB 0-670-53797-7. Age levels: 7–10. Grade levels: 2–5.

SUMMARY: *The Paper Airplane Book* combines information on aeronautics and aerodynamics. Such principles as lift, thrust, and gravity are presented within the context of instructions for how-to experiments. One can start with a simple folded paper design of airplanes and progress to more challenging models. The author suggests improvising on the designs in the book, and holding contests to display imaginative and creative innovations. While the title could suggest the making of a toy, the author treats the subject in a scientific and technological manner.

EVALUATION: This is a good resource book for children who have a particular interest in flying paper airplanes. It tells the reader how to build a variety of paper airplanes with clear verbal and visual instructions. In addition, each design is accompanied by instructions for further alterations and adaptations. There are many designs, and the black-and-white illustrations help to simplify the text. Some safety instructions are included; however, the book should contain a consistent reminder that an adult should supervise airplane-making activities, particularly when it is suggested that a razor blade be used. Multiple editions attest to the popularity of this book; the basic scientific principles contained herein do not change over time. Age ranges printed on the back cover may be misleading as the scientific principles themselves are rather complex. Thus this book could also be of interest to older readers.

□ □ □

Simon, Seymour. *Turtle Talk: A Beginner's Book of Logo.* Illus. by Barbara Emberley and Ed Emberley. Crowell, 1986. 32 pp. ISBN: LB 0-690-04522-0; pap. 0-06-445051-1. Age levels: 5–8. Grade levels: K–3. Series: Let's-Read-and-Find-Out Science.

SUMMARY: Through this book kindergarten and primary grade children will learn that by using little robots that look like turtles, they can create programs to make the turtle move in different directions. Logo is a language of a computer, and by learning logo the computer operator can tell the turtle what to do. Through a step-by-step process, *Turtle Talk* explains how to program a turtle to move in different directions, and gives shortcuts for direction-giving so an operator doesn't have to type an entire word in every instance. Simon also presents a program involving the creation of a star, showing how to put the star in memory so the turtle can draw it on a single command. The last few pages present eight additional turtle programs.

EVALUATION: It is difficult to imagine young children understanding this book without some prior experience with computers. Older readers may enjoy using it as a guidebook when an experienced programmer reads it with them and helps them through the processes. Lack of repetition is a problem; there is too much to recall with only one presentation. However, when the book is used as a reference, computer-interested students would enjoy the challenge of learning turtle talk. Upon mastering the movements of the turtle and the creation of the star, students can try the suggested challenges at the end of the book.

The illustrations are a distraction at times, because they create a busy appearance on the pages. Also, the conversation bubbles in the pictures have cursive captions. Most children do not learn to write in cursive until they are seven or eight, and it seems to take even longer for students to be able to read someone else's cursive writing.

□ □ □

Srivastava, Jane Jonas. *Computers*. Illus. by James McCrea and Ruth McCrea. Crowell, 1972. 32 pp. ISBN: LB 0-690-20851-0. Age levels: 6–9. Grade levels: 1–4. Series: Young Math.

SUMMARY: The purpose of a flowchart is the real focus of this book about computers. An introductory phrase addresses what computers are and what they do. Data are defined, and ways data are processed by a computer are explained. The speed with which computers process and convey information is stressed.

EVALUATION: The information included here is seriously dated, but children in the first three school grades might have fun revealing what they know about computers that this book does not tell. The saving grace of this title is the examples of flowcharts and the invitation to readers to create flowcharts of their own. The illustrations help here because they clearly aid the concept of loops and paths used in charting.

□ □ □

Stephen, R. J. *Cranes*. Illus. by Rhoda Burns and Robert Burns; photos. Watts, 1986. 32 pp. ISBN: LB 0-531-10183-5. Age levels: 5–9. Grade levels: K–4. Series: Picture Library.

SUMMARY: This book features a variety of cranes and describes what they can do and how they operate. After a brief introduction, there are very short chapters on mobile cranes, fixed cranes, and floating cranes. Three kinds of mobile cranes are presented: truck cranes, which include the tow truck; traveling cranes that run on tracks; and the crawler cranes that are able to travel over rough terrain. Most derricks and tower cranes are fixed machines. Floating cranes lift and move objects at harbors, docks, and at sea.

Cranes that load and unload freight, lift heavy materials and equipment in high-rise building construction, and help set up offshore oil and gas platforms are among the machines shown. Most cranes operate by using a system of pulleys and depend on electric or diesel motors; a diagram of the traveling crane covers a two-page spread. There is a short two-page history of cranes from primitive times to the present. A "Facts and Records" page includes pictures of the most powerful cranes in the world, the record lift of a single mobile crane, and a flying crane—a helicopter. Also included are a table of contents, an index, and a glossary of 12 entries.

EVALUATION: Bright, colorful photographs and descriptive captions provide much of the information about cranes in this book. The machines are shown at interesting work sites and from different perspectives. The text, which is simple and brief, conveys pertinent details in a straightforward manner. Young readers will readily learn about cranes from this visual reference book.

□ □ □

Vogt, Gregory. *Space Satellites.* Photos. Watts, 1987. 32 pp. ISBN: LB 0-531-10141-X. Age levels: 9–13. Grade levels: 4–8. Series: Space Library.

SUMMARY: Vogt's book *Space Satellites* explores the brief history of satellites, beginning with Newton's discovery of gravity. The launch of *Sputnik* by the Soviet Union and the United States' unsuccessful attempt to launch its first rocket are covered. Children will be interested to know that *Sputnik 2* carried a dog named Laika into space, and that Laika survived only a few days because of the satellite's insufficient life-support system. Readers will learn the many uses for satellites including scientific experimentation; study of the oceans, weather, and pollution; communication via telephone and television; and surveillance. The book contains a list of important dates, a table of contents, a glossary, and an index.

EVALUATION: The layout of this book is appealing because of the positioning of the illustrations and text: photo sizes and positions are varied on the pages. The text is enjoyable but too technical for children, although the glossary of terms is extremely helpful. Often the pictures are a story in themselves; they do not always correspond directly to the text, but give additional information. "Satellite Scoreboard" is an interesting feature that lists the various countries that have sent satellites into orbit and tells how many satellites are in orbit and how many are now obsolete.

□ □ □

West, Robin. *Far Out: How to Create Your Own Star World.* Illus. by Priscilla Kiedrowski; photos. Carolrhoda, 1987. 72 pp. ISBN: LB 0-87614-279-X; pap. 0-87614-463-6. Age levels: 7–10. Grade levels: 2–5.

SUMMARY: This book is about space creatures, vehicles, and so on, that kids can make. Each creation is designed to perform a function, as well. The book describes how to make and decorate these creations and gives other ideas for creative activity. The beginning of the book lists all materials needed for all projects. These materials are either common around the home, or can be purchased at a modest price. For each of the 17 space creations (for example, Hot Rod Transporter, Astro Shuttle, Meteors, Solar Saucer) there is a description of what is to be created, a list of materials needed, directions how to construct the item with line drawing illustrations, and a color photo of the finished product.

EVALUATION: This book will be popular with students interested in space, crafts, or both. The detailed, step-by-step, illustrated instructions are easy to understand and follow. The color photo of the finished product helps the reader understand what the creation will look like. The suggestions for additional ideas will help spark the children's creativity for modifying each project and personalizing it.

□ □ □

Westman, Paul. *Jacques Cousteau, Free Flight Undersea.* Illus. by Reg Sandland; photos. Dillon, 1980. 48 pp. ISBN: LB 0-87518-188-0. Age levels: 8–11. Grade levels: 3–6. Series: Taking Part.

SUMMARY: This biography of Jacques Cousteau covers his life from birth to age 70. The early years focus on his skill as a swimmer and his education. As his interest in the sea became more intense, and as he used equipment for underwater exploration, he realized that there was a need to improve on what was available for observing and diving. Thus began his contributions to the world of undersea exploration. Cousteau invented and perfected the Aqualung and the diving saucer. His work as an inventor, explorer, and conservationist is discussed. A short synopsis of Cousteau's major contributions and awards is given at the beginning of the book. Brief sketches of both the author and illustrator of this biography can be found at the back of the book.

EVALUATION: This biography is thorough with its explanations of events and circumstances leading up to Jacques Cousteau's inventions and contributions to oceanography. Drawings and photos of his inventions as well as other technological devices used in undersea exploration are included. The photos have captioned explanations. A photo or illustration sometimes fills an entire page, but generally the pictures occupy a much smaller part of the book than the text. The text print is quite large and fairly easy to read, except for the French names and places.

The book offers motivation for young people to achieve. It may encourage those trying to overcome some physical difficulties through its emphasis on Cousteau's long recuperation from an automobile accident when he was a young man.

□ □ □

Westman, Paul. *John Glenn: Around the World in 90 Minutes.* Illus. by Cliff Moen; photos. Dillon, 1980. 48 pp. ISBN: LB 0-87518-186-4. Age levels: 8–11. Grade levels: 3–6. Series: Taking Part.

SUMMARY: Starting with Glenn's birth, the author highlights many events and experiences that may have influenced his career. His schooling, flight training in the service, and love for airplanes are mentioned. He was a leader in mission attacks in World War II, became a test pilot after the war, and then joined NASA's Project Mercury program to become an astronaut. The author concentrates on Glenn's career as an astronaut, his piloting the *Mercury* flight, and his orbiting of the Earth. The descriptions of these events are told in detail. The book ends with an account of Glenn's involvement in politics.

Most of the illustrations are black-and-white photographs; drawings are used primarily to illustrate Glenn's childhood. The print is large and the text reads quite easily, except for some technical words pertaining to space exploration or the government. The last page contains short biographical sketches of the author and the illustrator.

EVALUATION: This book contains biographical information about Glenn as an astronaut and many interesting things that he and other astronauts experienced on their space missions. Readers should find Glenn's descriptions—dust storms blowing across the desert, brush fires, using the night sky for navigation, strange unexplained lights, the activities within the space capsule itself, and a problem with the capsule's heat shield—extremely interesting. They may also be surprised to see how much of his life, from boyhood on, actually prepared the way for his work.

Several photos accompany the text. The first page contains a short excerpt about Glenn's life and a reference to Jules Verne's book *Around the World in 80 Days,* used as a comparison to Glenn's orbiting the earth in 90 minutes.

□ □ □

Westman, Paul. *Neil Armstrong: Space Pioneer.* Photos. Lerner, 1980. 64 pp. ISBN: LB 0-8225-0479-0. Age levels: 8–11. Grade levels: 3–6. Series: Achievers.

SUMMARY: This book is a biography of Neil Armstrong, who was the first person to set foot on the moon. This historic accomplishment occurred on July 20, 1969. The *Apollo 11* space mission had a crew of three men, with Neil Armstrong heading the team as the commander, and Michael Collins and Edwin E. Aldrin, Jr., serving as pilots. When the lunar module, *Eagle,* landed on the moon near the Sea of Tranquility, TV viewers and radio listeners on Earth heard Neil Armstrong's voice speak memorable words that have become as indelible as his footstep on the moon's surface. "Buzz" Aldrin became the second person to step on the moon. Together they planted an American flag, collected lunar rocks and soil, took photographs, and undertook several scientific experiments. After they completed their tasks, they rejoined Mike Collins, who was orbiting the moon miles above in *Columbia*. Splashdown was in the Pacific Ocean, and after the 18-day isolation period ended, a ticker tape parade in New York City awaited them.

Neil Armstrong became a hero after many years of dedication, training, and hard work. At the age of two, he was taken to air races by his father; he first flew in a plane at the age of six. An interest in airplanes and astronomy continued throughout his childhood. By earning and saving money in his teens, he was able to take flying lessons and obtain a student pilot's license before he received his driver's license.

Combat missions during the Korean War, earning a flight engineering degree at Purdue University, and experience as a research and test pilot were some of his activities before he entered the NASA astronaut program. He was the command pilot on the *Gemini 8* mission, which accomplished the historic first space docking. Neil Armstrong displayed piloting skills, courage, and intelligence not only during intensive training but also during the space docking and lunar missions.

A list of the U.S. manned space flights through *Apollo 17* is supplied at the end of the book. The names of the crew members, the dates of the missions, the length of time in space, and the number of orbits around Earth are included.

EVALUATION: This biography of Neil Armstrong is interesting in that it covers all the aspects of his life that led directly to his becoming a space hero. The selection of events from his life includes not only successes but also family tragedy and close calls in flying. The influence of his family, a teacher, and a neighbor in nurturing and fostering Armstrong's interests in aviation are also mentioned. His talents, schooling, flying abilities, and experiences are recounted in simple text, large print, and a straightforward fashion. The black-and-white photographs are appropriate and help document Neil Armstrong's career, space projects that helped make the moon landing possible, and the first exploration of the moon itself. The reader will feel the excitement of the lunar landing and become acquainted with someone who ventured where no one had gone before.

□ □ □

White, Jack R. *How Computers Really Work*. Photos. Dodd, Mead, 1986, o.p. 112 pp. Age levels: 9–13. Grade levels: 4–8.

SUMMARY: Readers are told that before 1965, computers were much bigger and inconceivable for home use due to their high cost and excessive operational expense. Only businesses and government could afford computers then.

Now they are smaller in size and less expensive because of the use of the microchip; consequently the cost of computers has been steadily declining. Modern computers are built on the same techniques as those designed by the British genius Charles Babbage.

Readers are reminded that computers only do what they are told, and cannot think for themselves. Every computer must be able to receive input, store it in memory, follow a program of instruction, do arithmetic, and output information. Throughout the book details are given about computer languages, hardware and software, the displaying of letters and pictures, modems, and future "friendlier" computers.

EVALUATION: Information about computers can be very technical. However, this book manages to present some facts in an understandable way by using examples drawn from everyday situations and familiar objects. For example, the program counter is compared to an odometer, and transistors work like light switches: they are either on or off. Nevertheless, a fourth grade reader would probably have to struggle through this detailed book. If it were assigned as optional reading, it is doubtful that the average nine- or ten-year-old would complete it. In most cases this book will appeal to the serious computer whiz, without regard to specific age level. The book contains a few pictures and diagrams, primarily in the form of blowups of the inside of a computer.

□ □ □

Witty, Margot, and Ken Witty. *A Day in the Life of a Meteorologist.* Photos; illus. Troll, 1981. 32 pp. ISBN: LB 0-89375-450-1; pap. 0-89375-451-X. Age levels: 7–10. Grade levels: 2–5.

SUMMARY: This is a true-life story of a meteorologist who is also a television weatherman. Photographs detail the activities that occupy his workday. Joe Witty is shown using technical equipment and making predictions about the weather. Via computer, he remains in contact with the latest information from the National Weather Service, which collects data from thousands of locations. High-altitude balloons, ships, airports, and weather satellites supply some of the facts that are considered in making local weather forecasts.

Joe Witty's day begins with the knowledge that a severe storm might hit his city in a few hours. There is a special urgency as he tries to ensure that people are notified in sufficient time to prepare for any emergencies. His prime concerns are to be as accurate as possible in his reports, to broadcast warnings as needed, and to advise his listeners about safety precautions to be taken. During this long workday, Joe Witty tracks a major storm and reports current and probable conditions to his television audience.

EVALUATION: Sequential photographs record what meteorologist Joe Witty does during a day at work as he gauges constantly changing weather conditions. The informative text explains that some television weathermen are hard-working professionals, possessing training that includes much mathematics and physics. The functions of sophisticated equipment are presented in a concise and straightforward manner. This book for young readers provides an interesting behind-the-scenes look at how a dedicated television meteorologist determines short- and long-range weather forecasts.

Suggested Reading

Cullinan, Bernice E. *Literature and the Child.* 2nd ed. Orlando, FL: Harcourt Brace Jovanovich, 1989.
Children's literature textbook. Chapter 10, "Nonfiction." Evaluation criteria for nonfiction, and teaching ideas and activities for primary and intermediate grades.

DeLuca, Geraldine and Natov, Roni, eds. *The Lion and the Unicorn* 6 (1982).
A theme-centered annual issue featuring children's science books and science writers.

Elleman, Barbara. "Current Trends in Literature for Children." *Library Trends* 35 (Winter 1987): 413–426.
Genre revolution and quality literature in informational books.

Huck, Charlotte S.; Helper, Susan; and Hickman, Janet. *Children's Literature in the Elementary School.* 4th ed. Orlando, FL: Holt, Rinehart, 1987.
Chapter 11, "Informational Books." Textbook of children's literature. Criteria for evaluation, types of informational books, and curriculum use.

Lasky, Kathryn. "Reflections on Nonfiction." *The Horn Book Magazine* 61 (Sept.-Oct. 1985): 527–532.
Personal comments about nonfiction writing by informational book author.

Lima, Carolyn W., comp. *A to Zoo: Subject Access to Children's Picture Books.* 3rd ed. New York: Bowker, 1989.
Many scientific and technological headings in a comprehensive guide of over 12,000 picture books.

Lukens, Rebecca J. *A Critical Handbook of Children's Literature.* Glenview, IL: Scott, Foresman and Company, 1976.
Chapter 10, "Nonfiction." Literary elements in informational books.

Norton, Donna E. *Through the Eyes of a Child: An Introduction to Children's Literature.* 2nd ed. Westerville, OH: Merrill Publishing Company, 1987.
Children's literature textbook. Chapter 12, "Nonfiction: Biographies and Informational Books." Informational books, guidelines, and science curriculum activities for understanding science-related literature.

Podendorf, Illa. "Characteristics of Good Science Materials for Young Readers." *Library Trends* 22 (April 1974): 425–431.
Commentary by science book author on trade books and science curricula.

Radebaugh, Muriel Rogie. "Using Children's Literature to Teach Mathematics." *The Reading Teacher* 34 (April 1981): 902–906.
Math concepts and recommended picture and storybooks.

Sadker, Myra Pollack and Sadker, David Miller. *Now Upon a Time: A Contemporary View of Children's Literature.* New York: Harper & Row, 1977.
Chapter 10, "Spaceship Earth: Ecology in Children's Literature." Societal issues.

Smardo, Frances A. "Using Children's Literature to Clarify Science Concepts in Early Childhood Programs." *The Reading Teacher* 36 (Dec. 1982): 267–273.
Annotated list of books for third grade reading level and below. Science activities suggestions.

Sutherland, Zena and Arbuthnot, May Hill. *Children and Books.* 7th ed. Glenview, IL: Scott, Foresman and Company, 1986.
Survey of children's literature. Chapter 14, "Informational Books." Criteria for evaluating informational books, informational books in five broad categories, and major informational book authors.

Wilms, Denise M., sel. *Science Books for Children: Selections from Booklist, 1976–1983.* Chicago: American Library Association, 1985.
Best of science trade books from *Booklist* reviews.

Wolff, Kathryn and others, comp. *The Best Science Books for Children: A Selected and Annotated List of Science Books for Children Ages Five through Twelve.* Washington, DC: American Association for the Advancement of Science, 1983.
Annotated listing of over 1,200 selected science books.

Author Index

Title Index

Illustrator Index

Subject Index

Readability Index

The books in this text were evaluated according to the Fry Readability formula (see Preface). The readability levels in this index range from NA (Not Applicable) to 12.

READING LEVEL 1

READING LEVEL 2

READING LEVEL 3

READING LEVEL 4

READING LEVEL 5

READING LEVEL 6

READING LEVEL 7

CITRUS GROWING IN FLORIDA

Fifth Edition

Frederick S. Davies
&
Larry K. Jackson

UNIVERSITY PRESS OF FLORIDA

Gainesville/Tallahassee/Tampa/Boca Raton
Pensacola/Orlando/Miami/Jacksonville/Ft. Myers/Sarasota

Printed in the United States of America. This book is printed on Glatfelter Natures Book, a paper certified under the standards of the Forestry Stewardship Council (FSC). It is a recycled stock that contains 30 percent post-consumer waste and is acid-free.

14 13 12 11 6 5 4 3 2

Library of Congress Cataloging-in-Publication Data

Davies, Frederick Stanley, 1949–

Citrus growing in Florida/Frederick S. Davies and Larry K. Jackson.—5th ed.

p. cm.

Previous edition entered under Larry K. Jackson.

Includes index.

ISBN 978-0-8130-3409-6 (alk. paper)

1. Citrus—Florida. I. Jackson, Larry Keith, 1939– II. Title.

SB369.2.F6D38 2009

634.'304097599—dc22 2009020362

The University Press of Florida is the scholarly publishing agency for the State University System of Florida, comprising Florida A&M University, Florida Atlantic University, Florida Gulf Coast University, Florida International University, Florida State University, New College of Florida, University of Central Florida, University of Florida, University of North Florida, University of South Florida, and University of West Florida.

University Press of Florida
15 Northwest 15th Street
Gainesville, FL 32611-2079
http://www.upf.com

Contents

Figures

Plates

Plates follow page 142

Plate 22. Leaf notching caused by adult blue-green weevil feeding on citrus leaves.

Plate 23. Adult blue-green weevil. Weevil larvae feed on citrus tree roots.

Plate 24. Adult Diaprepes weevil.

Plate 25. Citrus rust mite (CRM) late-feeding damage on orange fruit called russetting.

Plate 26. Citrus red mite feeding on citrus leaves.

Plate 27. Citrus canker bacterial disease symptoms on citrus leaves.

Plate 28. Citrus canker bacterial disease symptoms on a citrus stem.

Plate 29. Greening bacterial disease symptoms on a mature grapefruit tree.

Plate 30. Orange dog caterpillar feeding on a citrus leaf. This is the larval stage of the swallowtail butterfly.

Plate 31. Sooty mold fungus on citrus leaves. This fungus is often found in association with insect feeding.

Plate 32. Sooty mold fungus on citrus fruit (black regions only).

Tables

Preface

Ten years have passed since this book was last revised. The Florida citrus industry has undergone some very significant changes during this time. In 1990, Florida citrus was produced on about 732,000 acres. Acreage and production increased significantly to 856,000 by 1996, but acreage stands at about 575,000 acres in 2009 and continues to decline. The reintroduction of citrus canker, new introduction of citrus greening, and the devastating hurricanes of 2004–2005 were largely responsible for this significant decline in production and equally important changes in cultural practices and production and marketing philosophies.

Citrus-growing technology has continued to develop and improve, and this latest edition addresses the many changes that have occurred over the past 10 years. Some new problems have emerged and most of the old ones are still with us. Among the new problems are citrus canker and greening bacterial diseases, which have dramatically changed cultural operations. Groundwater contamination concerns have brought about changes in fertilizer recommendations and resulted in the loss of certain herbicides for use in areas of the state prone to leaching. New methods of fertilizing and spraying are being developed to minimize environmental concerns without reducing yields or fruit quality. Problems continue with citrus blight, citrus tristeza virus (CTV), Diaprepes weevil, foot rot, and other disorders, but research is uncovering better ways to deal with them. Some new early orange cultivars are available to growers and several new rootstocks are being developed to adapt trees to various soil types and disease problems. Overall, a great deal of new information has been developed over the past 10 years and this revision attempts to summarize and incorporate all the latest technologies and strategies for growing citrus in a competitive world market.

As many readers know, this book was originally written by Louis W. Ziegler and Herbert S. Wolfe. Dr. Ziegler passed away in 1976 and Dr. Wolfe in 1991. Larry K. Jackson extensively revised the work of Ziegler and Wolfe for the third edition. In doing so, many of the chapters were completely rewritten while others needed only minor revision, which resulted in a style disparity among several of the chapters. Frederick S. Davies serves as the senior author and Dr. Jackson (retired) as the second author for the fifth edition, and the authors have continued to address format issues while bringing the material up to date. The new edition also includes color plates of some of the major citrus pests and diseases to help with their identification. We hope that the new version will be useful to citrus growers and homeowners alike.

Acknowledgments

The technical accuracy of this text is ultimately the responsibility of the authors, and we regret any errors or omissions. In order to produce the most usable and accurate text possible, we requested the help of various experts in the many aspects of citrus culture covered in this text. We gratefully acknowledge their contributions, criticisms, and suggestions.

W. S. Castle, Professor of Horticulture, Citrus Research and Education Center, University of Florida, Lake Alfred, Florida

S. H. Futch, Multi-County Extension Agent, Citrus Research and Education Center, University of Florida, Lake Alfred, Florida

J. Chapparo, Assistant Professor of Horticulture, Department of Horticultural Science, University of Florida, Gainesville

M. Kesinger, Chief, Bureau of Budwood Registration, Florida Department of Agriculture, Winter Haven, Florida

G. A. Moore, Professor of Horticulture, Department of Horticultural Science, University of Florida, Gainesville

R. P. Muraro, Professor of Food and Resource Economics, Citrus Research and Education Center, University of Florida, Lake Alfred, Florida

T. Obreza, Professor of Soil and Water Science, University of Florida, Gainesville

M. Ritenour, Associate Professor of Horticulture, Indian River Research and Education Center, University of Florida, Ft. Pierce, Florida

M. Rogers, Assistant Professor of Entomology, Citrus Research and Education Center, University of Florida, Lake Alfred, Florida

R. E. Rouse, Associate Professor of Horticulture, Southwest Florida Research and Education Center, University of Florida, Immokalee, Florida

T. Spann, Assistant Professor of Horticulture, Citrus Research and Education Center, University of Florida, Lake Alfred, Florida

J. P. Syvertsen, Professor of Horticulture, Citrus Research and Education Center, University of Florida, Lake Alfred, Florida

We also very much appreciate the cooperation and assistance from the University Press of Florida. Their helpful suggestions in preparing and editing have improved the overall quality of the text. In addition, we appreciate the fine efforts of Katrina Vitkus of IFAS Educational Media and Services in producing some excellent drawings.

We extend our deep appreciation to Kathy Snyder for her help with typing and editing the text and for her patience during the many revisions. She certainly has made the entire task much easier and more enjoyable.

We also thank our wives, Lisa Davies and Julie Jackson, for their love and support during this project.

F. S. Davies
L. K. Jackson
Gainesville, 2009

ONE

World, United States, and Florida Production

EARLY HISTORY AND ORIGINS

The origin of most citrus species was probably on the warm southern slopes of the Himalayas in northeastern India and adjacent Burma. Many years ago primitive, ancestral forms of the orange, shaddock (a grapefruitlike type of citrus), and some mandarins were transplanted by natural dispersal or humans over the Himalayan passes into western China. Secondary distribution centers developed in southern China and Indochina. Lime, lemon, and citron had origins south of the mountains, as did some of the mandarins, and spread southward into India and eastward into the Malay Archipelago. The trifoliate orange (*Poncirus*) and the kumquat (*Fortunella*), which are closely related to citrus, originated in the eastern and northern parts of China. *Microcitrus* and *Eremocitrus*, also part of the true citrus group, originated in Australia, probably from an ancient citrus relative.

The sweet orange likely developed in southwestern China. Like the shaddock, it is not known anywhere in the truly wild state. Quite possibly it originated from other citrus species under cultivation. It is impossible to say how long it has been cultivated in China because written records do not go back far enough. The earliest mention of oranges occurs in a book compiled in the sixth century B.C., the *Shu-ching*, which purports to have been taken from ancient records going back to before 2000 B.C. Chinese historians have long been aware, however, that any records dated earlier than 1000 B.C. (except inscriptions on bones used in divination) are highly suspect and that this particular reference cannot be credited to a period earlier than 600 B.C.

In the *Chou-li*, which deals with the Chou regime around 700 B.C. but was actually compiled some 400 years later, a distinction is made between sweet and sour oranges, whereas the earlier *Shu-ching* speaks only of oranges. There are other reasons to think that the sour orange only reached the Yangtze

Valley around 400 B.C., whereas sweet oranges may have been cultivated from much earlier times. Mandarins are first mentioned in Chinese literature about 200 B.C. The shaddock was included in the earliest mention of oranges, but citron was apparently not known until the fourth century A.D., and lemons not until the 10th century.

India seems to be the region where lemon, citron, and some species of mandarins developed. Sanskrit literature of the eighth century B.C. mentions both citron and lemon, but no Sanskrit name is known for the shaddock, and that of the orange dates only from the first century A.D. However, oranges had reached India long before this. The Sanskrit name for the orange, *nagarunga*, evolved into *naranj* in Persian, *aurantium* in Latin, *narañja* in Spanish, and *orange* in English.

The citron and lemon have greater written evidence of ancient origins than the orange, although they have not necessarily been cultivated longer. References to citrons being present in Egypt in 1500 B.C. have come into question. The lime is known in the wild only in the Malay Peninsula; it apparently spread westward into India and eastward into the Pacific islands.

The first citrus fruit present in Europe was undoubtedly the citron, which is prized for its fragrant rind. Naturalists who accompanied Alexander on his conquest of Persia and northern India brought back descriptions of the citron tree and its culture, which were made widely known by Theophrastus, the Father of Botany, around 300 B.C. This was not, however, the first experience of the Greeks with citrons. At least 50 years before this time, citrons were being imported into Athens from Persia. Since the Greeks initially called it the Median apple, it seems probable that they had known about it when the Medes ruled Persia, before the sixth century B.C. Greek colonists apparently introduced citron trees into Palestine about 200 B.C. Eventually the citron replaced the cedar cone as the fruit of the tree authorized for use in the Jewish Feast of Tabernacles. Tolkowsky, a noted citrus historian, advanced cogent arguments supporting this substitution which was ordered by Simon the Maccabee in 136 B.C. The significance of this substitution lies in the probable explanation of the origin of the word citrus. The Greek word *kedros* originally meant "the cedar," but apparently as the result of the substitution of citrons for cedar cones in this ceremony, the Palestinian Greeks began to use the term "cedar apple" (*kedromelon*) instead of "Median" or "Persian apple." The Greek *kedros* became Latinized as *cedrus*, which was changed to *citrus*. By the time of Pliny (A.D. 70), citrus was synonymous with the *citron* tree. Our word *citron* is derived from *citrus*, and *citreum, malum citreum*, or *malum medicum* were Pliny's names for the fruit. Cultivated in Asia Minor and Italy in the first century A.D., the citron was being grown in Greece a century later.

The fruit seems to have been widely used in Muslem medicine in Egypt and Spain before A.D. 900, and perhaps even long before this time.

The sweet orange must have reached India from China well before the beginning of the Christian era, for in the first century A.D. sweet oranges were called "Indian fruit" by the Romans. Certainly by the end of that century, oranges were being imported into Rome, probably from Palestine and Egypt, where plantings had been likely established by seeds from India. The time required for the journey from India to Italy was much too long for orange fruit or trees to survive. There is evidence that orange trees grew in southern Italy at that time, as suggested by frescoes found in Pompeii. By A.D. 300, oranges were being grown in elaborate orangeries, which protected them from cold damage. In the Dark Ages following the fall of Rome, culture of citrus trees declined or vanished.

In Moorish gardens of southern Spain around A.D. 900, oranges were quite common, but it is not certain whether these were sweet or sour oranges. However, sweet oranges were certainly abundantly available in Baghdad and Cairo, and Spanish Moors imported a great variety of exotic plants from Iraq and Egypt to make their gardens more spectacular. Therefore, it seems likely that Spaniards brought in both sweet and sour oranges.

Sweet oranges were undoubtedly cultivated in southern Europe long before Vasco da Gama reached India in 1498. The most direct evidence is a letter written in 1483 in which Louis XI of France asked that sweet oranges be sent to him from Provence, but there is also considerable indirect evidence. It is well known that Columbus took orange, lemon, and citron seeds from the Canary Islands with him to Hispaniola on his second voyage in 1493. While there is no statement as to whether the oranges were sweet or sour, we do have Oviedo's testimony that barely 30 years later, sweet orange trees were widely distributed in Hispaniola. Furthermore, on da Gama's return from India, his men reported that the sweet oranges they found were superior to those they had at home. These superior sweet orange types were brought back to Portugal and were more widely cultivated than their present selections. Consequently, the sweet orange was cultivated throughout Europe by the 17th century, primarily in orangeries. There is evidence of citron culture in houses covered with mica glazing in Italy as early as the first century A.D. By the 14th century, orangeries had become popular in northern Italy. The Portuguese finally introduced oranges directly from China to Europe in 1640; these fruit had much better quality than those brought from India. Oranges originating in China soon replaced Portuguese types as the preferred fruit throughout Europe.

Citrus and its relatives, with the exception of grapefruit, are not indigenous to the New World. They were brought to Hispaniola by Columbus on his second voyage in 1493. Columbus also is credited with introducing pigs to the New World. By 1525, oranges and shaddocks were widely grown throughout what was then known as the Americas. Spanish explorers were responsible for rapid dispersal of citrus to Central America (Juan de Grijalva, 1518), Bermuda (Juan Bermudez, 1522), and Florida (unknown origin, 1513–65).

WORLD PRODUCTION REGIONS

Most of the world's citrus production occurs between zero (the equator) and 40 degrees north-south latitude. Citrus fruits are produced in more than 140 countries according to FAO (Food and Agriculture Organization) statistics, although the top 10 countries produce over 80 percent of the world output. Citrus acreage and production have increased significantly from about 23 million metric tons in the early 1960s, to 48 million in 1975, to over 75 million in 1991. Moreover, the trend continued, with production increasing from 89 million tons in 1996 to 92 million tons by 2003. Figure 1.1 shows citrus production for the major citrus-producing countries for the 2006–2007 season (source—Citrus Summary, 2006–2007, Florida Agricultural Statistics Service, Orlando, FL).

Over the past 30 years, the relative position of the major producing countries has also changed. In 1975 the United States was by far the largest producer, followed by Brazil, Japan, Spain, Italy, Mexico, Israel, India, Argentina, and China. Currently, Brazil is the largest producer, followed by the People's Republic of China, then the United States (fig. 1.1). Collectively, Brazil and the United States produce most of the world's sweet oranges, so they also control the frozen concentrate juice market. Brazil was able to overtake the United States in production during the 1980s because of strong government support of its citrus program and severe freezes during the 1980s in the United States, especially in Florida, which greatly reduced United States production. Currently (2009) citrus canker and citrus greening diseases and continued urbanization will likely further reduce citrus production and acreage in Florida.

With the gradual shift toward capitalism, citrus production in China has increased substantially, with China becoming the number two citrus producer worldwide for the first time. China is by far the largest producer of mandarins, primarily satsuma. It is also a major producer of oranges. Most production is for fresh fruit and local consumption. However, the Chinese are interested in increasing their share of the Hong Kong and export market

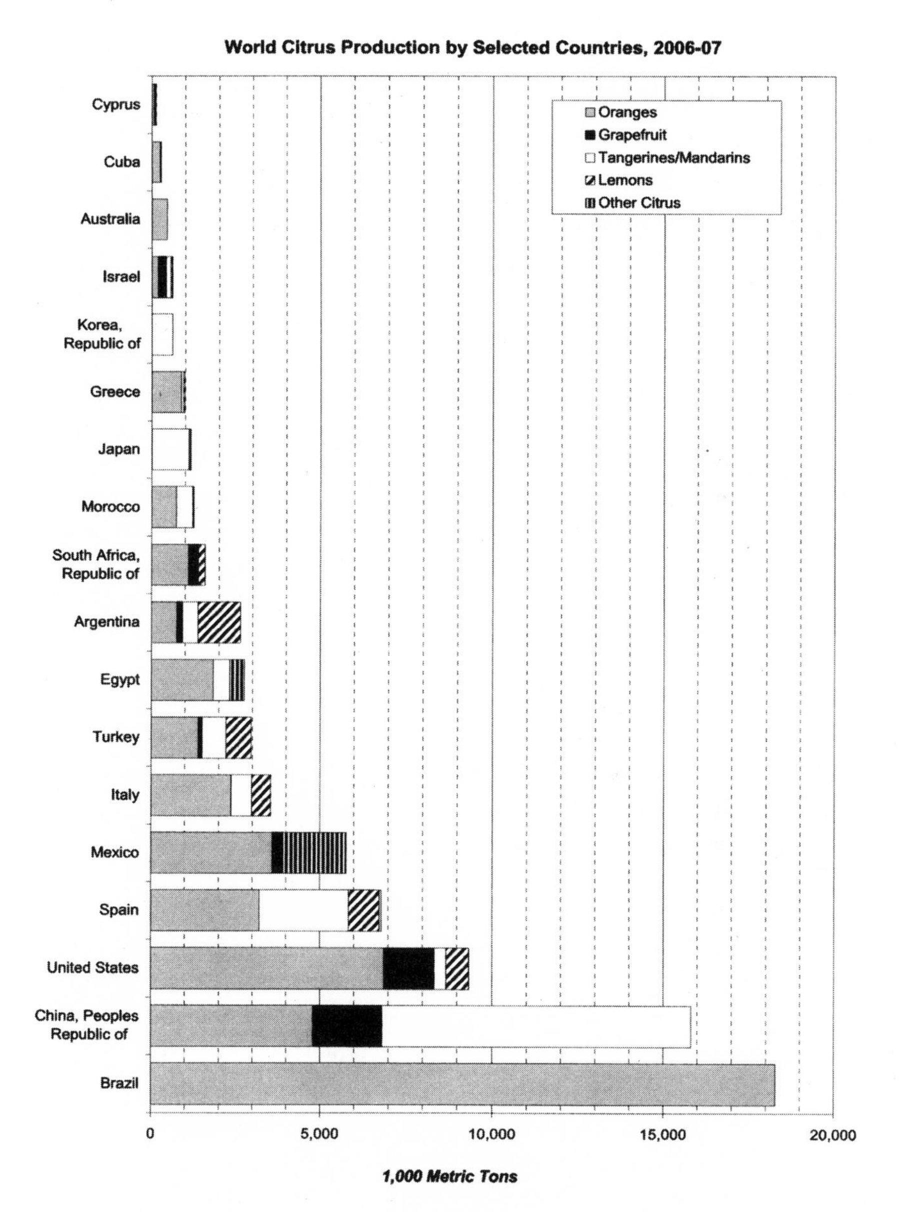

Figure 1.1. World citrus production by selected countries, 2006–2007. Source: U.S. Department of Agriculture, Florida Field Office, Citrus Summary.

where their fruit will compete with fruit from Australia, South Africa, and the United States and in increasing their processing capacity. Major shifts in government policy and low production costs have been largely responsible for these large increases in production.

Production in Spain has increased somewhat over the last five years with gains in orange at the expense of lemon production. Spain is a major producer of fresh oranges, mandarins, and lemons, primarily for the European market. It has remained competitive through innovative marketing and development of new cultivars. Many of the groves are small and have been family owned for hundreds of years. Competition from other countries in the Mediterranean region such as Morocco, Italy, Greece, Turkey, Egypt, and Israel is very intense. Mexico has moved into the fifth position, largely due to its government's move to expand citrus acreage and increase exports of juice, especially to the United States, following passage of the North American Free Trade Agreement (NAFTA). Previously most of Mexico's production was for the fresh fruit market and domestic consumption. Mexico is a major producer of oranges and limes. Other notable citrus producers include Italy, Turkey, Egypt, Argentina (the world's major lemon producer), South Africa, and Morocco.

Decreases in production have occurred in Japan, Israel, and Cuba. The economic situation in Japan and Cuba has forced a reduction in acreage, while in Israel competition from its Mediterranean neighbors and severe water shortages have caused a decline in production.

It is unlikely that world production will increase significantly for the foreseeable future. Overproduction has produced low prices and intense competition in the international markets. The intense competition among producers has forced some countries, such as South Africa, to develop successful niche markets for their citrus fruit. Consequently, it is likely that the relative position of producers will continue to change, with the exception of the three top countries—Brazil, China, and the United States—which will probably maintain their dominant positions.

UNITED STATES PRODUCTION REGIONS

Commercial citrus production in the United States occurs primarily between 25 and 35° north latitude, although within these regions, low winter temperatures render many areas unsuitable for citrus production. United States production accounted for 28 percent of world production in 1975 but declined to 15 percent of the worldwide crop in 1990–91. The United States share of world production remained around 15% through 2006. Severe freezes of 1983, 1985, and 1989 significantly reduced acreage and production in Florida and Texas, virtually eliminating the 1985 and 1989 crops in Texas. In 1998 a moderate freeze caused considerable damage to South Florida groves. Freezes in 1990 and 2007 severely damaged citrus crops and decreased production in California. Rampant urbanization has markedly decreased production in Arizona. Production in Texas has recovered since the 1989 freeze and has been stable through 2006.

The situation for Florida is one of confusion at this time (2009). Three hurricanes ravaged the state in 2004, followed by yet another in 2005. This caused a 40 percent reduction in production in the 2005 and 2006 seasons. Recovery from the storms is likely for many groves, but diseases now are posing a significant threat to future production. Minor outbreaks of citrus canker became major outbreaks as the bacteria were spread by the hurricanes of 2004 and 2005 and the disease has rapidly spread statewide. Eradication is now impossible and canker will soon likely be found in every area of the state where citrus is grown. Canker control measures will increase production costs and reduce yields, making the crop unprofitable (or too difficult to grow) for some growers, with a resultant decline in acreage and production (Table 1.1).

Soon after the hurricanes of 2004 and 2005, citrus greening disease was found in South Florida. The bacterium causing the disease is spread by the Asian citrus psyllid, which has become widespread at this time (2009). The disease is currently found in most of the state's citrus counties and is expected to spread to all citrus counties in the near future. Greening is a debilitating disease which produces tree decline followed by death. No cure is known at this time and none is on the horizon. This disease will no doubt bring about a greater decline in acreage and productivity for Florida for the foreseeable future.

While the new diseases will likely decrease Florida citrus production, the state should continue to have the greatest United States production, followed by California, Texas, and Arizona. Louisiana and Hawaii also produce minor amounts of citrus.

Citrus Production in California

HISTORY

Citrus seeds were probably brought from Mexico to California by Franciscan monks around 1769. The first orchard was likely planted around 1803 by Father Tomas Sanchez in San Diego. During the 1830s several other commercial groves were planted, all probably consisting of seedling trees. The California industry developed more slowly than the Florida industry, and H. J.

Table 1.1. Citrus production for the major citrus growing states in the United States (in thousands) 2006–2007

	Fresh	%	Processed	%	Total Boxes[a]	Total Tons
Oranges						
Florida	6,438	5.0	122,562	95.0	129,000	5,805
California	26,300	58.4	18,700	41.6	45,000	1,688
Texas	1,472	74.3	508	25.7	1,980	84
Arizona	182	60.7	118	39.3	300	11
Grapefruit						
Florida	10,962	40.3	16,238	59.7	27,200	1,156
California	3,500	87.5	500	12.5	4,000	134
Texas	3,462	48.8	3,638	51.2	7,100	284
Arizona	100	100.0	0	0	100	3.75
Tangerines[b]						
Florida	3,072	66.8	1,528	33.2	4,600	219
California	2,200	75.9	700	24.1	2,900	110
Arizona	178	59.3	122	40.7	300	13
Lemons						
California	9,900	61.8	6,100	38.2	16,000	608
Arizona	1,895	75.8	605	24.2	2,500	95

Pounds/box	FL	CA/AZ	TX
Oranges	90	75	85
Grapefruit	85	67	80
Tangerines	95	76	n/a
Lemons	n/a	76	n/a

Source: Citrus Summary, 2006–2007, Florida Agricultural Statistics Service, Orlando.

Webber, a noted California researcher, stated that there were "17,000 orange trees, 3700 lemon trees in 1867." The industry began expanding with the introduction of the navel orange (the most widely planted citrus type in California) in 1870. The industry moved northward in the late 1800s and further expanded as the transcontinental railroad permitted shipment of citrus to other areas of the United States.

The California citrus industry continued to expand during the 1900s, generally avoiding severe freeze losses as experienced in Florida and Texas. However, citrus tristeza virus (CTV) was discovered in 1939 and caused the loss of millions of trees on sour orange rootstock. The California industry addressed the CTV problem by establishing rigid budwood registration programs and transferring to CTV-tolerant rootstocks such as Troyer and Carrizo citranges. Citrus production has varied between 2.2 and 3.7 million tons over the past 20 years, with the exception of the severe freeze years of 1990–91 and 2007. California is a leading producer worldwide of lemons and navel oranges for the fresh market and also produces Valencia oranges, grapefruit, tangelos, and tangerines.

CURRENT PRODUCTION

California citrus is produced in five distinct growing regions, including the desert (Riverside, Coachella, San Diego), coastal (Ventura), and central valley (San Joaquin) regions. Citrus was grown on just over 250,000 acres in 2006–07, which produced 2.5 million tons of citrus. Because of California's Mediterranean-type climate, about ¾ of its production goes to the fresh fruit market.

Long-term projections suggest that acreage and production in California will decrease at a slow rate over time. The most likely causes for decline are urbanization resulting from increased population, increasing production costs including land and water, and more stringent pesticide regulations. As in many other citrus regions, much of the most productive land has been developed and any new plantings have to be made on more marginal sites. Recently, California has introduced several new tangerine and tangerine hybrids that have been widely planted and show considerable potential for the fresh fruit market.

Citrus Production in Other States

Citrus was introduced into what is now Arizona in about 1707 by Father Kinno. The industry grew very slowly and for many years consisted of small plantings in southern and central Arizona. By 1965 there were 40,000 acres of citrus located in the central (Salt River Valley) and Yuma areas. Acreage increased to 61,200 by 1975 but has steadily declined since then, primarily due to urbanization. In 2006–07 there were about 23,000 bearing acres in production. This acreage produced 121,000 tons of citrus in 2006–07. Yuma is a major production region for desert lemons but also produces mandarins and oranges. The central region produces grapefruit, oranges, mandarins, and some lemons. Arizona citrus production is very much directed toward the fresh market.

The commercial citrus industry in Texas is relatively new; production began in the lower Rio Grande Valley in the 1880s on the Laguna Seca Ranch near Edinburg. Early plantings consisted of satsuma mandarin on trifoliate rootstock, but these failed because the rootstock was not adapted to the alkaline soils that are so prevalent in Texas. Sour orange was first used as a rootstock by Charles Volz in 1908 and has remained the primary rootstock in Texas. The principal cultivar grown at that time was Louisiana Sweet orange. Commercial production began in 1910 and became sizable by the 1919–20 season.

During the 1920s the industry expanded considerably and several new cultivars were introduced, primarily through discovery of mutations rather than due to introductions as was the case in Florida. Important new cultivars included Marrs sweet orange, which is an early-maturing navel type that originated as a limb sport of Washington navel near Donna, Texas, in 1927; and several red-fleshed grapefruit cultivars—Hudson, Star Ruby, Redblush, Ruby Red, Henninger, Ray Ruby, and Rio Red (see chapter 4 for details). Valencia, Hamlin, and Washington navel oranges have been introduced into Texas from Florida over the years and are also grown to a limited extent. Production reached a peak of over 1 million tons from 116,000 acres in 1948 but decreased significantly due to the freezes of 1949, 1950, 1951, and 1962. Production again gradually increased to 808,000 tons by 1981–82, only to decrease to 0 in 1984–85 and 1990–91 due to the freezes of 1983 and 1989. In 2006–07 there were 18,500 bearing acres of grapefruit, 8,800 acres of oranges, and some limited plantings of tangerines. Production of high-quality red grapefruit marketed as Ruby Sweet and Rio Star will continue to be the mainstay of the Texas citrus industry.

The citrus industry in Texas is located in the lower Rio Grande Valley. Soils are relatively fertile, ranging from sandy loams to finely textured clays. Sandy loams with a pH range of 7.0–8.0 are most suitable for citrus.

Citrus has also been grown at various times in several Gulf Coast states including Louisiana and Alabama. Freezes and, more recently, hurricanes (Katrina, 2005) have largely eliminated the production in these areas. Very limited crops of navel oranges and satsuma mandarins are still produced on some of the fertile alluvial soils of southern Louisiana. There has also been some interest in reintroducing satsuma mandarins to the Gulf Coast of Alabama. Citrus is also produced to a limited extent in Hawaii.

Citrus Production in Florida

HISTORY

Citrus was brought to Florida by the Spanish between 1513 and 1565. By 1563, many citrus groves had been established around St. Augustine and along the St. Johns River as far south as Orange Lake. Apparently, Native Americans planted seeds, particularly sour orange, in clumps as they moved from location to location. These clumps of trees came to be known as "groves"—a term that has survived the test of time, especially in Florida. Present-day citrus plantings are, strictly speaking, orchards, not groves, since they consist of an orderly pattern or design, but the term "orchard" is rarely used in Florida.

Although citrus was grown throughout the northeastern part of Florida for many years, commercial production did not commence until 1763 when Jesse Fish established a grove on Anastasia Island near St. Augustine. Fish was among the first of many transplanted New Yorkers. He moved to Florida due to his wife's "frivolous and unfaithful" behavior and made his first shipment of oranges to England in 1776. Unfortunately, citrus growers experienced their first taste of politics shortly after this, and shipments were suspended with the advent of the American Revolution.

Growers also became painfully aware of a perennial problem, the freeze, as early as 1835. (In fact, the earliest freeze on record occurred in January 1766.) Fortunately, Captain Dummitt, a prominent grower at that time, had the foresight or good fortune to choose a warm location, Merritt Island, for his grove. Dummitt's trees survived the freeze of 1835 and provided budwood for many early groves in the Indian River area. Following the freeze, there is evidence that sour orange seedlings were topworked with sweet orange. Up to this time, only sweet orange seedlings were grown. Grapefruit had been introduced previously from Barbados by Don Phillipe in 1809, and mandarins had been brought from China in 1825. Other early and subsequently well-known groves were the Speer Grove at Sanford, the Mays Grove on Drayton's Island, the Dancy Grove at Orange Hill, the Hart Grove at East Palatka, the Carney Grove near Lake Weir, and the Sampson Grove at Orange Lake. All of these groves were located near transportation, specifically lakes or rivers, and all were budded on sour orange rootstock instead of being grown as seedling trees (The significance of a budded two-part tree is discussed in detail in Chapter 5). In 1870, General H. S. Sanford bought 40,000 acres of land near Sanford, where he established the first citrus research program of its kind. At this time, citrus production was also becoming established on the infertile sands of the Central Florida Ridge, largely due to the introduction of rough lemon rootstock which was adapted to the sandy soils found there.

The Florida citrus industry in 1890 consisted of approximately 115,000 acres and the leading counties in fruit production included Orange, Alachua, Volusia, Marion, Lake, Putnam, Hillsborough, Pasco, Brevard, and Polk. Alachua County accounted for about ⅓ of the state's citrus production in 1889–90. Production was an estimated 6 million boxes (90–pound equivalents) in 1894–95. However, the nemesis of Florida citrus production once again struck a cruel blow to the fledgling industry. The freezes of December 1894 and February 1895 virtually eliminated the industry, killing 90–95 percent of the trees and reducing production to only 147,000 boxes. The industry had been reduced to 48,200 acres by 1895, of which 97 percent was nonbearing. Orange, Lake, and Brevard counties accounted for more than half of the acreage remaining after these freezes. At that time, Polk County, which later became the leading county in production, was 10th on the list. Citrus growers had yet to develop Polk County's large expanses of rolling hills and sandy soils. Furthermore, a severe freeze in February 1899 brought some of the lowest temperatures experienced in the state prior to that time and caused major damage to the remaining industry. Subsequently, several freezes provided impetus for moving the industry southward and to the coast to locations less vulnerable to low winter temperatures. These freezes occurred in 1917, 1927, 1934–35, 1957–58, 1962, 1977, 1981–83, 1985, and 1989.

The late 1800s were also times of innovation and introduction of new cultivars into Florida. Of the current major cultivars, only Valencia was introduced directly from abroad. This important cultivar, originally known as Hart's Tardif because of its lateness, was brought to Federal Point by Hart from the Portuguese Azores. Parson Brown, for many years the most important early-season orange in Florida, was propagated by Captain Carney in 1874 from a seedling tree located on the property of Parson Brown. Carney purchased fruits from the original tree for one cent each. The trees, located near Webster, Florida, had been planted 14 years earlier from seeds obtained from a Georgia traveler and probably were imported from China. Hamlin orange, Florida's most important early-season orange, originated as a chance seedling in a grove planted in 1879 near DeLand, Florida, owned by Judge Isaac Stone. Similarly, the original Pineapple orange trees were propagated from nine seedlings brought from Charleston, South Carolina, by J. B. Owens. Buds were collected by P. P. Bishop from these trees, located near Citra, and propagated by the Crosby-Wartmann Nursery. Washington navel orange, one of the most widely planted navel cultivars, was introduced from Brazil via Washington, D.C., in 1870, probably for the second time. Navel had previously been introduced from Brazil in 1838 by Thomas Hogg, whose grove was later destroyed during the Second Seminole War.

During the Florida land boom of 1922–28, large-scale plantings were made in the central parts of the state and along the east and west coasts. These plantings established the future direction of the industry for many years. Much of this boom-developed acreage received inadequate attention, and some of the groves were forced into a semi-abandoned condition during the Great Depression in the 1930s. However, the Great Depression produced leaders in research and production who were instrumental in reestablishing the citrus industry.

Further plantings were made, and abandoned acreages were brought back to production with the resumption of demand for citrus fruit during the 1940s. By 1955 there were 522,000 acres of citrus in Florida. The leading counties in production included Polk, Lake, Orange, Hillsborough, St. Lucie, Pasco, Indian River, Highlands, Brevard, and Volusia.

The great influx of new residents from the North beginning in the 1950s exerted a strong influence on the distribution of plantings and relative importance of the citrus industry. These new residents began to settle in the best-drained areas along the coasts and in the central portions of the state; counties such as Orange and Pinellas began to lose many acres of citrus to urbanization, and land values and taxes began to increase to the economic detriment of citrus growers. New citrus plantings were forced onto less desirable locations. An attempt was made to expand citrus plantings along the upper west coast, but freezes in 1957–58 and 1962–63 interrupted this expansion.

Consequently, there was movement of citrus to the south onto soils that had earlier been avoided largely because of drainage problems. This move began after the 1962 freeze and reached a climax in 1969, at which time approximately half of the citrus acreage (and an even greater percentage of citrus trees due to closer planting schemes) was established on land with varying degrees of impeded drainage. These mineral soils of the flatwoods and marshes of central and south Florida presented growers and researchers many new problems quite distinct from those of the sandy Ridge area. By 1969 the better soils with adequate drainage had been developed, slowing the tempo of new planting. Furthermore, the federal income tax reforms of 1969 forced the capitalization of all expenditures during the first four years of a new planting, making such planting far less desirable to developers and speculators. Although new plantings continued, they did so at a reduced level until after the freezes of the 1980s.

During the expansion of the 1960s, large acreages of citrus were planted in the poorly drained flatwoods areas of the Indian River counties of Brevard, Indian River, St. Lucie, Martin, and Palm Beach, and in the flatwoods soils in the southern part of the peninsula. By the late 1960s, there were nearly

1 million acres of citrus in Florida. However, in the early 1970s, the new plantings did not keep pace with tree replanting; thus, in 1971 there were approximately 877,000 acres, with the principal producing counties being Polk, Lake, St. Lucie, Orange, Indian River, Hardee, Hillsborough, Martin, Highlands, and Pasco. Unfortunately, freezes of December 1983, January 1985, and December 1989 devastated the northern citrus areas. Nevertheless, new plantings increased tremendously, especially in southwest and southeast Florida following these severe freezes.

CURRENT PRODUCTION AND DISTRIBUTION

The general citrus-growing region of Florida lies within the limits bounded on the north by a line through McIntosh and Palatka with a small extension northward along the St. Johns River (although this area is very sparsely planted after the freezes of the 1980s) and on the south by the Everglades with extensions on the east coast to Homestead and on the west coast to Naples. The principal area of lime production was along the lower east coast and the southern part of the Ridge, predominantly in Dade county; lemon production was primarily in the southern third of the peninsula, particularly in Palm Beach, Martin, and Lee counties. Limes and lemons are more freeze sensitive than other commercial species and thus production was limited to more southerly regions. Unfortunately, lime production has been eliminated due to citrus canker disease and lemon production has been severely reduced because of freezes and limited demand.

Approximately 30 of 67 counties in Florida grow and ship citrus fruits in commercial quantities and all of them lie within the area described above. To the north, low winter temperatures limit commercial citrus growing, while counties at the southern extremes frequently have problems with either urban competition for resources or extremely poor soil drainage. The citrus crop is of major economic importance in many of these counties and, by far, the leading source of income in some. Polk, Lake, and Orange counties, in the center of the state, were for years the three leading citrus producers. However, with increased plantings beginning in 1959 in the western portions of the counties, and following the freezes of the 1980s, St. Lucie and Indian River counties became leading producers. By the late 1990s, citrus acreage in Hendry, Highlands, and DeSoto counties had increased tremendously. In the 2006–07 season the five leading counties in citrus production included, in descending order, Polk, Hendry, Highlands, DeSoto, and St. Lucie. Tables 1.2 and 1.3 show how counties compared in citrus acreage and production in the 2006–07 season.

Table 1.2. Citrus-producing counties in Florida by fruit type and acres as of 2006–2007

County	Oranges	Grapefruit	Specialty fruit[a]	Total
Polk	76,430	3,564	6,404	86,398
Hendry	73,583	3,324	2,819	79,726
Highlands	59,116	1,311	2,244	62,671
DeSoto	59,869	438	776	61,083
St. Lucie	25,672	23,073	2,642	51,387
Hardee	43,487	477	1,120	45,084
Indian River	18,144	20,877	1,170	40,191
Martin	32,138	2,113	787	35,038
Collier	31,189	1,061	1,144	33,394
Manatee	17,749	433	366	18,548
Lake	11,191	899	3,108	15,198
Hillsborough	13,848	180	755	14,783
Osceola	10,532	1,067	571	12,170
Charlotte	10,180	1,013	690	11,883
Lee	9,569	695	394	10,658
Okeechobee	7,832	1,022	368	9,222
Glades	8,127	175	253	8,555
Pasco	7,645	97	448	8,190
Brevard	4,289	404	387	5,080
Orange	3,998	98	452	4,548
Palm Beach	154	426	1,088	1,668
Sarasota	1,070	417	165	1,652
Volusia	993	156	82	1,231
Marion	967	41	177	1,185
Hernando	827	11	83	921
Seminole	382	19	128	529
Putnam	106	4	72	182
Citrus	113	20	12	145
Other Counties	41	4	8	53
TOTALS	529,241	63,419	28,713	621,373

Source: Citrus Summary 2006–2007, Florida Agricultural Statistics Service, Orlando.

Note: a. Includes temples, tangelos, tangerines, lemons, and others.

Table 1.3. Citrus production by county (2006–2007) in thousands of boxes

County	Oranges	Grapefruit	Tangerines and hybrids	Total
Polk	19,361	1,637	1,372	22,370
Highlands	15,563	757	424	16,744
DeSoto	15,389	245	198	15,382
Hendry	19,253	1,430	731	21,414
Hardee	11,392	265	346	12,003
St. Lucie	3,587	9,470	280	13,337
Indian River	2,933	9,184	163	12,280
Manatee	5,152	203	84	5,439
Lake	2,180	285	118	2,583
Collier	7,793	372	225	8,390
Martin	6,126	624	80	6,830
Hillsborough	3,810	78	239	4,127
Osceola	3,033	507	92	3,632
Pasco	1,886	42	62	1,990
Charlotte	2,415	391	190	2,996
Lee	2,180	285	118	2,583
Orange	915	54	84	1,053
Glades	2,214	59	99	2,372
Brevard	650	179	42	871
Sarasota	251	207	29	487
Marion	185	20	28	233
Hernando	166	5	10	181
Palm Beach	63	214	233	510
Volusia	157	66	7	230
Seminole	73	7	24	104
Other Counties	50	16	10	76
TOTALS	129,000	27,200	5,850	162,050

Source: Citrus Summary 2006–2007, Florida Agricultural Statistics Service, Orlando.

From 1920 to 1960 there were six areas of production: the Indian River, the lower east coast, the lower west coast, the upper west coast, the north central, and the south central. These areas were developed through the community interest of growers and the similarity of problems in each of the areas. The cohesiveness of these areas has changed considerably in the last few decades, with the change in the distribution of acreage and the expansion of organizations with plantings in various parts of the state.

The Indian River Citrus Area is defined by state law (Acts of 1941), which specifies that the name be used only for fruit grown on land adjacent to the Indian River and lying totally within the area described by law, along the east coast. It includes all or part of Brevard, Indian River, St. Lucie, Martin, Volusia, and Palm Beach counties. Since fruit grown in this area, especially grapefruit, has a recognized superiority in the market, as evidenced by demand and prices, it is accorded separate regulations under the marketing agreement. Fruit grown in all other sections of the state is designated as interior fruit for purposes of marketing in the fresh fruit channels.

Currently there are five citrus-producing areas as defined by the Florida Agricultural Statistics Service for recording acreage and production of citrus each year (fig. 1.2): the Indian River District, Southern, Central, Western, and Northern areas. The freezes of the 1980s killed virtually all the citrus grown in Alachua, Putnam, Marion, Citrus, and Hernando counties and very little of this acreage has been replanted.

County lines, as political boundaries, leave much to be desired in describing areas of similar climatic, soil, and biotic environments. One area grades almost imperceptibly into another, and within any area there will be islands of differences in these factors. Note that the East Coast Area includes all of the Indian River Citrus Area except for southeastern Volusia County. The East Coast, Upper Interior, and West Coast areas have few large differences within their boundaries, but the Lower Interior Area includes the well-drained soil of Polk and Highlands counties among those that generally have poorly drained soils. Extensive planting took place during the late 1970s and 1980s in the southwest counties of Collier, Hendry, DeSoto, Highlands, and others, which is reflected in the acreages shown in table 1.2. The actual citrus-producing belt of Florida, by legal designation, extends north and west to the Suwannee River.

There is yet another arbitrary delineation of certain regions within the state of Florida by marketing areas. There are four such areas in the state—the Indian River area (which has been described earlier), the Gulf area of Southwest Florida (Hendry, Lee, Collier, Charlotte, and Glades counties), the Peace River area (DeSoto, Hardee, Sarasota, and Manatee counties), and the Highlands County Citrus Grower Association and marketing area. Fruit from these areas are marketed by name such as "Indian River Marsh Grapefruit" or "Gulf Valencia Oranges." This gives each area a brand name that is useful for marketing purposes.

Environmental conditions, types of rootstocks and scions, general cultural practices, and marketing opportunities vary considerably in different parts of the commercial citrus belt. The prospective investor in grove properties

Figure 1.2. Major citrus-producing regions in Florida.

should study these factors in the area of interest. Potential investors may obtain help from established growers and production managers, extension and research centers, and representatives of allied industries and marketing agencies. Long experience has shown the value of considering land values and taxes, availability of production and market services, and general levels of production costs and returns. Studies of climatic, soil, and biotic conditions, rootstock and scion selections, and general conditions of grove properties with respect to tree size for its age, uniformity of tree size, uniformity of plantings, and incidence of inherent problems such as nematodes, viruses, frosts and freezes, and soil drainage will be most valuable for people who will be involved in the industry. This is especially true for people in a managerial position. For example, when soil drainage is a problem, producers must consider overall area drainage (through drainage district or otherwise) in order to understand the drainage of a particular site or location. There are areas with suitable soils that could likely produce citrus successfully if drainage and cold protection are considered.

The study of a particular area is a valuable exercise. However, each particular site or location within an area presents a unique set of conditions, ranging from best to poorest of the area, which determine its value to the investor. Once satisfied with the general conditions of the area, the investor will seek to optimize these conditions: natural cold protection; soil depth and drainage; land or grove cost (original, preparation, and continuing costs especially when concerned with soil drainage); and tax structure. Small differences with respect to conveniences may be considered as plus or minus factors but should never take precedence over the first four site considerations. Purchasers must have in mind their long-term objectives as an investment for citrus growing or for land speculation possibilities or both, which will greatly assist them in making a proper choice.

COMMERCIAL USES OF CITRUS

Citrus has a variety of uses around the home, although the most trees by far are planted for commercial utilization. All species have attractive foliage, flowers, and fruit, and they may be used in warm areas for landscaping where other evergreen trees of similar size would be suitable. The fruits are an added bonus and may be consumed fresh, juiced, or made into various types of preserves, marmalades, or crystallized fruits.

Commercial citrus production until 1920 was concerned only with fresh fruit production. Canning of citrus fruit began in 1921 and ultimately utilized a considerable part of the grapefruit crop as well as a small part of the orange crop. The advent of frozen concentrate processing in 1945–46 led to spectacular changes in the utilization of oranges, limes, lemons, and, to a lesser extent, mandarins. More recently, the introduction of ready-to-serve juice products has become the dominant method of supplying citrus products to consumers. Some juices are prepared from concentrate and others are simply fresh juice products that have not been concentrated and are sold as "not from concentrate," or "NFC" juice. The convenience and quality associated with NFC products has increased consumer interest and demand and this trend will continue in the future.

The processing of citrus fruits created a problem of what to do with the peel, expressed pulp, and seed residues. This problem has been solved by researchers, and valuable by-products are now derived from these former waste products. The principal by-products are dried citrus pulp (actually consisting of mostly peel) for cattle feed, molasses from the waste juice, and citrus peel oil from the oil glands of the peel; less important are citrus seed oil, pectin from the inner peel (albedo), bland syrup, and feed yeast. Ethanol production from by-products is proceeding on an experimental basis as of 2009.

Recently, the anticancer and other (antioxidant) health benefits of citrus have been recognized and promoted. The use of citrus fruit in recipes as part of a low-fat diet has also been promoted as a means of expanding markets.

TWO

Climate and Soils

"Climate" refers to the general or average conditions of a given area in regard to the various atmospheric phenomena, while "weather" refers to the transient conditions of atmospheric environment for this area, that is, weather is the climate of the moment, and climate is the average long-term weather. Both the weather and the climate are of importance to the citrus growers. They may make the growing of some or all kinds of citrus fruits difficult in some areas, while in others the effect is very subtle. No major citrus-growing region of the world is without some climatic problems. Citrus growers must recognize limits set by climate; in fact, more time will probably be spent worrying and talking about the weather than in administration of the production program.

CLIMATIC REGIONS

Tropical Growing Regions

Tropical regions are located between the equator and 23.5 degrees north and south latitude. They are bounded the Tropic of Cancer in the northern hemisphere and the Tropic of Capricorn in the southern hemisphere. Average annual temperature is usually greater than 65°F, and day and night temperatures are very similar. The tropics can be further subdivided into the lowland (sea level to 3,000 feet), midland (3,001–6,000 feet), and the highland (above 6,001 feet) altitude regions. Freezes rarely occur except at very high altitudes. There are distinct microclimates within each region that differ in temperature, rainfall, wind, or sunshine patterns. Citrus growing is also affected in different ways in each microclimatic zone.

The lowland tropics have a wet, humid climate during most of the year or have distinct wet-dry seasons in some areas. Average temperature is high and fairly constant throughout the year. Rainfall and relative humidity are high either throughout the year (wet tropics) or seasonally (wet-dry tropics).

Generally, orange and mandarin trees do not produce well and fruit quality is poor. In contrast, grapefruit and lime fruit are high quality in this region. However, disease, pest, and weed pressures are often severe in many of these regions.

Middle-altitude tropical regions generally have pleasant climates and are desirable places to live. They have lower average temperatures than the lowland tropical regions but also may contain high rainfall and humidity (wet regions) or have distinct wet-dry seasons. Day, night, and seasonal temperatures vary somewhat in this region. Citrus production is usually higher and fruit quality of oranges and mandarins is better than in lowland tropical regions. However pest, disease, and weed pressures are also great. Cloudiness in some areas reduces fruit quality since sunlight is necessary to produce high sugar content in the fruit.

High-altitude tropical regions are not important in commercial citrus production, but many of these areas have developed local selections of citrus that are specifically adapted to these regions. Generally citrus trees do not grow and yield well at altitudes greater than 7,500 feet. Seasonal and day-night temperatures are on average lower than in other tropical regions. Temperatures are also more variable seasonally and diurnally in the high than in the low or mid tropics. There is frequently cloud cover or fog that interrupts sunlight, thus reducing sugars in the fruit. In addition, ultraviolet radiation tends to be high in these regions, a factor that may decrease tree growth.

Subtropical Growing Regions

Subtropical regions that produce citrus include the area from 23.5° north-south latitude to about 40° north-south latitude, although not all areas within this region are suitable for citrus production. For example, 40° north latitude is near Philadelphia on the east coast of the United States, which is, of course, too cold to grow citrus. Subtropical climates have at least one month with average temperatures less than 65°F, with eight or more months with average temperatures above 50°F. These regions are also subject to frosts or in some cases severe freezes. The subtropics are further subdivided into arid or semi-arid and humid growing regions. Climatic differences among these regions have significant effects on all aspects of citrus production.

Dry subtropical regions include Mediterranean-type climates and are characterized by hot, dry days and cool nights. There is often a large difference between day and night temperatures due to radiational cooling and low humidity. Average rainfall is low in these areas (generally, 15–30 inches per year or less) and usually comes in the winter. Relative humidity is very low, often less than 20 percent. Some of the major producers in this region include the Mediterranean area (Spain, Italy, Morocco, Greece, Turkey, Israel, Egypt), California, northwest Mexico, Australia, and northern South Africa, among others. Within these areas, there are also arid regions such as those of the southern California or Arizona deserts, which often average less than 10 inches of rain per year. Due to the dry climate, pest and particularly disease pressures are less than in more humid regions.

Humid subtropical regions, which include Florida, Brazil, central China, Japan, and coastal areas of Mexico, have high average temperatures, but day-night temperature differences are not as great as in arid regions. Relative humidity is usually very high, and rainfall averages 48–60 inches per year, although variation among regions and years is very high.

CLIMATIC CONSIDERATIONS

Five components of climate are of interest because of their influence on tree growth, yields, and fruit quality. They include temperature, rainfall, relative humidity, wind, and sunlight. These factors, as they relate to citrus production in Florida, will be discussed in the following sections.

Temperature

Florida has a humid subtropical climate. During the period from April to October, temperatures are moderately high, but summer temperatures do not reach levels as high as those experienced in arid climates or even in the northern and central United States. Daily maximum temperature in summer is 90–95°F, with temperatures rarely reaching 100°F. The highest summer and lowest winter temperatures generally occur in the central areas of Florida. The Atlantic Ocean and the Gulf of Mexico moderate both summer maxima and winter minima temperatures in coastal regions.

Temperatures may vary significantly according to latitude and proximity to the Atlantic Ocean or the Gulf of Mexico. Minimum temperatures are generally lower from southern to northern Florida, and from the coast inland if elevation does not change. Locally, temperatures are influenced by such variable factors as elevation, soil type, air channels, bodies of water, surrounding vegetation, size of grove areas, presence of windbreaks, presence of hammock trees within a grove, and cultivation practices. Even though there is no known way to prevent periodic freezes in Florida, the grower can minimize their effects by proper selection of site, cultivars, and rootstocks, and by following satisfactory cultural programs (see chapters 8, 9, 10).

The temperature range for growth of most commercial citrus cultivars is 55–100°F. Within this range, optimum growth occurs at 80–90°F if other factors such as water are not limiting. Fruit maturation, including production of sugars and development of peel color, is optimal in the lower portions of this growth range. Winter temperatures of 35–50°F are best for developing freeze-hardy trees. Temperatures above 100°F rarely occur in the field in Florida but may occur in greenhouse nurseries. Trees appear to grow well even at these high temperatures if sufficient water is available. Canopy growth is slowed considerably or stops at temperatures less than 55°F.

From October through March, average and minimum temperatures are lower than at other times. Temperatures below 32°F occur nearly every year in some portions of the citrus-growing region. Usually, minimum temperatures are lower in northern versus southern or coastal areas and freezes occur for longer durations. This is not always the case as each freeze has different characteristics. For example, minimum temperatures were nearly the same in Gainesville (north central) and Labelle (south central) during the January 1997 freeze. Only the Florida Keys are completely free of freezes.

Low temperatures that will damage or kill citrus trees occur as a frost or a freeze. In both cases, temperature of the air or plant tissues falls below 32°F, but for different reasons. The methods of preventing freeze injury will be somewhat different under the two conditions (see chapter 9). Frosts are local occurrences with wind speeds less than three miles per hour. The cooling of the air and plants results from radiation heat loss from soil and plants into the atmosphere. As the soil cools, air above it releases heat to the soil and so is cooled slowly, with lowest air temperatures occurring next to the ground. Rapid loss of heat by radiation occurs only on clear, cloudless nights, for clouds reflect back much of this heat. On a calm, still night the air forms a series of layers which become progressively warmer from the ground up. This stratified condition is termed "temperature inversion." If there is a light breeze, these temperature strata become mixed and the warmer upper air mixes with the colder air next to the ground, thus increasing the temperature at ground level. Frost damage is most likely to occur on still, cloudless nights, and frost may occur anywhere on the mainland of Florida. Muck soils radiate heat faster than sandy soils, and, for the same latitude, frost danger is greater on muck. Elevated areas are less subject to frost than adjacent low areas because the heavier, colder air drains into the low areas from high ground. Even small differences in elevation can produce significant differences in cold damage.

Freezes are always general, not local, occurrences because they result from the advection of large masses of air at subfreezing temperatures. Thus, the term "advective" or "windy freeze" is used in this instance. These masses of frigid air often follow certain channels related to topography. These "cold pockets" commonly occur along the hilly Ridge area of Central Florida.

Because the air is near the same temperature from top to bottom of the moving mass, there is usually little difference between temperatures on high and low elevation, at least on the first night of a freeze. Whereas, a frost usually follows a warm, sunny afternoon, when the soil and trees have stored considerable heat, the afternoon before a freeze is usually cold and windy and may even be cloudy. Wind speeds of 15–25 miles per hour may occur during an advective freeze. A frost usually lasts only for a night (although it may recur the next night), but a freeze usually lasts for two or more days.

Freezes usually occur in Florida any time after November 15 until March 15 in most of the citrus belt. Conditions that result in a lack of tree freeze hardiness such as several days of warm weather preceding a freeze will usually produce more severe freeze damage. The most severe injury results when an early fall or winter freeze is followed by a period of warm weather sufficient to initiate new growth, and this in turn is followed by a second freeze in the same winter. Such was the pattern of the freezes of 1894–95. This "Big Freeze" killed many trees in the northern citrus areas, causing the industry to move south and to coastal areas. Trees were defoliated and fruit was frozen, but wood damage was not severe from the first freeze in early December of 1894. During January the weather was mild and trees began growing. Growers felt that they had come through the experience with little damage. However, many trees were killed to the ground by a second freeze in early February 1895 because the trees had become less freeze hardy in the interim. In contrast, in January 1940 a freeze of several days' duration caused considerable loss of fruit and injury to the branches; yet, because there were no subsequent freezes that winter, the new growth in February following the freeze developed normally and the trees were practically back to normal condition that summer. Freezes in the winter of 1957–58 occurred as repeated cold waves alternating with periods of warm weather and renewed growth; again freeze damage was severe in many areas. The "Big Freeze" of December 1962 was the most damaging of this century until freezes in the 1980s.

In the 1962 freeze, low temperatures occurred from December 11 to December 16, with hard freezes on December 13 and 14. Other freezes have produced lower minimum temperatures in low-ground locations; the advective windy freeze of 1962, however, resulted in low temperatures and severe damage even on high-ground locations. Less cold damage occurred in the

Indian River area and in southeastern Polk County and Highlands County than in more northerly areas.

Moderate to severe freezes occurred in Florida in 1977, 1981, and 1982. However, back-to-back freezes in December 1983 and January 1985 exceeded all previous records and severely damaged the northern citrus areas. Nearly 20 percent of the state's acreage was lost due to the 1983 and 1985 freezes. Yet another serious freeze occurred in December 1989. This freeze destroyed many of the newly replanted areas in the northern citrus region and brought extensive damage into all of Polk County and much of Highlands County. Since 1986 citrus production has moved southwest and to the east coast, and production reached record levels (see chapter 1) until citrus canker, greening, and hurricanes severely reduced acreage and yields in the 1990s and 2000s. Production in the northern area of the state will likely never again reach that attained before 1980.

FREEZE AND FROST DAMAGE

Freeze damage from low temperatures usually begins when the air temperature falls below 28°F. However, under radiative conditions (wind speed less than three miles per hour), the temperature of fruit and leaves may be less than air temperature and injury may occur earlier. Cold, dry winds may also injure tissues by desiccation while the air is still above 32°F. However, only very tender foliage or flowers are injured by such temperatures. Commercially, it is likely that fruit will be damaged after four hours at air temperatures of 26°F, but individual fruit damage will vary widely, depending on size, location in the tree, peel thickness, and sugar content. Obviously, as temperature continues to decrease, the amount of damage increases until fruit become completely frozen at 20 to 22°F. The primary source of freeze damage is through the loss of water through the peel, a reduction in juice content, and segment drying.

Fully expanded leaves may begin to show injury (evident as water-soaked, darkened leaves) after being exposed to air temperatures of 27°F or lower for two-four hours. Water-soaking does not always indicate severe leaf damage or loss, however. Location on the tree influences foliage injury because of temperature gradients within the tree. For example, during radiative freezes, minimum leaf temperature and thus injury is greater in the top versus the inside of the canopy due to these temperature gradients. Mature orange trees have survived 10 hours below 25°F and minimum temperatures of 20°F, with only 10 percent leaf loss. Previous temperature conditions (acclimation) play an important role in the freeze tolerance of citrus trees (see next section).

Nongrowing twigs and limbs may be injured at temperatures of 16–27°F, depending on size and maturity of wood and length of exposure to cold as well as on previous temperatures. Because their tissues are less mature, their bark is thinner, and their heat capacity is lower (because they have less mass), young trees may be killed outright by a single severe freeze. Older trees, even though not usually killed entirely, may have sufficient limb damage that they are no longer economically viable. Often serious damage to trees from freezing temperatures is evidenced by frost or cold cankers on scaffold branches and trunk, particularly in the crotch area. This occurs when cambium growing actively at the time of freeze is killed while the surrounding tissue is not damaged. As a consequence, the tree may be left in such a weakened condition that it is not capable of producing commercially satisfactory crops. Moreover, large cankers on the upper side of scaffold branches may cause the branches to break under the weight of fruit. Even though the cankers do not become progressively larger, there is no way to eliminate them except by removing the affected branches.

Factors Affecting Freeze Damage

The amount of freeze injury will vary with several factors: species and cultivar of scion and rootstock; condition of the tree with respect to vigor and active growth; minimum air temperature and duration; and the presence or absence of overhead or ground cover. Other things being equal, the vigor or health of the tree, resulting from previous cultural practices, plays an important role in minimizing cold damage. Trees in good health, with no major mineral deficiencies and free from serious pest or disease problems, are more freeze hardy than weakened, diseased trees.

If trees are in a nonapparent growth stage (the tree is not visibly producing growth flushes), trifoliate orange is the most freeze hardy of any of the commercial citrus types, followed in descending order by kumquat, calamondin, sour orange, mandarin, sweet orange, grapefruit, shaddock, lemon, lime, and citron. The range in freeze hardiness reflects the temperate origin of trifoliate orange compared with the tropical origin of citron. When trees are actively growing, however, there is little difference in freeze hardiness among the various kinds of citrus.

The amount of freeze acclimation or deacclimation, which depends largely on temperature, also has a significant effect on freeze hardiness. If nighttime minimum temperatures preceding a freeze or frost are in the 40s and 50s, tree tissues will become acclimated and survive many freezes with little damage. A few days of warm weather just before a freeze will usually decrease freeze hardiness. A prolonged drought may cause water stress, which makes trees more subject to freeze damage than those with adequate moisture content, although severely stressed trees can have greater freeze tolerance than nonstressed trees. Different citrus species tend to acclimate and deacclimate to cold at different times and rates during the winter months. For example, trifoliate orange, which is hardy in most winters, may be seriously injured by early, moderate freezes when temperatures in the early fall have been unusually high. Apparently, trifoliate orange needs lower temperatures to cease growth than are needed by other kinds of citrus, which may be much less injured by the same freeze. On the other hand, trifoliate orange requires higher temperatures to initiate growth than most other kinds of citrus. Therefore, budbreak in the spring occurs much later than other citrus types. Thus, freeze damage from spring frosts and late winter freezes is less likely for trifoliate orange than other citrus species. Similarly, kumquats have late budbreak in spring and so also tend to escape freeze damage in late winter and early spring. Conversely, lemons, limes, and citrons resume growth very readily in mild weather and may have flowers, new foliage, and young fruits at almost any time in the winter. This lack of freeze acclimation, along with new growth, makes them quite subject to winter injury.

Even when these species are used as rootstocks, freeze susceptibility must be considered because of rootstock influence on scion growth. Generally, scions budded on Swingle citrumelo, sour orange, and mandarin types impart more freeze hardiness to the scion than those on Carrizo citrange. Trees on lemon types such as rough lemon, *C. macrophylla*, or *C. volkameriana* are least hardy. However, following freezes when trees are not killed, scions on the latter rootstocks regrow much faster (see chapter 5).

Duration of subfreezing temperatures is as important as minimum temperature reached. An automated weather station, which records temperatures continuously, will indicate better the true temperature situation. Minimum temperature may cause little or no injury when it persists for only half an hour but may do serious damage when it continues for several hours. This difference in the amount of damage is because it takes time for the large limbs and trunks to cool to critical, damaging temperatures, whereas leaves and flowers may reach critical temperatures very rapidly.

Two factors that influence freeze tolerance by their effect on the temperature are overhead and ground covers. Citrus trees that are somewhat shaded by trees or other structures are usually less injured by low temperatures than trees fully exposed to the sky. In fact, because this difference had been observed, slat shade or canvas covers were constructed during the freezes of the late 1800s over some orange groves. The covers reflect back much of the heat radiated by the citrus trees and ground, which otherwise would be lost to the atmosphere, preventing them from reaching temperatures as low as trees fully exposed to the sky. On the other hand, freeze damage is usually greater when the ground under citrus trees is covered by grass, weeds, or mulch than when the soil is clean-cultivated. In this case, cultivated soil stores and radiates more heat to the trees than soil covered with vegetation, thus preventing the trees from reaching damaging temperatures as soon.

Windbreaks may be helpful or detrimental during freezes. A windbreak on the northwest side of a grove slows down the movement of cold air into the grove. The stream of cold air flows over the top of the windbreak and only slowly descends to ground level again, leaving a relatively undisturbed area on the other side of the windbreak. Heating the grove is more effective in this case since the heated air is not removed as rapidly as in a grove with no windbreak on the windward side. On the other hand, a windbreak downslope from the grove may prevent air drainage on a radiative-freeze night and thus cause more freeze damage than if it were not there. If the windbreak is both on the northwest side and lower on the slope, occasional openings in it will permit slow drainage of cold air while still giving considerable freeze protection.

HIGH TEMPERATURE

High temperatures alone rarely limit where citrus can be grown worldwide and in Florida but may reduce yields and growth. In March and April, high temperatures, coupled with lack of soil moisture and hot, dry winds, increase evapotranspiration (water losses from the tree and the soil combined) and may cause severe water stress in the trees. Desiccation damage often appears on leaves as darkened lesions. When such conditions are prolonged into May, even if not serious enough to cause severe water stress, an excessively heavy "June drop" of fruit may occur, which can decrease yields. "June drop" in Florida typically occurs in May. The term was originally coined to describe this drop period in deciduous crops in the northern United States. Moreover, normal high summer temperatures in June and July, with high humidity and intermittent cloudiness, may accentuate sunscald and spray burn damage. Warm weather during October and November, particularly if nights are warm and rainfall is much above normal, usually results in delayed development of peel color and an increased tendency for fruit drop. Warm weather, as stated previously, also increases susceptibility to freeze injury during the winter.

Rainfall

Rainfall is an important consideration in any citrus production program. The total amount of rainfall, the distribution of the rainfall during the year, its seasonal fluctuations, and its intensity all have an impact on citrus growth, yields, and fruit quality.

Average annual rainfall within the Florida citrus belt is approximately 52 inches, but amounts may vary considerably within seasons and areas in the state. The greatest amount of precipitation occurs during the summer rainy season from late May through September. During this period, rainfall usually supplies the water requirements of citrus trees and often provides an excess above what is needed. From October through April, and occasionally through May or early June, rainfall is often insufficient to meet the water requirements of the trees.

While the average annual rainfall is about 52 inches, the amount that falls in any individual year may vary from 37 to 100 inches. In addition, the proportion of the annual precipitation that falls in any given month also varies from year-to-year. This annual and monthly variation gives a dynamic quality to rainfall. There is also tremendous variability in rainfall from site to site. The static climatic picture gives the impression that predicting rainfall amounts should be easy, but dynamics of weather make it very difficult to do so even with new, advanced computer models.

The intensity of rainfall—that is, the amount falling in a particular time period (usually daily or monthly)—is also important. This amount may vary from a trace (less than $\frac{1}{100}$ inch) to more than 20 inches! A monthly rainfall of six inches may all come in one heavy shower or may represent a dozen rainy days with $\frac{1}{2}$ inch each. These variable patterns affect tree growth and yields. Rainfalls of $\frac{1}{10}$ inch or less provide little usable water to citrus trees, because the water evaporates from the soil surface with minimal effects on soil moisture. Rainfall of $\frac{1}{2}$ inch moves into a sandy soil to a depth of about six inches, but the effect is temporary because soil moisture at this depth is also readily lost by evaporation. Rainfalls of from one to three inches are ideal since they wet the soil deeply enough to supply moisture for a long period, yet are unlikely to provide excess water that percolates beyond the root zone and may cause leaching of nutrients or herbicides. Of course, several consecutive days of $\frac{1}{2}$-inch rainfall, with no intervening sunny days that increase evaporation, may also provide satisfactory soil moisture conditions. Very heavy rainfall accompanying tropical storms or hurricanes, which may run as high as 20 inches in one 24–hour period, are potentially very detrimental because they supply far more water than sandy soils can hold and consequently cause serious leaching of soluble nutrients and pesticides. Moreover, water may accumulate in poorly drained soils causing low oxygen conditions and root death. Even during moderate rainfall, trees in poorly drained areas can be stunted and have reduced yields. These low spots are frequently observed in groves that have not been properly leveled and bedded before planting. Generally as few as three days of flooding, especially in the summer, can cause root death. Rainfall deficit is less important to citrus production in Florida than in the past because much of the industry has irrigation.

The three periods in the annual growth cycle of a citrus tree when it is most sensitive to soil water deficiency are in early spring, when the new flush of growth is tender and flowering and fruiting occur; in May, during "June drop"; and in early summer, when fruit size is rapidly increasing. Water stress during fruit set may cause abnormally heavy abscission of young fruit. In addition, tender new shoots easily wilt if soil moisture is in short supply, due to their nondeveloped cuticle (waxy covering over the leaf) and poor stomatal (pores for gas exchange on the underside of the leaf) control of water loss via transpiration. Soil moisture deficiency in May and June may reduce subsequent fruit size. Rainfall shortages during October and November are not likely to prove critical unless the fruit is desiccated at the time of harvest, or the leaves become excessively wilted. Low average temperatures during the winter decrease tree transpiration (water loss through the leaves), thus decreasing tree water requirements.

Relative Humidity

The ratio of the amount of water vapor in the atmosphere at any given time to the amount that the atmosphere would contain at the same temperature if saturated is termed the "relative humidity" and is always expressed as a percentage. Relative humidity does not indicate the amount of water actually present in the air. The relative humidity in Florida varies from nearly 100 percent at night and early in the morning to an average low of 40 percent in midafternoon on clear, sunny days. During some dry periods, relative humidity may reach as low as 20 percent. There is some seasonal variation in relative humidity, but the daily average of 72 percent is fairly constant throughout the year.

This high average relative humidity may be advantageous for tree growth because it decreases transpiration for any given temperature and thereby increases water use efficiency by citrus trees in comparison with regions with low relative humidity. However, with properly controlled irrigation, tree growth and yields may be similar in humid and arid subtropical regions.

High relative humidity has some disadvantages, too, related to incidence of fungal diseases that cause extensive damage to fruit and trees in Florida. Citrus scab, melanose, foot rot (alga), greasy spot, postbloom fruit drop, and the bacterial disease, citrus canker, are much more serious under humid conditions than under arid ones (see chapter 10). High relative humidity, especially in combination with high temperatures, causes poor peel textural quality including excessive puffiness and a reduction in peel color.

Wind

Three types of winds occur with some regularity in Florida that may be detrimental to production and quality of citrus fruits. Southeasterly winds that occur in spring (especially March and April) increase transpiration and decrease the available soil moisture by increasing the rate of surface evaporation. Spring is typically a dry time, and new leaves transpire faster than older leaves. In addition, high winds cause leaf desiccation, often producing dark-colored lesions on the leaf. Wind scar also occurs during early fruit development before the cuticle is fully developed. It is a major cause of peel blemishes in Florida and worldwide.

The hurricane season extends from June 1 to November 30 and is a period of strong winds. Summer heating of tropical and Gulf waters produces unstable air masses that move across Florida, causing summer storms. Winds of hurricane velocity exceed 74 miles per hour and are capable of causing severe physical damage to buildings and equipment as well as to the grove by uprooting trees and shaking fruit to the ground. For example, in 1992, Hurricane Andrew uprooted thousands of trees and caused severe damage to the lime industry in Dade County.

Three hurricanes struck Florida in 2004 (Charley, Frances, and Jeanne) causing widespread damage and tree losses statewide. These storms spread citrus canker disease great distances by high winds and rain and essentially ended attempts to eradicate the disease in Florida. Yet another hurricane struck Florida in 2005 (Wilma), causing considerable damage to citrus in southwest sections of the state.

Citrus production volume in Florida has changed considerably since 2004–2005 when these hurricanes occurred. Total statewide citrus production was 291,800,000 boxes in the 2003–04 season and was reduced to 169,250,000 boxes in 2004–05, and 174,800,000 boxes in 2005–06.

Advective freezes (described previously in this chapter) move to Florida via cold, northwesterly winds, occurring any time between November 15 and

March 15. Large masses of high-pressure air move southward from Canada and the northern United States, unimpeded by natural barriers such as mountain ranges.

Sunshine

Florida's nickname, "The Sunshine State," suggests that sunshine is not a limiting factor to citrus production. In fact, the many hours of sunshine help to produce the high-quality juice fruit that is characteristic of Florida. Sunshine contains several components, notably heat, light, and ultraviolet rays. Light, all-important to green plants for photosynthesis, is rarely limiting for tree growth in Florida except where trees have not been properly pruned and shading occurs. Excessive heat is also a rarity, as discussed previously. Ultraviolet rays are absorbed by the atmosphere to a much greater degree under the humid conditions of Florida than in more arid or high tropical regions; they do not limit tree growth or yields.

SOILS

Soils provide anchorage for plants and provide the water and mineral nutrients required for plant growth. The native soils of Florida vary widely in the way they serve these functions. These soils have been grouped into series on the basis of origin, color, profile development, and other characters. Soils are further subdivided into classes according to the texture (particle size) of the surface layer.

Horticulturists have long recognized the extreme importance of certain soil characteristics and have selected planting sites accordingly. Drainage is perhaps the most important of all soil characteristics. In the early years of the Florida citrus industry, great emphasis was placed on the type of vegetation growing naturally on different soils, as this provided the experienced grower with a good idea of the nature and properties of the soil.

Historical Classification of Soils by Native Cover

The early citrus grower noted that the native flora not only served as an index of soil drainage but also gave indications of the natural fertility of the soil and the relative likelihood of potential cold damage for that location. In well-drained areas, the following types of land were distinguished; they are listed here in descending order of desirability for commercial citrus production.

1. High hammock land is characterized by dense stands of hardwoods such as live oak, magnolia, hickory, and dogwood.

2. High pine land sustains good stands of longleaf pine, often with some scattered red oak and post oak and little underbrush.
3. Blackjack oak land supports turkey oak (sometimes called "blackjack") and scattered pines. It represents a rather poor grade of high pineland.
4. Scrub, comprising areas of coarse sand with low organic matter, reflects its low soil fertility via poor stands of sand pine, turkey oak, and other shrubby oaks.

High hammock and high pine lands proved very well suited to citrus culture, while blackjack oak land was never very suitable. The scrub probably always will remain in its native condition except for real estate development potential for it has little value for agriculture. Scattered through the high pine lands are limited areas of coarse white sand that have the same characteristics as scrub land. These small areas, called "sandsoaks," are not suitable for citrus production.

On poorly drained soils, the following types of land were recognized, again in order of decreasing suitability for commercial citrus production.

1. Low hammock land is characterized by heavy growth of hardwoods, especially live oak and cabbage palmetto. Difficult to clear and drain properly, it has nevertheless produced some exceptionally good coastal groves. The old, so-called hammock groves were established with the underbrush cleared out, but many of the larger hardwoods and palmettos left in place.
2. Flatwoods land is low and level, with scattered longleaf pine predominant in the northern part of the state and slash pine in the southern part. Frequently saw palmetto and wire grass cover the ground and a hardpan often underlies the surface, making drainage very poor. These soils are suitable for citrus production if trees are grown on beds and proper drainage is used.
3. Bayheads are areas of standing water; although subjected to flooding, they originally supported stands of bald cypress and other trees often associated with well-drained soils at low elevations. They are not suitable for citrus culture.

The present-day horticulturist is rarely able to make use of only natural vegetation as an index of suitability of soils for citrus planting. Most areas that originally supported tree growth have been clear-cut, especially on soil types reasonably well adapted for citrus culture. There are areas of clear-cut blackjack oak and scrub land that have not changed greatly from their native condition, but these are some of the least desirable soils for planting citrus trees.

CURRENT SOIL CLASSIFICATIONS

Most Florida citrus production is on soils with inherently low fertility and low cation exchange capacity (CEC). Cation exchange capacity represents the ability of negatively charged soil particles to attract and hold positively charged cations such as calcium, ammonium, magnesium, and potassium. Thus, Florida soils in general are vulnerable to leaching of soil-applied agricultural chemicals. Soils on the central Ridge of Florida are primarily Entisols, which are deep, well-drained sands. Most of the state's citrus was produced in this area prior to the freezes of the 1980s. Lowland soils with a sandy surface that overlies fine-textured marine deposits (often calcareous) are termed "Alfisols." Sandy, acidic soils containing a spodic horizon (organic hard pan) are called "Spodosols." The distribution of these soils throughout the state is presented in figure 2.1. Many of the early citrus groves along the southerly coasts of Florida were planted on Alfisols. Recently, even more extensive plantings have been made on Spodosols. The spodic horizon frequently is strongly acidic and the overlying surface horizons may need to be limed (see chapter 9). The effects of liming appear to be limited to the depth of lime incorporation. In contrast, bedded citrus groves planted on Alfisols usually have alkaline materials present in the rooting zone. The latter soils initially were sought for citrus production because of their high organic matter and clay content, which provided greater CEC and water-holding capacity. However, the calcareous nature of these soils presents several grove fertilization problems, particularly with micronutrient deficiencies (see chapter 9).

Soils are also classified based on their water and air drainage characteristics. Knowledge of these soil types and their characteristics is important to the citrus grower. Based on drainage characteristics, soils may be grouped as follows:

WELL-DRAINED SOILS

Well-drained soils are found in the area extending north-south through the central part of the Florida peninsula as far south as Highlands County, with lateral extensions into Pasco, Hillsborough, and Pinellas counties. The area from Leesburg to Lake Placid is often called "the Ridge." Isolated areas of well-drained soils are found along both coasts and even on some of the coastal islands. These soils belong principally to the Astatula association. Astatula

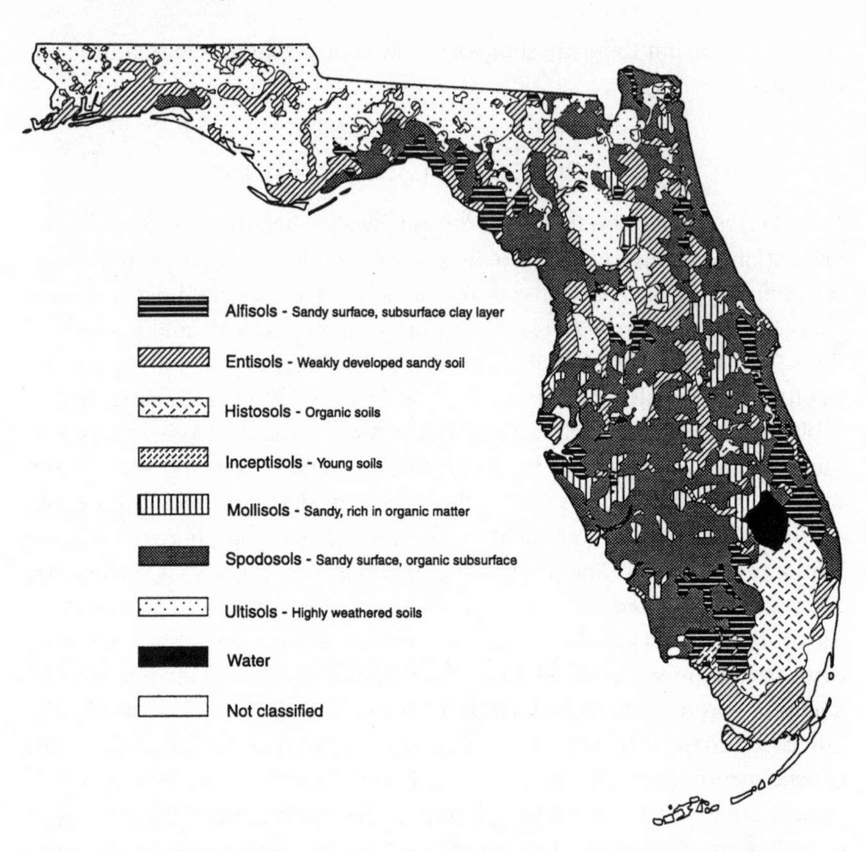

Figure 2.1. Soil types in Florida. Source: Mary Collins, University of Florida. Map by Katrina Vitkus.

fine sand is the predominant soil for citrus plantings throughout the central citrus area of the state. Its general characteristics include sand throughout the profile (although there may be a small amount of clay at varying depths) and surface three to four inches of brownish-gray fine sand with a subsoil of yellow fine sand. The water table is usually well below the soil surface, although during the rainy season it may temporarily move upward. The soil is acidic under natural conditions, ranging from pH 4.8 to 5.4, although in most groves pH is maintained at 5.5–6.5. Moreover, the organic matter is very low (0.5–1.5 percent), and the CEC is poor.

The main advantages of well-drained soils include excellent water drainage and deep rooting. Flooding rarely is a problem, although drainage may sometimes be slow because of compacted soil layers. Although regions having these soils are relatively frost free because of their elevation, frost pockets frequently are present in the low areas between hills. Soil uniformity, lack of a hardpan, and a generally low water table allow for deep, well-ramified, extensive root systems. Tree size is typically much larger for trees growing on the Ridge than in flatwoods areas due to the extensive root system.

The disadvantages of these well-drained soils include their inherently low levels of mineral nutrients and organic matter and their limited water-holding capacity. Excessive leaching, together with the low CEC, is responsible for the low nutrient levels and the potential for nitrate and pesticide leaching into groundwater. Exchange capacity is based on the amounts of clay and organic matter in the soil, and both are low in these soils. In addition, sand holds less water than clay, due to larger particle sizes. Fertilization programs are adjusted to account for this low CEC. As suggested above, development of a deep, extensive root system can compensate for low water- and nutrient-holding capacity of these soils.

POORLY DRAINED SOILS OF THE COASTAL AREAS

The low hammock lands are found principally along the east and west coasts, although they also occur in isolated areas in the central portion of the state. They are represented by the Myakka, Oldsmar, Immokalee, Riviera, Winder, and other series. The first three soils are usually quite acidic and have an acidic hardpan. Riviera and Winder are similar, but they may occasionally have marl in the profile, making them somewhat less acidic. The organic matter content of these soils is high (at least for Florida), ranging from 3 percent to 8 percent in their native state. Usually the water table is close to the surface and in the natural state it may often be above the soil surface. These soils must be drained and bedded before they can be used for citrus planting.

The principal advantages of these soils are their high natural nutrient supply, their relatively high CEC, and their relatively high water-holding capacity compared with Ridge soils. The higher CEC permits better utilization of the nutrients in the native soils or added via fertilization. The high water table may also supply water via capillary movement to the root system.

These soils also have disadvantages for use in citrus production that include poor air drainage, poor water drainage, shallow rooting depth, and occasionally high pH in the subsoil. Although these soils are usually located geographically where the freeze probability is relatively low, they are by no means free from frost, and their low elevation does not allow for drainage of cold air. Canals or ditches are necessary for adequate water drainage. In times of very heavy rainfall, however, such as may occur in the hurricane season, pumps are needed to remove excess water. Costs of removing water add greatly to total production and land preparation costs. Three of Florida's five water management districts (St. Johns River, Southwest Florida, and South Florida) oversee and regulate water movement on and off commercial citrus groves (see chapter 6). In most flatwoods groves, a reservoir must be constructed so that water may be discharged in a controlled manner so as not to exceed the rates allowed.

Shallow depth to water table limits root development to about 12 to 18 inches, so only a small volume of soil is present for storing available water. In periods of drought, the moisture supply in the root zone is quickly exhausted, so these soils are droughty in spite of their high water table. Irrigation water must be applied in frequent light applications, which can make irrigation more expensive than on the deep, well-drained soils. Often, in large plantings typical of this area, it becomes difficult to maintain adequate soil moisture due to limitations on irrigation capacity.

The alkalinity (high pH) of the subsoil in some sites causes micronutrients such as copper, zinc, iron, manganese, and boron to be less available to the tree. Therefore, it is often necessary for the grower to supply most micronutrients as foliar sprays to enhance their absorption. Moreover, potassium and magnesium must be applied in high amounts because of competition with calcium for root uptake, and in extreme cases, these elements may need to be supplied in nutritional sprays.

As the result of the shallow root system and erratic soil moisture supply, citrus trees tend to be smaller and lower yielding in the flatwoods compared with the Ridge areas. Root systems of citrus trees are often further restricted by temporarily high water tables. These prevent the efficient use of the maximum theoretical depth for rooting, which is already shallow. Practices that lower the water table and increase rooting depth help to improve tree size and crop yields.

FLATWOOD AND MARSH SOILS OF CENTRAL AND SOUTH FLORIDA

Flatwood and associated swamp soils are found in many areas of the citrus belt outside the elevated central Ridge. For years, these soils were used for open range for cattle and hogs; then, improved pastures were developed. In the past, they were considered unsuitable for citrus plantings. Typical soil types in the area include Myakka, Immokalee, Oldsmar, Pineda, and Wabasso, among others.

As urban and suburban populations have increased during the last two decades, a premium has been placed on well-drained soils with expansion of existing communities and development of new population centers and urban sprawl. This has forced new citrus plantings onto a whole range of soil types, with varying levels of pH and drainage. With the establishment of drainage districts or large-scale area drainage by citrus corporations, and the installation of ditches, dikes, and pumps, it is possible to provide adequate drainage for commercial citrus production.

Groves in the flatwoods are planted on double- or multiple-row beds, and spodic or hardpan layers are sometimes broken up by deep plowing or use of a dragline. Often limestone is incorporated into those soils to increase the pH. Tiles, open-ended pipes seated on pervious gravel, are placed across the beds in various locations within the grove to determine fluctuations in water table depth. Aside from problems with high water levels, accumulation of hydrogen sulfide due to low oxygen levels may cause root dieback. The smell of rotten eggs is a good indication of a poorly drained area.

OOLITIC LIMESTONE SOILS

One soil series in particular, Rockdale, occurs in Dade County and is used for production of various citrus fruits, especially limes. A shallow layer of clay or sand overlies a very porous limestone formed by the cementing together of the shells of tiny marine organisms. To provide adequate depth for root development, this oolite must be broken up by bulldozers or scarifying plows. Since the water table is shallow, rooting depth is shallow and soil moisture is limited. As a result, many lime trees with shallow root systems were blown over during Hurricane Andrew in 1992. In addition, the water may be above the soil surface for some time following periods of very heavy rainfall, for while the soil is porous, drainage is slow because the land elevation is near sea level.

In summary, during the early period of citrus growing in Florida, the grower could select land by observing native flora. Beginning about 1930 this was no longer feasible in most areas because the native vegetation had been removed. Since about 1950 there has not been enough land with naturally good drainage to take care of expanding demand. Thus, production has moved to poorly drained coastal and southwest flatwoods areas. Land preparation costs, including bedding and creation of wetlands and reservoirs, have greatly increased in recent years, and likely will continue to do so (see chapter 7).

Important Factors in Choosing Soils for Citrus Planting

Several factors should be considered in appraising a given site for suitability for citrus planting. Two of them, water drainage and freeze susceptibility, are of major importance; a third one, depth for rooting, is related to drainage.

No single factor is more important than adequate drainage. Accumulation of water in the root zone results in poor aeration (low oxygen) and causes reduced uptake of water and nutrients. This condition, termed "anoxia," will cause progressive injury and death of the roots, leading to decreased tree vigor. Such trees are stunted, unproductive, and short-lived. Water damage may be chronic (affecting the tree throughout its life) or acute (affecting the tree for only a brief period due to temporary lack of drainage). Poor drainage also produces conditions favorable for infection by foot rot and root rot (*Phytophthora spp.*) and other soil-borne diseases.

In addition, susceptibility to freeze or frost damage is a major factor to consider in site selection as discussed previously in this chapter. Adequate air drainage is important both in Ridge and flatwoods areas. Low-lying soils are usually colder than elevated soils because there is no place for cold air to drain during radiative freezes. Cold pockets, valleys in the hilly areas, are usually problematic on frosty nights when cold air flows down into them from the adjacent elevated areas, forcing the warm air up above tree height. After a severe radiative freeze, there is usually a visible bench line around the sides of such a pocket, below which trees are damaged and above which they are undamaged.

If a site meets the above criteria, such factors of secondary importance as water- and nutrient-holding capacity and soil pH are considered. Soils ranked high in these favorable characteristics lower grower costs while achieving the same degree of tree vigor and productivity. Growers must also consider proximity to roads and processing or packing facilities and availability of labor in isolated areas. Increased labor (due to post-9/11 immigration policies) and hauling costs (due to increased fuel costs) can have considerable influence on overall profitability of the grove, especially in times of low fruit prices.

Water-holding capacity is influenced by the texture (particle size) of the soil and especially by the content of colloidal material, such as clay and organic matter. In addition, the volume (depth) of soil occupied by the roots plays an important role. Each soil has a characteristic field capacity (the percentage of soil moisture remaining after saturation and free drainage) and wilting percentage (the amount of moisture remaining after plants have reduced the water content until they have become permanently wilted). The difference between these two percentages represents the available water that can be used by the plant. The higher the percentage of available water, the greater the potential supply in the root zone. Furthermore, the greater the depth of the soil, the greater the total reservoir of water to sustain the tree during drought. The amount of available water in a given depth of soil can be computed in inches of water, equivalent to inches of rainfall, and as such gives a valuable clue as to how much irrigation will be needed during periods of extended drought (see chapter 9).

The nutrient supply in most of the soils used for citrus production in Florida does not vary greatly in the natural state because Florida's sandy soils have been leached by high rainfall for millennia and all have inherently low levels. However, soils of relatively high organic content have somewhat greater nutrient levels than those of very low organic content. More important is that soils of high water-holding capacity and those with relatively high content of clay and organic colloids also hold mineral nutrients applied as fertilizers much more effectively against leaching losses than soils low in colloids. Mineralization of soil organic matter also provides nutrients. In some soils, mineralization may provide 50 pounds of nitrogen per acre per year. Mineralization also accounts for extensive soil losses in some areas of the state.

Florida soils are classified as acid or alkaline, a condition customarily expressed as pH, where 7.0 represents neutral pH, values below 7.0 indicate increasing acidity, and values above 7.0 indicate increasing alkalinity. The most favorable soil pH for citrus trees on sandy soils is 5.5–6.5, or slightly acidic conditions. Most sandy soils are naturally more acidic than this and require periodic applications of lime to maintain the desired pH. Most alkaline soils cannot be acidified to reach this desired pH level and require specialized management practices, notably the application of microelements, which are made unavailable to the tree at high pH. In the case of moderately alkaline soils, regular application of acidic fertilizer materials such as ammonium sulfate will help to reduce soil pH over the long term.

Certainly, climate and soils have a major effect on the long-term profitability of a citrus grove. Careful attention to climatic factors, in particular minimum temperature, is well worth the effort. Soil characteristics can be determined by observing natural flora and by soil testing within a given area (see chapter 9). Important information concerning soils can also be obtained from the U.S. Natural Resources Conservation Service.

THREE

Systematics and Botany

The term "systematics" was introduced in 1735, but only recently has supplanted the terms "taxonomy," which is the science of classification and its basis, principles, rules, and procedures, and "nomenclature," which involves the correct naming of taxa (groupings). Systematics is a much more inclusive term than these two and consists of four basic elements: identification, description, nomenclature, and classification. The science of systematics has changed drastically in recent years with the advent of molecular biology techniques and the development of phylogenetic (evolutionary development) associations. This chapter will discuss some of the history of systematics, taxonomy, and nomenclature and emphasize specific characteristics of citrus botany including growth and development.

HISTORY

Theophrastus (310 B.C.), the "Father of Botany," developed one of the first taxonomic systems for plants. He classified all plants as shrubs, sub-shrubs, trees, or herbaceous plants. Future generations of taxonomists greatly refined these groupings based on morphological characteristics such as leaf size and shape, fruit structure, or, most important, flower structure. These taxonomists realized that there were natural relationships among species and grouping. Carolus Linnaeus, a Swedish taxonomist (sometimes called a systematist), was responsible for systematically arranging thousands of species into a binomial (two-name) system in 1753. Each plant was then identified by using a genus and a species, for example, *Citrus sinensis* (sweet orange). Taxonomists continued to classify plants using this descriptive system until the mid 1900s, when new biochemical techniques were also used. Several new systems based on phylogenetic relationships were developed in the early to mid 1900s. Then in the 1980s, molecular biology techniques further allowed systematists to classify plants at the genome level. The new techniques have caused major restructuring of the existing taxonomic systems which were based primarily on morphological characters, particularly in the case of citrus.

Classic taxonomy has guidelines but no exact rules for placing a species into one category or another. For example, in the early 1920s, Tanaka, a well-known Japanese taxonomist, divided the genus *Citrus* into 162 species whereas, Swingle, a prominent American citriculturist and mentor of Tanaka, believed there were only 16. Incredibly, modern molecular biology techniques suggest that most commercial scion cultivars are interspecific hybrids of only three citrus species! Each system has its supporters and detractors, depending on the observer's own preferences and prejudices.

In contrast, nomenclature involves a system of properly and legally naming a plant and differs from taxonomy in that it has a strict set of rules set down in the International Code of Botanical Nomenclature. These rules are necessary to avoid confusion in naming and identifying a plant. Imagine if each systematist were permitted to name or rename a species without constraints. Every species would have a myriad of different names and identifying plant species would become very difficult and confusing. A proper botanical name consists of a genus, species, and authority (the person who named the species). It is this authority—or a future person who may reclassify the species—that must adhere to international rules to name each species properly. For example, the proper name for sour orange is *Citrus aurantium L.* The "L." is an abbreviation for Linnaeus, who first gave sour orange a binomial designation. In addition, all scientific names are based on a "type" species, which must be carefully described in Latin and properly published in an accepted journal to avoid redundancies in naming and identification. The advantage to such a system is that every plant species has a Latin name which is used to identify it independent of local common names or language differences.

CURRENT SYSTEMATICS

Systematics is based on an inverted pyramid system which begins with broad, large categories, eventually terminating at the species or cultivar level (table 3.1). Unlike classical taxonomy, plants are arranged into clades (groups) that have similar genetic traits. Similarities among species in each group become closer as one moves through each taxon (grouping). For example, citrus species are classified within the rosid clade (140 plant families) which contains the eudicot clade, which is further subdivided into the eurosid II (malvid)

Table 3.1. Systematic classification of important citrus and related species based on molecular biology and morphological characteristics (adapted from D. Soltis, University of Florida)

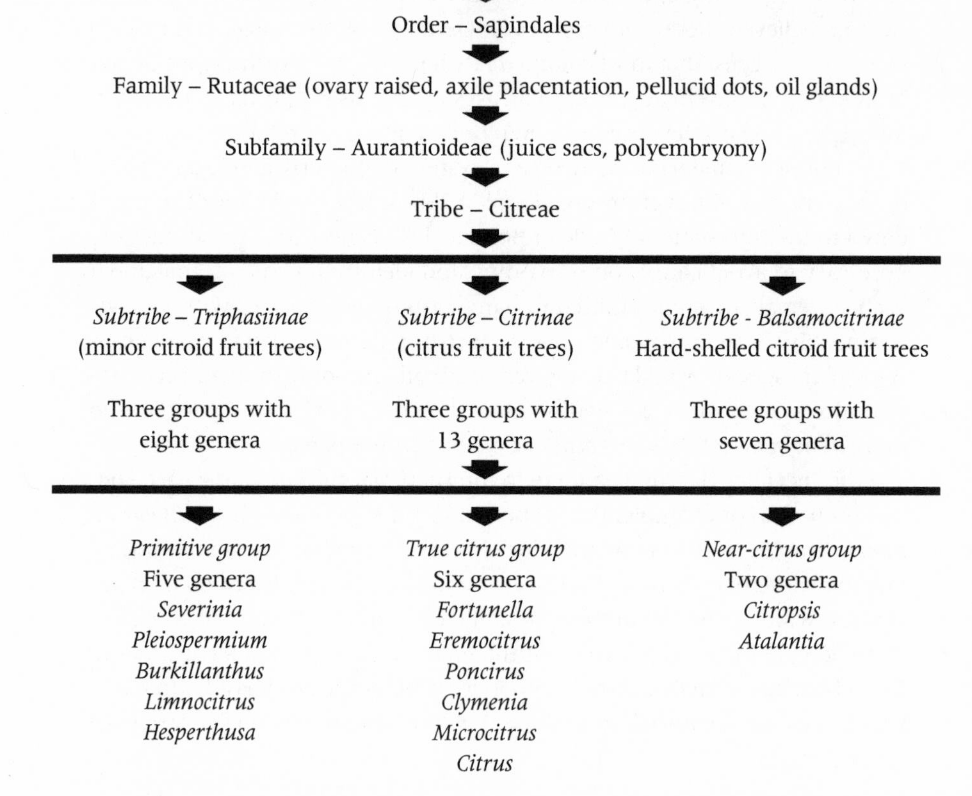

clade (34 plant families). This clade is further divided into orders (Sapindales) and families within this order such as the Rutaceae. The new system is based on molecular-genetic relationships among species unlike the classical taxonomic systems of the past. Plants in the Rutaceae also include the cork tree (*Phellodendron*), orange jessamine (*Murraya*), and prickly ash (*Zanthoylum*). Important family characteristics of the Rutaceae include presence of oil glands, presence of pellucid (translucent) dots in leaves, an ovary raised on a nectary disc, and axile placentation. The nectary disc contains sugars and aromatic compounds that attract pollinators to the flower. Axile placentation refers to the arrangement of the seeds around the central axis of the fruit. The area of seed attachment to the central axis is the placenta. Therefore, most but not all species in this family will have these basic characteristics.

The next subordinate taxon to family is the subfamily, Aurantioideae. Plants in this subfamily have unusual characteristics including presence of juice sacs in the fruit and polyembryonic seeds (see next section on botany). Fruits in this subfamily are called "hesperidium berries," which are typified by a single enlarged ovary surrounded by a leathery peel. A hesperidium berry differs from other true berries, like tomato or blueberry, due to the presence of the leathery peel.

The Aurantioideae was divided by Swingle into two tribes, the Clauseneae and the Citreae, the latter of which includes citrus and citrus-like genera. The tribe Citreae is further divided into three subtribes, of which the Citrinae contains the true citrus group. This in turn is separated into three groups: the primitive citrus group, including the ornamental shrub Severinia; the near citrus group including Citropsis and Atalantia; the true citrus group, which contains six genera. Three of the six genera (Clymenia, Microcitrus, and Eremocitrus) are of little or no commercial importance, while the other three include all of the commercially grown citrus fruits.

DESCRIPTIONS OF THE TRUE CITRUS GROUP

The three commercially unimportant yet botanically interesting genera of the true citrus group occur only in Australia and some of the southwest Pacific islands. *Eremocitrus,* native to southern Queensland and northern New South Wales, is the only distinctly xerophytic species in the subfamily and is called the "desert lime" because it is adapted to semiarid and arid regions. It has strap-shaped leaves with thick cuticles (the waxy layer covering the leaf) to retard water loss. Additionally, it has great tolerance of saline soils and may be of potential value as a rootstock for citrus trees in areas where the soil has a high salt content. It also has been used as a parent in some rootstock breeding programs in an attempt to improve drought and cold tolerance of the progeny. *Microcitrus* species are also native to eastern and northern Australia, and one species is native to New Guinea. Some species show drought tolerance, but others are found in areas of high rainfall. Leaves, as the name implies, are quite small, although the tree itself grows very large. The fruits are known as "wild limes" and are quite acid. Some of these species also may have value as rootstocks for commercial citrus fruits. The third genus, *Cly-*

menia, is known only from the island of New Ireland in the Bismarck Archipelago, northeast of New Guinea, and has not been extensively studied or propagated. It was geographically isolated from other citrus types millions of years ago. It has very large unifoliate leaves but differs morphologically from true citrus, as would be expected.

The three genera of true citrus of commercial significance are *Poncirus* (trifoliate orange), *Fortunella* (kumquat), and *Citrus*. These will be discussed in more detail below.

Trifoliate Orange (*Poncirus trifoliata* [L.] Raf)

The genus *Poncirus* was separated in 1815 by Rafinesque from the genus *Citrus*, but the Linnaean name *Citrus trifoliata* persisted until Swingle transferred it to another genus in 1915. It has only one species, *Poncirus trifoliata*, and is distinguished from all other true citrus fruits by being deciduous and having trifoliolate leaves. It is also distinct in having flower buds which form during the summer previous to their spring opening (as is common in apples and peaches) and in having the fruit covered with fine hairs (pubescence). Since all other citrus species and their close relatives are of tropical or subtropical origin, it seems likely that this species is derived from a primitive tropical form that migrated northward long ago when the climate became warmer and managed to adapt to the cool climate of what is now northern China by developing a deciduous growth habit. The trifoliolate leaf is likely a very primitive trait, and it is somewhat surprising that no additional species have evolved. *Poncirus* was cultivated for centuries as an ornamental tree in China and in Japan (which received it from China in the eighth century). It came to the United States in 1869 as an introduction by William Saunders for the United States Department of Agriculture. *Poncirus trifoliata* is the most freeze hardy of the true citrus group. It can be grown outdoors as far north as Long Island, New York, and will withstand temperatures of 0°F when fully dormant. This freeze hardiness seems to be a dominant character, which may be transmitted to hybrids of less hardy species such as orange and grapefruit. The common name, trifoliate orange, is a misnomer since the fruit are similar to the sweet orange only in their general shape. The pulp is not only acid but also contains very bitter oil, which makes it inedible. Nevertheless, both immature and mature fruits were utilized, along with powdered rhinoceros horn, in classical Chinese medicine. The fruit is small, round, yellow at maturity and has a distinct pubescence and numerous seeds.

The trifoliate orange is further categorized into two very similar forms, distinguished only by the size of the flowers (large-flowered and small-flowered). *Poncirus* is used primarily as a rootstock, particularly in Japan and China. It has also been used sporadically in Florida since 1892 as a rootstock for citrus because it is freeze hardy and confers some degree of freeze hardiness to the scion. In Florida, as in Japan and China, it is the primary rootstock used for satsuma mandarins and the one often used in the northern part of the state for kumquats, oranges, or mandarins. Scion cultivars on this rootstock apparently are very cold-hardy when they are fully acclimated because of the tendency of this rootstock to prevent growth during the period when *Poncirus* is "dormant." It is unclear whether this is true dormancy (endodormancy) as exhibited in crops like apples and peaches, or dormancy that can be interrupted by environmental conditions (ecodormancy). In addition, *P. trifoliata* tends to stop growing at higher temperatures than other rootstocks such as rough lemon, thus imparting increased hardiness to the scion. When scions on this rootstock are actively growing, however, they are no hardier than on other rootstocks, which is a distinct disadvantage in years of early cold weather following a warm fall period. The trifoliate orange is susceptible to citrus exocortis viroid (CEV), a disease that seriously reduces tree vigor.

Poncirus readily forms hybrids with species of *Citrus*. Most important of the hybrids are those resulting from crossing trifoliate orange with sweet orange and grapefruit. The first *Poncirus* by sweet orange hybrids were produced by W. T. Swingle in 1897, working at Eustis, Florida, and named "citranges" by Webber and Swingle in 1905. The citranges usually grow true-to-type from seed, because they rarely develop a sexual embryo but form several nucellar embryos. Nucellar embryos arise from the nucellus within the seed and are genetically identical to the mother. Citranges show a considerable range of variation in degree of deciduousness, leaf morphology, and freeze tolerance, being somewhat intermediate between the two parents in these characters. They are deciduous during some years, trifoliolate, and usually more cold-hardy than sweet oranges.

Unfortunately, citranges usually inherit the acidity of trifoliate orange but are less bitter. The fruits are juicy and presumably as rich in vitamin C as lemons. The cultivars Rusk (from the first crosses in 1897) and Troyer (from a cross by Swingle in 1909 at Riverside, California) attracted attention as possibly valuable rootstocks. Neither has been widely used in Florida, but Troyer is an important rootstock in many other citrus regions. The Carrizo citrange, which is a sister seedling to Troyer, is currently very popular in many areas including Florida to replace lemon-type rootstocks that are highly susceptible to blight or freeze damage. Several other citranges are also being investigated as rootstocks, but they are not widely used at present in Florida with the exception of Kuharske citrange (see chapter 5). Several *Poncirus* by grapefruit hybrids (citrumelos) have been produced, most of which have been of little commercial value. However, Swingle citrumelo has been the most widely used rootstock in Florida for the past 19 years due to its freeze hardiness, blight and foot rot tolerance, and high fruit quality it imparts to the scion.

As mentioned earlier, the trifoliate orange has been used as an ornamental for centuries in northern China and Japan. It has white flowers, and the twigs and thorns are dark green when the leaves have fallen, making the tree visually attractive in all seasons. The long, sharp thorns make a formidable hedge. The thorns also help make this an excellent species for wildlife protection. Thickets of trifoliate orange trees give excellent cover for ground-nesting birds, rabbits, and other animals and have often been planted as game refuges by wildlife associations.

Kumquat (*Fortunella* sp.)

The kumquat has been cultivated in China for at least a thousand years and in Japan for several centuries, but somehow it remained unnoticed by early European plant explorers. Ferrari, in his account of citrus fruits in 1646, mentioned kumquats as Chinese fruits of secondary importance, and Kaempfer included them briefly in his 1712 account of plants cultivated in Japan. In 1846, Robert Fortune brought oval kumquat trees back to England from China, where he had been collecting plants for the London Horticultural Society. Attempts may have been made earlier to introduce the kumquat by seed, but it does not grow well on its own roots, and such introductions probably died before they fruited. Thunberg named the common oval-fruited species *Citrus japonica* in 1784. This name was used until 1915, when Swingle reassigned *Fortunella* to a different genus in honor of Robert Fortune.

A specimen of oval kumquat was brought to America within three or four years of its reaching England and probably was introduced into Florida very soon afterward. In 1885, however, both the Glen St. Mary Nursery and the Royal Palm Nursery made importations directly from Japan of the oval and small, round kumquats under their Japanese names, Nagami and Marumi, respectively. Most of the species growing in Florida are derived from these introductions.

Kumquats are shrubby evergreen trees, rarely over 10 feet high, with dense branching and small, apparently simple leaves with much reduced petiole wings. The underside of the leaf has a silvery appearance. Most kumquats are thornless. The fruits are small, ovoid, or globose in shape, less than two inches long or wide, and have a very thick, sweet peel and mildly acid pulp. The seeds are distinctive in having green rather than white cotyledons (seed leaves), as observed in oranges and grapefruit. Kumquats are more cold-hardy than any commercial citrus species, and are able to survive temperatures of 10°F when fully dormant, especially on trifoliate orange rootstock. This freeze hardiness is due in part because trees have less tendency to start growth during warm periods of the winter than do citrus species. Moreover, kumquats typically do not bloom until late in the spring (May), long after danger of frost is past.

The fruits have a brilliant orange peel color and are very attractive. The name "kumquat" is an English form of the Chinese words for "golden orange." Kumquats are currently one of the top selling citrus-type ornamentals in Florida. The kumquat is slow growing and very suitable for landscaping. Since the fruit color is fully developed by December, the kumquat also makes an attractive living Christmas tree.

Three cultivars of kumquat are grown most commonly, Nagami, Marumi, and Meiwa. Nagami, the oval kumquat, is a species, *F. margarita,* separate from the Marumi, or small round kumquat, *F. japonica*. The large round kumquat, Meiwa, was not introduced to Florida from Japan until 1911. Swingle named it *F. crassifolia* in 1915, but in 1943 decided that it was a hybrid between the other two species and withdrew the specific name. Nagami is the most popular kumquat due to its vigor and productivity. Its fruits are oval, from 1¼ to 1¾ inches long and about ⅔ as wide, and have two to five seeds. It has a pleasant, but somewhat acid flavor. Meiwa has globose fruits, 1 to 1½ inches in diameter, and is often nearly seedless. Its peel is much thicker than that of the other two species, and it has a somewhat sweeter taste. Marumi fruits are also globular but small, rarely over an inch across, and have one to three seeds. The flavor is good but often is considered inferior to that of Nagami; the peel is thinner, and the trees are slightly thorny instead of practically thornless. Marumi appears to be a little more cold-hardy than Nagami, which is slightly hardier than Meiwa. Nagami is the species best known in China, while Meiwa and Marumi are known primarily from Japan, although undoubtedly they were introduced from China.

The fruits are used chiefly for decorating gift boxes of oranges and grapefruit and for making preserves. Sprigs consisting of twigs and fruit are cut for gift boxes. Judicious pruning methods are used by gift fruit shippers to assure continued fruit production. Kumquats are excellent for making marmalade or candied fruits, but most cultivars contain rather large seeds. Fruit may be eaten fresh including the peel or used as an attractive garnish. Kumquats are also popular as decorations for buffet and dinner tables.

The superior freeze hardiness of kumquats suggested they would be useful in breeding programs to develop cold-hardy acid fruits, since most acid citrus fruits such as lemons and limes are more cold sensitive. Swingle first made such crosses in 1909 using the small round kumquat and the Key lime; he coined the name "limequats" for the resulting hybrids. Limequats are sometimes used in dooryard plantings in Florida.

Citrus

Commercial production of citrus fruits is limited to certain species within the genus *Citrus*, which had its primary center of origin in northeastern India and a secondary center in southern China but became widely disseminated throughout southeastern Asia and the adjacent island chains. An amazing diversity of species and natural hybrids has developed over thousands of years, and authorities differ dramatically regarding which of these forms are true species as described previously in this chapter. A number of different types undoubtedly have arisen as the result of extensive crossing among the original species. Some botanists and systematists recognize these resultant forms as true species, although they are not. Most of the varied forms of *Citrus* are known only as cultivated plants. As mentioned previously, Swingle recognized only 16 species. In contrast, Tanaka recognized 162 species of *Citrus*, accepting in this category most (but by no means all) distinctive forms, regardless of the theoretical possibility of a hybrid origin so long as such origin was not proven. Swingle's taxonomic classification is narrow and Tanaka's is broad in the number of species they assign. Tanaka made a particularly strong case for recognizing several species of mandarins instead of only one.

Recent work using molecular biology techniques suggests that many commercial citrus scions and rootstocks originated as hybrids involving only three true species: the pummelo (*C. maxima*); the citron (*C. medica*), and the mandarins (*C. reticulata*). Eureka and volkamer lemons and sweet lime are hybrids of citron and sour orange; rough lemon and Rangpur lime are hybrids of citron and mandarin; sour orange and sweet orange are hybrids of mandarin and pummelo; and grapefruit is a pummelo-sweet orange hybrid (fig. 3.1).

Nevertheless, the following morphological characters are common to all *Citrus* species.

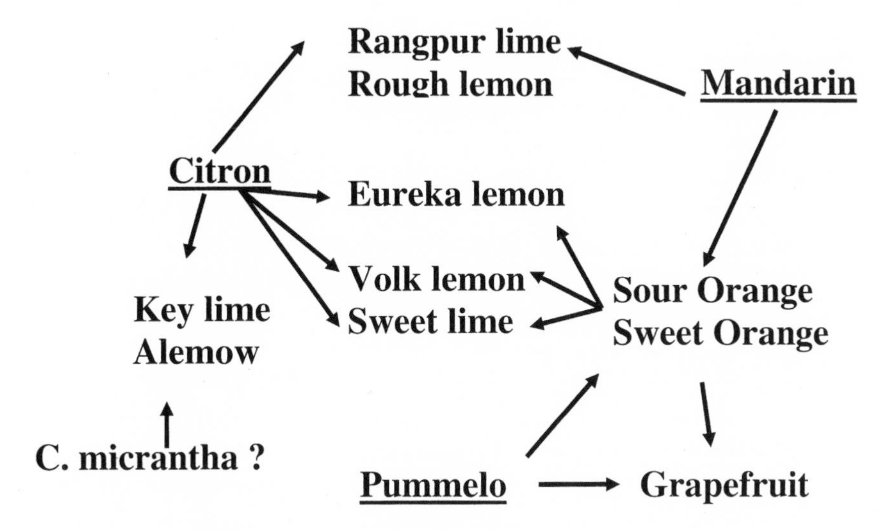

Figure 3.1. Interspecific origin of commercial citrus. Source: J. Chapparo, University of Florida.

1. Trees are evergreen, small to medium in size, and sometimes thorny, depending on age and species. For example, young trees tend to have more thorns than older ones, and lemons and limes have more thorns than sweet oranges and mandarins.
2. Leaves appear unifoliolate (although plants are trifoliolate in primitive species), the petioles may be with or without wings, and the blade is jointed to the petiole (except in the citron). Leaves persist for two or more years in Florida and their main drop period is in the spring.
3. The flower buds do not have protective scales and are formed just prior to a growth flush, normally in late winter or early spring, and occasionally in the summer. Flowers are usually large and fragrant, and petals are white except for citron, lemon, and Tahiti and Key limes, which have some pink or purple coloration. Both stamens (male) and pistil (female) are normally present in the same flowers (perfect flowers). All floral components are normally present, including sepals, petals, stamens, and carpels (complete flower). Carpels are foliar units that collectively comprise the ovary. They roughly are equivalent to the fruit segments.
4. The typical citrus fruit, the hesperidium berry, is spheroidal, oblong, oblate, or prolate (lemons) in shape, with a leathery peel, which is green when immature and at maturity is green, yellow, orange, or red depending on species. The peel possesses abundant oil glands, which are developed

by the separation and degradation of some of the subepidermal cells. These glands contain essential oils which differ characteristically among species. Oil from the peel of orange, lemon, and lime is commercially important for use in soft drinks, perfumes, soaps, hand cleaners, and other products. The inner portion of the peel (mesocarp) is a whitish, spongy material known as the "albedo," which often has a bitter favor. (Any time peel zest is used in cooking, care should be taken to remove only the flavedo and not the albedo.) The outer, colored portion containing oil glands and pigments is called the "flavedo" (exocarp). The interior of the fruit, within the peel (endocarp), is divided into several segments (most citrus fruit have 8 to 18 segments) by thin, membranous walls; the number of these segments can be used to distinguish among species in some cases. Each segment contains juice vesicles, except for the space occupied by the seeds (when present); these vesicles normally have thin, easily ruptured walls.

5. The juice in the vesicles contains over 400 different compounds, including sugars (sucrose, glucose, fructose), organic acids (including citric, malic, and ascorbic acid, or vitamin C), pigments, glucosides (which are often characteristic of species), carotene (vitamin A) in some species, and inorganic salts as well as traces of other organic substances. Taken together, these constituent compounds are termed "total soluble solids" (TSS) of the juice; sugars form by far the largest part of the soluble solids in juice, about 75–85 percent.
6. Seeds often contain embryos formed from the nucellus in addition to the usual embryo formed by sexual fertilization following pollination. These nucellar embryos are derived wholly from the mother plant and thus produce seedlings that are genetically identical to the maternal tissue. Apparently the stimulus of fertilization is needed for nucellar embryos to begin development, and the sexually derived embryo may fail to develop in many cases. Seed number varies from none (Tahiti lime), to few (lemons), or many (Duncan grapefruit). There may be from 0 to 12 seeds in each segment. The cotyledons (seed leaves) are usually white but are pale green in mandarins.

Although citrus systematics is complex and often confusing, the genus *Citrus* is most conveniently divided into five primary groups, which are easily identified by growers, brokers, and consumers. They include sweet oranges, mandarins (tangerines), grapefruit, lemons, and limes. The characteristics of the most commercially important species and cultivars will be discussed in chapter 4.

BOTANY AND LIFE CYCLE

Species in the genus *Citrus* are evergreen perennials which may live for hundreds of years. The oldest documented age is 421 years for a citrus tree growing in an orangery (enclosed area) in France, but this is obviously an unusual and artificial situation. Citrus trees may live for more than 200 years under commercial conditions in some citrus regions, but the oldest trees in Florida are about 100 years old. Freeze, pest, disease, and urbanization pressures over time are the primary causes of reduced tree longevity.

The life cycle of citrus trees is typical of that of most seed plants (fig. 3.2). It begins with seed germination, followed by growth of the vegetative seedling (the juvenility period). The tree then becomes mature when it flowers and fruits consistently. The adult phase may last many years, although productivity typically decreases for very old trees. Various phases of the citrus life cycle are discussed in the following sections.

Seed Germination, Seedling Growth, and Juvenility Period

Citrus seeds are often morphologically distinct from one another and representative of a particular species. The seed consists of a seed coat surrounding one to many embryos each with two cotyledons (fig. 3.3a). The characteristic of having several embryos (polyembryony) is unusual in the plant world and occurs primarily in the Aurantioideae subfamily. Seed germination occurs when temperatures and moisture are favorable. Germination occurs in one to two weeks after planting at optimum temperatures of 85–95°F with emergence of the radicle (primary root).

The primary root helps to stabilize the new seedling in the soil or planting medium. The epicotyl (primary shoot) then emerges, and both begin to elongate (fig. 3.3b). The seed remains in the soil or medium (hypogeous germination) versus moving upward out of the soil as the shoot grows, as is the case with beans (epigeous germination). The primary root will form the tap root of the seedling and acts to anchor it in the soil (fig. 3.3c). It produces lateral roots which further proliferate into the soil, finally forming feeder roots which are important in nutrient and water uptake (fig. 3.3d). The primary shoot grows from a specialized group of cells at the tip called the "apical meristem" (AM). All of the new cells in this shoot originate from the AM. With time, buds (resting shoots) form in the leaf axils, and branching occurs via lateral shoot formation (fig. 3.3d).

The production of these vegetative structures represents the juvenile phase

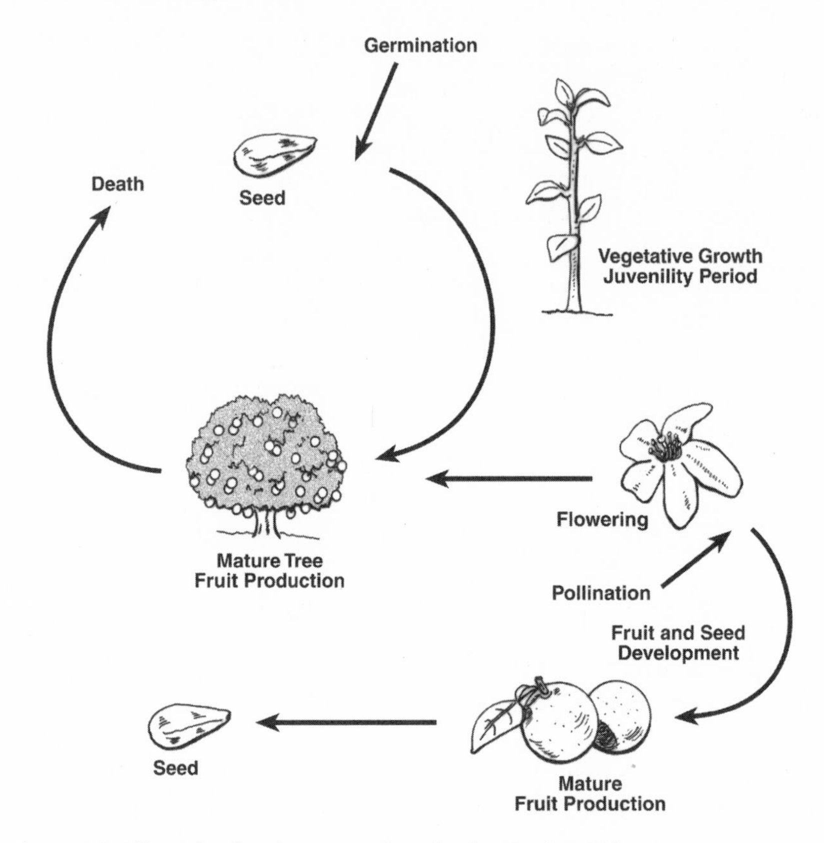

Figure 3.2. Life cycle of a citrus tree. Drawing by Katrina Vitkus.

of seedling growth. During this period, plants grow very vigorously and have many thorns. The juvenility period is the time until the first consistent flowering occurs. Many of us have eaten a citrus fruit and planted seeds in pots along the window sill. The resulting young seedling initially may produce a few flowers, but the production is not consistent over the long run. In fact, the protracted juvenility period is one reason most commercial citrus is not produced from seed. The juvenility period in citrus ranges from one to two years for limes to as long as 15 years for some sweet oranges! The juvenility period is directly related to growth rate and may be shortened by high temperatures and rainfall and very favorable growing conditions. Thus, the juvenility period is much shorter in lowland tropical than in subtropical regions. For example, a two-year-old tree growing in Costa Rica may be the same size as a four-year-old tree growing in Florida. The tree enters the ma-

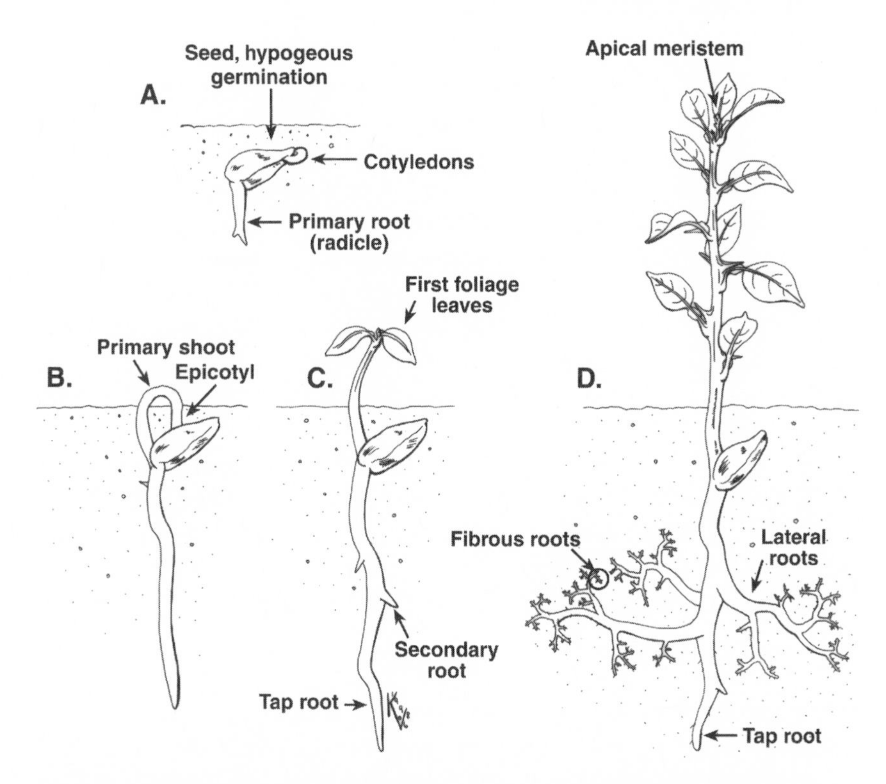

Figure 3.3. Developmental sequence of citrus from seed through flowering: (a) seed, (b) young seedling primary root and shoot development, (c) young seedling development of first foliage leaves, (d) young seedling lateral and fibrous root development. Drawing by Katrina Vitkus.

ture (adult) phase when it is capable of flowering and fruiting consistently. During this phase, the tree stores large reserves of carbohydrates (starches), nutrients, and water in the trunk and branches.

Vegetative growth occurs primarily in distinct flushes from buds located in the leaf axils, while roots grow from apices covered by a root cap. In Florida's subtropical climate, there are typically three growth flushes for mature trees which occur in late winter and spring, summer, and fall. During the late winter and spring flush, root growth precedes shoot growth. The root and shoot growth flushes overlap during the summer- and fall-flush periods. The spring flush produces many short shoots; the summer flush has fewer, longer shoots; and the fall flush has a small number of short shoots. Young trees may have three to as many as five growth flushes per season.

Shoots elongate with growth originating from the terminal bud and ceasing at various times, producing a wide range of shoot lengths within a growth flush. The terminal bud then abscises and subsequent growth occurs from lateral buds, producing a distinct zig-zag growth pattern termed "sympodial" growth (fig. 3.4). As the new shoot elongates, leaves and new buds are produced and develop along it, and the process repeats itself. Shoot-growth rate is dependent on temperature, water, nutrient status, and genetic factors. For example, a lemon tree naturally grows much more rapidly than a mandarin tree under the same environmental conditions. Growth also may occur from adventitious buds which are hidden underneath the bark in branches and the trunk. Adventitious bud break is apparent after severe freeze damage or pruning occurs and helps to revitalize and renew the tree.

Root growth occurs from root apices from cells just below the root cap. The major types of roots produced are the tap root, the pioneer (structural) roots, and the fibrous (feeder) roots (fig. 3.4). The tap and pioneer roots serve as support and storage organs, and the fibrous roots, with their tremendous surface area, are important for water and nutrient uptake.

Flowering

As mentioned previously, the citrus flower typically has five sepals, five petals, 20–40 stamens, and a single ovary composed of 8–18 fused carpels. A carpel is roughly equivalent to a segment (fig. 3.5a). The flower is perfect: It has both male (androecium) and female (gynoecium) parts and is complete in having all four types of organs (sepals, petals, stamens, and ovary). Flowers develop from vegetative buds. The bud is transformed in the fall and winter during periods of low temperature or water stress and develops into a flower prior to opening (anthesis). The new shoots may produce only flowers without new leaves (bouquet or leafless bloom) or flowers with new leaves (leafy bloom). Fruit set is usually higher for leafy than bouquet blooms. The flowers develop along a central axis in a distinct branched pattern called a "cyme," or they may develop singly in leaf axils (fig. 3.5b). The terminal flower is the largest and opens first followed by the most basal lateral flower on the shoot. The other lateral flowers then open sequentially from the base to the apex of the shoot.

In Florida's climate, most flowers are produced in the spring. Large citrus trees may produce over 100,000 flowers. Occasionally, flowers are also produced in May and June, which is termed the "off" or "June" bloom. Fruit developing from these flowers usually is of poorer quality and matures later than that from the spring bloom. However, off-bloom fruit development

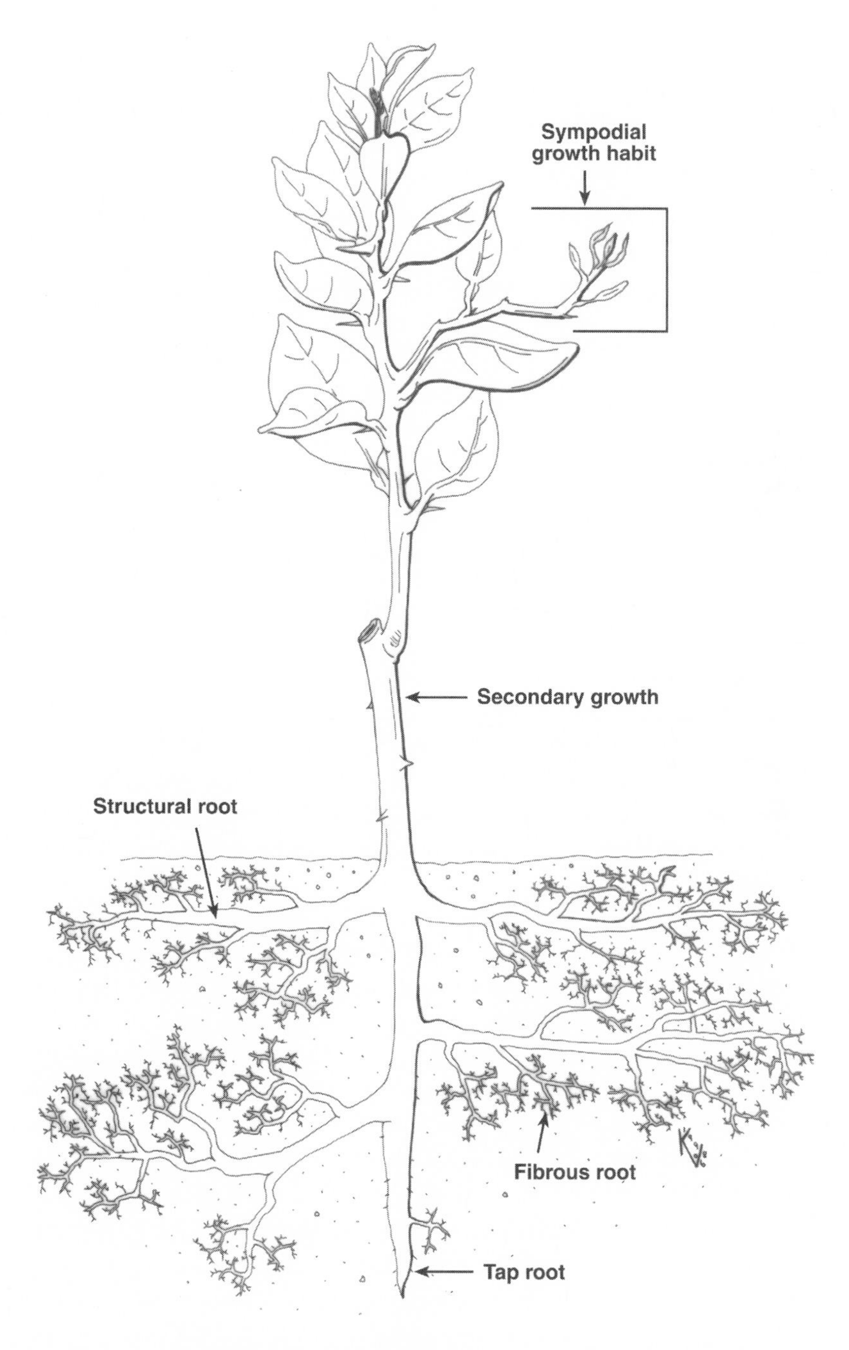

Figure 3.4. Vegetative growth of a budded citrus tree showing sympodial and secondary growth. Drawing by Katrina Vitkus.

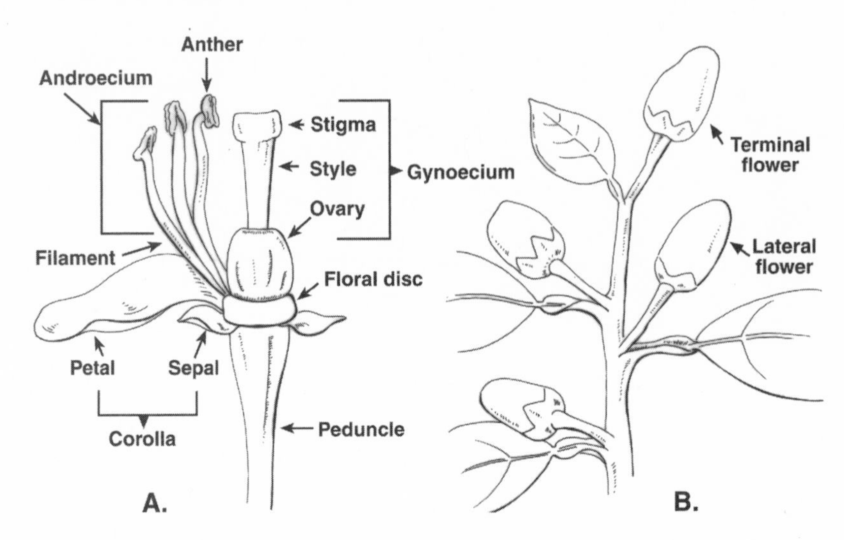

Figure 3.5. Diagram of a (A) a citrus flower and (B) a flowering shoot called a cyme. Drawing by Katrina Vitkus.

from limes may bring a superior price and is a desirable characteristic. Very few (if any) flowers are produced in the summer and fall flushes. In lowland tropical regions, some coastal areas of California, and for lime trees in Florida, flowering occurs in small numbers over the entire year, with several different-sized crops produced on the same tree. Some cultivars of lemons are termed "everbearing" because of this year-round production of flowers and fruit.

Fruit Drop

Citrus trees produce far more flowers than they can mature into a crop. For example, if only 10 percent of flowers became fruit, a single tree might produce over 10,000 fruits! In most cases, production of several thousand fruit per tree is considered a good yield. Citrus trees have evolved control mechanisms using the process of abscission (fruit drop) to reduce crop load. Only 0.5–2.0 percent of flowers mature into a harvestable fruit. There are three major periods of fruit drop: postbloom, June, and preharvest. During postbloom drop, nearly 98 percent of the flowers and young fruitlets abscise. This natural drop removes weak or defective flowers and fruit. June drop (which generally occurs in May in Florida) causes a further reduction in crop load. Again, small fruits that cannot compete for nutrients and water are naturally thinned from the tree. However, sometimes June drop is extensive if trees are heat- or water-stressed, and yields are reduced significantly. Extensive June drop is of particular importance in arid regions such as southern California and Egypt, where daytime temperatures exceed 100°F. Fruit drop following this period is inaccurately termed "preharvest" drop because any fruit that drops does so preharvest. In fact, it is better termed "late-season" drop and can greatly reduce yields, especially for some cultivars such as navel and Pineapple oranges. Late-season drop also occurs when fruits are stored on the tree long past their normal harvest period, as is sometimes the case with grapefruit and some sweet oranges.

Pollination

Unlike some apple and blueberry cultivars, most citrus species do not require cross-pollination with a second cultivar in order to set a good crop. Species that do not set a crop when planted without cross-pollination are termed "self-incompatible." Oranges, grapefruit, lemons, limes, and most mandarins may be planted in large blocks containing only a single cultivar. Thus, citrus for the most part is self-compatible. Moreover, most cultivars do not need bees to produce an adequate crop, as is commonly believed. Citrus trees are the source of excellent quality, flavorful honey, however.

In contrast, some mandarins (tangerines) and mandarin hybrids do require cross-pollination with a compatible cultivar called a "pollinizer" (a plant) in order to produce adequate, consistent crops. They also require a pollinator (agent of pollination—in this case, the bee) to transfer the pollen among flowers. These include (among others) Orlando and Minneola tangelos, and Robinson and Sunburst tangerines. It is important to select a pollinizer that has certain characteristics to ensure consistent cropping:

1. The pollinizer and primary cultivars must bloom at the same time. Navel oranges (which produce non-viable pollen) tend to bloom earlier and kumquats much later than most other cultivars and thus would make poor pollinizers.
2. The pollinizer must produce a sufficient number of flowers and pollen each year to ensure cross pollination. In cultivars prone to alternate bearing, such as Murcott, few flowers are produced in the "off" light crop year.
3. The pollinizer must be of similar freeze hardiness to the primary cultivar to ensure sufficient flower production. Temple orange is a good pollinizer for many cultivars, but it is cold sensitive and may sustain damage during moderate freezes that reduces flower number. Therefore, it should not be used as a pollinizer in very cold areas.

4. The pollinizer must be self- or cross-compatible with the primary cultivar and must produce a salable crop. For example, Minneola is not compatible with Orlando, and Orlando is also incompatible with Minneola. Duncan grapefruit may be a possible choice as a pollinizer, but the fruits are of little value commercially.
5. The pollinizer and primary crop should have similar cultural practices. For example, Minneola and Temple are susceptible to citrus scab fungus disease, while Robinson and Sunburst are not. Consequently, the spray programs may differ, which may make management more difficult. Moreover, mixed plantings may cause harvesting problems because they reach maturity at different times of the season.
6. Both the pollinizer and primary cultivar must be attractive to bees, which are the primary pollinators. Hives must be placed in the grove during flowering, and growers must avoid application of toxic pesticides at this time.

Pollination schemes—the placement of pollinizers—can be done in several ways. The pollinizer trees should be placed no further than two rows from the primary trees to allow for effective bee movement and pollination. In typical schemes the pollinizers may be placed every fourth row (1:3); in groups of two rows (particularly where double beds are used); or every other row (1:1). In some cases, pollinizers are placed within the row about every third tree. This scheme provides for excellent transfer of pollen but may be difficult to manage during harvesting if fruit matures at different times.

Pollination is the transfer of pollen (male gamete) to the stigma (female), which may result in fertilization (fusion of the male and female gametes within the ovary) and seed formation. Pollen may be transferred within the same flower (self-pollination), from one flower to another on the same tree (also selfing), from one clonal tree to another (also selfing), or from one flower to another of a different cultivar or species on a different tree (cross-pollination).

During pollination the anther dehisces bright yellow pollen grains that are haploid. They contain a single set of 9 chromosomes, compared with 18, which is the diploid (normal) chromosome number. The pollen grains adhere to the stigma, which is sticky when receptive, and are carried in the pollen sacs of bees. Citrus pollen is not readily carried by the wind.

Fertilization

The pollen grain then germinates, sending the pollen tube down through the style. There one nucleus joins with the haploid (one set of chromosomes) egg cell, creating the zygote, which is diploid (two sets of chromosomes). A second nucleus joins with the polar bodies, creating the endosperm, which is triploid (three sets of chromosomes). This process is termed "double fertilization." The zygote (sexual) embryo develops within the seed and is a unique organism having the combined genomes of the male and female parents. In many citrus types, embryos also develop spontaneously from the nucellus. They are genetically identical to the diploid maternal tissue and are called "nucellar embryos." There may be one to as many as seven of these developing within the seed, and often they outgrow and crowd out the zygotic embryo. Nucellar embryos are very important to the citrus industry. They are widely used to produce clonal rootstocks. These rootstocks are true-to-type and free of most virus and viroid diseases, except occasionally psorosis. The final product of pollination and fertilization is the seed. In the mature seed the endosperm and nucellus are largely used up in providing nutrition to the developing embryos. The seed then begins a new life cycle, and the process of growth and development repeats itself.

Fruit Growth and Development

Following pollination the stigma and style abscise and cell division begins in the fruit. This is stage 1 of fruit development (fig. 3.6). In some cultivars, cell division occurs without sexual fertilization and fruit are seedless, a process termed "parthenocarpy." There are various degrees of parthenocarpy in citrus. Marsh grapefruit and Tahiti lime are examples of strongly parthenocarpic fruit and navel orange, and Orlando and Minneola tangelos are examples of weakly parthenocarpic fruit. Parthenocarpic (seedless) fruit such as navel oranges and Afourer and Clementine mandarins usually bring the highest prices in fresh fruit growing regions such as California and Spain. Most of the cells present in the mature fruit are produced during stage 1. These cells then differentiate into various tissue types. (Some researchers identify "differentiation" as stage 2.) The fruit cells then expand, and the fruit grows considerably during stage 2 of development termed cell expansion. Sugars and juice content increase and acids decrease during stage 2. In stage 3, fruit matures, the peel changes color, and growth rate and sugar accumulation slow down. The fruit is then ready to harvest when the minimum maturity standards are met (see chapter 11). The fruit consists of an exocarp (peel), mesocarp(white albedo tissue), and the endocarp (edible tissue). The length of each stage of development varies with climate and species. In lowland tropical areas, fruits of Valencia orange attain final size in 6 months after flowering, whereas the process may take 18 months in cool, coastal regions

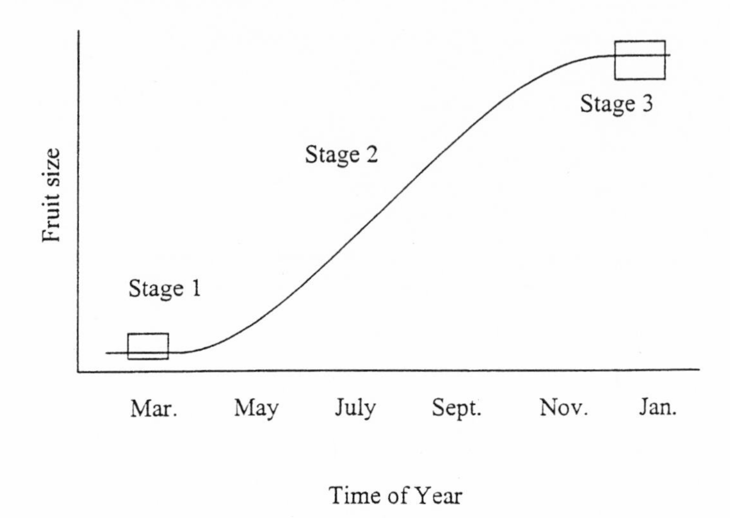

Figure 3.6. Diagram of a typical sigmoidal fruit growth curve for citrus (Hamlin orange): stage 1, cell division; stage 2, cell elongation; stage 3, cell maturation.

of California. In Florida, Valencia oranges attain full size about 10–12 months after flowering. Naturally small fruits (such as lemons or limes) attain final size much sooner than large fruits (such as grapefruit) under the same climatic conditions, and the duration of each growth stage differs.

The Mature Citrus Tree

The mature, bearing citrus tree can remain productive for many years, as discussed previously. Maximum, consistent production in Florida generally occurs 10–15 years after planting in the field. Yields in Florida range from 400 boxes per acre (95 pounds per box) for tangerines, to 1,000 boxes per acre (90 pounds per box) for Hamlin oranges, to more than 1,200 boxes per acre (85 pounds per box) for grapefruit. Often, especially for tangerines, these yields vary considerably from year to year. Many groves never achieve these yields due to production problems such as inadequate irrigation, fertilization, or disease and pest control. In addition, yields have been reduced considerably by debilitating diseases like citrus canker and greening in recent years (see chapter 10). As trees age, fruits are produced toward the outside and top of the canopy due to shading of the lower branches. Yields generally decrease, and trees have a large percentage of dead or unproductive wood. Eventually, trees become commercially nonproductive and are removed (pushed) or, in some cases, pruned to rejuvenate them (see chapter 9).

FOUR

Cultivars and Cultivar Improvement

For many centuries, citrus trees were grown from seed. Seedling trees were often vigorous but were slow to come into production. Furthermore, citrus was disseminated predominantly by seed which could be easily transported and stored. Many of the earliest plantings in Florida consisted of seedlings planted on hammocks. Trees were planted in clumps, which were later referred to as "groves," a term still used more widely in Florida than "orchard."

Currently, most citrus trees worldwide and in Florida are propagated as a two-part tree consisting of a scion (above-ground portion of the tree) and a rootstock (below-ground portion of the tree comprising the root system). The many advantages to such a system will be discussed in this chapter and chapter 5 on rootstocks. Certainly, choosing the proper scion cultivar and rootstock is one of the most important decisions in any citrus production program.

CULTIVARS AND HYBRIDS

Hybrids

The most commercially important types of citrus are divided into five groups: sweet oranges, grapefruit, mandarins (tangerines), lemons, and limes. These groups are easily identifiable in the marketplace, although they are certainly very similar genetically (see chapter 3). Other types of citrus are grown to varying degrees throughout the world. These include kumquats (the Chinese golden orange), pummelos (shaddocks), sour oranges, acidless oranges, and sweet limes. None of these are of commercial importance in Florida or elsewhere in the United States.

Most citrus and related species hybridize with one another, including species of the closely related genera *Fortunella* and *Poncirus*. Undoubtedly such hybridization had a part in the development of some of our present species. However, no one can say with certainty which species are primitive and which are derivative. There are several cultivars such as Temple orange, Murcott tangerine, Meyer lemon, and Tahiti lime, which are clearly hybrids. There is not, however, agreement on the exact species involved. New molecular biology techniques may be useful in the future in determining the origin of citrus hybrids (see chapter 3).

Many hybrids have been produced through controlled crosses in which both parents are known with certainty. The first hybrids of this kind were produced by Dr. W. T. Swingle for the U.S. Department of Agriculture in 1897. Most of the many early human-made citrus hybrids were the work of Swingle and his associates in the succeeding 15 years. No one can predict the fruit characters that will result from any given cross due to the inherent genetic variability in citrus. Each seed in a single fruit produced by hand pollination may develop into a seedling with fruit and tree characters different from the others. This is true, however, only if a true hybrid (zygotic) embryo matures. As stated previously in chapter 3, the primary embryos that mature in a high percentage of citrus seeds are nucellar ones, which are derived wholly from the maternal tissue and, thus, are genetically identical to it. In some crosses, the plant breeder may never be able to produce a true hybrid or must make hundreds of crosses and grow the seedlings to fruiting before a single hybrid is obtained. Thus, only recently has it become possible to produce hybrids using Dancy tangerine as a seed parent (Dancy produces only nucellar seedlings). This is accomplished by removing young developing embryos from immature seeds and culturing them on nutrient gels, a technique known as "embryo rescue." Another innovation in producing hybrids involves the combination of genetic material at the cellular level. This technique, known as "protoplasmic fusion," can be used to make hybrids between species that could not be made using conventional breeding methods. For example, it is difficult to make viable hybrids from crosses between rough lemon and sour orange. New rootstock cultivars are currently being developed using this method. Once a hybrid has been obtained, it can be maintained permanently as a cultivar by vegetative propagation.

Hybrids may display characteristics intermediate to those of their parents, some characteristics of each parent, new characteristics that are not exhibited by either parent, or characteristics once present in an ancestral form. Only hybrids that represent definite improvements over the parents in some quality factor should be propagated as new cultivars. Qualities that are considered in new cultivar (scion) development are fruit flavor and ease of handling and shipping, seedlessness, disease resistance, vigor, productivity, and adaptation to climate. Some hybrids have potential as rootstocks because of superior adaptation to soil conditions or climate; resistance to soil or systemic diseases, insects, or nematodes; and ability to produce high-quality fruit and high yields in the scion.

Citrus hybrids are divided into the following classifications:

1. *Intercultivar*. These represent crosses made between cultivars of the same species. They usually result in little or no more variation in seedling characters than those produced by self-pollination. They are of little importance in producing new cultivars, with the exception of three midseason oranges: Sunstar, Gardner, and Midsweet, which were selected from self-polllinated seedling populations.
2. *Interspecific*. These represent crosses made between species within the same genus. These hybrids are usually produced quite easily. Most of the interesting and economically important citrus hybrids have arisen in this way, notably, the tangelos and tangors, which will be discussed later in this chapter.
3. *Intergeneric*. These represent crosses made between different genera. These hybrids must usually be closely related and be within the same family. Intergeneric hybrids are more difficult to obtain than interspecific hybrids, and many attempts have failed. Nevertheless, some successful hybrids have resulted which are valuable as rootstocks. Citranges (*C. sinensis* x *P. trifoliata*) and citrumelos (*C. paradisi* x *P. trifoliata*) are good examples of intergeneric hybrids.
4. *Complex*. These represent crosses made between interspecific or intergeneric hybrids and another species or genus. Complex hybrids are the result of two or more crosses. The difficulty of making such hybrids is even greater than for simple bigeneric hybrids. Most of these complex hybrids are not of commercial importance but are curiosities.

Citrus hybrids are known by common names, such as "tangelo" or "citrange," which consist of portions of the name of each parent, though sometimes these names are difficult to recognize. In practically all instances, they were coined by W. T. Swingle to identify the types of hybrids he had developed. The relationship between and among citrus hybrids is given in figure 4.1.

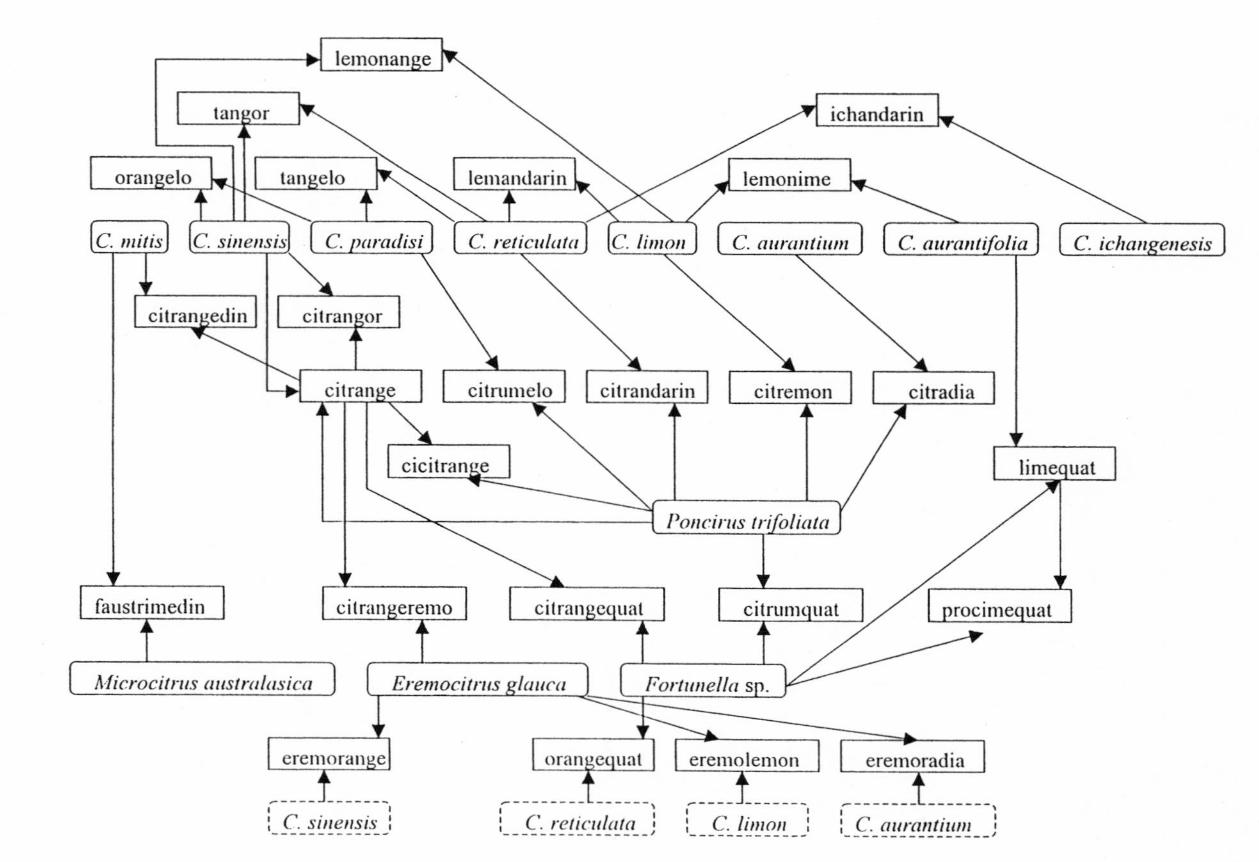

Figure 4.1. Interspecific and intergeneric hybrids of citrus and related genera.

SWEET ORANGE (*CITRUS SINENSIS* [L.] OSBECK) CULTIVARS AND HYBRIDS

General Characteristics

Both in Florida and worldwide, the sweet orange is the leading species based on total acreage, production, and dollar value (see chapter 1). Sweet orange trees have a moderately upright growth habit and branching patterns. The leaf blades (lamina) are medium in size among *Citrus* species, and the petioles have rather narrow wings which may flare out somewhat at the upper end but rarely overlap the blade. Fruits vary from oval to flattened-globose in shape and have thin, smooth, tightly adhering peels. Fruit are rarely bitter; the pulp is solid without a hollow core. Peel color at full maturity ranges from light to deep orange, especially under low temperature conditions. In regions where average temperatures are high and day-night temperature fluctuations are small, the peel may remain green (as is the case in lowland tropical regions) or become green again if previously orange. This regreening is especially a problem for Valencia oranges held on the tree after May in Florida.

The great popularity of the sweet orange is attributed to several factors. The flavor of oranges is appealing to a wide range of ethnic groups worldwide and many cultivars are available that allow for a long marketing period. Fruit are available during the entire year. The fruit are not only popular for fresh use but also can be processed into frozen concentrate juice (FCOJ), ready-to-serve single-strength juice, frozen and canned sections, frozen salads, marmalades, wines, and other products.

The sweet orange will likely remain the principal citrus fruit in Florida in the future as it has always been in the past. Most of the new citrus plantings in Florida are sweet oranges, in large part because net returns to the grower average higher than for most other citrus fruits with the exception of some tangerine cultivars.

The name "sweet orange" is used advisedly. While "orange" refers to the fruit of *C. sinensis* to most people, it is often applied to fruit of other species or hybrids that may resemble sweet oranges. "Trifoliate orange," "Temple orange," "King orange," "Orlando orange," "Satsuma orange," and "mandarin orange" are not true oranges. Only the term "sweet orange" refers unequivocally to this species and its cultivars.

Four groups of sweet oranges are recognized when the cultivars are classified by fruit characters: round, navel, blood, and acidless. The round, or common, orange has the characteristics previously listed but without development of a navel or red pigmentation in the flesh. Its cultivars constitute by far the largest part of the Florida citrus industry. Navel oranges are distinguished from round oranges by the development of a secondary or even tertiary fruit at the stylar end (located opposite from the stem end) of the primary fruit, called a "navel" for obvious reasons. Blood oranges develop red pigmentation due to anthocyanin pigments throughout the flesh or in streaks under certain climatic conditions. Sometimes they also have red coloration in the peel. Under Florida's humid, subtropical conditions, the red pigmentation usually does not develop extensively. Conversely, deep blood-red color develops under Mediterranean-type climates, with hot days and cool nights. Acidless oranges are a curiosity in Florida and have no commercial value. However, they are cultivated to some extent in some other countries such as Brazil and Egypt.

The Florida citrus industry also separates orange cultivars into three groups based on season of maturity: early, midseason, and late. These categories were developed for convenient separation of cultivars, but some overlap may occur. Early oranges are those maturing from late September through November; midseason oranges mature primarily in December and January; late oranges mature from February through August.

A great many cultivars have been selected and named in the past, only to be replaced by other more suitable cultivars. There are also many variant types, producing fruit slightly different from the parent cultivar or sometimes distinctly off-type. Citrus trees are quite subject to bud mutations, and off-types may arise in the field even years after planting. Many important cultivars are products of mutation—for example, Washington navel orange and most red grapefruit cultivars.

Sweet oranges are also classified as seedy and seedless. The latter group includes cultivars with only a few seeds (commercially seedless) as well as those with no seeds (strictly seedless). In general, if there are more than six seeds in the fruit, it is considered seedy; usually fruits classified commercially as seedless have six or fewer seeds. This situation differs from that in the European market, where only fruit with no seeds are considered seedless.

Any new cultivar that is to replace one of the established cultivars must represent a distinct improvement upon it, because usually much time and money have been invested in the established cultivars. A new cultivar cannot be a mere novelty, but must fill a production niche. For example, there is an interest in developing a high-quality, productive navel orange, an early-season sweet orange with highly colored juice, or an early Valencia orange to help fill processing needs between mid and late season oranges.

In selecting cultivars for large-scale planting, the grower is usually guided by market returns for the crop in the preceding year or, still better, over a period of years. Cultivars that consistently bring high returns and can be either shipped fresh or processed are most desirable. Another factor of equal importance is, of course, the size and regularity of the crop because total returns depend both on price per box and productivity. In addition, selections that are profitable one year may not be the most profitable in future years when the planting comes into full production. For example, Sunburst tangerines brought very high prices when first introduced, but prices declined during early 2000 and low prices continue today. The needs of the processing industry are not the same as those of the fresh fruit industry, although this difference has, in the past, had its primary effect on cultural practices rather than on cultivar selection. It is usually more advisable from an economic standpoint to plant new cultivars based on consumer acceptance and market demand than to plant on the basis of personal satisfaction with the performance of a particular cultivar.

In selecting new orange cultivars, the demands of both fresh fruit shippers and processors must be recognized. The commercial packinghouse and the gift box shipper desire oranges with attractive peel and flesh color, a smooth and thin peel, rather fine flesh texture, a good flavor balance between sugars and acids, good shipping qualities, and optimum fruit sizes. In contrast, the processing industries care little about peel color, texture, or fruit size but want high juice content with good color, high solids (sugars) content, moderate acidity, good flavor and aroma, and freedom from any off-flavors.

The commercially important sweet orange cultivars in Florida will be discussed in the following section.

Early Season Oranges

HAMLIN

This is the principal early-season orange grown in Florida where it originated. The parent tree was a seedling in a grove planted in 1879 by Judge Isaac Stone a few miles northwest of DeLand, Florida. Later this grove was bought by A. G. Hamlin. The cultivar was propagated under Hamlin's name when extensive budding was done following the severe freezes of 1894 and 1895 to take advantage of its unusually early maturity. Hamlin is usually harvested from late October to January. Fruits have less than six seeds and are thus commercially seedless. Fruit internal quality never equals that of Pineapple or Valencia, juice color is poor, and solids are relatively low. Under comparable conditions, Hamlin produces higher, more consistent yields per acre than any other sweet orange cultivar. Consequently, it was the most extensively planted cultivar during the 1980s and 1990s and is used predominantly for processing fruit. Nursery propagations of Hamlin, which is an indicator of future trends, were well below those of Valencia in the mid to late 1990s. However, from 2002 to 2006 they have been slightly greater than those of Valencia.

PARSON BROWN

Before 1920, Parson Brown was the leading early season orange cultivar, but it has been replaced by Hamlin and other early season oranges because of its seediness and low yields. The original tree was one of five seedlings growing at the home of the Reverend Nathan L. Brown near Webster, Florida. The seedlings had been given to him in 1856 by a man who said they originated from orange seeds brought to Savannah, Georgia, from China on an English ship. In 1874, Captain J._L. Carney was looking for a scion to bud into his wild sour orange trees on his island in Lake Weir. The fruit of one of the Brown trees caught his attention because it matured earlier than those on the other four trees. He bought the rights to the budwood of this tree and propagated it as the Parson Brown. The other four trees were also sources of budwood for many groves following the freezes of 1894–95 because people believed that all Parson Brown trees were genetically the same. However, the trees were not identical, and considerable variation in fruit and tree characteristics were observed as groves matured. The Carney selection represents the original Parson Brown cultivar and matures from October to December. Fruit has 10–20 seeds and a thick, slightly pebbled peel, which loses its green color very slowly. The quality is moderate, and the juice color similar to that of Hamlin, although many purport that juice color is superior to that of Hamlin. Parson Brown is used primarily for processing and is currently not widely grown or propagated in Florida.

RECENT EARLY SEASON ORANGE INTRODUCTIONS

In 1985, Dr. W. S. Castle brought seeds of Earlygold, Itaborai, Ruby, and Westin oranges to Florida from Brazil. He then propagated these on several rootstocks and compared yields, vigor, fruit quality, and time of maturity with the Hamlin orange as a standard. All of the new cultivars except Westin had superior juice color to Hamlin. They also generally produced higher cumulative pounds-solids (juice weight x total soluble solids) per acre than Hamlin. Juice color and pounds-solids are important factors in determining processing fruit prices (see chapter 11). Juice color development was earliest for Itaborai and Earlygold and somewhat later for Ruby and Westin. The earliest maturing cultivars were Ruby and Westin. Nursery propagations for Earlygold were much greater than for the other cultivars, peaked in 2001–02, and have since declined.

Midseason Oranges

PINEAPPLE

This was the leading midseason cultivar in Florida and has an obscure origin. About 1860, the Reverend Dr. J. B. Owens moved from South Carolina to Sparr, Florida. He planted orange seeds, which W. J. Crosby said he had obtained in Charleston from an English ship, probably from China. When the Reverend P. P. Bishop was looking for desirable trees to bud to his wild sour orange trees around Orange Lake in 1873, he was told about Owens's trees. He subsequently bought the tops of nine trees that Owens wanted to move. One of these trees had been named "the pineapple tree" by one of Owens's daughters. When Bishop's budded trees began producing, their fruit was superior to that of the nonbudded trees. He budded the rest of his large tract of sour oranges from this particular Pineapple tree. There is disagreement among people who were familiar with the original tree at the time it was first propagated as to the origin of the name "pineapple" in the Owens family. One account states that it was because the fruit had a pineapple flavor, another that the aroma of the fruit, not the flavor, was like pineapple. A third says it was the shape of the tree, with cylindrical sides and bushy top, which resembled a pineapple fruit. Since the fruit does not taste remotely like a pineapple and the marked aroma is noted only in a packinghouse full of this fruit (such as Owens never had), the third explanation seems the most reasonable. The Pineapple oranges from the Bishop Hoyt (later the Crosby-Wartmann) grove were well received in the market, but trees were propagated at a slow rate by other growers. After freezes of 1894 and 1895, however, when extensive replanting had to be done, Pineapple was widely planted. The fruit achieves premium quality during January and February and has an excellent and distinctive flavor. Internal juice color is dark orange, and total soluble solids are high, making Pineapple fruit excellent for processing. Moreover, peel color becomes deep reddish-orange at maturity, especially in the cool, northern regions of the state. Pineapple fruit is only slightly acceptable for the fresh market due to its seediness. Pineapple trees have moderate yields, but late season fruit drop is often a problem. Moreover, trees have a tendency toward alternate bearing, that is, the production of many fruit in one season followed by few fruit in the next season. The number of seeds varies from 10 to 20 and with modern processing equipment is not a concern to processors.

Some observations suggest that Pineapple orange trees are less freeze hardy than other sweet oranges. This may be because Pineapple is a midseason fruit and trees are sometimes harvested just before a freeze. Harvesting stresses the tree and may decrease cold hardiness. However, there is likely no inherent difference in hardiness among sweet orange cultivars.

ROBLE

This cultivar has become a somewhat popular midseason orange in recent years, although it was first introduced to Florida in 1851 from Spain and named after Joseph Roble. The cultivar has been planted sparingly and was only the 10th-ranked sweet orange cultivar propagated in 2005–06.

GARDNER, MIDSWEET, AND SUNSTAR

The USDA research station in Orlando introduced three midseason oranges: Gardner, Sunstar, and Midsweet. Gardner is an open-pollinated seedling of Sanford Mediterranean orange, Sunstar is an open-pollinated seedling of Berna orange, and Midsweet is an open-pollinated seedling of Homosassa orange. Initial testing suggested these may have superior juice color to Hamlin and are more productive than Pineapple. Midsweet, in particular, has produced high pounds-solids in some areas of the state. Gardner and Sunstar are not widely planted. Midsweet was planted to a limited extent in the early 1990s, but since then it has been the dominant midseason orange propagated in Florida and acreage of Midsweet will likely increase in the future.

Late-Season Oranges

VALENCIA

This is the most widely planted sweet orange cultivar in the world. Valencia oranges were initially sent to the Parsons Nursery in Flushing, New York, and General Sanford in Palatka, Florida, in 1870 by the Rivers Nursery, London, England. Rivers Nursery obtained their original plants from the Portuguese Azores. In 1877, E. H. Hart of Federal Point, Florida, announced that he had a tree with unusually late-maturing fruit. The tree was from the original importation by S. B. Parsons, who planted it and several other cultivars at his orange grove and other nursery at Federal Point in 1870. When the tree came into bearing, Hart noticed its unusually late maturity and found it had no identification label. The Pomological Committee of the Florida Fruit Growers Association examined specimens of the fruit, confirmed its lateness of maturity, and named it Hart's Tardif.

Another importation, also unlabeled, was made from Rivers Nursery to

California in 1876. After this tree had attracted attention, it was recognized by a visiting Spanish citrus expert as the cultivar known in Spain as *naranja tarde de Valencia*, and the name "Valencia Late" was soon adopted (1887) in California. It was not until 1914 that Hart's Late and Valencia were considered as the same cultivars in Florida, although as early as 1893 many citrus authorities in California had considered them identical.

Valencia produces premium quality fruit in Florida from late February until June depending on grove location. Moreover, fruits remain on the tree in fairly good condition all summer, although considerable fruit drop, regreening of the peel, and drying of the flesh may occur. There are usually less than six seeds per fruit. The internal quality is excellent. Fruits have high total soluble solids and superior flavor; juice color is deep orange at maturity and juice usually receives a premium price. Processors blend this high-quality juice with that of lower quality to attain a uniform product for frozen concentrate orange juice (FCOJ). Valencia orange juice also makes an excellent single-strength, ready-to-serve product. Several types of Valencias have arisen by bud mutation and are given cultivar names in regions such as California and South Africa; in Florida all types of Valencias are lumped together as a single cultivar. One exception is Rohde Red, a Valencia type with a somewhat superior juice color. It is likely a mutation that originated in the grove of Paul Rohde near Sebring in 1955. It was released in 1975 and has achieved some commercial popularity since good juice color is such a desirable characteristic. Valencia trees are less vigorous and productive than cultivars such as Hamlin. The fruit is excellent for fresh fruit as well as processing markets and Valencia production accounted for nearly 50 percent of the total Florida orange production in the 2005–06 season.

Navel Oranges

Navels as a group comprise a small fraction of the orange trees grown in Florida. The fruit often commands a premium price early in the season, but the major marketing times are around Thanksgiving and Christmas. Navel oranges are grown primarily for fresh fruit and generally are larger than most round orange cultivars. The juice contains limonin, which after processing may impart a bitter flavor to the juice, although processors now have methods to remove this substance. The fruit has a very distinct flavor, unlike that of other oranges. Navel orange trees have low-to-moderate yields and are more susceptible to water or heat stress than round oranges, likely due to the presence of the navel. Navels are probably mutations of Seleta orange from Brazil and were first introduced to Florida in 1838 by Thomas

Hogg. These plantings were later destroyed and Washington navel was introduced to Florida soon after it was grown successfully in California. Trees were introduced from Brazil via Washington, D.C., in 1870. Some trees seem to have been sent by the U.S. Department of Agriculture directly to Florida at the same time as to California, but real interest developed only after the California success in the 1880s. Fruit quality is good to excellent, especially when grown in cooler regions of the state. Unfortunately, because this cultivar often yields poorly under Florida conditions it has never been extensively planted. In addition, Washington navel tends to produce very large, coarse fruit with dry segments when the cropload is light, and it may have large unattractive navels.

Many old-line local selections of navel oranges have been made in the past which differ from the original Washington navel in time of maturity, yields, and fruit quality. These include Summerfield, Pell, Port Mayaca, Dream, and several others. The old-line cultivars have largely been replaced by improved nucellar types developed through the University of Florida and the Citrus Budwood Registration Bureau. These selections are numbered (e.g., F-60–13 and F-56–11) and have been selected for superior yields over old-line cultivars. For several years, Glen navel (S-F-56–11-X-E) has been the most widely propagated selection in Florida. Average yields for navel orange trees have steadily increased with the introduction of these new cultivars and use of improved cultural practices including better irrigation management and use of growth regulators (see chapter 9).

Cara Cara navel, also called red navel, was discovered at the Hacienda de Cara Cara in Valencia, Venezuela, and brought to Florida by Dr. A. H. Krezdorn. It has become a very popular navel selection in the last five years due to its deep red flesh color. It is an excellent fresh fruit cultivar.

GRAPEFRUIT (*CITRUS PARADISI* MACF.)

History

Grapefruit is considered a distinct commercial species (*Citrus paradisi* Macf.), although it is likely a natural hybrid of shaddock (*C. grandis* Osbeck, also known as "pummelo") and sweet orange which originated in the West Indies during the 18th century. It was first described in Barbados in 1750 under the name "forbidden fruit." In 1789 the forbidden fruit or small shaddock was reported to be common in Jamaica. In 1814 it was called "grapefruit," the name being given because the fruit hung in small clusters, like some grapes, instead of one to a twig, as is more common with shaddocks. The specific epithet paradisi reflects the early name "forbidden fruit." During the first quarter of the 20th century, horticulturists tried to have the name "pomelo" adopted instead of "grapefruit," but the latter name was more familiar to growers and the public. "Pomelo" is still the common name in Brazil.

Grapefruit originated in the Western Hemisphere and is not widely produced in the Mideast and East and Southern Hemispheres except in Israel, China, and South Africa. The United States, particularly Florida, is a major producer of grapefruit, although production decreased significantly following hurricanes of 2004 and 2005 (see chapter 2). The introduction of grapefruit to Florida took place in 1809, when Dr. Odette Philippe, a French count, settled near Safety Harbor on Tampa Bay, bringing with him seeds or seedlings of grapefruit and other kinds of citrus fruits from the Bahamas. All of our present grapefruit cultivars are likely descended from this introduction via seedling variability, mutations, or natural hybridization.

Grapefruit attained popularity very slowly in Florida. It was chiefly a curiosity until about 1885, usually being considered merely a variation of shaddock, although considerable quantities were imported from the West Indies (78,000 fruits came into New York in 1874). However, fruit flavor and texture differ considerably between grapefruit and shaddocks. Juice sacs of shaddocks are much larger and chewier than those of grapefruit. In 1875, Florida nurseries produced only grapefruit seedlings, as grapefruit were considered unsuitable for commercial cultivation. Yet in 1884, a New York fruit dealer was quoted in a Florida newspaper as saying there was a considerable demand for the fruit. Evidently some shipments were being made prior to 1884–85, for J. A. Harris reported he had shipped 1,500 barrels from his grove in Citra. Curiously, in 1889, W. S. Hart considered that there was still no commercial production. He noted that several improved cultivars were being produced by nurseries, but at the Florida State Horticultural Society meeting that year, there was a premium only for the best grapefruit, while premiums were offered for about 50 named orange cultivars. The first named grapefruit cultivars appear in the horticultural literature only in the 1897 list from the American Pomological Society, even though nursery catalogues had been listing some cultivars such as Marsh and Duncan for several years.

In spite of an auspicious beginning, grapefruit growers in Florida have had and continue to have many difficult economic times. Often during the 1950s and 1960s, the fruit was simply left on the trees (economic abandonment) because the prices were too low to pay for picking, hauling, packing, shipping, and selling, let alone return a profit to the grower. Low prices occurred because production far exceeded consumer demand. The tendency to market grapefruit too early in the season has not improved the situation.

For many years, few plantings of grapefruit were made in Florida because of the above-mentioned possibility of economic abandonment. In 1968, however, growers began to make new plantings with the realization that the older plantings were gradually declining. This renewal grew with the increased demand and better prices for grapefruit occasioned by consumer interest in the diet- and health-related value of the fruit. Planting continued in the 1980s and 1990s, especially in the Indian River area and the southern Florida flatwoods, again causing a tremendous oversupply and low prices. This situation was corrected by the hurricanes of 2004 and 2005, which sharply reduced supply with a concomitant large increase in prices.

Fruit on the original grapefruit trees in Florida had a yellow peel at maturity and pale ivory-yellow flesh, which has been termed "white." Infrequent natural mutations produced pink, red, and even dark red flesh color, due to the presence of lycopene pigments. The naturally occurring pink and red colors have appeared primarily as bud mutations, when a single bud on a tree develops into a branch bearing fruit of a color different from that on other branches. Buds taken from this mutated branch will typically produce trees with fruit exactly like that borne by this branch, and the new mutation can be propagated as a cultivar. One red cultivar, Star Ruby, is the product of intentional seed irradiation by citrus researchers and will be discussed later in this section.

The earliest mention of grapefruit in Florida by Atwood in 1867 refers to the red flesh of the fruit, although all the seedlings known in the 1880s were white fleshed. It was not until the winter of 1906–07 that a pink-fleshed sport, Foster, was discovered; not until 1929 and 1931 were the red-fleshed cultivars, Ruby and Redblush, discovered, respectively, in two different groves in Texas. While pink, or especially red, flesh may have eye appeal for the buyer and consumer, there is no correlation between color and taste, although fruit with red pigments may have superior antioxidant properties to white-fleshed fruit. Starting in about 1935, there was strong interest in Florida in planting pink seedless grapefruit such as Pink Marsh, followed in about 1945 with similar heavy planting of red seedless cultivars such as Redblush. After 1954, however, there was a notable decrease in enthusiasm for red grapefruit, and there had even been extensive topworking (see chapter 6) of red grapefruit to other citrus types, sometimes even before they were mature enough to bear. This was due in large part to the inability to produce a satisfactory juice product from pigmented cultivars with the technology available at that time. However, new technology and blending of red grapefruit juice with other juices has expanded the juice market. Use of glass and paper containers instead of metal cans has also helped to improve the flavor and shelf life of grapefruit juice.

Seedlessness (less than six seeds per fruit) has also arisen in grapefruit by mutation in the seed, not the bud. In this instance, the genetic change occurred during the formation of the embryo (see chapter 3) and all tissues are genetically alike. Cecily, Davis, Marsh, and Star Ruby cultivars are seedless mutations arising from seedy fruit. The mutation for pink or red color is independent of that for seedlessness. Pink and red mutations of both seedy and seedless types exist. Duncan grapefruit was likely a seedling of the original tree introduced to Florida in 1809 and is possibly the source of several other cultivars. However, many other grapefruit seedlings were present in Florida at that time, which also may have served as parents. The relationships among white-, pink-, and red-fleshed mutations are shown in figure 4.2.

Red seedy grapefruit have been propagated in Florida, but no cultivars have been described, and the possible relation to Foster (no other pink seedy cultivar has been recorded) cannot be determined. Hudson, a red-fleshed seedy bud sport of Foster, was named and described in Texas in 1930. Seeds of this cultivar were irradiated at Texas A&I University and planted in 1959, and Star Ruby was selected in 1970. It has dark, red flesh and is seedless. Additional red-fleshed cultivars arose from Pink Marsh. These originally included Redblush and Ruby Red, which were discovered as limb sports in Texas. Fruit with deeper red coloration (Ray Ruby and Henderson) also originated from limb sports and were selected and propagated in Texas. Rio Red originated as a limb sport from irradiated Ruby Red budwood. Flame grapefruit, which is nearly as red as Star Ruby, originated as a seedling of Henderson in Florida. For many years, red-fleshed grapefruit brought superior prices to white- or pink-fleshed fruit. Prices for all grapefruit plummeted in the mid-1990s due to overproduction and surged to all time highs in 2005–06 due to hurricane damage and significantly reduced yields. Red-fleshed cultivars continue to be favored in the marketplace and the relative percentage of red to white cultivars will continue to increase.

Many seedy types with white flesh are indistinguishable from Duncan. Some distinct cultivars have arisen by natural seedling variation. Others have appeared that lack the slightly bitter taste characteristic of grapefruit; while these seem otherwise to be true grapefruit, they are likely hybrids. These in-

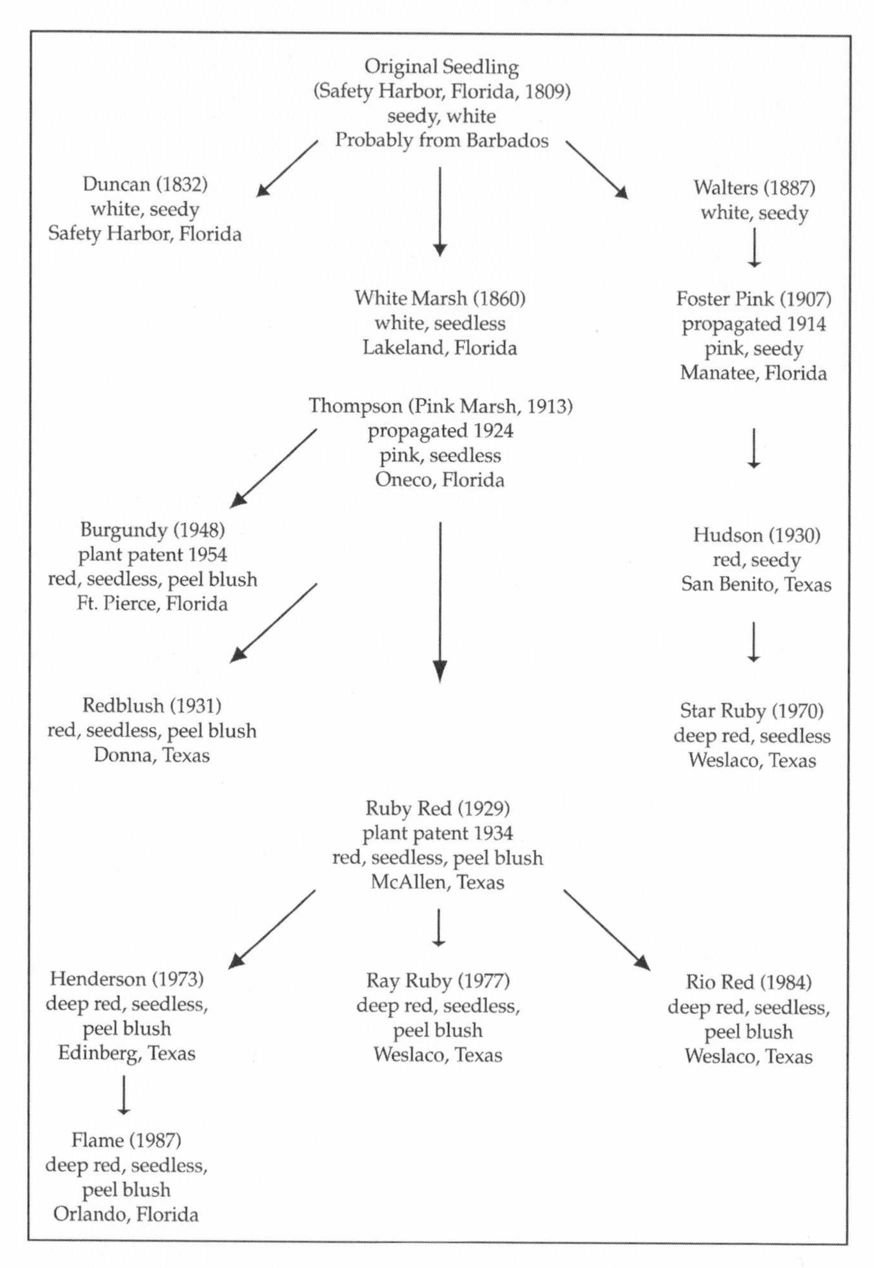

Figure 4.2. Origin of commercially important grapefruit cultivars. Numbers in parentheses indicate date of release. (Adapted from F. S. Davies and L. G. Albrigo, *Citrus*, CAB International, 1994)

clude the Royal and Triumph of Florida, the Imperial of California, the Isle of Pines from Cuba, and the Chironja of Puerto Rico (which appear to be a natural group of orange-grapefruit hybrids—orangelos). It is also possible that the loss of the bitter principle is due to a gene mutation.

Grapefruit Characteristics

Grapefruit cultivars, like oranges, may mature early, in midseason, or in some cases late when allowed to develop normally. Commercially, the differences in maturation time among cultivars are small, and most grapefruit are considered early to midseason, although fruits of some cultivars such as Marsh may be held on the tree until June. Considerable differences in maturity dates among similar cultivars are also due to bloom date and tree age, rootstock, and microclimate. Generally speaking, grapefruit grown in the southern or coastal portions of the state attain minimum maturity standards before similar fruit in the northern portion because they accumulate more degree-days, they usually bloom earlier, and acid levels in the juice (acids are important determinants of minimum ratio) decrease more rapidly. Degree-days are a measure of the number of hours or days that the average temperature for a region is above a threshold value. Generally, the higher the average temperature in a region the greater the number of degree-days. Degree-days accumulation affects tree vigor and fruit quality among other factors.

Grapefruit markets do not always look for the same fruit characteristics. The fresh fruit and gift box shippers demand a regularly oblate shape, which can be cut into two symmetrical halves and eaten fresh. Fruit should have a smooth and uniformly yellow or blushed peel, without any green color. Seedless fruit is preferred. Desirable sizes are large to medium, from 27 to 40 per standard carton (4⁄5 bushel).

Processors of canned sections and citrus salads are concerned with having excellent flavor and intact segments after processing. Seedy fruit (such as Duncan) have greater section stability than seedless fruit as well as higher solids content. However, seedy grapefruit production had virtually ceased by 2000.

For single strength and frozen concentrate juice, the important qualities are attractive juice color, high juice content, high solids, a moderate degree of acidity, and low bitterness. Seedy white-fleshed cultivars were preferred but currently are virtually unavailable due to low prices received in past years. Recent innovations in processing technology, including blending with other fruit juices, now allow the use of pigmented juices. Older methods of processing often yielded a brown, muddy-looking juice.

White-fleshed Grapefruit Cultivars

DUNCAN

The oldest commercial grapefruit cultivar, Duncan, was not the first to be named and propagated. Its parent tree was planted around 1830 near Safety Harbor and lived to be nearly a century old. About 1892 Duncan was introduced and propagated by A. L. Duncan of Dunedin. Trees are large and very productive when mature. Duncan has been the standard of quality in grapefruit and is probably the most cold-hardy grapefruit cultivar, although this is based largely on observation and may be the result of its dense canopy. The fruit has 30–50 seeds, matures normally from October to June, stores well on the tree, and has white flesh. Seediness detracts from the otherwise fine fruit quality. Duncan is no longer an important commercial cultivar in Florida due to its white flesh color and seediness.

MARSH (MARSH SEEDLESS)

The original Marsh tree was a seedling allegedly planted about 1860 near Lakeland (although Hume, a noted citriculturist, claims the original seedling was 60 years old in 1895). The tree was first noticed in 1886 by E. H. Tison, who began propagation of it in 1889 as a budded seedless grapefruit. A few years later, C. M. Marsh bought Tison's nursery and named this cultivar "Marsh's Seedless," which has been shortened to "Marsh." The fruit is smaller than that of Duncan and Foster, normally matures a little later, and can be stored longer on the tree without excessive drop. Flesh color is white, and seed number ranges from six to none. Marsh has long remained a very popular grapefruit cultivar in Florida and the world due to its high productivity, seedlessness, and long harvest season. However, Marsh acreage in Florida has declined over the past 10 years and has been largely replaced by red-fleshed cultivars.

Pink and Red-Fleshed Grapefruit

FOSTER

The first named cultivar with pink flesh was Foster, which arose as a bud sport of Walters white grapefruit and was first noted in 1907 by R. B. Foster of Manatee in the old Atwood grove near Ellenton. Foster trees were first propagated by the Royal Palm Nursery at Oneco in 1914. Its peel is also pink, so that the fruit was easily identifiable. Normal maturity occurs from November to December. Fruit quality is good, but the number of seeds is equal to, or greater than, those in Duncan and it was replaced by redder seedless cultivars from Texas and Florida. Foster has not been grown in Florida for many years.

THOMPSON (PINK MARSH)

This cultivar originated in 1913 as a pink-fleshed bud sport of a white Marsh tree discovered by S. B. Collins in the grove of W. B. Thompson near Oneco, Florida. "Thompson," also called "Pink Marsh," was not propagated until the Royal Palm Nursery of Oneco introduced it in 1924. It is seedless and has the same internal quality and external appearance as Marsh. There is no peel blush. Thompson matures earlier than Marsh but holds fruit well on the tree for several months, although the pink color fades with time. It has been replaced by other redder-fleshed cultivars.

BURGUNDY

The parent tree of Burgundy was found about 1948 in a grove of Pink Marsh owned by H. J. McReynolds near Ft. Pierce, Florida, probably originating as a bud sport. It was patented in 1954 and introduced commercially in 1956. The fruit is red-fleshed, seedless, and late ripening with white albedo similar to that of Pink Marsh. Burgundy does not have a red peel blush. It was not widely planted because its flesh color fades early in the season, and it has been replaced by cultivars with superior red color.

REDBLUSH AND RUBY RED

The first grapefruit with red flesh coloration originated in a planting of Pink Marsh made at McAllen, Texas, in 1926 by A. E. Henninger. When the trees first fruited in 1929, one branch had fruit with a red peel blush and red, not pink, flesh. Plant patent No. 53 was issued for Ruby grapefruit to Henninger in 1934. However, several similar mutations of Pink Marsh with pink to red flesh and peel, which cannot be distinguished from Ruby, have appeared in both Texas and Florida. Redblush, also called Ruby Red, is the most notable. It was first observed in a grove in Donna, Texas, in 1931. Many nurseries now simply refer to red-fleshed fruit as "Red Seedless" or "Red Marsh." The name "Redblush" was used in Texas, where the cultivar originated, and is more descriptive than "Ruby" due to the peel blush and occasional red albedo. Red grapefruit are usually harvested between October and March because flesh color may fade late in the season. Fruit quality is good, and yields are usually quite high. These early red-fleshed cultivars have largely been replaced by superior, redder fleshed cultivars like Ray Ruby, Rio Red, Rio Star, and Flame.

STAR RUBY

This cultivar was developed by Dr. R. A. Hensz of the Texas A & I University Citrus Center (currently known as Texas A & M University Citrus Center) from irradiated seeds of Hudson grapefruit. Star Ruby differs from the seedy, red-fleshed Hudson in being commercially seedless, with zero to six seeds per fruit. Similar to its parent, it often shows a reddish tinge in the wood and bark at the cambial layer, and the flesh is the deepest red of all grapefruit. Leaves are often narrower than those of other grapefruit and are typically variegated, having yellow to nearly white streaks in them. The tree is bushy and compact, blooms profusely, and bears some fruit in clusters. The original tree produced substantial crops of good-sized fruit with a red albedo as well as flesh. However, fruit size is often small and production is erratic. Fruits are suitable for sectioning and juicing since, at single-strength, juice retains its color well. Introduced in Texas in March 1970, it was officially released in Florida in 1973. The previously mentioned problems with leaf color and size, which are probably a result of genetic changes due to the radiation treatment, make this cultivar more difficult to grow than most other grapefruit cultivars. Star Ruby trees are also more sensitive to foot rot, herbicides, and cold damage than other cultivars. However, a premium price has been paid in the past due to the deep red fruit and flesh color. It, too, has been replaced by superior, more easily managed red-fleshed cultivars.

FLAME

This cultivar originated in Florida as a seedling of Henderson grapefruit planted in 1973 and was released in 1987. Flame has deep red flesh and peel color, although not to the extent of Star Ruby. In some areas, trees have become chlorotic and may be more difficult to grow than white-fleshed cultivars. In addition, Flame bears heavy crops of fruit in tight clusters. This clustering prevents development of oblate fruits, which are very desirable in the fresh market. It may also produce small fruit size when heavy crops are produced. Flame has remained a widely propagated and planted cultivar in Florida.

RAY RUBY

This cultivar is a mutation of Ruby Red grapefruit which originated in Weslaco, Texas, in 1977 and is very similar to Henderson which originated in Edinburg, Texas, in 1973. Both cultivars have a peel blush and red flesh color far superior to that of Ruby Red. The fruit are seedless and are used primarily for fresh and gift fruit, but they are also processed. Ray Ruby has been the most widely propagated and planted grapefruit cultivar in Florida and has replaced Star Ruby because of its excellent fruit quality and lack of production problems.

RIO RED

This cultivar is also a mutation of a seedling of irradiated Ruby Red grapefruit which was selected in 1976 and released in 1984 from Weslaco, Texas. The peel blush is similar to that of Ray Ruby, but the flesh color is redder. The fruit are seedless. Yields are very good, although sometimes are variable, which is unusual for grapefruit. Rio Red along with Ray Ruby has also been widely propagated in Florida and has replaced Star Ruby and other red-fleshed cultivars due to superior yields and fruit quality.

THE MANDARINS (TANGERINES)

History

The name "mandarin" or "mandarin orange" is applied to several citrus fruits that have similar common characteristics distinguishing them from oranges. These include a peel that separates easily from the flesh, green seed cotyledons, and flowers that form singly or in clusters that rarely form a branched inflorescence. The term "kid-glove orange" was used at one time because it was stated that "a woman could eat one while wearing kid gloves (goatskin) without getting them wet." Mandarins have also been marketed as "zipper-skinned oranges" or "soft citrus." The mandarin group as a whole is one of specialty fruits with a rather short season and difficulties with handling and shipping because the soft, loose peel is easily injured, with resultant development of diseases and postharvest disorders.

The name *Citrus reticulata* Blanco was originally applied to the Ponkan mandarin of Taiwan and the Suntura mandarin of India. Tanaka, a noted Japanese citriculturist, limited this name to these mandarins and divided the remaining mandarins into several species. For example, differences between Dancy tangerines and Ponkan mandarins are much greater than those between Parson Brown and Valencia oranges, justifying placement into separate species based on morphological characteristics alone. However, genetically these species are very closely related, based on molecular biology techniques (see chapter 3).

Some tangerines are native to India, some to Malaysia, and several to China. All types grown in Florida are marketed as tangerines. However, until about 1945, the only economically important tangerine was Dancy, because it was the only cultivar that had been marketed for a long period in large quantities. Several other tangerine types are currently grown and marketed commercially as tangerines. These are natural and human-made hybrids that closely resemble tangerines and usually have tangerine parentage. These types are discussed in the section on mandarin hybrids. At one time, tangerines were extensively planted in Florida. However, new plantings of tangerines dwindled from 1930 until the 1980s, when renewed interest in specialty fruit occurred following serious freezes in 1983, 1985, and 1989. Growers in the northern part of the citrus belt in particular planted mandarins and mandarin hybrids because they are freeze hardy and are usually harvested before freezes occur. Some of the citrus hybrids (tangerine types, tangors, and tangelos) that will be discussed later have some of the desirable qualities of tangerines without some of the serious disadvantages. Citrus hybrids are often marketed fresh in competition with oranges rather than tangerines, although they are specialty fruits.

Tangerine Characteristics

Tangerines have long been popular as fresh and gift fruits. The attractive peel and flesh color, the ease of removing the peel and separating the segments, and the excellent flavor make them desirable as dessert fruits.

As fruits for large commercial production, tangerines have several disadvantages in comparison with oranges or grapefruit. If left on the tree after becoming mature, segments dry out. With shrinkage of the segments, the space between flesh and peel increases, causing puffiness. Puffy fruits are easily damaged during harvesting, packing, and shipping. Tangerines tend to overbear, producing a large crop of uneconomically small fruit in one year and a very light crop of large fruit the following year. This cycle of alternate bearing is then repeated. In fact, Murcott tangerine (tangor) actually produces so heavily in the "on year" that the root growth is often severely reduced. In time the tree may become very weak and die (Murcott collapse). Tangerine fruit produced inside the canopy is much less highly colored than that borne on the outside and may also have poorer internal quality due to low light levels. Adequate light levels are needed to produce sugars in the fruit. Moreover, loss of green peel color and development of red-orange color is often slow. Attractive peel color is an important factor in determining packout and price. Early in the season, fruit may be degreened in the packinghouse (see chapter 11) to attain acceptable peel color.

In the early days of the citrus concentrate industry, people believed that small, blemished, or poorly colored tangerines might be utilized as frozen concentrate juice and to enhance juice color of orange concentrate by blending. This has not occurred to a large extent, although some tangerine juice blends are available. Nevertheless, some tangerine juice is blended (up to 10 percent allowed by law) into orange juice for concentrate to increase color of the finished product.

The price per box "on tree" (before production costs are subtracted) for tangerines is often higher than that for oranges and grapefruit, but the number of boxes of marketable fruit per acre is lower and the net return per acre is often lower for tangerines. Thus,tangerine fruit prices and market conditions are very volatile and unpredictable.

Four species of mandarins have been identified by Hodgson, a prominent citrus taxonomist: the satsumas, the King or Kunenbo, the Willowleaf, and the common mandarins (tangerines). Swingle, in contrast, placed them all into one species, *Citrus reticulata* Blanco, but Tanaka divided the mandarins into seven distinct species. We now realize through molecular biology techniques that mandarins are a single species (see chapter 3). Nevertheless, from a horticultural, marketing, and consumer viewpoint, the four "species" of Hodgson are readily identifiable and this distinction forms the basis for the text that follows.

SATSUMAS (*CITRUS UNSHIU* MARC.)

The satsuma originated in Japan in about A.D. 160 from seed brought from China. Its Japanese name, *unshiu*, is a variation of Wenchow, the area of China where the seeds originated. In all probability, current satsuma cultivars are mutations of the original satsuma. In Japan, satsumas are the most important fruit grown on a commercial scale. The fruit is exported to the United States and Canada, although the bulk of the crop is consumed locally. There are more than 100 satsuma cultivars in Japan, differing in time of maturity and fruit characteristics. Of these, Owari is the main satsuma cultivar grown in the United States, although only to a very limited extent.

The Owari satsuma was first introduced to Florida as budded trees in 1876 by George Hall and reintroduced in 1878 by General George Van Valkenburg. Satsumas are appreciably more cold hardy than all other commercially important citrus species. They have been successfully cultivated in north Florida and around the Gulf Coast from Florida to Texas. Owari satsuma is cultivated almost entirely north of the main citrus belt. Fruit quality and yields are usually best when grown on trifoliate rootstock. The trees bear heavy, consistent crops; the fruit, which has a distinctive flavor, matures in October and November. Both peel and flesh are deep orange in color when fully mature and grown in a cool climate. However, peel color in Florida remains green early in the season even though fruit is mature and the flesh color is deep orange. If left on the tree after they reach maturity, the fruits become puffy, and peel-tearing (plugging) during harvest is a problem. Fruit are seedless. Satsuma is not commercially important in Florida and is better adapted to cooler climates such as Japan and China, although it is also grown in Spain and other Mediterranean countries and in southern South Africa.

KING MANDARIN (ORANGE) (*CITRUS NOBILIS* LOUR.)

The King Orange of Indochina and the closely related Kunenbo of Japan are the only representatives of this mandarin group. They may have arisen many hundreds or thousands of years ago as natural hybrids, but they have maintained distinctive characteristics and have not been duplicated in any human-made crosses.

King was introduced from Saigon to California in 1880 as fruit, and to Florida by J. C. Starin in 1882 as small seedling trees which fruited in 1884. The fruits are some of the largest, with the thickest peel, of any mandarins. King has never been important commercially because of a tendency to alternate bearing, with resultant limb breakage in the "on years." Sunscalding of fruit is also a problem because of King's upright branches with relatively sparse foliage. Furthermore, the rough peel is unattractive to consumers. The maturity season is March to June, and the fruit has several seeds that, unlike most other seeds of the mandarin group, have white cotyledons and are monoembryonic. King has been used in citrus-breeding programs and is a parent of Kinnow mandarin.

COMMON MANDARINS (TANGERINES, *CITRUS RETICULATA* BLANCO)

There is no distinction between the terms "mandarin" and "tangerine." The earliest use of "tangerine" (1841) was as a synonym for "mandarin." The name was originally spelled "tangierine" because fruits were exported from Tangier, Morocco, to England. At one time, cultivars with yellow-orange peels were called mandarins and those with reddish-orange peels, tangerines. But these characteristics do not always distinguish between the groups in Florida because of its hot, humid climate and lack of distinct peel coloration. "Mandarin" is the preferred term worldwide except in Florida and for some cultivars in Australia.

The first mandarins (tangerines) introduced to Florida probably arrived about 1825, since the village of Monroe on the St. Johns River changed its name in 1830 to Mandarin, possibly based on the town's experience with mandarins. There is little information about the source or the cultivar introduced, although the Ponkan mandarin was available for possible transport from England after 1805. A second unknown source of mandarin importation occurred in 1838. The tree came from China via Parsons Nursery on Long Island. Unfortunately, long scale(an insect, see chapter 10) was also introduced on the leaves and nearly destroyed all the citrus trees in the state. Presumably all or many of the tangerine trees were killed by the scale, because in 1877 Major Atway was given credit for introducing (or reintroducing) mandarins to Florida from Louisiana. The Read and Hartley groves at Mandarin survived the long scale infestation, but they may have had no tangerine trees. The type of mandarin that Major Atway introduced is unknown. Usually he is credited with bringing in Willowleaf, which originated in China and was brought to Italy before 1840. It was reportedly introduced to New Orleans from Italy between 1840 and 1850. The Atway grove at Palatka was bought by N. H. Moragne in 1843; his daughter stated that a mandarin tree was growing there at that time. This could have been the Willowleaf, although it is unlikely. The fruit that the Pomological Committee of the Florida Fruit Growers Association described from the Moragne grove in 1877 as Tangerine Orange (synonyms: Mandarin, Kid Glove, Tomato Orange) and ascribed to Major Atway's introduction was definitely not the Willowleaf. Instead it was very similar in appearance to Dancy tangerine. In 1885 Colonel Dancy, who was then a neighbor of Dr. Moragne, stated that the Moragne tangerine had been introduced from Tangier about 1850. Therefore, it appears that several different mandarins were introduced at various times into Florida during the 1800s.

WILLOWLEAF (*CITRUS DELICIOSA* TEN.)

The Willowleaf (Mediterranean) mandarin was placed in a different species by Hodgson because of its unique strap-shaped leaves and resemblance to a willow tree in appearance. The only Willowleaf tangerine known to have been in Florida in 1877 was one small tree in the grove of E. H. Hart at Federal Point. This had been obtained from England by S. B. Parsons in 1870. It almost certainly was the type introduced to Louisiana in the 1840s. If Major Atway brought it to Florida, it must have died before 1877, for Hart certainly knew of the trees in the small Moragne grove, and his tree was unique. Some mystery surrounds the supposed Atway introduction. The first mention of Willowleaf by name was by A. H. Manville in 1883, and in 1887 Pliny Reasoner erroneously made it a synonym of the Moragne tangerine. Willowleaf is grown primarily in the Mediterranean region and is not adapted to Florida's climate.

The commercially important mandarin and mandarin-hybrid cultivars will be discussed in the following section.

DANCY (*CITRUS RETICULATA* BLANCO)

This cultivar was by far the most widely planted tangerine in Florida until 1945. By the 2002–03 season, Dancy had been replaced by mandarin hybrids such as Fallglo, Murcott, and Sunburst due to its small fruit size, poor fruit color in some years, susceptibility to *Alternaria* fungus, and alternate-bearing tendencies. The original Dancy tree was discovered in the grove of Colonel F. L. Dancy at Orange Mills from seeds of the Moragne tangerine. The cultivar is typical of the tangerine type native to the Foochow area of southern China and introduced to Japan in the 16th century, where it is known as *Obeni-mikan*. Dancy reproduces practically 100 percent true-to-type (nucellar seedlings) from seed unless a mutation occurs, which is very rare. The first mention of Dancy occurred in the report of the Pomological Committee in 1877, which considered this a new fruit, similar but slightly superior to Dr. Moragne's fruit. Various opinions have been expressed on its origin, but Colonel Dancy himself stated in 1885 that it arose from a seed of the Moragne tree planted in 1867. Commercial propagation began about 1890 by the Rolleston Nursery at San Mateo. Dancy trees and fruit have characteristics typical of most tangerines and also have the previously mentioned defects. It matures in December and January and has from 7 to 20 seeds.

Mandarin Hybrids

TANGELOS

These are interspecific hybrids resulting from crossing tangerines with grapefruit (*C. reticulata* x *C. paradisi*). When the name was coined in 1905, horticulturists were trying to establish "pomelo" as the accepted name for grapefruit. Thus, the names tang(erine) and (pom)elo were combined into tangelo to identify this hybrid. The first tangelos, Sampson and Thornton, were named and released in 1905 and were quickly planted on a small commercial scale. They did not prove very successful because of poor fruit quality and are not grown commercially. In 1931, Lake (Orlando), Minneola, and Seminole tangelos were released. Only Minneola and Orlando are still commonly found today. Early attempts were made to promote high fruit quality and stimulate market demand. The Tangelo Act of 1955 placed these cultivars under maturity regulations for the first time (see chapter 11) and was a further aid to orderly marketing of good quality fruit.

Orlando

Orlando (Duncan grapefruit [female] x Dancy tangerine [male]) has little resemblance to either parent. It was initially accepted by the market as an orange under the name "Orlando orange," which replaced its original name, "Lake tangelo." But today tangelos are placed in their own category, and this cultivar is known commercially as "Orlando tangelo." Size and shape are similar to those of the tangerine, but the color and texture are more orange-like. Peel and pulp color are deep orange at maturity, and the peel adheres firmly to the pulp, much like an orange. It is variably seedy (depending on the amount of cross-pollination) and matures from November to January. It was planted on a relatively large commercial scale, totaling about 20,000 acres in 1970, but by 2006–07 less than 6,000 acres remained due to consistently low prices and declining consumer demand in recent years. Temple orange and/or Robinson or Sunburst tangerine trees are usually interplanted with Orlando to produce satisfactory crops since it requires cross-pollination in most locations to attain maximum fruit size and production. It is also susceptible to *Alternaria* fungus disease. Orlando trees sometimes produce an acceptable crop without cross-pollination when grown under very favorable conditions. Fruit are produced parthenocarpically (without seeds). Gibberellic acid (a plant hormone) application during bloom also produces an adequate crop of seedless fruit (see chapter 9).

Minneola

This cultivar resulted from the same cross as Orlando but shows much more evidence of its tangerine parentage, although it also does not have the loose-skin character of mandarins. Minneola has distinctive flavor and aroma when fully mature, with a dark reddish-orange peel, dark orange flesh, and high juice content. Unfortunately, it often produces poorly under Florida conditions, although the tree is very vigorous like its grapefruit parent. It is also susceptible to citrus scab and *Alternaria* (see chapter 10), and the fruit shape does not lend itself to easy packing. The fruit has a ridged neck at the stem, giving rise to another common name, "Honeybell." It is also variably seedy, depending on degree of cross-pollination, and fruit mature in late season, January to March. The fruit is grown primarily for fresh and gift fruit. Like Orlando, acreage as well as production, have declined in recent years.

Seminole

The third cultivar, Seminole, from the cross-producing Orlando and Minneola, has not been as widely planted as Minneola, which it resembles greatly. It differs from Minneola chiefly in being easier to peel, having less susceptibility to scab, a much larger number of seeds (22), and a later season of maturity (February to April). Flavor, aroma, and juice content are good, although fruit are sometimes tart. Seminole is grown commercially in only a few places in Florida.

Sampson

This cultivar is the result of crossing Dancy with an unnamed grapefruit in 1897 by W. T. Swingle. Sampson was the first tangelo developed. Like the tangelos described previously, it is very juicy, highly colored, and moderately seedy (about 15 seeds), but it is quite tart. It reaches maturity from February to April. There is also a trace of the bitterness characteristic of grapefruit, which is absent in the previously described tangelos. The combination of high acidity, very thin peel, and easy bruising, together with high susceptibility to citrus scab, has prevented Sampson from being grown commercially in Florida.

Thornton

This cultivar resulted from the same type of cross as Sampson but was made two years later. It differs markedly from all the tangelos discussed above in having a thick peel which is easily separated from the flesh. The peel becomes quite loose when overmature, producing puffy fruit. Peel and pulp are both light orange and there are about 15 seeds. The flesh is soft, juicy, and sweet, becoming insipid when overmature. The season of maturity is December to February. Thornton has never been planted on a commercial scale in Florida.

K-Early

This cultivar was a product of some of the early breeding work of Swingle and Webber and was initially not released because of poor fruit quality. K-Early more closely resembles a grapefruit than a tangerine. However, the early maturity (after September) and freeze hardiness has prompted limited planting in spite of the fruit quality problems in an attempt to market fruit early when prices are high. K-Early production has virtually disappeared since 2001.

COMPLEX MANDARIN HYBRIDS

Robinson

This hybrid is known commercially as a tangerine. It is a Clementine mandarin (female) x Orlando tangelo (male) hybrid that was released by the U. S. Department of Agriculture in 1959. Clementine mandarin originated in China and is widely grown in the Mediterranean region, with some production in California and South Africa. It was widely used as a female parent in crosses because the seed is monoembryonic and produces only zygotic (sexual) progeny. Robinson fruit is harvested from September to December, making it among the earliest fruit packed in Florida. Fruit quality is very good, having deeply colored flesh, high solids, and a thin peel. Fruit size is unacceptably small in some years. Maximum production occurs with adequate cross-pollination using Orlando, Temple, or Sunburst as pollinizers. As with other mandarin hybrids, seed number varies with the degree of cross-pollination. The wood of the tree is brittle and may break with heavy crop loads. It is also susceptible to limb and twig dieback problems. The tree itself is moderately vigorous and suitable for planting in cold areas. Robinson acreage and production have declined considerably since the mid 1990s and represented a small portion of tangerine production and acreage in 2006–07.

Osceola

Another Clementine x Orlando citrus hybrid, Osceola, was released in 1959. The fruit most closely resembles a tangelo rather than a tangerine. It has a highly colored fruit with 15–25 seeds and matures in October to November. It is scab and *Alternaria* susceptible and has a rather tart flavor. It has never achieved significant commercial production levels.

Nova

This cultivar, released by the U.S. Department of Agriculture in 1964, is a hybrid of Clementine mandarin (female) and Orlando tangelo (male). It resembles Orlando in several respects and likewise requires a pollinizer to attain maximum productivity. Additionally, seed number is a function of degree of cross-pollination. The cultivar has not been widely accepted in Florida but is a popular fresh fruit in parts of the Mediterranean region and South Africa.

Lee

Like Nova, Lee is a hybrid of Clementine and Orlando. This cultivar was released by the U.S. Department of Agriculture in 1959. While the fruit most closely resembles a tangelo, it has been described (and sold) as both a tangelo, and at times, a tangerine. Lee matures from November to December and is slow to develop adequate peel color, often leading to degreening problems. It has never achieved significant commercial importance, but is an excellent quality fruit and a good pollinizer for Orlando.

Page

This is a Clementine mandarin (female) x Minneola tangelo (male) citrus hybrid, most closely resembling an orange. The cultivar was released by the U.S. Department of Agriculture in 1963. Seed number varies with degree of cross-pollination, and fruit quality is one of the best of all citrus cultivars. The fruit peel and flesh are deep orange, and the flavor is very rich. The fruit has a characteristic ring around the stylar end. However, Page is plagued with a problem of commercially unacceptable small fruit sizes and has not been grown commercially. The fruit does have potential for dooryard use.

Sunburst

Commercially marketed as a tangerine, Sunburst is a Robinson tangerine (female) x Osceola (male) hybrid; it most closely resembles the former. Sunburst was released in early 1979 and became fairly popular and widely planted in the early 1980s. Prices for fruit were excellent in the 1980s but dropped considerably in the late 1990s and early 2000s. Production has steadily declined over the past five years. The cultivar produces larger crops when cross-pollinated. Fruit quality is excellent. Interestingly, Sunburst appears to be resistant to citrus snow scale but highly susceptible to citrus rust mite damage. The main period of fruit maturity is October to January.

Fallglo

This 1987 release was made by crossing Temple orange (female) with Bower citrus hybrid (male). It most closely resembles Temple but matures two months earlier (September to November). Fallglo trees have a distinctive appearance, appearing very bushy in the field. Fruit are difficult to degreen and may develop peel-related problems during shipping. The fruit is moderately seedy. Acreage has declined steadily because of the shipping and handling problems beginning in the 2002–03 season.

Ambersweet

This orange hybrid was released in early 1989 by the U. S. Department of Agriculture. It is a complex hybrid of a Clementine mandarin (female) x Orlando tangelo (male) hybrid, which was further crossed with a seedling mid-season sweet orange (15–3) in 1963. Ambersweet is technically considered an orange for commercial purposes. It is an early cultivar, maturing most years from September to January. Fruit size is quite large, and juice color is better than early-season oranges. The original intention was to use Ambersweet as both a fresh and processing fruit. Unfortunately, during its first several years in the field, production was low and variable, and fruit quality was unacceptably poor. Fruits tend to be large and dry, having up to 15 seeds per fruit when grown in mixed plantings. In addition, the tree is not as freeze hardy as expected. As a result, many growers have topworked Ambersweet to other cultivars or removed trees entirely. Ambersweet will probably not be planted in the future.

TANGORS

Tangors are interspecific hybrids between tangerines and sweet oranges, the name being a combination of "tang(erine)" and "or(ange)." Of the human-made tangors, only the Murcott has attained any commercial importance. Several natural tangors have occurred by chance under cultivation. Notable among these is the Temple orange. Some authorities believe the King mandarin is also a natural tangor, but its parentage is unknown. In general, the flesh texture and flavor of tangors are intermediate between those of the parents with peels that separate fairly readily from the flesh (though less so than that of true mandarins). Sugar content of juice and juice content of tangors are often greater than is usual for oranges, but fruit cannot be held on the tree as long as oranges.

Temple

The most widely planted tangor is marketed as "Temple orange." It originated in Jamaica sometime late in the 19th century but was first noticed in 1896. A fruit buyer named Boyce, who usually obtained his fruit for northern markets in the Oviedo area, went to Jamaica for oranges following the severe freezes of 1894 and 1895 because there were few fruit in Florida. He came across a seedling orange that impressed him greatly, and he sent budwood to friends at Oviedo. Several trees were propagated, and budwood was shared with others. The fruit caught the attention of W. C. Temple, formerly manager of the Florida Citrus Exchange. Temple contacted M. E. Gillett, his associate in the exchange and a leading citrus nurseryman. After testing it on several rootstocks, Gillett signed a contract in February 1916 for the exclusive propagation of this unnamed cultivar. It was Temple's intention to have it called "Winter Park Hybrid," although Edgar Wright, editor of the Florida Grower, urged that it be named "Temple." After propagating a large supply of nursery trees, Buckeye Nurseries, of which Gillett was president, first announced this cultivar publicly in May 1919 under the name "Temple" as a posthumous tribute.

Temple was not the overwhelming success that had been anticipated, in part because of considerable propagation on rough lemon rootstock, which produced poor quality, often dry, fruit. The fruit also proved somewhat hard to ship when given the same packinghouse treatment as oranges. Fruit quality was very good on other rootstocks, such as Cleopatra mandarin, and often commanded a premium over that of oranges. Temple fruit have a deep orange or reddish-orange peel at maturity, with about 20 seeds. Temple fruit and leaves are very susceptible to scab and *Alternaria* fungus diseases, and trees are freeze sensitive. Fruit reach maturity from January to March. Temple planting was heavy from 1940 until 1970 when there were more than 20,000 acres. Freezes in the late 1970s and early 1980s and low prices reduced tree acreage to just under 2,500 in 2007. In addition, prices have continued to be low in the 2000s and future planting is unlikely.

Murcott

This cultivar has been planted extensively only since 1952. The parent tree was sent for field trials from the U.S. Department of Agriculture nursery at Little River (Miami), Florida, to R. D. Hoyt of Safety Harbor about 1913. Undoubtedly it was a tangor from the breeding work being carried out by Swingle and his associates at that time, but the identification label was lost. A neighbor, Charles Murcott Smith, obtained a bud of this tree from Hoyt and propagated a few trees about 1922. Small-scale commercial propagation was undertaken in 1928 by the Indian Rocks Nursery under the name "Honey Murcott," and several other nurseries on the west coast propagated the cultivar to a limited extent prior to 1950. In 1944, J. Ward Smith (no relation to C. M.) became interested in commercial production and made the first planting near Brooksville. He used the name "Smith Tangerine," apparently unaware that the name "Murcott" had previously been approved by C. M. Smith. The cultivar is also known as the "Honey" tangerine in the fresh fruit trade.

The fruit is yellow-orange, smooth, and glossy with an oblate shape. The flesh and juice are deep orange, and the flavor is excellent at maturity. Seed number ranges from 18 to 24 and fruit are some of the latest maturing of the mandarin types (December to April).

Murcott is somewhat susceptible to scab and *Alternaria* fungi, and cachexia viroid. The tree is bushy, with willowy branches, and wind-scarring of fruit is often a problem because fruit is borne terminally on the branches. This orientation also makes fruit more subject to freeze injury and sunburn. Like Temple, the Murcott has higher soluble solids than sweet oranges of comparable maturity. The trees tend to bear very heavily in alternate years, which may result in poor root growth, reduced starch accumulation, and low potassium levels. As indicated previously, trees may actually die (Murcott collapse) due to this heavy seasonal production. Careful attention to pruning, irrigation, and nutrition is essential to successful cultivation.

Historically, prices for Murcotts have been the highest of all tangerines, but prices and acreage have declined over the past several years (2002–07). W. Murcott Afourer is a natural tangor imported from Morocco in 1985. It is very similar in quality and appearance to Murcott but has few seeds if it is not cross-pollinated. Lack of seeds is very important in the fresh fruit market and Afourer and Tango, a California selection, may have future potential in Florida.

OTHER HYBRIDS

Orangelos represent a group of cultivars that have the general appearance of grapefruit without its bitterness and aroma and are considered to be natural interspecific hybrids of *Citrus paradisi* and *C. sinensis*. Triumph, the first named cultivar, and the Imperial and Royal should probably be classified as orangelos.

Limequats are intergeneric hybrids produced by pollinating Key limes (*Citrus aurantifolia*) with kumquats (*Fortunella* spp.). Two of these hybrids from crosses made by Swingle in 1909 have been grown sparingly in Florida. It was hoped that they would extend lime culture further north by adding some of the freeze hardiness of kumquats to freeze-sensitive limes. This has been achieved to some degree, but the hybrids are more like Temple in hardiness than sweet oranges. They are quite prolific fruit producers that closely resemble limes, but they never became very popular and are planted solely as dooryard trees. Lakeland and Eustis are the most important cultivars.

Citrangequats and citrangedins are examples of the numerous classes of complex hybrids created by Swingle and his associates, none of which has any commercial importance. Citrangequats are crosses of kumquat (*Fortunella* spp.) with citrange and, thus, are trigeneric hybrids. Citrangedins result from crossing citrange with calamondin and, thus, are bigeneric, although complex. The Thomasville citrangequat and the Glen citrangedin have both been described and propagated to a very limited extent for use as lime substitutes in areas like Georgia that are too cold even for calamondins.

LEMONS (*CITRUS LIMON BURM.* F.)

Historical

The true lemon appears to have originated in northwestern India. The term "true" may not be appropriate since lemon is likely a hybrid of citron and sour orange based on molecular analysis. It has not been identified in the wild and may have arisen thousands of years ago under cultivation. Lemon trees reached southern Italy by A.D. 200 but may have failed to survive the political and economic turmoil of the Dark Ages. Tolkowsky, a noted citrus historian, has shown that the lemon was reintroduced to Sicily before A.D. 1000 by the Moors, if it was not still surviving there. It was being cultivated in Iraq and Egypt by A.D. 700. With nearly a millennium of culture in Sicily, it is not surprising that Sicily has traditionally been the source of commercial cultivars. Apparently the true lemon did not reach China until around A.D. 1200. The lemon mentioned in early Chinese literature was a related species, *C. limonia* Osbeck. By A.D. 1100, lemons were widely grown in Italy, Spain, and Portugal.

Lemon seeds were brought to Hispaniola by Columbus in 1493, and while they are not specifically mentioned, the importance of lemons in the Spanish diet suggests strongly that they might well have been grown in 1579 in St. Augustine. Little mention is made of lemons in colonial Spain, but in 1839 Williams noted that lemons increasingly were being planted in northeastern Florida. Because they were grown only as seedlings, lemon trees would have survived the periodic freezes of this area by regrowing from the roots following freeze damage.

Commercial lemon culture in Florida was begun after 1870 and was spurred by the extensive importation of lemons to the United States from Sicily. Florida and California orange growers felt that this market could be supplied profitably with homegrown fruit. They recognized, however, that the seedling types already growing in this country could not compete with the superior types coming from Sicily. Thus, they began searching for superior local cultivars. Several cultivars resulted simply from planting seeds of Sicilian lemons with favorable characteristics and selecting among the seedlings for superior quality and performance. Budwood was also imported in a few cases. An industry rapidly developed in both Florida and California, but trees in Florida were severely damaged by a freeze in 1886. In addition, in the humid subtropical climate of Florida, lemon scab (a fungus) was difficult and costly to control, and proper curing of lemons to attain suitable commercial peel color was difficult. Curing consists of harvesting lemons when green and storing them at 55°F under high humidity. This process increases juice content and lessens peel damage during handling. Peel color also changes from green to yellow. After the freezes of 1894 and 1895, lemon culture was almost wholly abandoned on a commercial basis in Florida, although lemons were grown for home and local use.

After 1953 there was renewed interest in growing lemons in Florida, and several large plantings were established primarily to produce fruit for making frozen lemonade concentrate. It was not necessary to cure fruit, and control of lemon scab was not a severe problem with the new fungicides and spray equipment then available. Florida growers were confident that they could produce lemon juice at a lower cost than their competitors. Additionally, cured Florida lemons were well accepted at that time, and the demand was large. Fruits were spot picked for the fresh market early in the season (August to September), while fruits harvested later went to processing plants. These fruits were generally too large for the fresh market but acceptable for processing. The value of processing lemons is based on high juice and acid content, the absence of an unpleasant aftertaste, and very important, peel oil content. Lemon peel oil is used in a variety of products including soft drinks and cosmetics.

Lemon plantings in the southern portion of the state gradually increased to about 10,000 acres, and for a time there was continued interest in further development. Most of the acreage was located near Lake Okeechobee in Martin and Palm Beach counties, an area that is less likely to have damaging freezes. Smaller plantings were also located in Hillsborough County and other locations that could afford active cold protection such as grove heaters. Production exceeded 1 million boxes in the early 1970s. The freezes of 1977 and the 1980s reduced lemon plantings to about 1,000 acres by 1999 with production becoming essentially nonexistent by 2005. Some important lemon cultivars are discussed in the following section. These cultivars are not grown commercially in Florida but may be grown as dooryard trees.

Lemon Characteristics

There are basically only two types of commercial true lemons; these were apparently selected long ago in Sicily and Portugal. The various named cultivars are selections within these two types, represented by Eureka and Lisbon. As suggested by H. J. Webber, the Eureka-type represents a structurally sound tree, characterized by an open, spreading tree habit with relatively few branches and twigs, and leaves that are dark green and rather blunt pointed. In contrast, the Lisbon type makes a rather dense tree with many slender, upright branches, and leaves that are light green and acute at the tip.

EUREKA

This is the principal lemon cultivar of California. It is not grown under that name in Florida to any extent. It originated as the best seedling from seeds of a Sicilian lemon planted in California in 1858 and was introduced as a cultivar in 1878 by T. A. Garey of Los Angeles as "Garey's Eureka." Within two years, however, the name had been shortened to "Eureka." Thorns are few, fruit acidity is high, and seeds are few, but fruit is mostly borne on the tips of the branches, so that it is somewhat subject to wind and sun damage. The tree is more subject to cold injury than most other lemon cultivars, probably due to its morphology rather than to inherent genetic differences. Trees flower throughout most of the year and are harvested in winter, spring, and early summer in regions like California. Fruits mature in August to December when grown in Florida.

In California, the Eureka is highly productive but tends to have a short bearing life because of susceptibility to shell-bark and dry-bark diseases. H. B. Frost developed a nucellar type that is free from these viral infections, and it has been widely planted. Many clonal selections are now being used in California.

LISBON

This cultivar remains the second most widely planted cultivar in California. It originated in Australia from the seed of lemons from Portugal and was introduced to California from Australia in 1874 as budwood. Like Eureka, it was first propagated commercially by Garey around 1880. Trees are much thornier than Eureka trees but are more vigorous and productive. Fruit acidity is as high, there are usually more seeds, and the fruit tends to be produced mostly inside the tree canopy, where it is better protected from sun, wind, and freeze damage than in Eureka. Selections of outstanding trees for propagation by nurserymen in California have given rise to many clonal types.

PONDEROSA

Ponderosa has been used in home plantings since around 1900. It originated around 1886 as a seedling of unknown source grown by George Bowman of Hagerstown, Maryland, and was first propagated as a greenhouse plant under the name "American Wonder lemon." The fruits resemble ordinary lemons except for being much larger and are grown more for their large size than for their juice quality. The fruits are very juicy and have many seeds. The tree is small and fairly thorny. This is probably not a true lemon, but experts disagree on its parentage. It is more freeze susceptible than true lemons, although not more so than limes.

MEYER (*CITRUS MEYERI* TANAKA)

This cultivar is most likely not a true lemon but is commonly called a lemon because the juice tastes much like lemon juice. It was introduced from China in 1908 by Frank N. Meyer for the U.S. Department of Agriculture; he found it cultivated near Beijing (then Peking) as an ornamental plant. The mature fruit greatly resembles an orange in both external and internal appearance, although it may possess a small nipple at each end. It is very juicy, with a smooth, thin peel and about 10 seeds. Juice quality is good but acid content is lower than the true lemon, and the peel lacks the true lemon aroma. The inferior character of the peel oil is the biggest handicap to more widespread use of Meyer for concentrate. Even a small amount of Meyer peel oil gives an undesirable flavor to the juice, although this can be masked, if not too intense, by adding peel oil of true lemon. Some fruit matures throughout the year, but the main crop is harvested from December to April. The tree is small and bushy, nearly thornless, and more cold hardy than any true lemon. It was planted on a large commercial scale in Texas and on a small scale in Florida beginning in 1930. Because Meyer carries the citrus tristeza virus (CTV), it should not be planted on sour orange rootstock or in areas where CTV is a problem. It is prohibited by law in some areas of California. It is popular as a lemon tree for dooryard use since it produces heavily, needs a limited amount of space, and is reasonably cold hardy. Nursery propagations of Meyer lemon have increased in the 2000s to fill this homeowner niche. Undoubtedly Meyer is a hybrid, but the other parent besides lemon is unknown, although the fruit color suggests orange or mandarin.

LIMES (*CITRUS AURANTIFOLIA* [CHRISTM.] SWINGLE)

History

Las Casas, a Spanish historian, does not specifically include limes among the seeds brought by Columbus to Hispaniola in 1493. However, Oviedo reported that limes were plentiful there in 1520; therefore, it is quite probable that Columbus did bring lime seeds from the Canary Islands. The lime soon became naturalized on some of the West Indies, on the coast of Mexico, and ultimately on some of the Florida Keys. The first mention of limes in Florida was by Williams in 1839, who noted that planting was increasing. In 1838 Henry Perrine planted a few lime trees, obtained from the Yucatan on Indian Key and perhaps some adjacent islands. Naturalized lime trees found on many of the Keys at the turn of the century have often been considered the result of his plantings. However, in 1876 his son was unable to find any trace of those plantings still in existence. Throughout the 19th century, the common lime—also known as "Key," "Mexican," or "West Indian"—was primarily a dooryard fruit in Florida, although there was some limited commercial production by 1883 in Orange and Lake counties.

.Pineapple production was abandoned on the Keys in 1906 after the depletion of soil organic matter and the 1906 hurricane and a lime industry slowly developed in its place. Plantings increased rapidly after 1913 on the Keys and on islands near Fort Myers, with production reaching a peak in 1923. The hurricane of 1926 eliminated or severely damaged most of these plantings.

Meanwhile, the Tahiti or Persian lime was being reintroduced to Florida. Tahiti is not a true lime but is probably a hybrid that was classified as *Citrus latifolia* by Tanaka. Undoubtedly it was a hybrid that originated under cultivation because it was known only as a local selection appearing in California about 1875. From 1850 to 1880, oranges and limes were imported in great numbers to San Francisco from Tahiti, and seeds of these fruits were sometimes planted. The parent tree of Tahiti lime must have arisen from a seed that resulted from a chance cross with some unknown cultivar, possibly citron. The cultivar has a triploid chromosome number (it has three sets of chromosomes), a condition producing sterility. There is little information about the original tree. The first record concerning it is Garey's statement in 1882 that the Tahiti lime was "worthless" in California. The next year, Rooks noted that the Tahiti lime was growing in Florida and that he also knew of the Persian lime. For many years, Tahiti and Persian appeared as distinct types, with one sometimes rated inferior to the other, but the two names have long been considered synonymous. The name Persian lime probably refers to a second area of propagation in Iran (Persia), although this is not well documented. A seedling selection of Tahiti lime is called "Bearss lime" in California. It appears that Tahiti, Persian, and Bearss are the same cultivar.

Pliny Reasoner reported in 1887 that the Tahiti lime was planted in Lake County, but lemons were considered more commercially promising acid fruits, and plantings of Tahiti lime increased slowly. With the decline of Key lime plantings in the late 1920s, much interest developed in planting Tahiti lime in southern Florida. Since 1930, Tahiti lime has been the primary type of lime grown in Florida. Most production was limited to the calcareous soils of Dade County and at one time there were more than 6,000 acres in Florida In 1992, Hurricane Andrew significantly reduced Tahiti lime acreage in this area. Many trees were toppled over and later reset in their original position. In the early 1990s, citrus canker was found in Dade County. Trees that were infected and all neighboring trees (a 1900–foot radius around the original tree) were removed and burned and the lime industry virtually disappeared. It is likely that the lime industry will not return to Florida.

Lime Characteristics

Key limes are much smaller and seedier than lemons and are easily distinguished from lemons. Tahiti limes are seedless, as Eureka lemons tend to be; they are similar in size and shape to lemons, so they may be easily confused for lemons. Lime growers have preferred to market Tahiti fruit green as an easy means of separating them from lemons. Like lemons, Tahiti limes are largely picked by size. Very few Tahiti limes are grown in California because of the cool climate during the winter in the major lemon-growing regions. Tahiti limes grow and produce well in hot, humid climates such as found in lowland tropical regions. Tahiti limes, like Key limes, mature throughout the year, but peak production is during the summer. Limes produced in the off-season during the spring bring much higher prices than those harvested during summer.

Three other citrus fruits are sometimes referred to as limes or resemble them greatly: the sweet lime, the calamondin, and the Rangpur lime. None of these is a true acid lime, and none is of commercial importance in the United States. The Palestine sweet lime (*Citrus limettioides* Tanaka) is common in India and parts of the Middle East. It has been used as a rootstock in Israel, although susceptibility to cachexia (viroid) has been a problem in the past. It has been tested as a rootstock in Florida but has not been widely used. The Palestine sweet lime is not the same as the Mediterranean sweet lime (*C. limetta* Tanaka) of Spain and Italy. The latter species was responsible for British sailing ships of the 18th century being known as "lime-juicers" and British sailors as "limeys." The lime juice carried to prevent scurvy was not the acid West Indian lime but the sweet lime, which has a higher ascorbic acid (vitamin C) content.

The calamondin (*Citrus madurensis* Lour.) is a small seedy lime type with a thin, orange peel and soft, juicy, orange flesh. Native to the Philippines, it was named "Citrus mitis" by Blanco, but Loureiro had seen and named it earlier in Madura, an island near Java. It is sometimes designated as "*C. microcarpa* Bunge," but this is a different fruit, the musk lime. The calamondin was introduced to Florida as an acid orange in 1899 by Lathrop and Fairchild from Panama and was initially called "Panama orange." It had come to Panama from Chile and to Chile from China, where it has long been cultivated both as an ornamental and as a rootstock for mandarins.

The calamondin is very cold hardy; the tree is bushy and is quite attractive as an ornamental for months when the bright orange, mature fruits are produced. It is ever-bearing, with the heaviest crop occurring in winter. The acid content is high, the flavor is distinctive, and juice color is attractive. Seed propagation is used because very little seedling variation has been observed. Various hypotheses have been put forward as to the possibility of the calamondin being a hybrid of lime and mandarin or lime and kumquat, but these are highly speculative.

The Rangpur lime and its near relative the Kusaie lime were considered by Tanaka as lemon relatives and belong to *Citrus limonia* Osbeck. Both are erroneously called limes and are used like true limes and calamondins. Rangpur especially is used in some areas of Central America as a substitute for true acid limes. Fruit have thin peels easily separated from the soft, juicy flesh, the peel and flesh both being red in Rangpur and yellow in Kusaie. These are probably hybrids of mandarin with lime or lemon. Rangpur was introduced to Florida as seed from India in 1887 by the Reasoner brothers. Fruits mature in the fall and winter. It has been sparingly grown as an ornamental dooryard tree but has been of more interest in recent years as a rootstock. Rangpur lime is currently a very important and widely used rootstock in Brazil and Argentina. Kusaie lime is not cultivated in Florida.

CITRONS (*CITRUS MEDICA* L.)

The citron is probably the first citrus fruit known to Europeans and has been grown in the warmest areas around the Mediterranean Sea for at least 2,000 years. It is one of three true citrus species (see chapter 3). Commercial culture today is mostly limited to Israel, Sicily, southern Italy, Corsica, Crete, and a few small Greek islands in the Cyclades. The flesh and juice are not used, but the thick peel is preserved in brine for use in making candied peel for fruitcakes and confections.

Citron was among the citrus species brought to the New World by Columbus, but it has never been extensively grown here. It was likely introduced to Florida when St. Augustine was settled, but it also probably did not survive winters there. There is no commercial production in Florida today, although a few specimen trees may be found in citrus collections. No cultivar is grown as such, although the common large-fruited type grown commercially in Italy is similar to what has long been seen in Florida. It is propagated readily by cuttings or may be budded. A few seedlings are available of the small-fruited type with persistent styles called "Etrog," which is used in the Jewish ceremonies of the Feast of Tabernacles. Etrog citron is also used as an indicator plant for citrus exocortis viroid (CEV). The fingered citron, which has very distinct, separate carpels, is grown as a curiosity in some citrus collections.

CULTIVAR IMPROVEMENT

For centuries, citriculturists have tried to find or develop new and improved cultivars and rootstocks. Favorable scion characteristics include tree yield and vigor, fruit quality and disease and pest resistance, and cold hardiness. Rootstock characteristics include adaptation to soil texture, pH and salinity, flooding and drought, and resistance to soil-borne diseases and pests. There are three basic methods for finding or developing superior cultivars of citrus: selection of favorable mutations or natural hybrids; classical breeding; and use of molecular techniques.

Selection of Mutations and Hybrids

Most of the commercially important citrus cultivars have resulted from selection of superior material in the field. These include most of the red grapefruit cultivars, Valencia and navel oranges, and many mandarin cultivars. Many local selections have been made through long-term observation of trees and fruit in a localized region.

Citrus and its relatives spontaneously produce bud, limb, fruit, or entire tree mutations. Navel oranges and satsuma mandarins among others are very prone to produce mutations. Some citrus also form chimeras, which are two or more genetically distinct tissues in the same plant or plant part. Periclinal mutations occur when one genetically dissimilar tissue surrounds another. Sectorial mutations occur throughout an entire botanical region often including an entire fruit section from the seed to the peel. Mericlinical mutations are similar to sectorial mutations but do not extend through an entire region. These mutations are particularly prevalent in fruit. It is common to observe fruit in packinghouses and processing plants that have distinct variations in peel color or texture within one or more segments.

Citrus also naturally hybridizes in the wild as described in chapter 3. Grapefruit, Temple orange, and Ortanique tangor are natural hybrids that were selected for superior characteristics.

Traditional Citrus Breeding

Citrus breeding programs began in Florida in 1892 with the establishment of a Subtropical Laboratory in Eustis. W. T. Swingle and H. J. Webber were hired to study diseases such as blight (which is still unsolved today) and pests but also to begin a targeted breeding program based on genetic principles and specific goals. They made many interspecific and intergeneric hybrids, most notably the citranges and citrumelos, which have been important rootstocks.

Traditional or classical breeding is based on the selection of parents with favorable, heritable characteristics and the production of literally thousands of progeny. Often female parents are chosen that produce only a sexual embryo to ensure that a sexual hybrid has been made, for example, Clementine mandarin or Robinson tangerine. Citrus breeding is difficult and extremely time consuming because many parents do not readily transfer genes or are incompatible, and few important traits have single gene inheritance patterns. Moreover, often nucellar embryos arise that are identical to the maternal parent (see chapter 3).

Breeding is accomplished by transferring pollen from one parent (male) to the stigma of another (female), usually by hand. Sexual fertilization then occurs and the resulting seeds are planted out in large numbers. Seedlings are allowed to fruit, often requiring 10 years (juvenility period), and fruit and tree characteristics are evaluated. Seedlings are also evaluated for pest, disease, and nematode tolerance. Seedlings with favorable characteristics are budded onto a rootstock and transferred to another field where they are evaluated for another five years. Superior selections are in many cases placed into growers' fields and further evaluated. The entire process may take 15–20 years (or even more) before a new cultivar is released.

Molecular Biology Approaches

Molecular biologists have made major strides in isolating and identifying many genes that are associated with important disease (CTV) and pest (nematodes) problems as well as potential resistance to freeze damage. These individual "markers" can be isolated and characterized for individual crosses when the plants are very young. This process saves time and money by eliminating the need to screen each seedling for various diseases as is the case in traditional breeding. Molecular biology also provides an accurate genetic assessment of similarities among species (see chapter 3). Plant breeders can then avoid making crosses among closely related species or hybrids which often grow poorly. Traditional approaches then can be combined with this advanced knowledge to shorten generation times.

FIVE

Rootstocks and Rootstock Improvement

GENERAL CHARACTERISTICS

Citrus trees were grown from seed for many years. But seedlings of many citrus types often have long juvenility periods (time until first consistent flowering), are susceptible to soil-borne diseases and nematodes, and may produce inferior quality fruit. Therefore, most citrus trees in Florida and worldwide are propagated and grown as two-part trees, consisting of a scion (aboveground portion) and a rootstock (below-ground portion). Fortunately, many species used as rootstocks come true-to-type from seed. Thus, they are genetically similar to the maternal (seed) parent (see chapter 3). This makes the propagation of rootstocks much simpler and more consistent than for rootstocks of other tree crops, which are generally not true-to-type from seed. The propagation of rootstocks is discussed in chapter 6.

Every rootstock has its advantages and disadvantages. Therefore, it becomes important to choose a rootstock based on its major limitations first, followed by less-important characteristics. In general, several different qualities are important to rootstock selection. These include:

1. *Compatibility with the scion.* Most species within the true citrus group are graft (bud) compatible. However, these species are generally incompatible with their remote relatives. Often incompatibility symptoms do not occur for many years after budding.
2. *Nursery adaptability.* Desirable traits include a high percentage of nucellar embryos, seedy fruit and good availability of seed, and ease of growing seedlings in the nursery. Ideal seedlings are vigorous, not too thorny, have straight trunks and few branches for ease of budding, and are not susceptible to diseases and pests.
3. *Soil adaptability.* The rootstock should grow vigorously in a variety of soil textures, structures, depths, water-holding capacities, and salinity and pH levels.

4. *Influences on fruit and tree characteristics.* The rootstock should favorably affect vigor, yields, fruit quality, and cold hardiness of the scion.
5. *Biotic adaptability.* The rootstock reduces the susceptibility of the tree to soil-borne pests (nematodes, weevils), diseases (foot rot, root rot), or to viruses or viroids present in the tree (CTV, CEV, cachexia).

Rootstocks can be divided into general categories based on their overall characteristics. There are the lemon types, sour orange types, sweet orange types, mandarin types, and trifoliate types. The general characteristics of commonly available rootstocks in each group are presented in table 5.1. The table is intended to give general, overall characteristics based on experimentation and observation. However, expression of these traits is not always consistent and may be affected by local soil or climatic factors.

LEMON-TYPE ROOTSTOCKS

Rootstocks in this group in general impart great vigor to the scion, are drought tolerant, produce poor-quality fruit, and decrease freeze hardiness of the scion. Most lemon-type rootstocks are susceptible to burrowing nematode and citrus nematode damage but vary in susceptibility to the major citrus viruses, viroids, and diseases. Rootstocks in this group include rough lemon, *C. volkameriana* (Volkamer lemon), *C. macrophylla* (alemow), Palestine sweet lime, and Rangpur lime.

Rough lemon (*Citrus jambhiri* Lush.)

Apparently only sour orange and sweet orange were used in Florida as rootstocks prior to 1865. Rough lemon attracted attention as a rootstock due to faster growth and earlier bearing for oranges and lemons budded on this rootstock. As the high pineland soils were planted more extensively in the southward development of the citrus industry, rough lemon increasingly replaced other rootstocks or seedling trees.

Rough lemon rootstock was used on about 70 percent of the citrus acreage of Florida in groves planted from the 1920s to the 1960s. However, in 1964, at Wauchula (Hardee County), a six-year-old grove of Pineapple orange on rough lemon began to show symptoms of a decline that was subsequently called "young tree decline." During the intervening years, this decline (discussed more fully in chapter 10) reduced the vigor and appearance of many young trees in the flatwoods areas of Florida. A similar type of decline, called "sandhill decline," was observed in the well-drained areas in the southern portion of the citrus belt. Because this decline appears most preva-

Table 5.1. General characteristics of rootstocks for Florida citrus

	Characteristic															
Rootstock	Wet soil	Clay soil	Drought	Freeze tolerance	Phytophthora tolerance	Blight tolerance	Tristeza tolerance	Exocortis tolerance	Cachexia tolerance	Burrowing nematode tolerance	Citrus nematode tolerance	Yield/Tree	Juice quality	Fruit size	Tree size	High pH tolerance	Salt Tolerance
Rough lemon	I	I	G	P	S	S	T	T	T	S	S	H	L	LG	LG	G	I
Volkamer lemon	G	I	G	P	I	S	T	T	T	S	S	H	L	LG	LG	G	I?
C. macrophylla	G	I	G	P	T	P	S	T	S	S	S	H	L	LG	LG	G	I
Palestine sweet lime	G	I	G	P	S	P	T	S	S	S	S	H	L	LG	LG	G	P
Rangpur	?	?	G	P	S	P	T	S	S	S	S	H	L	LG	LG	G	G
Sour orange	G+	G	I	T	T	T+	S	T	T	S	S	I-H	I+	I	I-LG	G	G
Smooth Flat Seville	(G)	G	G	I	T	(T)	(T)	(T)	(S)	(S)	(S)	I-L	L-I	LG	LG	G	G
Sweet orange	P	I	P	I	S	T+	T	T	T	S	S	I	I	LG	LG	I	(I)
Cleopatra mandarin	P	G	I-G	G	S	G	T	T	T	S	S	L-I	H	SM	LG	I	G
Sun Chu Sha	P	G	I-G	(G)	S	?	T	T	T	S	S	L-I	H	SM	LG	I	I+
Trifolate orange	G	G	P	G	T+	(S)	T+	S	T	S	T+	L-I	H	SM	SM	P-	P-
Carrizo citrange	I	P	G	G	I	I	T	S	T	T	T	H	I-H	I-LG	LG	P	P
Kuharske citrange	?	?	(G)	(G)	T	?	(T)	(S)	(T)	T+	(T)	(H)	I	I-LG	LG	(P)	(P)
Swingle citrumelo	(G)	P	I-P	G	T+	T	T	(T)	T	S	T	I	H	I	I	P	P

Source: Castle et al., Florida Citrus Rootstock Selection Guide. SP-248, 1998, 2006.

Notes: a. G = good, P = poor, H = high, R = resistant, I = intermediate, S = susceptible, L = low, SM = small, LG = large, T = tolerant, ? = inadequate information available, ( ) = expected rating, (+)(-) = relative ranking.

lent on rough lemon budded with sweet oranges, it was also called "lemon root decline." Further observations indicated that other rootstocks were also affected, but growers were reluctant to use rough lemon following tree losses of more than 80 percent in some groves. This decline has caused significant economic loss for the citrus industry, and no solution is available. These declines are now known to be so similar that they share a common name, "blight." The gravity of this problem is evidenced by the fact that citrus nurseries have reduced propagations on rough lemon rootstock from 60 percent of total nursery inventories in 1960 to less than 10 percent in 1973, and less than ½ percent in the 2000s.

Rough lemon is especially well suited to deep, well-drained, sandy soils, where it produces large trees in a relatively short time. The vigorous growth of trees on rough lemon also make scions on it very susceptible to cold damage as illustrated in the December 1962 freeze. It is tolerant of citrus tristeza virus (CTV) and citrus exocortis viroid (CEV). Trees budded on rough lemon form a normal bud union with all scion cultivars, develop a deep root system, and produce high fruit yields. Longevity of trees on this rootstock is limited in many areas by blight, and the rootstock is susceptible to foot rot (an alga). In addition, fruit quality (soluble solids and acids) is lower on this rootstock than on sour orange, Carrizo citrange, or Swingle citrumelo. Often the higher yield produces more pounds-solids per acre than other rootstocks. However, fresh fruit quality is low, which may affect time of maturity, peel quality, and market.

Citrus volkameriana (Ten. and Pasq.)

Citrus volkameriana, or "Volk," as it is often called, is a lemon hybrid having many of the same effects on the scion as rough lemon, including poor fruit quality (low sugars and acids) and freeze hardiness, and blight susceptibility. It is not susceptible to CTV, CEV, or cachexia but it is susceptible to citrus and burrowing nematode damage. Nevertheless, in long-term rootstock studies, Volk consistently produced more pounds-solids due to its higher yields than most other rootstocks. More important, this even occurred with considerable losses due to blight! Consequently, there is some interest in planting trees on Volk for processing oranges. More than 1 million trees were propagated on Volk from 2000 to 2005, indicating a renewed interest in this rootstock.

Alemow (*Citrus macrophylla* Wester)

Macrophylla, or "Mac," as it is sometimes called, is possibly a hybrid of *C. celebica* and pummelo. Most citrus, especially lemons and limes, are quite vigorous and productive when budded on Mac. As with other lemon types, fruit quality of oranges and mandarins budded on Mac is poor, particularly for young trees. Generally juice sugars and acid are very low compared to other rootstocks such as sour orange. Trees on Mac are more tolerant of foot rot than those on rough lemon, and CEV does not cause tree stunting. Trees on Mac are very susceptible to CTV and cachexia, and scions on it are very susceptible to cold damage. Therefore, with the exception of limes, macrophylla is not often used as a rootstock in Florida and probably will not be, due to concerns about CTV.

Palestine Sweet Lime (*Citrus limettioides Tan.*)

This rootstock was mentioned previously as an important species of sweet lime. It, like Volk and Mac, is also probably a natural hybrid. Vigor of scions budded on it is slightly less than that of rough lemon, but fruit quality is slightly better. Citrus trees on Palestine sweet lime are susceptible to foot rot, may be stunted by CEV and cachexia, and are possibly susceptible to CTV. They are susceptible to burrowing and citrus nematode, and scions budded on Palestine are cold susceptible. It has not been used extensively as a rootstock in Florida, primarily being confined to rootstock studies.

Rangpur (*Citrus reticulata hybrid*)

This hybrid is likely not a true lime, as indicated by its scientific name, but it is used as a lime substitute in some countries. Trees budded on it have characteristics similar to those of sweet lime, except those on Rangpur are susceptible to blight and are stunted by CEV and cachexia. Scions on Rangpur are tolerant of drought and high-salinity soils and reportedly tolerant of CTV, although its CTV tolerance has sometimes been questioned. Like sweet lime, Rangpur is rarely used in Florida, but it is the major rootstock in Brazil because of its reported CTV- and drought-tolerance. Trees on Rangpur are susceptible to burrowing and citrus nematode, and scions budded on it are cold susceptible. Rangpur will likely not be widely used as a rootstock in the future in Florida.

SOUR ORANGE-TYPE ROOTSTOCKS

Sour Orange (*Citrus aurantium* L.)

In Florida, sour orange was the preferred rootstock for the low hammock and flatwood soils with fine texture and high water tables, where its tolerance to foot rot gave it a distinct advantage over other rootstocks. It imparts freeze hardiness to the scion, and trees on sour orange are standard size (in spite of a slight tendency for the scion to overgrow the rootstock) and produce moderate crops of excellent quality fruit (high sugars). Trees on sour orange rootstock are not affected by cachexia and CEV and are blight tolerant. Sour orange is, unfortunately, adversely affected by CTV, with almost all scions except lemons. The citrus industry of Brazil, with nearly all trees on sour orange, was literally wiped out in the 1940s by a severe strain of CTV. Similarly, CTV has caused significant tree losses in California, South Africa, Australia, and Venezuela in the past. Several strains of CTV are present in Florida in some areas and CTV is now widespread. It is spread through infected budwood and by aphid vectors and represents a serious threat to trees on sour orange rootstock. In the 1990s, more severe strains of CTV (stem pitting) were discovered in Florida, and the brown citrus aphid, which is a very efficient vector, was inadvertently introduced from Central America. The use of sour orange as a rootstock in Florida has declined precipitously in the 1990s and 2000s, and probably will continue to do so in the future.

Smooth Flat Seville (*C. aurantium x C. grandis*).

Smooth Flat Seville, also called Australian orange, is likely a hybrid of sour orange and pummelo and appears to be tolerant to CTV. Otherwise, it has the same general characteristics of sour orange and many trees were propagated on this rootstock from 2000 to 2005. It will probably continue to be planted at low levels, especially for grapefruit. However, some growers have experienced problems with low sugars in the fruit from scions on Smooth Flat Seville, and fruit may be slow to reach minimum maturity standards for harvest (see chapter 11).

SWEET ORANGE ROOTSTOCKS (*CITRUS SINENSIS* [L.] OSBECK)

Sweet orange rootstock was extensively used in the first big expansion of the citrus industry after budding became standard in the early 1880s, but it was extremely susceptible to foot rot. This was particularly a problem on low-lying, fine-textured soils but was encountered even on high pineland areas. By 1890, sweet orange was no longer used as a rootstock; when the second big expansion took place on the sand hills of southern central Florida after the big freezes of 1894 and 1895, rough lemon was the predominant rootstock. Sweet orange rootstock produces orange, grapefruit, and tangerine trees larger than those on sour orange, although not as large as those on rough lemon. Fruit quality is almost as good as that on sour orange. The rootstock is compatible with most scions, but trees are delayed slightly in coming into bearing as compared to those on rough lemon. The rootstock is resistant to CTV and tolerant of cachexia, CEV, and moderately tolerant of blight; it is of intermediate freeze hardiness between Cleopatra and rough lemon rootstocks. Sweet orange is very susceptible to drought, foot rot, and burrowing and citrus nematodes with the exception of Ridge Pineapple orange, which has tolerance to burrowing nematode.

MANDARIN-TYPE ROOTSTOCKS

Cleopatra Mandarin (*Citrus reticulata Blanco*)

This species came into use in Florida rather late compared to rough lemon and sour orange, having been first used by Reasoner about 1920 as a rootstock for Temple oranges on excessively drained, sandy soil. It achieved limited success as a rootstock for Temple oranges and tangelos on sandy soils and as a substitute for sour orange with kumquats on poorly drained soils. However, it remained a very minor rootstock until the discovery of CTV in Florida in 1952. Cleopatra (Cleo) was perceived to be a possible substitute for sour orange and is moderately popular among Florida rootstocks. It imparts nearly as much freeze hardiness to the scion as sour orange and is fairly tolerant to foot rot, but it is very susceptible to root rot. Feeder root regeneration following root rot damage is very slow for trees on Cleo. Cleo can be used on a wide range of soil types and is particularly good for high pH, calcareous soils found in the flatwoods.

Scions on Cleopatra rootstock are also tolerant of CTV, cachexia, and CEV. Cleo is compatible with many citrus scions and produces high-quality fruit. Some cultivars on Cleo come into bearing slowly and fruit size is usually small. However, by the time the trees are 10 years old, they have made up the difference in both size and yield, even on deep sandy soils. Valencia on Cleo rootstock has a bad reputation primarily because of relatively low yields of nucellar Valencia budlines on Cleo accompanied by considerable tree vigor. However, old-line Valencia selections with lower vigor have satisfactory yields. In many ways, Cleo seems to combine several of the good features of both rough lemon and sour orange and may represent a good compromise for some growers, especially those growing mandarins or mandarin hybrids.

TRIFOLIATE ORANGE-TYPE ROOTSTOCKS

Trifoliate Orange (*Poncirus trifoliata [L.] Raf.*)

In the past, trifoliate rootstock was not very well suited for use on deep sands because of its limited root system. However, the widespread use of micro-irrigation systems has largely solved this problem. In addition, trees on it do poorly on calcareous soil, but it is well adapted to poorly drained, fine-textured soils. It is resistant to CTV and cachexia and is immune to foot rot. However, it is the most susceptible of all rootstocks to CEV and may be severely stunted by this viroid. It is resistant to the citrus nematode, although susceptible to the burrowing nematode. Blight may also be a problem for trees on this rootstock, but information on its susceptibility is limited. The trifoliate orange rootstock always overgrows the scion markedly, producing a bench. Some species of citrus are dwarfed on trifoliate, while other species produce standard trees in spite of what seem to be poor bud unions. Infection with CEV almost always induces dwarfing, and in fact, trees are sometimes deliberately inoculated with CEV to control tree size in regions such as Israel and Australia. Trees on trifoliate orange rootstock produce excellent quality fruit. Yields are good for the tree size attained. In the past, trifoliate has been used little in Florida except for satsuma mandarins and kumquats; now it is used for other types of citrus in restricted numbers. Trifoliate orange imparts a great degree of freeze hardiness to the scion in north Florida and when trees become fully acclimated during the winter. Trifoliate orange does not confer exceptional freeze hardiness to the scion in central and southern Florida. Warm weather in early winter may delay the development of hardiness of the rootstock, and the scion may be even more injured by cold than on other rootstocks. In California and Australia, it has been valuable on certain fine-textured soils for oranges. Because CEV infection can easily be avoided by careful budwood selection, the resistance to CTV and the high fruit quality make it of interest as another possible replacement for sour orange in soils having a pH less than 7.5.

Carrizo Citrange (*Citrus sinensis* [L.] Osbeck x *Poncirus trifoliata* [L.] Raf.)

This intergeneric hybrid was developed by Dr. W. T. Swingle in 1909 and is one of many citranges. Trees budded on Carrizo are vigorous and productive, and fruit quality is very good. Carrizo rootstock was originally intended to replace rough lemon in the Ridge area of Florida on deep, sandy soils. Carrizo imparts superior freeze hardiness to the scion compared with rough lemon;

unfortunately, it also is susceptible to blight, and losses to this malady have been great in some areas of Florida. Moreover, trees on Carrizo grow poorly on high pH or highly saline soils typical of many flatwoods areas, and trees are stunted by CEV. Nevertheless, Carrizo is still widely planted and was the second most widely propagated rootstock from 2000 to 2005 next to Swingle citrumelo because it is tolerant of CTV, it is productive, and it produces high-quality fruit. In addition, Carrizo is one of the few rootstocks that is tolerant of burrowing and possibly citrus nematodes. Thus, it is well suited for use in Ridge areas where burrowing nematodes and deep sands are prevalent and in flatwoods areas with soil pH less than 7.5.

Other Citranges

Several other citranges have been tested as potential rootstocks of citrus. Of these, Troyer is by far the most widely planted and originated from the same hybrid cross as Carrizo (navel orange x *Poncirus trifoliata*). It is rarely used in Florida but is widely used in California and parts of the Mediterranean region. Its effects on the scion are similar to those of Carrizo, with the exception of its lower tolerance of burrowing nematode. Kuharske citrange was widely propagated from 2000 to 2005, although it has not been extensively field tested. All early indications are that it is very similar to Carrizo in its characteristics but has superior burrowing nematode tolerance. Research from Dr. W. S. Castle at the University of Florida/IFAS, Citrus Research and Education Center, suggests that it is a good choice for Earlygold oranges. C-32 and C-35 citranges, which are siblings, are also being field tested; C-32 produces vigorous trees and C-35 moderate-sized trees. Both are productive, but their blight tolerance is unknown. They have potential as rootstocks for oranges and grapefruit. Rusk and Morton citranges also have been used as rootstocks in Florida to a limited extent. Rusk has performed quite well as a size-controlling rootstock in some field trials and may have some potential for use in high-density plantings in the future.

Swingle Citrumelo (*Citrus paradisi Macf. x Poncirus trifoliata [L.] Raf.*)

The original intergeneric crosses that produced many citrumelos were made by Dr. W. T. Swingle in 1907. One of these crosses, Swingle citrumelo, was released in 1974. Tree size on Swingle ranges from moderately small for most oranges to large for grapefruit and navel oranges. Yields, particularly for oranges, are considerably less than those on rough lemon. Nevertheless, Swingle has become the most widely planted rootstock in Florida. Nearly 40 percent of all nursery trees produced in Florida from 2000 to 2005 (46 percent in 2006) were on Swingle rootstock. Swingle's popularity stems from its tolerance of CTV and blight, immunity to damage by foot rot (although some damage may occur in association with Diaprepes weevil feeding), tolerance to citrus nematode, and the excellent freeze hardiness it imparts to the scion. Fruit quality for scion cultivars on Swingle is also excellent, although peel color development with most scions is delayed compared to that on Carrizo. In a typical planting, tree losses are much lower where Swingle is used as the primary rootstock compared with those on sour orange (CTV) or lemon types and Carrizo (blight). Nevertheless, during the 1990s and 2000s, growers began to realize that Swingle also has its problems, such as poor growth in high pH or clay soils, which occur frequently in east coast and southwest flatwoods areas. Some scions on Swingle are also stunted by low levels of calcium carbonate in the soil, or in soil series like Winder and Riviera with distinct clay layers or sand spots. Trees on Swingle may also require more rigorous nutrient management than those on other rootstocks. In addition, Swingle sometimes forms an incompatible union with Murcott tangerine or Roble orange, and trees begin declining after six to eight years in the field. Some research suggests that long-term yields of oranges on Swingle are less than those on more vigorous, productive rootstocks because of low per-tree yields despite the small percentage of tree losses. In contrast, there are several very productive groves of Hamlin and Valencia oranges on Swingle that are nearly 20 years old. It appears, however, that Swingle will continue to be popular in Florida despite these limitations.

Other Citrumelos

Many other citrumelos besides Swingle have been created through breeding programs worldwide but few have become commercially important. The F-80 series, including F-80–7, 80–9, 80–14, and 80–18, has shown some promise in field trials with Valencia oranges and Marsh grapefruit. It is assumed that many of their characteristics will be similar to those of Swingle. Some of these may not have decline problems associated with Swingle and warrant further study.

OTHER ROOTSTOCKS

The citrus industry is perpetually in search of the perfect rootstock. Literally hundreds of different rootstocks have been tested worldwide, but the previously mentioned ones remain the most commonly used in Florida. Other rootstocks with potential value in Florida include Sun Chu Sha (Sun), Kinkoji mandarin hybrid, and Gou Tou sour orange type. Sun Chu Sha has similar characteristics to Cleopatra but may perform well in areas with severe blight or high pH. Gou Tou is tolerant of CTV and produces large fruit. However, initial yields and fruit quality for trees on Gou Tou are inferior to those on sour orange, yet trees have other favorable qualities of sour orange. Kinkoji also has shown some promise in field trials due to its favorable performance relative to sour orange without being susceptible to CTV. It has not produced well, however, in some flatwoods areas. A Cleopatra x trifoliate hybrid (x639) has also produced well in several field studies and has many characteristics similar to Cleo. It is well adapted to some soil series in the flatwoods and has potential as a rootstock for this region. Information on many of these newer rootstocks is still being compiled.

NEW ROOTSTOCK DEVELOPMENT

The development of new rootstocks is a very long-term and expensive process. Several projects are under way at the USDA and the University of Florida to find new and improved cultivars. These new rootstocks are being created by traditional breeding or through molecular or cellular biology techniques.

Traditional Breeding

Traditional rootstock breeding relies on large-scale hybridizations and planting of many seedlings as described for scion breeding in chapter 4. The seedlings are then screened for the major problems associated with rootstocks as mentioned in the introduction to this chapter. The successful seedlings must then be field tested using a variety of different scions and vigor, yields, fruit quality, disease, and pest resistance further evaluated.

For example, the USDA has been testing several grapefruit x trifoliate (citrumelo) hybrids. These are listed in table 5.2. As expected, there is considerable variation in yields, tree size, and fruit quality among selections. Of these, US-1215 and US-1209 had superior yields and US-1213 produced the highest soluble solids (sugars) in the juice. Many other hybrids are being tested for *Phytophthora* foot and root rot and *Diaprepes* weevil/ P. palmivora complex tolerance. These include US-852 (Changsha x trifoliate), US-812 (Sunki x trifoliate), US-802 (pummelo x trifoliate), and US-897 (Cleopatra x trifoliate). Although these are long-term studies, the potential exists for finding the next important rootstock for Florida.

Table 5.2. Citrus hybrids currently being tested by USDA in Florida

Rootstock	Availability[a]	Parentage
US-1203	USDA	*Citrus paradisi* 'Duncan' x *Poncirus trifoliata* 'Kryder 5–5'
US-1205	USDA	*Citrus paradisi* 'Duncan' x *Poncirus trifoliata* 'English Large'
US-1208	USDA	*Citrus paradisi* 'Foster' x *Poncirus trifoliata* 'Kryder 43–3'
US-1209	USDA	*Citrus paradisi* 'Foster' x *Poncirus trifoliata* 'Kryder 43–3'
US-1210	USDA	*Citrus paradisi* 'Foster' x *Poncirus trifoliata* 'Kryder 5–5'
US-1211	USDA	*Citrus paradisi* 'Foster' x *Poncirus trifoliata* 'Ronnse'
US-1212	USDA	*Citrus paradisi* 'Foster' x *Poncirus trifoliata* 'Ronnse'
US-1213	USDA	*Citrus paradisi* 'Foster' x *Poncirus trifoliata* 'Ronnse'
US-1214	USDA	*Citrus paradisi* 'Foster' x *Poncirus trifoliata* 'Ronnse'
US-1215	USDA	*Citrus paradisi* 'Foster' x *Poncirus trifoliata* 'Ronnse'
US-1216	USDA	*Citrus paradisi* 'Foster' x *Poncirus trifoliata* 'Ronnse'
US-1217	USDA	*Citrus paradisi* 'Foster' x *Poncirus trifoliata* 'Ronnse'
Carrizo	C	*Citrus grandis* 'US-145' x *Poncirus trifoliata* 'Yamaguchi'
Swingle	C	*Citrus sinensis* x *Poncirus trifoliata*
		Citrus sinensis x *Poncirus trifoliata*

Source: Bowman and McCollum. 2006. Proceedings Florida State Horticultural Society 119:124–27.

Molecular Biology Approaches

PROTOPLAST FUSION

In the 1990s, researchers at the University of Florida/IFAS Citrus Research and Education Center developed the technique of protoplast fusion for circumventing traditional breeding methods. Protoplasts are cells with their cell walls removed. They are extracted from potential rootstock parents and combined in the laboratory under controlled conditions. The resulting hybrids have four sets of chromosomes (tetraploids) and are designated as parent A + parent B rather than as parent A x parent B as with traditional hybrids. There is no male or female plant as in traditional breeding and plants are called "somatic hybrids." The best of these have been field tested as rootstocks for several years. The long-term performance of somatic hybrids has been variable and further testing is warranted. Types of somatic hybrids are listed in table 5.3.

Table 5.3. Rootstocks used in Murcott tangor trials near Loxahatchee (1) or Sebring (2)

Trial	Common name	Scientific name
1	Benton citrange	*C. sinensis* (L.) Osb. 'Ruby' x *P. trifoliata* (L.) Raf.
1	*C. amblycarpa*	*C. amblycarpa* (Hassk.) Ochse
1	C-35 citrange	*C. sinensis* 'Ruby' x *P. trifoliata*
1	C-65–165	Mandarin hybrid (identity lost)
1	Calamandarin	Probable hybrid of *C. reticulata* (Blanco) and *C. mitis* Blanco
1, 2	Carrizo citrange	*C. sinensis* x *P. trifoliata*
1	Changsha mandarin	*C. reticulata* Blanco
1	Changsha x English Large TF	*C. reticulata* x *P. trifoliata*
1	Cleo x Rubidoux TF	Cleopatra mandarin (*C. reshni* Hort. ex. Tan.) x *P. trifoliata*
1	Cleopatra mandarin	*C. reshni*
1	F80–3 citrumelo	*C. paradisi* Macf. x *P. trifoliata*
1	F80–7 citrumelo	*C. paradisi* x *P. trifoliata*
1	F80–8 citrumelo	*C. paradisi* x *P. trifoliata*
1	F80–18 citrumelo	*C. paradisi* x *P. trifoliata*
1, 2	Flying Dragon TF	*P. trifoliata*
1	Kinkoji	*C. obovoidea* Hort. ex. Tan.
2	Milam	Putative *C. jambhiri* Lush. hybrid
1	Morton citrange	*C. sinensis* x *P. trifoliata*
1	Murcott	Tangor (putative mandarin x sweet orange)
1, 2	Rangpur x Troyer citrange	*C. limonia* Osb. x (*C. sinensis* x *P. trifoliata*)
2	Shekwasha mandarin	*C. depressa* Hayata
1	Smooth Flat Seville	Putative pummelo-sour orange hybrid
1, 2	Sour orange	*C. aurantium* (L.)
1	Sun Chu Sha mandarin	*C. reticulata*
	Sunki mandarin x Flying Dragon TF	*C. reticulata* x *P. trifoliata*
2	Sweet orange	*C. sinensis* 'Ridge Pineapple'
1	Swingle citrumelo	*C. paradisi* x *P. trifoliata*
2	Uvalde citrange	*C. sinensis* x *P. trifoliata*
1	Volkamer lemon	*C. volkameriana* Ten. & Pasq.
1	W-2 citrumelo	*C. paradisi* x *P. trifoliata*
1	Yuzu	Probable sour mandarin hybridized with *C. ichangensis*

Somatic hybrids			
1	Cleopatra + Flying Dragon	ClFD	*C. reshni + P. trifoliata*
1	Hamlin + Flying Dragon	Ha+FD	*C. sinensis + P. trifoliata*
1	Hamlin + Rangpur	Ha+Rg	*C. sinensis + C. limonia*
1	Hamlin + Rough lemon	Ha+RL	*C. sinensis + C. jambhiri*
1	Valencia + Rough lemon	V+RL	*C. sinensis + C. jambhiri*

Source: Castle and Baldwin, 2006. Proceedings Florida State Horticultural Society 119:137.

TRANSGENIC TREES

In traditional breeding, thousands of genes are combined and the desired characteristics of the progeny are evaluated over time. In contrast, molecular biologists select genes that are involved with certain desirable traits such as CTV tolerance or cold hardiness. These genes, which may not even originate in citrus, are inserted into a plasmid vector (carrier) housed in the bacterium, *Agrobacterium tumefaciens*, using enzymes, and the gene sequence of the insert is checked for accuracy.

Agrobacterium is a pathogen of many plants and is readily taken up by the plant. Stem pieces of the species to be "transformed" (citrus in this case) are exposed to the vector by inoculation or infection and the new gene is taken up by the plant. The remaining bacteria are then destroyed and the transgenic plant is checked to see if it has been transformed. The transgenic plants are then cultured in artificial media in the lab, after which they are transferred to the greenhouse and grown under carefully controlled conditions. The plants are then transferred to the field for evaluation of the desired characteristic. This method is being tested as a means of producing sour orange rootstocks that are tolerant of CTV. The potential exists to genetically engineer citrus trees that are adaptable to a variety of situations. While molecular biology offers more precision in gene transfer over traditional methods, it is a difficult procedure for woody crops such as citrus, and undesirable, unwanted changes in the tree may also occur. Citrus trees produced via transgenics will require rigorous and long-term field evaluation before being released to growers.

SIX

Nursery Production and Propagation

While seedling citrus trees were seen as satisfactory in Florida for many years, it became evident over time that there were some very good reasons for growing two-part citrus trees consisting of a rootstock budded with an appropriate scion cultivar for the top part of the tree. Such trees were a little more difficult to produce but they had several quite compelling advantages over seedling trees. Chief among them was reduction in the length of the juvenility period common to most citrus species produced from seed and grown as seedling trees.

Seedling orange and grapefruit trees exhibit lengthy juvenility periods (often up to 10 years or more) during which trees grow rapidly and produce little or no fruit due to an absence of flowering. Lemons and limes, in contrast, have much shorter juvenility periods. Seedlings are also characterized by thorny growth that makes harvesting difficult; they tend to have vigorous upright growth patterns, and when fruit production begins, the fruit are produced in the upper canopy (another character of juvenility).

On the other hand, two-part trees with scions budded onto rootstocks have abbreviated juvenility periods when budwood is selected from mature, fruiting trees. Strictly speaking, the juvenility period is not actually shortened because the budwood has already gone through this period, but the time until the first crop is produced is much shorter. It is now becoming clear that budded trees will bear fruit more quickly than seedling trees, will not grow as tall as quickly, and usually have fewer (if any) thorns—with the exception of lemon and lime trees.

Another advantage of the two-part (budded) tree is that rootstocks upon which the tree is budded can be selected to better adapt the tree to growing conditions of the planting site. For example, some rootstocks do better in sandy soils while others might require less irrigation. Rootstocks can also affect the cold hardiness of the scion. Fruit quality can also be affected by rootstock choice, with some rootstocks enhancing and others lessening quality. Tree yields and vigor may also be affected by rootstock choice. Thus, a budded tree can provide the grower with an opportunity to select a rootstock that will enhance tree performance under a given set of conditions. Specific rootstock characteristics are discussed in chapter 5.

While advantages of budded trees are obvious, there can also be a downside. The two-part tree can be subject to certain virus and virus-like diseases that can seriously compromise tree growth and productivity. Such diseases cause problems associated with the bud union; seedling trees have no such union and, consequently, no problems from these diseases. Fortunately, only certain rootstocks are susceptible to these diseases and by using disease-free budwood, many problems can be avoided when susceptible cultivars and rootstocks are used. At least one major virus, citrus tristeza virus (CTV), can still be a problem for susceptible rootstocks even when clean budwood is used since it can be transmitted by aphids. For example, sour orange is an excellent rootstock for most citrus, but it has fallen from popularity dramatically because it is susceptible to CTV which is found all over Florida, often in a latent form in trees that are budded on nonsusceptible rootstocks.

PROPAGATION

The first citrus trees grown in Florida were no doubt seedlings, since seeds were far easier to transport on long sailing voyages than budded trees. For 250 years thereafter, it seems unlikely that any method of propagation other than by seed was practiced here. Budding was apparently standard practice in Europe during the 16th and 17th centuries, but during the 18th century growers realized that seedling orange trees were larger and more productive, and budding became unfashionable in Spain and Italy. This observation was undoubtedly largely responsible for the delay in using budding to propagate citrus trees in Florida.

Historical Background

The first recorded use of budding in Florida was in developing the Dummitt grove on the upper end of Merritt Island in 1830. Zephaniah Kingsley had imported budded orange trees from Spain in 1824 for a planting later known as the Mays grove at Orange Mills, a few miles north of Palatka. Buds from this grove were taken, according to E. H. Hart in 1877, for budding wild sour orange trees in the Dummitt grove. Thereafter, groves were occasionally budded, but far more were planted by seed.

The phenomenon of nucellar embryony in citrus seeds was another potent factor for delaying recognition of the superiority of budded trees. The term "polyembryony" refers to the production of extra embryos within a seed (see chapter 3). Most seeds contain a single embryo (one grain of corn, for example, contains a single embryo), but some plants and most citrus plants produce seeds with multiple embryos. A sexual embryo is usually formed following pollination and sexual fertilization, but within the developing seed, extra vegetative embryos may arise from nucellar tissue (nucellar embryos) found within the seed. Thus, a single seed may give rise to several plants, most from nucellar embryos with occasional ones from sexual embryos. Plants produced from nucellar embryos are very similar in appearance to the maternal parent, since they have been formed vegetatively and not by sexual fertilization. Consequently, a high percentage of many citrus seedlings arise from nucellar embryos. There are always some variant trees in a seedling grove, however, and all trees are usually thornier when grown from seed and require a much longer time to come into commercial production as mentioned previously.

In the 1870s, large-scale development of orange groves occurred in Marion County around Orange Lake and Lake Weir by budding wild sour orange trees with budwood carefully selected from a very few trees. Citrus growers began to be interested in clonal cultivars derived from one parent rather than in seedling types, and budded nursery trees of various named cultivars were imported from Europe. Most of these came from the Rivers Nursery in England, but some were brought by General H. S. Sanford from the Mediterranean area. In 1877, the short list of such cultivars with descriptions was drawn up by a committee for the Florida Fruit Growers Association. Thereafter, the cultivar list grew rapidly and more groves were developed from budded trees than seedlings, although many growers continued to argue the merits of using seedlings. Around 1880, several articles on orange culture appeared in both Florida and California, all stressing the superiority of budded trees; thereafter, few seedling groves were planted.

Any propagation method other than by seed will provide the advantages of early bearing, few or no thorns, and uniform fruit maturity and quality that distinguish a budded tree from a seedling. While stem cuttings or air-layers may also be used, only budding or grafting can give the additional advantage of having a root system that is better adapted to soil conditions than the scion roots are, or that can confer greater freeze hardiness to the scion.

Grafting can also be used for vegetative propagation but takes longer to do than budding, requires more vegetative material, and is more expensive. It is more commonly used for topworking trees, which will be discussed later in this chapter. Limes and citrons are sometimes propagated by cuttings or air-layering. Nevertheless, nearly all citrus trees produced in Florida are budded.

RECENT NURSERY PRODUCTION TRENDS

Recent nursery propagation trends provide information on future industry trends. The top 10 scion and rootstock nursery propagations from 2003–07 are presented in table 6.1. Swingle citrumelo has been the most highly propagated rootstock for the past five years, a position it has held for the past 19 years. It is followed in popularity by Carrizo citrange, Kuharske citrange, Cleopatra mandarin, and volkamer lemon. Recently, there has been increased interest in Kinkoji, US 812, US 802, X 639, and US 897 rootstocks. The leading scion propagations from 2003–07 included Hamlin, Valencia, Midsweet, Rohde Red, and Vernia oranges and Ray Ruby and Ruby Red grapefruit.

NURSERY ROOTSTOCKS

Rootstocks produced for budding scion cultivars are practically always grown from seed because fruit production and juvenility periods are not a concern. In the case of a cultivar that produces few seeds, such as Rusk citrange, cuttings may sometimes serve in place of seedlings. Tissue culture has also been used with a limited number of cultivars. Thanks to the high percentage of seedlings that develop from nucellar embryos, there is a high degree of uniformity in the most commonly used seedling rootstocks. The few seedlings that are obviously more or less vigorous can be rogued out (removed) easily. Such seedlings usually originate from sexual embryos and are not true to type. Several different species or hybrids have been used as rootstocks, each having some merit under certain conditions. Commercially, only a small number have proven desirable. A satisfactory rootstock must be compatible with the scion budded onto it; that is, the two must form a union that permits good growth, longevity, good yields, and favorable scion fruit quality. Dwarfing may be an indication of incompatibility or of the presence of a systemic disease such as citrus exocortis viroid, although yield may be adequate relative to tree size and fruit quality may be excellent. Some rootstocks produce normal, moderately productive small trees such as Rusk, Swingle, and trifoliate orange.

Table 6.1. Nursery propagation of citrus cultivars with five-year trends

2007		2006	2005	2004	2003	Five-year average	
Top Twenty Scion Cultivars (Rankings)							
1	Valencia	2	2	2	2	1	Hamlin
2	Hamlin	1	1	1	1	2	Valencia
3	Midsweet	4	6	5	3	3	Midsweet
4	Rohde Red Valencia	7	12	6	4	4	Ray Ruby Grapefruit
5	Ray Ruby Grapefruit	3	3	3	9	5	Rohde Red Valencia
6	Ruby Red Grapefruit	11	8	8	7	6	Vernia
7	Glen Navel	8	11	11	14	7	Ruby Red Grapefruit
8	Pineapple	9	10	12	11	8	Earlygold
9	Earlygold	12	7	8	6	9	Rio Red Grapefruit
10	Vernia	21	13	4	5	10	Pineapple
11	Murcott	6	9	13	15	11	Flame Grapefruit
12	Marsh Grapefruit	14	17	18	8	12	Glen Navel
13	Rio Red Grapefruit	17	4	9	13	13	Murcott
14	Minneola Tangelo	15	15	14	10	14	Minneola Tangelo
15	Key Lime	18	27	25	17	15	Marsh Grapefruit
16	Hirado Buntan Pum	13	34	36	39	16	Ruby Sweet Orange
17	Fallglo	19	38	31	33	17	Sunburst
18	Meyer Lemon	24	14	21	18	18	Meyer Lemon
19	Cara Cara Navel	23	22	30	37	19	Star Ruby Grapefruit
20	Sunburst	10	19	16	16	20	Meiwa Kumquat
Top Ten Rootstocks							
1	Swingle	1	1	1	1	1	Swingle
2	Carrizo	2	3	2	2	2	Carrizo
3	Kuharske	3	2	3	3	3	Kuharske
4	Kinkoji	6	10	11	10	4	Cleopatra
5	Cleopatra	5	4	4	6	5	Volkamer
6	Volkamer	8	6	6	5	6	X-639
7	US-812	10	11	16	14	7	Smooth Flat Seville
8	Sour Orange	14	13	14	20	8	Sun Chu Sha
9	Sun Chu Sha	7	7	8	8	9	Kinkoji
10	US-802	11	12	17	-	10	Benton

Source: Table provided by Michael Kesinger, Chief of Bureau of Citrus Budwood Registration in the Annual Report 2007 (FY 2006–2007).

In addition to scion compatibility, the rootstock must be adapted to the soil and environmental conditions it will be placed in if it is to grow successfully. Nursery adaptability and ease of handling from the nursery's standpoint is also very important. These two factors include ready availability of seed, production of a high percentage of nucellar embryos, rapid growth, relative tolerance of pests and diseases, and a smooth, straight, thornless stem.

Some rootstocks are superior in some qualities but inferior in others and none is superior on all counts. The same rootstock may be superior for one scion under given environmental conditions and inferior for another. The choice of a rootstock, therefore, must be made for a particular scion-rootstock combination in a particular environment. Detailed information on rootstock characteristics is found in chapter 5.

BUDDING PROCEDURES

Two important phases of successful budding are choice of a suitable rootstock and selection of satisfactory budwood. Since the factors influencing choice of rootstock were discussed in chapter 5, this section will deal only with budwood selection. Under the Citrus Budwood Registration Program of the Division of Plant Industry, it has been possible for many years to obtain budwood (and rootstock seed) that has been tested for most of the viral diseases and inspected for trueness to type for the cultivar. Florida's citrus budwood registration program began in 1953 with voluntary participation. After a new mandatory Citrus Budwood Protection Program was initiated in 1997, growers and nurseries began working even more closely with the Division of Plant Industry to refine regulations to protect the Florida citrus industry. All commercial citrus nurseries were required to become participants in the Citrus Budwood Protection Program. Own-use nurseries that were propagating a small number of trees that individual growers use for their own resets were also required to register for and follow program rules. Own-use nurseries were also required to register or certify their source trees and have them tested annually for citrus tristeza virus (CTV). Propagative material used for topworking in a citrus grove must also be either registered or certified and in addition be grown in a greenhouse. Regulating the use of this propagative material lessened the potential for spreading various vectored pathogens to neighboring groves.

Several things have happened since 1997 that have brought about significant changes in the Florida citrus nursery industry. First was the continuing spread of citrus canker within the Florida citrus industry; this spread was greatly exacerbated by three serious hurricanes (Charlie, Jeanne, and Frances) that pummeled the state in 2004 and a later one (Wilma) that occurred in 2005. Not only did these storms cause tree damage and death and significant yield reductions from high winds, but they spread citrus canker disease widely over previously uninfected areas of the state.

Citrus greening disease, first detected in 2005, also made its way into the Florida citrus industry, threatening the industry's very existence if left uncontrolled. The disease is transmitted by citrus psyllid insects which arrived several years in advance of the disease. Greening represents an even greater threat than canker because it has the capability to stunt trees, reduce yields and fruit quality, and ultimately cause tree death. The disease cannot be controlled once the tree is infected.

In order to help the Florida citrus industry recover from hurricanes and canker, and deal with the new problem of citrus greening, the Florida Citrus Plant Protection Committee, made up of growers, nurserymen, scientists, and regulators, developed new guidelines for the safe production of citrus nursery stock. The Florida Department of Agriculture, Division of Plant Industry, Bureau of Budwood Registration wrote new rules and regulations for citrus nurserymen called the Citrus Nursery Stock Certification Program (Florida Department of Agriculture Rule 5B-62) from the recommendations of the Plant Protection Committee. Full details of the rule are beyond the scope of this book, but additional information can be obtained from the Bureau of Budwood Registration, 3027 Lake Alfred Road, Winter Haven, FL 33881–1438 or from their Web site, www.doacs.state.fl.us/pi/budwood/index.html.

Among the many rules contained in 5B-62 are standards which must be met or exceeded for nursery site approval and for structures used to grow citrus plants including seedlings, budded trees, and budwood. Standards for decontamination and sanitation are also detailed. Rigid rules for handling all propagated materials and development of detailed records are clearly spelled out and must be followed from seed selection and planting through tree sales. Prospective nursery owners will find full details on all nursery procedures in the publication titled "Citrus Nursery Stock Certification Manual," available through the Budwood Office.

All citrus trees produced in Florida have to be produced in approved greenhouse structures at least one mile away from commercial citrus groves to help prevent the transmission of greening or other vectored diseases (fig. 6.1). Existing citrus nurseries were exempted from the one-mile distance requirement but all new nursery sites must conform to this standard. All propagating material will need to meet the same standards as the nursery production

Figure 6.1. Approved greenhouse structure for Florida citrus nurseries. Source: Steve Futch, University of Florida.

houses. Nurseries are under increased inspection for insects, diseases, sanitation, and structural integrity. These new rules for nursery production will help protect the citrus industry and represent significant changes for nursery growers. The Florida citrus industry has gone from primarily field nurseries, to a mix of field and greenhouse nurseries, to greenhouses exclusively for the production of all citrus trees.

Growers should understand that two types of citrus trees, registered and certified, will be legally available to purchase for grove planting. Both registered and certified trees will be tested for the graft-transmissible pathogens of citrus found in Florida. Registered trees are those that have fruited and been validated as true to type. Certified trees may be younger and waiting for fruit verification before becoming registered. The nursery must indicate on the invoice or citrus nursery inspection tag whether the trees are registered or certified. Both registered and certified sources are recurrently tested for graft-transmissible pathogens.

In budwood selection, there are two important steps: the selection of the scion tree and the selection of bud sticks from that tree. Note that most budwood is selected from a block of clonally propagated parent trees known as an "increase block." Increase blocks must originate from program foundation or scion trees, and budwood is generally cut without seeing the fruit.

New cultivar selections in the budwood program are evaluated and tested before industry release. In the past, many successful new parent candidates were found in groves that improved yields of many different cultivars. Today, very few new parents are found in existing groves or in the wild, and the greatest cultivar improvements are now obtained through dedicated breeding programs.

The following points should be considered in selecting budwood source trees:

1. Trees should produce fruit that is characteristic of the particular cultivar and be free from chimeras. Chimeras occur when two or more types of foliage, fruit, or growth are present on the same tree and result from a mutation that produces two or more genetic constitutions.
2. Trees should have a record of satisfactory production over as many years as possible to even out any year-to-year differences due to alternate bearing.
3. Trees should be tested for systemic (viral) diseases such as psorosis and CTV, and viroid diseases CEV (citrus exocortis viroid) and cachexia. Absence of visible symptoms is not sufficient evidence of freedom from these diseases, for many scion cultivars may carry viruses (viroids) without showing any symptoms on one rootstock while showing clear symptoms when budded on another rootstock. This is where the above-mentioned Budwood Protection Program plays such an important role. Enclosed greenhouses allow even CTV-free budwood to be produced, although absence of CTV cannot be guaranteed once the tree is planted in the grove. Program policy allows the propagation of trees with mild strains of CTV but not the more virulent sour orange decline strain or stem-pitting isolates. Budwood should also be free of propagatable malformations. Trees must also be checked for symptoms of the bacterial diseases, canker and greening.
4. Trees should have attained maturity, be in good health and vigor, and have reached a normal size for their age, rootstock, and soil type.
5. Budwood will never be distributed from the initial parent tree, as shoot-tip grafting will be used to produce clean, virus and viroid free propagating material from which a foundation tree will be established.

Foundation trees are located outside the commercial citrus-growing areas of Florida. The protected greenhouse-grown foundation trees are used to provide citrus nurseries with source material to establish self-contained scion and increase blocks at nursery locations. Scion trees are evaluated by Division of Plant Industry inspectors and placed in protected, carefully arranged plantings. Scion trees are annually tested for vectored citrus diseases and routinely for other endemic graft-transmissible pathogens. Scion trees are used by the nursery to start increase blocks which allow the rapid multiplication of propagating material. It is in the best interest of every citrus nursery to become self-sufficient in regard to their own budwood supply and limit the risks of bringing outside material into their nursery sites.

In selecting budwood from the source tree, the following points should be taken into account:

1. The bud stick is usually cut from the second flush back from the tip of the twig from scion trees and should have five to nine good, healthy buds, which have not yet started to grow. *NOTE*: The last flush can be used if "hardened off." It is often used when producing "increase" trees.
2. The bud stick should consist of healthy, well-matured wood. The novice can cut buds from well-rounded twigs much more easily than from angular wood, although the experienced budder will have equally satisfactory results with either. Leaves should be cut off at once to conserve moisture in the bud stick.
3. Budwood, preferably, should be cut just prior to use and protected from drying out. However, for budding in early spring when buds may start to grow on the twigs as soon as the bark "slips" (bark slipping indicates that the cambial cells under the bark are actively growing), it is advisable to assure a supply of unsprouted buds by cutting budwood before any buds begin growth. The bud sticks should be put at once into polyethylene bags, such as those used for freezing vegetables, and stored in a refrigerator at 40–45°F. Budwood will keep one to two weeks or longer under refrigeration, but be sure to periodically monitor bags for moisture content or mold development.
4. Bud sticks must be labeled at once after being cut if more than one cultivar is used, for it is very difficult to identify the cultivar after it is cut from the tree. Each source tree is treated as an individual, and records are available to track/trace nursery propagation back to this source. This procedure is important if a disease or problem is identified in the future.

Techniques used to produce citrus trees are very similar for all nurseries since all trees are grown in similar, approved greenhouse structures. Formerly, a large share of citrus nursery trees were produced in field nurseries and sold as bare-root trees for planting or were dug and planted in large pots for later transplanting. Field nursery trees typically required two years to produce from the seedbed to digging and sale.

Containerized trees produced under controlled conditions in the greenhouse are often ready for sale in about one year from seed but are usually smaller than bare-root trees. The key to good performance for any planted tree is careful attention to planting procedures and aftercare. More often than not, poor growth, or even tree death, are the result of improper planting or care.

DEVELOPMENT OF A BUDDED NURSERY TREE

The development of a budded nursery tree, ready for planting in a grove or as a dooryard tree occurs in four steps described below.

1. *The Seedbed.* Seeds of the desired rootstock were formerly sowed directly into a specially prepared area outdoors in the field. Since only greenhouse propagation in approved structures on approved sites is permitted today, the seedbed usually consists of a well-drained shallow pan/tray/flat three to six inches deep that is designed to fit on top of benches. Some nurseries use undivided trays; others use compartmented pans whose cells will each contain one seed and later, one seedling rootstock. Others may use pans which hold interlocking tubes, each holding one seed and later, the emerged seedling.

The containers are filled with an appropriate medium to facilitate seed germination. Some nurserymen prefer their own proprietary medium, while others will use media from horticultural supply warehouses.

Seeds are planted about one inch deep and the seed flats are placed in the appropriate racks in the greenhouse. The soil mix should be kept moist but not wet until the seeds germinate, which should be in one to two weeks, depending on variety and greenhouse temperature. Thereafter, water should be applied as needed to prevent wilting. However, keeping the medium too wet can inhibit growth of the young seedlings and may encourage development of fungal diseases.

Citrus seeds are unlike seeds of deciduous fruits in that once they have become thoroughly dry, they are no longer able to germinate. It is necessary to

Table 6.2. Approximate seed numbers for cultivars used as citrus rootstocks

Cultivar	Approximate seeds per quart
Carrizo citrange	2,500
Troyer citrange	2,500
Swingle citrumelo	6,500
Trifoliate orange	3,500
Cleopatra mandarin	6,000
Citrus macrophylla	4,500
Palestine sweet lime	5,000
Rangpur lime	6,500
Milam lemon	5,700
Rough lemon	5,500
Citrus volkameriana	6,000
Sour orange	2,500
Smooth Flat Seville	2,500
Ridge Pineapple	2,500

use seeds that have not been out of the fruit for more than two or three weeks unless special storage precautions have been taken. Seeds planted within a few days after extraction have the highest germination. Freshly extracted seeds may be stored for several weeks by washing them, drying them in thin layers in the shade for a day or two, and then keeping them in a cool place. Dry seeds can be stored in a plastic bag in a cooler or refrigerator at around 40°F for six months to two years for some species such as sour orange. Seeds should never be allowed to freeze. A fungicidal dip or hot water treatment (120°F for 10 minutes) is recommended to minimize losses due to fungal and algal diseases. Citrus seeds are generally sold by the quart, and seed number and price may vary considerably (table 6.2). Seeds should be purchased from a reputable dealer who follows the above recommendations. Some growers produce their own seeds, and proper precautions should also be taken in this situation.

2. *The Liners.* In a traditional field nursery, the small developing seedlings that were transplanted from the seedbed into another area where they could grow satisfactorily and be ready for budding were called "liners." The term was coined because the seedlings were planted close together in straight lines. The term is sometimes still used today even though the procedure is different in greenhouse propagation. In this case, the usual procedure is to trans-

plant the small seedlings into larger and taller pots (one plant per pot) so as to develop larger seedlings that are suitable for budding. Only vigorous plants of about the same size should be transplanted; all weak plants or those with crooked or malformed roots should be discarded. Off-type variants should also be discarded since they may be genetically inferior and produce non-uniform trees in the grove.

Liners are grown in their new containers until they achieve the desired size for budding. Water, nutrients, and pesticides are applied as needed.

3. *The Budding Operation.* With outdoor field nurseries, budding was usually done in warm months during the fall and spring when liners were actively growing and the bark would "slip" which would allow the bud to be slipped into a cut made on the stem. Since all nurseries are now in climate-controlled greenhouses, the liners can be budded at any time the plants are deemed ready.

The type of budding most widely used for citrus in Florida is called "shield budding" or "T-budding" because the scion piece is cut in an oval or shield shape. This shield of bark plus a small sliver of wood contains the bud which is in the leaf axil. A T-shaped incision is made in the rootstock bark and the bud shield (bud-eye) is inserted. In most citrus-growing areas of the world, including Florida, the cut is made to form an inverted T so that the bud shield is pushed up from below. The procedure for inverted T budding is presented in figure 6.2. The upright T is used in California and some other citrus-growing areas, but there is no inherent superiority of either method. A budder can attain as high a percentage of success with one as the other if equally skillful in using both. There is some variation in different citrus areas related to the height of bud insertion. For many years in Florida, it was considered best to make the rootstock incision only two to three inches above the soil level. Research has shown, however, that foot rot will occur less on a susceptible scion such as sweet orange when the bud is placed at a height of six inches above the soil line, and most Florida trees are budded at that height today. In some tropical regions, trees are budded at 18 inches or more above ground level to avoid foot rot infection.

The initial vertical cut should be one inch long but may be smaller if small buds are used (fig. 6.2a). The shield should be about ½-inch long, cut from the bud stick so that the inner face is nearly flat and not tapering greatly toward either end from the center (fig. 6.2b). The corners of the bark are lifted at the bottom of the rootstock incision, and the bud is gently pushed up under the bark (fig. 6.2c) (taking care that the shield is not doubled back on itself) until it is entirely within the incision (fig. 6.2d). If the bark does not slip easily

(a) Initial incision into the rootstock

(b) Removing the bud from the budstick

(c) Inserting the bud into the inverted T incision

(d) Properly inserted bud ready for wrapping

(e) Wrapping the bud with plastic tape

(f) Final tying of plastic wrap

(g) Properly wrapped and tied bud. Note that the tape covers the entire bud.

Figure 6.2. Inverted T budding of citrus trees.

enough for the bud to enter the incision readily, it is often possible to lift the bark and open the incision for its whole length with the budding knife or a probe, and then to insert the bud. The cambium tissue of the rootstock is now in close contact with the cambium tissue around the wood sliver on the inner face of the shield. For successful union of the bud and rootstock, and to prevent drying-out of the bud, it is necessary to maintain this close contact, which is achieved by wrapping the rootstock tightly with a waterproof material. In the past, waxed cloth strips were used as wrapping material, but strips of polyethylene plastic are now often used. Wrapping is started at the base of the bud and continues upward in a spiral so that each successive turn overlaps slightly the one below it (fig. 6.2e). This will keep water out of the bud and incision area. At the top of the incision, the wrap is secured by slipping the free end under the last lap around the stem (fig. 6.2f). Excessive pressure from the wrap on the bud should be avoided, but there should be no slack in the wrapping material because the bark must be held tightly against the bud (fig. 6.2g).

Another method of budding that has rather recently gained some popularity in Florida is the use of a so-called hang-bud. Instead of the usual T-shaped incision, a piece of bark is lifted by a single upward cut with a knife held almost parallel to the bark surface. The lower half of this flap is cut off, and the usual shield bud is inserted under the bark of the rootstock. The bark flap should cover the shield as far down as the bud itself, which should not be covered. The pressure of the bark on the upper part of the shield holds it in place so that it hangs with the lower end free. Wrapping is done in the usual manner for standard shield budding. The advantages of this method are that making the incision and inserting the bud may be done slightly faster than for shield budding, and failure of the bark to slip is of less importance than in the standard method.

Chip budding may also be done when the bark is not slipping. A downward cut is made through the bark of the rootstock, leaving a flap of bark which will hold the chip bud. A partial bud shield is cut from the budwood and inserted into the flap. The bud is then wrapped as described previously.

Any time budwood is cut or budding is done, knives and pruning shears should be kept very clean to minimize transfer of viroids. Both exocortis and cachexia viroids can be transmitted by infected pruning or budding tools.

To minimize disease spread, a solution of household bleach (sodium hypochlorite) diluted to 20 percent by volume with water (one part bleach to four parts water) can be used. Prepare and use the solution daily as heat and sunlight can cause a reduction in the strength of this solution. Dip clippers, knives, and pruning tools in the bleach solution for several seconds. Dry and oil tools lightly before storing to prevent rust, but use freshly made bleach solution when pruning or budding. Clippers should be sterilized after use on each tree when cutting budwood, and budding knives should be sterilized frequently in the greenhouse.

Soon after budding (about two weeks), the wrap may be removed and the bud examined. If it is still green and callus tissue has formed, the bud has "taken." If the shield has turned brown and slips easily out of the incision, a new bud may be inserted at another location on the stem if the bark is still slipping.

4. *The Budded Tree.* As soon as the rootstock-bud union has formed, it is time to "force" the bud to grow. This is most often done by looping or bending the seedling by tying the top to the base of the tree. In time, the bud should sprout and the top of the original seedling (rootstock) can be removed.

Forcing buds by cutting tops off seedlings can also be used instead of looping, but it may take more time for buds to sprout. The looping method with subsequent top removal has become popular with most nursery growers. Young budlings which do not sprout in a reasonable length of time should be clipped as an attempt to force buds to grow.

Budded trees are usually staked so that the developing shoot can be supported and trained to an upright growth habit. Heavy galvanized wire (No. 8) is used for most stakes. When the new shoot has reached a length of four to five inches, it should be tied to the stake with soft, heavy twine, fastened loosely enough so that the expanding stem tissue will not be constricted. As the new shoot continues to grow, additional ties may be necessary. When the tree reaches an appropriate height, it may be topped to promote formation of lateral shoots. These will form the framework of the tree. Most trees, however, are sold as single stem, untopped plants which are trained to an upright position as a single leader.

Care of the young budded tree should be the same as already discussed for seedlings, with watering, fertilizing, and spraying performed as needed to produce healthy, vigorous growth. The young tree should be examined quite frequently for rootstock sprouts so that these may be removed before they have a chance to compete with the scion for nutrients and water. Young, tender sprouts can simply be rubbed off; older ones should be cleanly pruned off.

The plant is usually sold in the container including stakes as part of the completed package. Container-grown trees have an advantage over bare-root trees in that they can be watered and held in the field before planting without drying out.

TOPWORKING

The term "topworking" refers to the operation of changing the scion cultivar to another cultivar. Topworking is a rarely used, very tedious, costly, and time-consuming procedure which will become even rarer since regulations now require propagating material to be registered or certified and come from approved, enclosed structures. The process was mostly used to change an undesirable cultivar to a desirable one without losing the advantage of its established root system. The topworked tree usually came into production in two years and would be much larger than if the old tree were pushed out and a new, small nursery tree planted.

Topworking can be done in several ways, using budding or grafting as the situation warrants. The easiest way is to cut back the scaffold branches to short stubs and to bud the shoots that develop from these branches as if they were seedlings in the nursery. A second possibility is to insert buds into bark incisions on the branches just below the points where they will later be cut off after the buds have taken satisfactorily.

INARCHING

Sometimes the root system of a tree is seriously damaged by fungi, algae (*Phytophthora*), viruses (viroids), or burrowing animals so that tree loss is imminent. Usually it is best to remove such a tree and replace it with a healthy one budded on a rootstock more suited to the soil conditions. However, in the case of trees with special value, it is often possible to replace the diseased or injured roots by inarching one or more vigorous seedlings into the trunk. Seedlings of suitable size for budding or a little larger are planted as close as possible to the base of the damaged tree. As soon as the seedlings have become well established and have a stem at least two feet high, they may be grafted into the trunk of the damaged tree. One common technique is to make an incision in the trunk as if for shield-budding and to insert the obliquely sharpened end of the seedling stem into this. Another is to cut a slot in the bark of the trunk of the same width as the seedling stem and, after cutting away a thin slice of bark and wood from the latter, press it tightly in place so that the exposed cambium of the seedling and trunk are in contact. In either case, the cuts must be made with care to assure matching of the two cambia to be joined. The seedling stem must be held firmly in position using a small nail until union takes place. If all goes well, the new seedling roots will take over the functions formerly performed by the original root system, and the rejuvenated tree may survive. This method is not at all recommended for commercial use, however, as it is too slow and expensive to be profitable. As with topworking, the disease status and susceptibility of plant materials should be thoroughly investigated before this sort of operation is undertaken.

MARCOTTAGE

Marcottage, also known as "air layering," is yet another method of propagating citrus. It has been used primarily to propagate Tahiti limes in south Florida because it was easy to do, and marcotted trees generally came into production sooner than budded trees. However, marcotts also have shallow root systems and were easily blown out of the ground in the Homestead area during Hurricane Andrew in 1992. Marcottage is rarely, if ever, used to propagate other citrus species.

Marcottage consists of rooting branches while they are still attached to the mother tree. A one- to two-inch bark patch is removed 18–36 inches down from the tip of a shoot of a Tahiti lime tree. The exposed area is wrapped with moist sphagnum moss, which serves as the rooting medium. The moss is then wrapped tightly with aluminum foil or plastic which is tied at both ends. After two to four months, depending on time of year and temperature, the moss is unwrapped. New roots which arise from the stem tissue should fill the sphagnum by that time. Thus, marcotts do not have a taproot typical of citrus rootstocks grown from seed. The new tree is then cut off below the new roots, and the tree can be planted immediately or grown for a while in a container to increase tree size.

Propagation by marcottage should not be undertaken in Florida without first checking with the Citrus Budwood Registration Office. Strict interpretation of current (2009) regulations would mandate that trees from which propagation materials are obtained must be tested and certified disease free and that propagations would have to be made within the confines of an approved greenhouse structure.

To ensure freedom from disease and to grow trees with strong root systems, growers should consider using registered or certified trees from approved nurseries.

SEVEN

Site Selection, Grove Design, and Planting

This chapter considers factors involved in the development of a producing citrus grove. The first and arguably the most important decision when establishing a grove is location. Site selection is extremely important because the success or failure of the operation may depend upon having a satisfactory location. At the very least, the site will influence development and operation costs. Of course, site selection is a decision that cannot be easily modified, short of selling the land and moving elsewhere. After a site is chosen, the topography often determines the grove geometry or planting plan. Even when topography is not a complicating factor, there are usually several planting schemes which can be used. Spacing between trees, distance between rows, and row orientation are all important factors to consider. These factors are, for all practical purposes, irreversible short of removing planted trees and starting over. After selecting a site and planting plan, land preparation must begin, and tree planting should follow. These factors are also very important since mistakes made at this time will be difficult and costly to correct in later years.

Once the trees are planted on the prepared site, the grower embarks on a young tree care production program (discussed in chapter 8) to develop the young, nonbearing trees into healthy bearing trees. This process will, in most cases, take three to four years. During this period, the trees will receive cultural practices favoring rapid growth and development into efficient fruit-producing units. Fruit production in the first several years is very low (or nonexistent) and erratic, and fruit quality is typically poor. During the establishment period, emphasis is on producing a vegetative tree that will later become productive, and the amount and quality of fruit produced are of minimal importance. This establishment period is but another cost, in addition to the site, preparation of the site, and the purchase and planting of trees, which a citrus grower must accept before having a bearing grove that will be productive enough, ideally, to pay its way and produce profitable crops.

SITE SELECTION

The importance of site selection cannot be overemphasized. Many factors must be considered when choosing the site for a citrus grove. Some factors are more important than others and carry more weight in the decision-making process. Some factors are extremely important and outweigh all others. Other factors are not as important but must be considered in the total evaluation process since several such factors might combine to become significant.

Every location will have some features that make it less than perfect. In the site-selection process, it is important to maximize the positive attributes of a site while minimizing the negative ones. The process of elimination can be used when the factors affecting value of a given site are evaluated critically and the worst locations eliminated until the best site remains.

Site-Selection Factors

Weather can be the single most important consideration in selecting a site. The threat of cold damage has been the most important limiting factor in Florida in the past. In other parts of the world, many other weather-related factors can come into play. These include rainfall patterns, excessively high temperatures, strong winds, or the presence of cloud cover for extended periods.

Because of Florida's latitude and orientation on the continent, all commercial citrus sites have some degree of cold risk. Major cold fronts originate in Canada and the northern United States and move unimpeded across the central plains of the United States. This is why Florida and Texas have been more severely damaged by freezes than California, even though California is located at a more northerly latitude. (However, note that nearly the entire California citrus crop was eliminated during a January 2007 freeze.) Sites at northerly latitudes in Florida are generally at greater risk than those in the southern or coastal areas. Following the severe freezes of the 1980s, however, most of Florida's citrus production has moved out of the northern citrus areas (see chapter 1).

Other factors can also influence the risk of cold damage. Proximity to a heat reservoir can make a grove warm when a freeze occurs, especially if the heat source is to the north. Heat reservoirs are usually bodies of water, including the Gulf of Mexico, the Atlantic Ocean, or one of the many large lakes in the state. Some heat is also generated by large urban areas.

Site elevation is another factor that must be considered with cold risks. Usually areas of high elevation are warmer than areas with low elevation. This is because cold air is more dense than warm air and will settle into low areas. This, of course, occurs only during calm periods (radiation freezes) when there is no mixing of various temperature layers due to wind. Not all freezes, however, occur under calm conditions. Some of the historically worst freezes were windy (advective). In this instance, higher elevation sites can be damaged as much as or even more than locations with lower elevations.

Latitude, elevation, and proximity to heat reservoirs can be easily determined. However, long-term temperature records for the area and the specific site should be investigated and other nearby groves inspected for signs of previous cold damage. Just because a site has been free of freeze damage for several years does not necessarily guarantee successful citrus production in the long run. For example, Florida has not had a severe freeze for the past 19 years, 1989–2008, but as we saw in the 1980s, this does not eliminate the possibility of having several severe freezes in a short period.

Soil supports the tree and is the medium in which the tree is grown (see chapter 2, Soils). It should be adequately but not excessively drained and be of sufficient quality to grow citrus economically without expensive modification. Sites near the coasts and in the southern portion of the state tend to be quite level and are usually poorly drained. These sites need to be artificially drained to provide an adequate rooting zone for the trees. Land preparation increases grove development costs and drainage systems require additional maintenance over time. Trees planted in these areas are placed on raised beds, and a system of ditches removes excess water from the rooting zone.

Elsewhere, in areas with adequate drainage, trees are planted at ground level; and raised beds and additional drainage are not necessary. These groves are easy to plant and maintain, and development costs are low. Such sites are scarce since most of the better sites were planted many years ago or have been sold for real estate development.

Soil quality is determined by several factors including texture, organic matter content, depth of rooting, pH, and the presence of possible toxins or harmful salts. All these factors should be considered when assessing the suitability of a citrus-planting site.

Even though citrus can survive in impoverished soils with modern fertilization and irrigation practices, tree vigor and productivity are compromised and overall costs increased. Such sites should be avoided. Areas with a hardpan (restrictive layers) that inhibits satisfactory root growth should also be avoided unless artificial drainage can be provided. Areas in which sandy soils are predominant are usually infertile and excessively drained; they will likely require extra irrigation and fertilization and increased costs to produce economic crops.

Soil pH can be an important consideration, although citrus trees are grown in a wide soil-pH range. Acid soil conditions can be corrected with lime applications, but this represents an additional cost. Some soils, particularly in coastal flatwoods areas, may have high pH which cannot be easily and economically reduced. High soil pH may mandate application of certain heavy metal elements, usually in a spray or special chelated form, to keep trees productive. This represents additional costs. High soil pH will also eliminate some rootstocks from consideration (see chapter 5).

Salts can create problems for citrus trees. Therefore, it is wise to check both soil and irrigation water for elevated salt content. Copper levels are especially important if the site was previously planted in citrus or vegetables. Copper is often used to control plant diseases and may accumulate in the soil over the years. Sites with high copper levels should be avoided since tree growth can be compromised. Copper toxicity can also be lessened by maintaining soil pH above 6.5. At this level, copper becomes complexed in the soil and unavailable to the tree.

Citrus grove site costs consist of land acquisition and development. The price of land will depend on the real estate value, and some of the most desirable sites may be too expensive to produce citrus economically. These sites may have greater value for other purposes, such as shopping centers or residential communities. Interest rates also must be considered. For example, in the early 1980s interest rates were very high, and some groves did not generate enough income to make mortgage payments let alone produce a profit. In contrast, current interest rates for the early part of 2009 are at historic lows.

Development cost is the expense required to make the land suitable for planting citrus. Land clearing, leveling, mowing, and discing will be required for most sites. Areas prone to flooding will have to be drained. This will require construction of raised beds, water furrows, and subsurface drainage ditches. Reservoirs will usually need to be built so that drainage water can be released in a controlled manner. Roads and pumps and wells may need to be constructed or repaired and irrigation systems installed, repaired, or modernized. Legal costs of purchase and development may also be substantial. There may also be some set-aside requirements that will not allow the planting of all the purchased land area, and this will increase the per-acre cost of planted land. Typically only 70–75 percent (or less) of the acreage in flatwoods areas can be planted versus 90–95 percent on the Ridge. The number of net grove acres must be calculated and acquisition and development costs determined to be able to calculate the actual cost per acre of the planted grove.

Regulations may make otherwise very good sites more expensive by placing additional development costs on the landowner or, in a worse case, completely disqualify a site from consideration. Most of the regulations fall within the categories of site development, water management rules and regulations, and environmental concerns.

Site development restrictions include land use and zoning requirements. While modification of these requirements is sometimes possible, it is often expensive and time-consuming. Even where zoning restrictions may permit grove development, it would be wise to determine whether home construction will be permitted nearby. Many homeowners are intolerant of necessary agricultural operations such as spraying, and this can prove troublesome for the grower.

Water regulations are often complicated and are usually handled by one of the five water management districts or persons specializing in this area. Water management districts were legislated in 1972 to manage programs related to consumptive use, aquifer recharge, well construction, and surface water use. They are also involved with flood protection programs and regulating water use during droughts. The five districts—Northwest, Suwannee River, St. Johns River, Southwest Florida, and South Florida—are funded by tax dollars. Permits are required for well construction, and consumptive use permits are necessary to use the water that is pumped. These permits and their allocations will usually determine the type of irrigation system selected and how it is used. Potential hazards to water quality are also regulated, requiring backflow-prevention devices; the use of certain chemicals is restricted or not allowed in some areas where leaching might create a problem. In addition, water movement on and off the property is carefully monitored and regulated so as not to disrupt natural flow patterns or impinge on water rights of adjacent properties.

Environmental regulations are in place to protect the environment, and they vary greatly among locations. They can restrict drainage water discharge, often compromising the ability to effectively drain a given site. They also protect wetlands and can greatly restrict a property owner from utilizing all of the land in a given site. In general, wetlands cannot be disturbed, but

Plate 1. Scab fungus on Temple orange fruit and leaves.

Plate 2. Melanose fungus on grapefruit leaves. Lesions are raised on the leaf and rough to the touch.

Plate 3. Melanose fungus on grapefruit. Lesions are raised on the fruit surface and may occur in solid patches.

Plate 4. Greasy spot fungus on citrus leaves. Dark lesions occur on the underside of the leaf in mid-late summer and may cause defoliation.

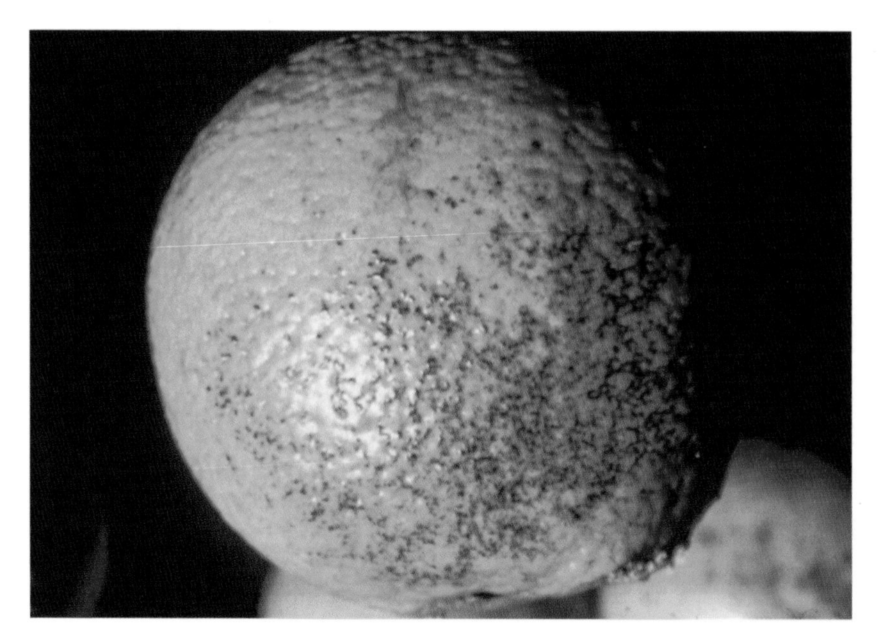

Plate 5. Greasy spot fungus on Temple orange fruit. This problem is also called greasy spot rind blotch when it occurs on fruit.

Plate 6. Alternaria leaf spot fungus on tangerine leaves.

Plate 15. Cottony-cushion scale feeding on citrus stems. This is a soft-bodied scale whose population is controlled by natural predators such as ladybird beetles.

Plate 16. Black scale feeding on a citrus stem. Black scale usually feeds at nodes and has a distinct H shape on its back.

Plate 25. Citrus rust mite (CRM) late-feeding damage on orange fruit called russetting. Note that undamaged areas were exposed to high sunlight temperatures, which discourages CRM feeding.

Plate 26. Citrus red mite feeding on citrus leaves. The adult has eight legs and can be seen with the naked eye. Note a reddish egg in the center of the photo.

Plate 27. Citrus canker bacterial disease symptoms on citrus leaves. Note the distinct brown lesions surrounded by a yellow halo. Symptoms are often confused with infection by some types of fungi.

Plate 28. Citrus canker bacterial disease symptoms on a citrus stem. Note the raised brown pustules.

Plate 29. Greening bacterial disease symptoms on a mature grapefruit tree. Symptoms include small misshapen fruit and shoot dieback. The disease is transmitted by the Asian citrus psyllid. Source: Jamie Yates.

Plate 30. Orange dog caterpillar feeding on a citrus leaf. This is the larval stage of the swallowtail butterfly. It is often described as looking like a bird dropping. Source: Jerry and Jan McDonald.

Plate 31. Sooty mold fungus on citrus leaves. This fungus is often found in association with insect feeding.

Plate 32. Sooty mold fungus on citrus fruit (black regions only). It generally does not harm the fruit and can be easily washed off with water. Citrus mealy bug (white area) or other insects are often found in association with sooty mold.

there is often a provision whereby "mitigations" are negotiated. When these compromises are possible, landowners create new wetlands to offset the use of others. For example, a small wetlands area in the middle of a grove might be filled and planted to citrus by creating a new wetland area of equal or greater size elsewhere on the site. (It should be noted that filled wetland areas seldom result in the production of satisfactory citrus trees.) Environmental regulations also protect endangered species and other wildlife but often trigger emotionally charged battles between environmentalists and growers.

Site infrastructure concerns include availability of water, utilities, labor, and housing and proximity to roads, processing and packing facilities, and markets. All of these become a part of site assessment. In the past, some of these items were of considerable concern. They are still a concern for growers developing properties elsewhere in the world or in isolated plantings in Florida.

Special problems may occur in some locations and have been previously discussed. Other possible problems, to name a few, include insect pests (such as *Diaprepes* or other weevils); disease and nematode problems on the property or nearby; water runoff and retention; and restrictions to air drainage, which might make the site colder during freezes.

Clearly, site selection is a complex process with numerous considerations to be made. However, the best technology available will never overcome the problems created when a poor site is chosen. It is important to remember when selecting a grove site that citrus trees must occupy the same area for many years to be profitable, and the crop must mature on the trees each year. In contrast, vegetable growers may lose a crop one year and recoup their loss later in the year or in the following year; citrus growers, however, must count on a consistent crop every year. The loss of an occasional crop from cold, wind, or pests and diseases means a permanent loss in profits and there is a limit to how many such losses can be taken without the grove becoming insolvent. The level of crop production is determined at planting time by site and rootstock and scion selection. A small, short-lived, nonproductive tree growing in a cold location on shallow, poorly drained soil is rarely an economically productive unit on a long-term basis.

PREPARING THE SITE FOR PLANTING

A carefully selected grove site must be cleared of all trees and shrubs. All large pieces of roots and as many small roots as possible should be removed from the upper foot or two of soil. The soil is then plowed, disced, and leveled or sloped slightly to facilitate drainage. In some areas of the world, the grove sites are on slopes steep enough to require contouring or terracing to prevent soil erosion. This planting pattern is rarely used in Florida because of the relatively flat topography.

Water drainage must be considered for any site in a poorly drained area. Surveys and engineering will determine the slope of the land for proper run-off of surface water. Today, most groves requiring drainage are planted on double-row beds with water furrows between beds at 45- to 55-foot intervals. Many older, single-row plantings still exist but often are plagued with drainage problems. Moreover, cultural practices such as spraying and harvesting are difficult since grove equipment will always be running in water furrows which move accumulated water to ditches at the ends of the beds. Commonly, the beds are from 660 feet to ¼ mile (1,320 feet) in length. The ditches carrying the water from the furrows then run to progressively larger and larger channels until the water is moved off site or into a retention area. Soil removed from the ditches is usually placed along the side of the ditches to construct roadways that facilitate movement within the grove. Groves should be designed to allow rapid drainage during periods of heavy rains; at the same time, water must be accessible for irrigation during dry periods.

Subsurface drainage pipes are often used within the grove, either as a substitute for water furrows or to further enhance drainage within the beds. Various types of perforated plastic pipes are buried under the soil surface, sloping down to lateral ditches at the ends of the beds. Water seeps into the pipes through the holes throughout their length and moves via gravity down the slope to the ditches.

Thorough planning before the grove is established is very important. Permits from all appropriate agencies must be obtained before grove construction commences. Consultations with extension personnel, engineers, environmentalists, and horticulturists may result in slightly higher preparation costs but will greatly increase future profitability of the enterprise.

GROVE GEOMETRY

The planting system refers to the regular arrangement of trees for a given land area. Formerly, several different systems were in vogue, each with its loyal advocates. These included rectangular, hexagonal, triangular, and quincunx patterns. Quincunx refers to a pattern of five trees arranged in a square with the fifth tree placed equidistant from the other trees in the middle of the square. As with other fruit crops, citrus trees are in most cases planted using the rectangular system. The other systems—triangular, hexagonal, and quincunx—sometimes permitted planting of more trees per acre for a given spacing but they required cultivation, spraying, and harvesting to be done along diagonals through the grove, which was often more troublesome and confusing than when cultivars were set in rows parallel to the boundaries of the grove.

The rectangular system places the trees at the intersections of lines that cross at right angles. The length of the sides may be the same (square) or the lengths of the opposing sides may differ. The system is easy to lay out, understand, and use in grove operations. Such groves can be cultivated in two directions (when young or if spacing is adequate), and cultivars can be planted in compact blocks of rows, making it easy to keep track of them during spraying or harvesting. Planting on the square was formerly by far the most common practice, but today it is desirable to have trees closer within than between rows. This optimizes land use, canopy bearing volume, and potential productivity. The distance between rows cannot be too close, however, because cultural practices such as spraying and harvesting become increasingly difficult due to restricted equipment movement as the trees mature.

The hedgerow system of planting was adopted by some citrus growers in Florida in the late 1950s and early 1960s. It permitted planting the largest number of permanent trees per acre in a pattern that was commercially manageable. Space between the rows allowed equipment movement, and space between the trees in the row was maintained as a hedgerow. While trees were small and had not yet become crowded, higher yields were possible than for other systems because there were more trees per acre. The rectangular system required sound fertilizer and irrigation practices (especially on sandy soils low in organic matter) to produce satisfactory results; however, the problem of crowding in later years forced some growers to abandon or modify the hedgerow system. Increased interest in early production, the development of sophisticated pruning machinery, and more intense production management have brought about a revival of interest in hedgerows. Tree density has increased rapidly over the years from an average of about 70 trees per acre in the 1960s to 140 or more trees per acre by the 1990s. Currently (2009), some growers are planting 200 or more trees per acre to obtain an earlier return on investment due to concerns about greening, canker, and shortened productive grove life.

In the contour system, the trees are planted in rows along lines of equal elevation, or contours. Its use is limited to slopes steep enough for soil erosion to be a serious problem when using conventional planting systems. If slopes are so steep that it is difficult to drive machinery along the contour lines, it may be necessary to make beds several feet wide for planting, so that machines can run on nearly level ground. Laying out contours requires surveying skill and takes much more time than laying out a rectangular system. A contour planting is difficult to maintain, and the rows along contours often will not all have the same number of trees. The system is rarely seen in Florida and should never be used unless it seems absolutely necessary for erosion control. Other methods of controlling soil erosion, such as the use of cover crops, should be given careful consideration before deciding to plant on contours. These groves may require rigorous and regular tree pruning and/or tree removal programs as they mature.

Regardless of the planting system used, it is necessary to decide how far apart trees should be placed. Spacing was originally determined chiefly by the ultimate tree size. Trees that are naturally small or that would remain small because of poor soil conditions or rootstock selection would be spaced more closely than trees which naturally reach large size and are growing in deep, well-drained soil.

Under similar conditions grapefruit trees grow much larger than orange trees, which in turn are usually larger than lime and most mandarin trees. In the early days of the citrus industry, spacing in the row averaged 10–12 feet for kumquats; 15–20 feet for satsuma mandarins, limes, and lemons; 20–25 feet for oranges and mandarins (except satsumas); and 30–35 feet for grapefruit. The effects of in- and between-row spacing on citrus tree density per acre are given in table 7.1. It was, however, desirable in commercial groves to have the drive middles at least 25 feet wide in order to use mechanized equipment such as sprayers, fertilizer distributors, and harvesting equipment. Many closely set early plantings developed into canopied groves. The bottom branches died out as a result of shading, and the fruit was borne in the upper canopy. Fruit was difficult to harvest, and yields were low due to the small amount of fruiting wood.

A large number of trees per acre does not necessarily produce a large yield per acre (except during the early years of production), but planting too few trees to utilize soil and light efficiently is certainly an uneconomical use of space. During the Florida boom of the 1920s, common spacings, on the square system, were 25 × 25 feet and 30 × 30 feet, giving about 70 and 48 trees per acre, respectively. On deep, well-drained soils, 25 × 25 feet eventually became too close for efficient grove operations, even for oranges. The same number of trees per acre was possible by using a 20 × 30 spacing to get a wider drive middle by crowding the trees slightly in the row. There was a

Table 7.1. Citrus tree within- and between-row spacing and plant densities

Spacing (ft.)		Trees/acre
In-row	Between-row	(Number)
6	15	484
7	15	413
7	20	310
10	10	435
10	12	363
10	15	290
10	20	217
12.5	15	232
12.5	20	175
12.5	25	140
15	15	192
15	20	145
15	22	132
15	25	116
20	20	108
20	22	99
20	25	87
22	22	90
22	25	79
25	25	70
25	30	58
30	30	48
35	35	36

suggestion in the early 1940s that the best spacing for grapefruit trees would be 35 × 35 feet, which would allow only 36 trees per acre. This tree spacing produced very large trees that were productive individually, but per-acre yields were often low.

Citrus and other fruit growers have long been troubled by the problem of how to use the grove space more fully while the trees are young without having them too crowded at maturity. The solution used during the 1940s was double planting. In theory, this close initial planting and later removal of alternate trees after a few crops was a good solution. It was feasible to transplant the temporary trees to a new grove site if further plantings were being made. Growers, however, postponed as long as possible the removal of the nonpermanent trees because trees were producing valuable crops; usually thinning operations were delayed until the yields of permanent trees had been reduced considerably from crowding. Gradual thinning partially solved the problem. Alternate trees were pruned sufficiently to leave a two-foot clear space between them and their permanent neighbors. Each year these same alternates could be further headed back so that they never shaded the permanent trees. After 12–15 years in a 15 × 25 planting, they had been reduced to primarily scaffold branches (skeletonized) and could be removed entirely, leaving the permanent trees spaced 30 × 25 feet. Too often growers refused to prune at all. Therefore, it became apparent that to sacrifice a little income during the early years of grove life because trees were widely spaced was better than to gain extra returns from close spacing of interplanted trees that had been left too long in double plantings.

After much trial and error, the concept of hedging began as a practice in the 1950s. It began as a remedial operation simply for the safe operation of equipment through the groves. It proved, however, to be a means of controlling tree size with maximum production per acre for closely spaced groves, with many advantages and few disadvantages, as discussed in chapter 9. Today, hedging is a necessary maintenance operation.

After grove geometry and spacing have been determined, the future tree locations may be marked by stakes. Boundary rows are first laid out and then parallel rows established across the field. Once the tree locations are staked on two sides of a field by careful measurement, and perhaps a central row similarly measured and staked, the rest of the locations can be quickly sighted in and staked. Sometimes the tree positions are marked initially by crosshatching the ground by means of an outrider on a tractor, with the trees set where the lines intersect. Many current groves are laid out using laser-guided transits to ensure proper tree spacing and grove orientation.

PLANTING YOUNG TREES

Florida citrus growers formerly could choose between field-grown trees which were usually sold bare-root or small greenhouse-grown trees sold in a pot or specialized container. Widespread canker incidence in the 1990s and 2000s and greening in 2005 has caused nurserymen to grow trees in greenhouses with special regulations regarding distance from known canker sites. Special precautions to prevent the introduction of canker and greening into the nursery greenhouse sites have been developed by the Florida Department of Agriculture (see chapter 6). Trees formerly produced by Florida citrus field nurseries consisted of one-year scions on two-year rootstocks (that is, the rootstock was a one-year-old seedling when budded, and the scion was a year old at digging). Such a tree actually took 24 months to grow but was considered one year old by nursery convention, which disregarded age of the rootstock. A desirable one-year-old tree would have a trunk diameter caliper of ½–⅝ inch at a point two inches above the bud union. The pruning cut made in removing the rootstock above the bud would be healed over and be hardly noticeable, and the trunk straight. Field-grown trees had extensive canopy branching and a well-developed root system.

Greenhouse-produced trees are now the norm (as required by law) and are considerably smaller than the previously described field-grown trees. These greenhouse-produced trees are sold in containers and may be only one year from seed because plants are grown intensively under controlled climate conditions and given optimum care. The trees produced in the greenhouse generally are trained to a single upright stem, while field-grown trees have considerable lateral branching. Research suggests that field-grown trees are initially larger than greenhouse-grown trees, but that there are no long-term differences in tree size or yields. Details on greenhouse production and other propagation information can be found in chapter 6.

Price is a fairly good indication of tree value if one is dealing with a reliable nursery. Greenhouse-produced trees should be vigorous with a good root system, yet not held until roots become pot bound. The cost of trees is a small part in the total cost of bringing a grove into production and is a very poor place to economize. Current market prices for good nursery trees can be found in advertisements in citrus magazines, online, or by speaking with other growers and citrus professionals. The price per tree in small lots will always be a little higher than it is in large quantities. Most citriculturists believe that the price for citrus nursery trees will increase due to the new regulations and limited supplies.

Citrus growers should deal only with reputable nurseries to assure the purchase of uniform, high-quality, true-to-type trees. Florida has many citrus nurseries that are careful to maintain a reputation for producing high-quality nursery trees. The transient nurseryman may also produce high-quality trees, but the prudent grower should carefully investigate before buying from an unknown source.

Handling and Delivery of Nursery Trees

The handling of bare-root and greenhouse-grown nursery trees is quite different. The first step in preparing bare-root nursery trees for transplanting was to prune the canopy since much of the root system was left in the soil. This loss in root volume called for a similar reduction in leaf surface area to reduce water losses via transpiration. Most trees were quickly placed into covered trailers with mist irrigation to keep plants cool and moist. If it was necessary to delay planting after trees reached the grove site, they could be held for several days by heeling them in, that is, by making a shallow trench and placing the roots in it while the trunk is inclined at an angle about 60° from the vertical. The roots were then covered with soil, which was kept moist.

Trees produced under controlled environment (greenhouse) conditions are sold in a container (usually with a stake to support the slender trunk) and are simply loaded into a vehicle and transported to the grove site for planting. Although these trees are smaller than the typical bare-root tree, the added pots and growing media constitute considerable bulk, making transportation somewhat difficult. Container-grown trees are less likely to dry out than bare-root trees if adequately watered before being transported to the field.

Some growers used to grow a few trees in bushel hampers to replace nursery trees that failed to survive transplanting. The cost per tree was considerably greater than for the average nursery trees, but there was no root or canopy loss when transplanting. Bagged and burlapped trees, which contain the entire root system and the surrounding soil, are commonly used in several other citrus-growing regions, but not in Florida.

Planting Season

Citrus trees in Florida are probably best planted during the winter months (December to February) while they are not actively growing. Roots are able to start growing before budbreak because soil temperatures are often higher than air temperatures during this time. By the time the new growth flush is present and demand for water increases, new fibrous roots are ready to meet the demand. Trees planted in winter also have the longest possible growing season but may face considerable risk from cold injury in northern locations in some seasons. Spring plantings (March to April) are also successful, especially now that solid set micro-irrigation systems are widely used. Summer planting is sometimes less successful but is practiced by some growers. The high summer temperatures place stress on trees and tree planters alike. In addition, the growing season is not long enough for the tree to develop proper tissue maturity before cold weather occurs in northerly citrus-growing areas. This is less of a concern in southern and coastal regions where much of the citrus is currently grown.

Planting Methods

In general, citrus trees respond well to many different planting methods if the following criteria are met:

1. The tree must be set in the grove at the same depth at which it grew in the nursery. It may be set slightly higher if settling of the tree is expected. Planting too deeply negates the effectiveness of the rootstock and can subject the scion to soil-borne diseases such as foot rot.
2. The roots should be spread out. Spreading out the roots of a container-grown tree may seem unnecessary, but experience has shown that removing at least half the artificial media from such trees results in better tree growth. The media removal exposes young roots to the new soil of the planting site and affords more rapid development of roots, and subsequently, the top of the tree. Media that is removed is mixed with soil from the planting site; this alleviates the great disparity in soil water holding capacity between the highly organic media the containerized tree grew in and the light sandy soil surrounding the new tree. This difference in soil-media type causes interface problems that often result in less than adequate water movement even if plants are heavily irrigated.
3. The soil should be returned and packed into the planting hole without leaving any air pockets around the roots.
4. Sufficient water must be supplied to thoroughly wet the entire soil volume surrounding the roots.

The hole may be dug with a hoe, shovel, or mechanical auger. The tree may be set in place with or without use of a planting board, and water may be applied after each of several increments of backfilling or simultaneously with the soil in a mudding-in process. A planting board is placed across the hole to act as a guide for proper planting depth. Cost of planting will vary a little from one method to another, but low cost cannot be economically justified if the trees do not grow-off well. In some cases, trees are not mudded-in at planting if sufficient irrigation is available, particularly if a large number of trees are being planted.

One method previously used on a large commercial scale greatly decreased hand labor at planting. Holes were dug by an auger operated as a tractor attachment, and the soil was formed into a water ring by machine. Then the nursery tree was jetted in using a stream of water under considerable pressure from a spray tank or from an auxiliary pump. The water was directed at the taproot by one person, while another placed the tree in the proper position. Then the soil was filled in around the roots by washing it down from the sides by the jet of water, assuring the absence of air pockets.

Some growers plant container-grown trees using posthole diggers without watering-in the trees. There are examples where this has worked, but often trees grow-off poorly using this system. In most recently developed groves, water rings are not necessary. Instead, in-place irrigation is used to irrigate newly set trees. Trees are usually irrigated every few days until they become established.

It is not necessary to apply fertilizer at planting since most young trees have adequate nutrient levels from the nursery. If the fertilizer is applied in the planting hole and is not very carefully distributed, tender new roots may be injured by coming in contact with a high concentration of fertilizer salts. Sometimes soluble fertilizers are dissolved in irrigation water and applied, or controlled-release fertilizers are applied at planting time. However, the use of any chemical at planting time should be done very judiciously. Applications of fungicides, nematicides, and herbicides may also be made but are usually best applied after the tree has become established and after several irrigations have been made. There is no need to apply these materials initially unless the grove has a previous history of problems with fungi, algae or nematodes.

BANKING OR WRAPPING NEWLY PLANTED TREES

When winter or fall plantings are made in northern regions of Florida, it may be necessary to protect the new trees from possible cold injury since they are very susceptible at this time. Cold protection can be accomplished by banking soil or applying insulating wraps around trees to a height of a foot or more, preferably up to the scaffold limbs. The important thing is to have the bud union well covered so that new shoots can develop from the scion even if the top is damaged. If the bud union remains covered by a properly made soil bank, it will survive most freezes that occur in Florida. It is very desirable to make soil banks free from decaying vegetation or wood particles, as decaying vegetation attracts ants and wood particles attract termites. Both insects may feed on the tree under the bank. Wind and rain may remove some of the banked soil, even to the point of exposing the bud union, so periodic inspection and rebanking may be necessary. Normally the banks remain in place until the last danger from frost is past in spring; but in an extremely dry winter, wilting leaves may indicate the need for irrigation, and the banks must be removed. Of course, they should be quickly replaced. While soil banks work very well, they are difficult to construct and maintain as labor is often in short supply. Soil banks are not widely used in Florida except in some northern regions of the Ridge or by some homeowners. They are not used in flatwoods groves because shallow root systems may be damaged.

Insulating wraps may provide a satisfactory substitute for soil banks. While the degree of cold protection provided is not as great as that of a well-made soil bank, wraps do afford certain advantages. Most tree wraps, unlike soil banks, can be attached in the fall and left on the trees throughout the year or even for several years. Tree wraps also inhibit sprouting and protect trunks from herbicide and mechanical damage. Some wraps retain water and predispose trees to foot rot. Additionally, some wraps harbor ants which may damage trees or provide a potential hazard to workers.

Selection of the proper tree wrap for a particular grove depends on a number of factors, including cost, ease of installation, and probability of freeze damage. For example, growers in northern regions of the state should choose wraps with good insulating qualities, while growers in warmer southern locations may opt for less costly, thinner wraps.

Wraps used in combination with microsprinkler irrigation have proven to be extremely successful in protecting young trees from cold damage. To ensure success, microsprinklers should be placed on the north or northwest side of young trees, and microsprinklers that concentrate the water on the tree (for example, 90° or 180° emitters) should be used. Microsprinklers should apply at least 10 gallons per hour per emitter, although 15–20 gallons per hour provides even more cold protection. Irrigation should commence before the temperature reaches freezing, so that irrigation lines and emitters do not freeze; it can be discontinued when temperatures rise above freezing, especially under calm conditions.

If top of the tree is killed or damaged by a freeze and warm weather ensues, it may be necessary to remove the banks or wraps because new shoots may start growth on the uninjured stem under the bank or wrap. If the tree is still covered under these conditions, the bark may become infected by soil fungi or algae, become spongy, and slough off, with eventual death of the tree. A tree unbanked or unwrapped for this reason, with tender new growth starting, is exceptionally sensitive to cold and must be rebanked or rewrapped when further cold weather is forecast.

EIGHT

Cultural Practices for Young, Nonbearing Groves

By convention, citrus trees are classified as nonbearing during the first three years after they are planted. Although they may bear a few fruits as early as the second or third year, all efforts during this time are directed toward optimum tree growth, and any fruit production is incidental. The cultural program for these nonbearing trees is different from that for bearing trees. In the following discussion, it is assumed that trees have been planted as one-year-old nursery trees when they are not actively growing. The slight adjustments in the production program that are needed to fit this program to trees planted in spring or summer will be obvious and easily made.

UNBANKING/UNWRAPPING

If the young trees have been banked or wrapped, when danger of frost is over (February 15 in the southern area, March 15 in the northern area of Florida), they should have their banks or wraps removed. Some growers may elect to leave wraps in place to reduce sprouting and protect the young trees from wind-blown soil or misplaced herbicide sprays. Soil banks, which are not commonly used today in Florida, are removed with a large hoe or tractor attachment, removing the soil carefully so as not to injure the bark of the tree, and spreading the soil around the tree to make a basin about a yard in diameter for holding water. Where permanent irrigation systems that can deliver adequate water to the root zone of the young tree are in place, construction of basins is not necessary. The last remnants of soil adhering to the tree should be removed by hand to avoid scarring the trunk with a hoe. Then the basin should be smoothed by hand and any exposed roots covered with soil. Watering young trees regularly during the spring dry season is far more important than fertilizing them, as many growers have learned the hard way.

Trees lacking adequate mineral nutrients may be stunted temporarily, but young trees without sufficient water are likely to die.

IRRIGATION

Irrigation is critical and may be handled by transporting water to the trees by water wagon, or more commonly today, by in-place irrigation systems (usually microsprinklers or drip irrigation). The historical method of irrigating is discussed first and the contemporary method second.

Immediately after unbanking the trees and forming the water basins, apply about 10 gallons of water per tree. The purpose of the basin is to hold the water in the limited root zone. Water must be poured into the basin slowly enough that it does not overflow the sides and run off into the middles. During the spring and until summer rains begin, water should be supplied to prevent the soil in the root zone from drying out. As a rule of thumb, if no rain occurs for a week, trees should again be given 8–10 gallons each, with more frequent applications being made during protracted dry periods. The basin is retained as long as watering is needed. Growers should try to water all of their young trees within a two-week period, even if they have large acreage, so that they can repeat the process within 14 days or less of the previous watering. In some seasons, the spring rainfall may make watering unnecessary, but this is unlikely; growers must also be prepared to deal with severe drought. Fortunately, most new groves are established with irrigation systems in place, which will greatly simplify irrigation.

Watering in the fall is not as critical because by that time the tree root system has expanded and roots have access to a much larger reserve of water than they had in spring. In addition, decreasing temperatures reduce the rate of water loss (transpiration) from the tree. Except in extreme drought, it is advisable to decrease active growth by allowing the soil moisture content to become a little low in the fall, particularly in cold locations. Even though summer is the rainy season, there are sometimes periods of two or three weeks without rain. Young trees in their first season may not have adequate root systems to carry them through such dry periods, and watering may be required.

Where micro-irrigation such as microsprinklers or drippers is used, the irrigation scheme obviously requires some modification. If no rainfall occurs, two to four drip irrigation applications or one to two microsprinkler applications per week may be necessary from November to February. During the warmer months of March to October, three to seven drip irrigations or two

Table 8.1. Citrus microsprinkler irrigation schedule for young (age = 1–3 years) trees under average conditions

Month	Allowable soil water depletion (%)	Duration (Hours)	Interval (Days)
January	50	3–4	7–8
February	25	2–3	3–4
March	25	2–3	3
April	25	2–3	2–3
May	25	2–3	2
June	25	2–3	2
July	25	2–3	2
August	25	2–3	2
September	25	2–3	2–3
October	25	2–3	3
November	50	3–4	5–6
December	50	3–4	7–8

Source: Management of Microsprinkler Systems for Florida Citrus, http://edis.ifas.ufl.edu/HS204.

Notes: Irrigation efficiency was set at 75%.
Irrigation frequency and duration depend on soil type, root depth, climate, spray volume, and diameter.

to three microsprinkler applications may be necessary during weeks with no rain (table 8.1). The recommended intervals assume that drippers are run three to six hours and microsprinklers two to three hours per application. The length of time actually required varies with emitter output and soil type. Some growers irrigate based on allowable soil water depletion. This number is based on the water holding capacity of various soil types (see chapter 9), the rooting depth, and the daily cumulative water loss via evapotranspiration. "Depletion" refers to the amount of water that is lost from this soil reservoir. Several University of Florida bulletins are available to help with this calculation.

Irrigation is very important in the first year in the field. By the second growing season, the young tree is larger with a more expansive root zone, but irrigation is still required to attain maximum growth. Throughout the first three to four years in the field and even after trees come into full bearing, regular inspection should be made during periods of prolonged drought. If leaves are wilting, it has been too long since the last irrigation, although during periods of very high temperatures leaves may wilt in the afternoon even with adequate irrigation. It is preferable to measure soil moisture using tensiometers or capacitance devices to determine when to irrigate rather than to rely on visual symptoms. Trees are typically irrigated when soil moisture tensions fall to 10–20 centibars. When it is evident that watering is necessary, enough water should be applied to replace soil moisture that has been lost to the atmosphere. Where irrigation systems are used, the same guidelines will apply. The ease with which water is applied using in-place irrigation, however, will often result in superior growth since trees will not be exposed to the stresses of their hand-watered counterparts. Care must be taken to ensure that over-irrigation does not occur. Besides leaching away nutrients and increasing weed growth, over-irrigation can actually damage citrus roots by flooding the root zone.

Watering scattered resets requires extra effort because these young trees will usually have to compete with adjacent, larger trees for water, nutrients, and possibly light. Moreover, a mature tree irrigation regime may not provide frequent enough irrigations for resets.

FERTILIZATION

Citrus trees differ little from other evergreen trees or shrubs in their need for mineral nutrients, and fertilizer mixtures with a 1N-1P-1K ratio are generally used with satisfactory results for young citrus trees. A fertilizer with an analysis such as 6N-6P-6K or 8N-8P-8K with micronutrients is often used. The numbers on the fertilizer label refer to the percentages of nitrogen (N), phosphorus as (P_2O_5) and potassium as potash (K_2O) in the fertilizer. A good deal of research, however, suggests that micronutrients are usually not needed until leaf-deficiency symptoms arise (and this is especially true with resets). High-analysis fertilizer mixtures can be used at a lower rate per tree with proper care to assure even distribution. Dry, water-soluble fertilizer should be spread evenly over the ground in a circle whose diameter is twice that of the spread of the tree canopy. For nonbearing trees, especially in the first year or two, the fertilizer is usually spread by hand, although some operators with large acreages of young trees use specially modified fertilizer spreaders. Uneven distribution of fertilizer may result in a relatively high concentration of soluble salts in some places, causing root injury under those areas. Liquid fertilizer should be applied directly to the same general area, usually via fertigation (application of liquid fertilizers through the irrigation system)using the existing irrigation system. Specific recommendations will be discussed in the next section.

Table 8.2. Recommended nitrogen rates and minimum number of annual applications for nonbearing citrus trees

		Lower limit of annual application frequency		
Year in grove	lbs N/tree/year (range)	Controlled-release fertilizer	Dry soluble fertilizer	fertigation
1	0.15–0.30	1	6	10
2	0.30–0.60	1	5	10
3	0.045–0.90	1	4	10

Dry fertilizer is applied four to six times annually for the first three years in the field. Six applications are used the first year, five in the second year, and four in the third. The first application is made as soon as the banks or wraps have been removed or after the tree has become established in about two weeks after planting. The other applications follow at regular intervals until late October. In the northern part of the citrus area, the first application will be a few weeks later, and the succeeding applications will be at shorter intervals. It is often considered risky to apply fertilizer after November 1 in areas north of the Orlando-Tampa line because favorable temperatures in winter may cause trees to grow, thus decreasing freeze hardiness. In more southern regions of Florida, trees are fertilized almost year-round to prevent leaves from becoming chlorotic (yellow).

The quantity of fertilizer to apply at each application increases each year as the tree grows and is based on the amount of nitrogen per tree, not per acre (table 8.2). For example, applying 0.30 pounds of N annually per tree would require 3.75 pounds of an 8 percent N analysis fertilizer but only 3.0 pounds of a 10 percent N analysis fertilizer. A range of rates is used because of differences in soil types, climatic conditions, and tree size throughout Florida (fig. 8.1). The number of applications of dry soluble fertilizer decreases with age, reflecting the increase in both tree and root zone size. In contrast, controlled release fertilizer can be applied only once per year, which reduces application costs and nutrient leaching into groundwater.

Fertigation (defined previously) is widely used for irrigation and fertilization in many new plantings. It is a convenient and cost-effective method of fertilizing young trees. The same rates of nitrogen and other elements should be applied as in conventional fertilization. However, at least 10 frequent light applications per year should be used and the amount applied per tree per application reduced accordingly. Some growers using the open hydroponic

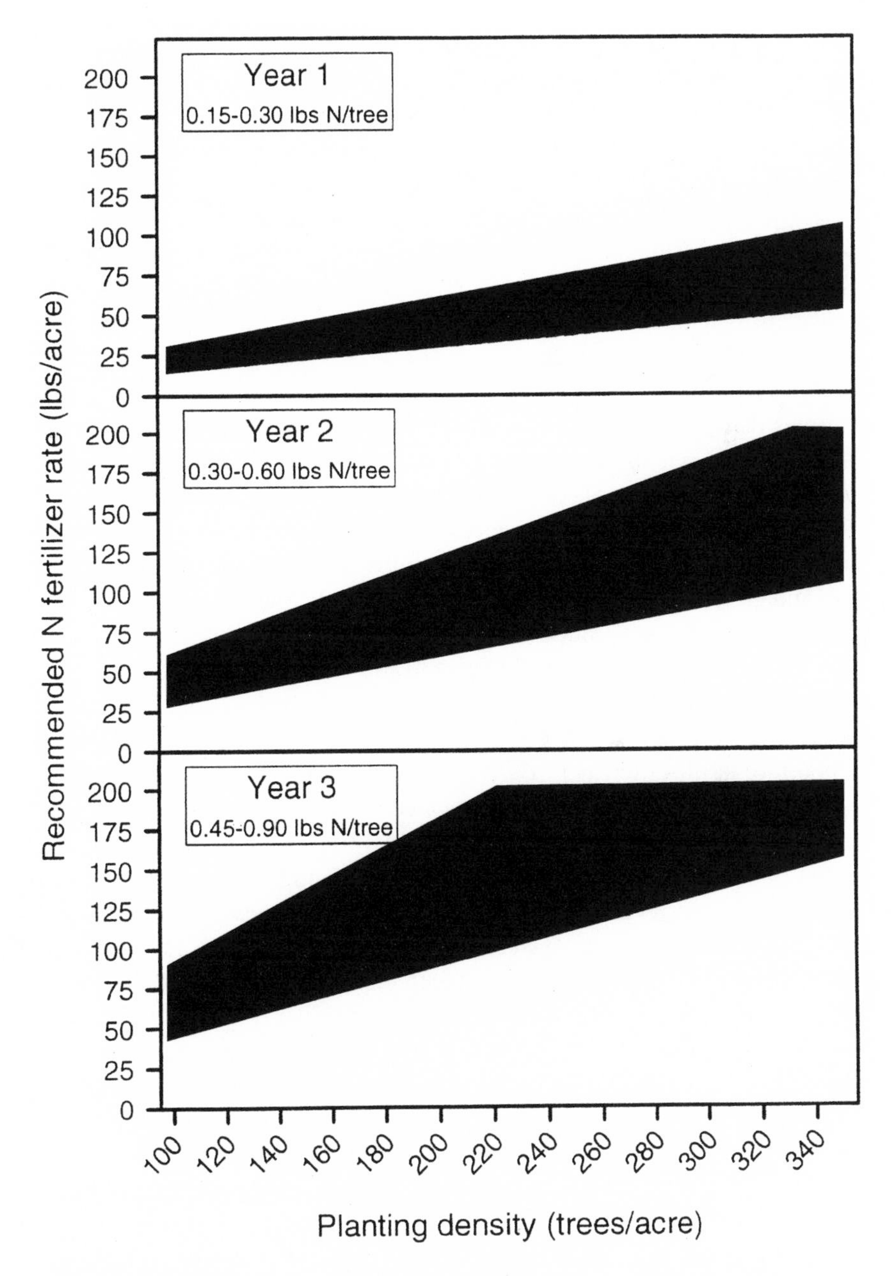

Figure 8.1. Recommended nitrogen rates for nonbearing citrus trees. Source: *Nutrition of Florida Citrus Trees*, edited by T. Obreza and K. Morgan, University of Florida Cooperative Extension, SL 252, 2008.

System (OHS) fertigate much more frequently than 10 times per year (see chapter 9 for details). Making frequent applications of fertilizer reduces the potential for nutrient leaching. Some research suggests fertigation also increases tree growth, while other studies find no increases in growth related to frequent applications of liquid fertilizer over less frequent dry applications.

As the quantity of fertilizer per application is increased, the area covered by the fertilizer should increase proportionally so that the amount of fertilizer per square foot of soil surface remains fairly constant. Thus, with twice the quantity applied in August the second year, the diameter of the application circle should be half again as large the second year; and in the third year, twice as large as in the first year.

Trees planted in spring follow a modified schedule the first year in that each application is made about six weeks later and the fall application omitted. The second year trees can be fertilized according to a regular schedule. Trees planted in summer will receive only the first two or three applications. In the spring of the next year, trees will initially receive one pound of fertilizer and gradually be given larger amounts, until by the end of summer they are receiving almost as much fertilizer as trees planted in winter.

It is not practically worthwhile to try to adjust the amount of fertilizer applied to each tree to take account of small differences in size. Uniformity in size is desirable and will most nearly be attained by applying the same amount of fertilizer to each tree as used for the average tree. New precision agriculture techniques are being developed that may someday allow for precise fertilizer application based on individual tree requirements. Some cultivars, however, grow more rapidly than others, and on some soils a given cultivar grows faster than on others. Such differences between cultivars and locations may well warrant a small variation in the amounts of fertilizer applied, increasing amounts when average tree size of a particular block is above normal and decreasing amounts for markedly smaller trees.

The third or fourth year is a season of transition from the vegetative to the fruiting tree. Growers often wonder whether to begin the distribution of fertilizer by mechanical spreaders, which are commonly used in fertilizing bearing groves of uniform tree size. As a rough guide, mechanical spreaders will prove satisfactory when the radius of the crown (the distance from tree trunk to the dripline) is at least ¼ the width of the drive middle in which the spreader operates. After the fourth year, the trees are considered to be of bearing age. Production practices for bearing trees are described in chapter 9.

SPRAYING

The extensive spray programs needed in bearing groves are not required during the first three to four years, since little or no fruit is produced and spray programs are chiefly concerned with keeping fruit free of pests. There are, however, some insects and mites that may feed on the young trees and cause considerable injury if not controlled, and there is always the threat of infection by citrus canker and citrus greening bacterial diseases. Perhaps the most troublesome of these pests are the Asian citrus psyllid and citrus leafminer. Serious leaf damage may take place from such infestations, resulting in a reduction in metabolites for plant growth and tissue maturity, since leaves are the primary organs responsible for photosynthesis. Psyllids can transmit greening and leafminer tunnels can harbor canker and exacerbate the canker problem. The trees should be regularly inspected several times a year so that infestations may be detected before pest populations have become seriously large. These same pests also occur in bearing groves, and methods of controlling them are discussed in more detail in chapter 10.

There are several scale and mite pests that can damage young trees, and regular inspection and control where needed are important. Orange dog caterpillars and grasshoppers can defoliate small trees; aphids damage leaves and may require control. Recently, *Diaprepes* weevils, which feed on roots, have caused severe damage to young trees in some areas and must be controlled. Several kinds of ants make their nests in grove soils and feed upon leaves or even the bark of young citrus trees. These pests may injure the tree and may be agents for the dispersal of foot rot. In making inspections for other pests, lines of ants are sometimes observed moving up and down the trees. Ants are often associated with other problems such as scales. Ant control is important not only for tree health but for worker safety as well, since the imported fire ant is one of the most common and most serious ant pests. Trees that are wrapped or banked should be checked regularly, as the wraps often provide ideal habitats for ants. Current recommendations for control of all the above pests can be obtained from extension or industry personnel.

Fungal and algal diseases are not much of a problem in young groves, except for foot and root rot caused by the alga, *Phytophthora* sp. This disease is very devastating in some areas and causes tree stunting and death. It can be controlled by choice of rootstock (see chapter 5), use of certain chemicals, and by planting trees at the proper depth so that the susceptible scion is well above soil level. Citrus scab, a fungus, causes distortion and malformation of leaves and tender shoots, and fruits of grapefruit, Temples, satsumas, lemons, and certain tangerine hybrids, but it does not affect oranges or tangerines. While rarely as injurious to the trees as some of the insects and mites, severe scab injury may retard growth. Again, control methods are given in chapter 10.

Minor fertilizer elements (microelements), especially zinc and copper, may not be available in the soil in sufficient quantity (especially at high pH) to meet the needs of the young trees, and they may be more readily assimilated through the leaves if applied as a foliar spray. It is better never to let serious deficiency symptoms appear than to correct them after they have retarded growth. Transient deficiency symptoms, however, are usually no cause for concern in a rapidly growing young tree. Persistence of deficiency symptoms is cause for concern and should be corrected using microelement sprays applied in the spring. However, there is no need to routinely apply foliar sprays to young trees. Details concerning these sprays can be found in chapter 9.

PRUNING

Undesired sprouts frequently appear on the trunks of young citrus trees. Trunks should be checked several times during the growing season so that sprouts may be removed before they have become large enough to compete with desired shoots for water and nutrients. Various types of trunk wraps and growth regulators such as NAA have been used to reduce sprout formation. While the metabolites produced by sprouts may have some effect on trunk and root growth, the nutrients and water that they divert from the permanent branches more than offset any benefit, and the earlier the sprouts are removed, the better. Sprouts develop chiefly either from the rootstock below the bud union or from the first six to eight inches of the scion above the union; any sprout arising less than 12 inches from the ground should be removed. While sprouts are still only a few inches long, they are easily broken off the stem. If they are allowed to grow larger and to form woody tissue, pruning shears will be needed. During the first year, the young grove should be inspected every two months from March to November for sprout removal. In succeeding years, the interval between inspections may be lengthened until, in the fourth year, a single inspection in summer should suffice.

Pruning of permanent branches is usually not practiced during the first three years in the field for young trees in Florida, although it is done in other citrus-growing regions of the world, such as Japan and China. Thus, citrus growers are not required to perform the careful tree training required of apple and peach growers. In the fourth year, however, just prior to the bearing period, it is well to examine the trees for weak limbs, branches that are parallel to and just above stronger branches, and limbs that lie across and rub against others. Such branches should be pruned out to strengthen the bearing framework. The objective is to develop a well-shaped canopy with evenly spaced scaffold limbs and space for branches to develop properly without excessive competition from other branches for light. Only limited thinning should be done, and heavy pruning should be avoided; while it is easy to remove branches, it is very difficult for the tree to replace branches unadvisedly removed. This shaping and thinning operation should be entrusted to an experienced person and not to a novice as a means of supplying experience.

Special pruning procedures may need to be used on young trees if mechanical harvesting is a possibility in the future. Lower limbs should be removed to a height that will accommodate the harvesting machinery.

Painting pruning wounds is unnecessary and may actually delay wound healing. It is, however, just as important with small limbs as with large ones to make pruning cuts nearly flush with the surface of the parent limb or trunk. Removal of large limbs at the trunk may allow wood-rotting fungi to enter the trunk. These limbs will become structurally weak over time and may break due to fruit load, wind, or mechanical damage. Some growers top young trees before they have filled their allotted space. There seems to be no advantage to this practice, except possibly with young Murcott trees, which tend to produce alternating large and small crops. Most young trees in Florida are not pruned and are allowed to develop their natural shape (except the trees grown with mechanical harvesting in mind). Selective hand pruning and training of young trees is labor intensive and costly.

CULTIVATION AND WEED CONTROL

Competition from weeds is a much more serious problem for young trees than for mature ones because of the small, shallow root volume and the lack of shading which inhibits weed growth under mature trees. Consequently, weed control is important for rapid and vigorous tree growth in the young grove. Weeds used to be controlled by hoeing or discing, but now chemicals are used almost exclusively. All weeds should be removed in an area extending from the trunk to at least a foot beyond the spread of the branches.

Chemical weed control has been the subject of much research. Several herbicides are now available for use in the young, nonbearing grove. These can be used within the tree rows to reduce to a minimum the need for hand labor and mechanical cultivation. Their proper use reduces mechanical root injuries and incidence of foot rot, improves the health and growth of the tree, and can reduce the need for water during drought periods. Proper use includes avoiding herbicide contact with the tree, especially when systemic chemicals are used. No specific recommendations for chemical control of weeds are suggested here due to the rapid development of new materials and techniques. Growers should seek the advice of extension personnel or industry representatives regarding available chemical agents and their proper usage.

Probably the best weed-control management technique for young citrus consists of an herbicide-treated strip within the tree row, extending out from the trunks to just beyond the branches. Weeds in the middles between the rows do not affect tree growth and can then be controlled by mechanical means, usually by mowing during the summer.

Instead of permitting the natural growth of weeds in the middles, the grower may wish to plant a leguminous, nitrogen-fixing cover crop, such as hairy indigo or perennial peanut, which may increase organic matter in the soil. In September or October, tall cover crops (whether natural or planted) are often disced under. Of course, discing is not advised in bedded groves due to the danger of erosion and damage to the shallow tree roots. Mowing or treating with low rates of herbicides are the preferred methods for middles management in bedded groves. A standing cover crop is a potential fire hazard, and it may increase cold injury to young trees by preventing air drainage and insulating the soil from the heat of the sun during the day. On the other hand, wind erosion is much more severe in hilly ground without plant cover. For this reason, if the middles are cultivated in autumn, it should be only enough to assure freedom from risks of fire or cold damage, leaving the cover-crop roots undisturbed at a depth of a couple of inches to reduce erosion and sandblasting of young trees during windy periods.

NINE

Cultural Practices for the Bearing Grove

The care of mature, bearing citrus groves involves many different operations. This chapter will deal with cultural practices, and chapter 10 will cover pest management strategies. Both chapters are further subdivided by major cultural operations. These subdivisions are convenient, but it is important to keep in mind that many of the cultural operations affect one another. For example, irrigation can impact fertilizer, weed control, and pest management programs. Irrigation not only reduces drought stress but is also the most popular form of cold protection. The skilled citrus grower has to successfully integrate all the production practices into one comprehensive program that considers all factors and their interactions.

It is difficult to say which cultural practice is most limiting, so it is important not to assign significance to the order in which the operations are presented. It is the total package of all production operations and their associations with one another and the environment that is important.

FERTILIZATION

Most Florida soils are low in both native fertility and the ability to retain nutrients that are applied to them as fertilizer. Consequently, trees can obtain an adequate supply of nutrients for satisfactory growth and commercial fruit production only if they are periodically provided appropriate fertilization. Fertilization represents about 15–20 percent of the total production costs, and all other cultural practices may be compromised if the nutrition program is managed improperly. To a large extent, tree vigor and fruit quantity and quality are dependent on the fertilization program. Therefore, understanding proper citrus grove fertilization is of great importance to the grower.

Complicating the fertilization process is the inability of many Florida soils to retain applied nutrients, which may result in contamination of groundwater, particularly by nitrates. Groundwater contamination has been a problem in some locations with very sandy soils on the central Florida Ridge where trees have received high levels of fertilization. Modifications to fertilization practices in areas with vulnerable soils have been developed through research. Best Management Practices (BMPs) is a program initiated in the 1990s to reduce nitrate levels in groundwater. The Florida Department of Agriculture and Consumer Services (DACS) has developed voluntary, economically viable BMPs for Polk, Highlands, and Lake counties as well as the Indian River, Gulf, and Peace River areas. Growers who comply with the BMP regulations are not subject to financial penalties if minimum groundwater standards are not met. Growers who use excessive fertilizer may be subject to legal action.

On the positive side, in areas where salinity can be a problem, the same inability of most Florida soils to retain applied nutrients (along with the state's heavy rainfall) assists in the rapid removal of soluble salts by leaching. Otherwise, soil salinity might reach toxic levels as it does in many other citrus-producing areas of the world.

The Essential Nutrient Elements

Seventeen chemical elements are essential to satisfactory growth and functioning of citrus trees and to most other green plants. Of these, three are adequately provided in any natural environment suited to tree growth and are largely beyond the control of the grower so far as their nutrient functions are concerned. The other 14 are mineral nutrient elements supplied in fertilizer or by natural breakdown of soil organic matter and minerals.

The three elements provided by nature are carbon, hydrogen, and oxygen, which together make up about 95 percent of the dry weight of a tree. Carbon and oxygen are taken into the leaves as carbon dioxide from the air, and there they combine with hydrogen, taken up as a part of the water molecule by the roots, to synthesize carbohydrates. Since this process requires energy from light, it is called photosynthesis and can take place only in cells containing chlorophyll. Thus, light energy is stored in the carbohydrates. These carbohydrates, together with proteins, fats, and other organic compounds derived from them, are the building blocks of the plant; they are used by the plant to make new tissues, to provide energy for growth, and most important to growers, to produce fruit.

The 14 remaining nutrient elements are rarely present in adequate supply in Florida soils and must be provided by the grower. Some of them are used by the tree in making amino acids, proteins, and other compounds while others seem to have some regulatory functions as components of enzymes. When any of the nutrients are not present in sufficient amounts, tree function may be affected. The severe shortage of a mineral nutrient element usually produces a characteristic visible deficiency symptom exhibited by the leaves (or other organs), and this symptom usually persists until the deficiency is corrected. Often, however, two or three elements can be deficient in varying degrees simultaneously, and the resulting symptoms do not permit the specific deficiencies to be so easily recognized. Conversely, excessive amounts of some elements may be present in the soil and may prevent the tree from functioning properly. It is important that the fertilizer elements be present in proper balance as well as in certain minimal quantities.

The fertilizer elements are divided into two groups, often termed "major" and "minor" but probably better known as "macronutrient" and "micronutrient" elements. The adjectives "major" and "minor" imply that one group is of greater importance than the other, but the elements in both groups are essential for plant growth and fruit production. Within both groups, the amount of any element needed may be affected by the amount of the other elements. The macronutrient elements are those needed by the plant in relatively large amounts, while the micronutrient elements are needed only in very small amounts. Micronutrients are also sometimes called "trace elements."

Macronutrients

The macronutrients are usually divided into two groups: the primary elements (nitrogen, phosphorus, and potassium) and the secondary elements (calcium, magnesium, and sulfur). The distinction between primary and secondary reflects the opinion of early plant scientists of the 19th century that the elements of first and greatest importance for plant growth were nitrogen, phosphorus, and potassium. If these elements were deficient, they produced the largest increases in plant yield when added as fertilizer, whereas the other three elements seemed to exert a lesser effect and therefore, were deemed secondary. We recognize now that these responses merely reflected the degree of deficiency commonly encountered. Today the primary and secondary labels have no significance as to the relative importance of these six macronutrients. For most crops, the fertilizer mixture often contains the intended amounts of only the primary elements unless it is a complete fertilizer.

In Florida, magnesium is sometimes added as a supplement, but calcium and sulfur are not considered in most citrus fertilizer programs, although they are important in connection with control of soil pH.

Micronutrients

The micronutrients of interest to citrus growers are iron, copper, manganese, zinc, boron, molybdenum, chlorine, and nickel. In general, they are needed by plants in only small amounts equal to about 1⁄100 or less of the amounts of the macronutrients. Until about 1920 none of these elements except iron was known to be needed by plants because it was difficult to purify sources of the macronutrients sufficiently that they did not contain the micronutrients as impurities. Today all of the above micronutrients can be applied to citrus trees in Florida for satisfactory growth and fruiting. They are not always needed as fertilizer components, however, since they are present in some soils in sufficient and available supply. Many microelements are present in organic matter added to the soil or even in certain foliar applied sources used to supply macronutrients.

In general, the micronutrients are needed by the plant to form organic catalysts, or enzymes, in combination with a protein. Only minute amounts are needed because very small numbers of enzymes control chemical reactions within plant cells between large amounts of plant metabolites or other compounds. Such catalysts or co-enzymes are not themselves metabolized in chemical reactions. Macronutrients may be involved in the composition of cell walls and stored metabolites, but micronutrients do not become part of these products; rather, they are repeatedly formed and degraded to become available again.

Micronutrient deficiencies usually result in rather distinctive visible patterns of leaf chlorosis (yellowing). The intensity of the chlorosis is proportional to the severity of the deficiency. Sometimes twigs and fruits may also exhibit characteristic deficiency symptoms. On the other hand, macronutrient deficiencies do not usually produce characteristic symptoms in leaves, twigs, or fruits of comparable distinction unless the deficiency is quite acute. Excessive amounts of the macronutrients are often tolerated by the plant but can result in nutrient imbalances, in lower fruit quality, or in the case of nitrogen, reduced yields. An excess of the micronutrients often results in toxicity and will interfere with absorption of the other elements, which may also cause lower yields and reduction in fruit quality.

As a final important contrast, there is usually a relationship between the amount of the macronutrients available (up to an optimum) and the amount of tree growth and fruit production. For citrus trees in Florida, in soils that are very low in nitrogen, the applied nitrogen level is closely related to yields until leaf N levels approach sufficiency. There is no such relationship with the micronutrients and yields. A certain level is necessary to avoid deficiency symptoms, but sufficient amounts have no effect on tree performance until they reach toxic levels.

Soil pH and Nutrient Availability

The term "soil reaction" refers to the acidity or alkalinity of the soil and is expressed as pH (the negative log of hydrogen ion concentration). In Florida, soil pH values range from pH 3.0 (an exceedingly acid condition) to about pH 8.5 (a moderately alkaline condition). The neutral point on the scale, equivalent to pure water, is pH 7.0. Each unit increase in pH represents a 10-fold change in the number of hydrogen ions present in the solution. Thus, pH 5.0 is 10 times as acid as pH 6.0.

Two general situations are found in Florida with regard to soil pH. Groves on alkaline soils are found in many coastal and southern areas and in Dade County, but the great majority of groves are planted on acidic, sandy soils, called entisols such as those in the Ridge (central) area of Florida.

The alkaline soils of Florida are calcareous and contain large quantities of calcium carbonate. Leaching due to rainfall or irrigation has removed most mineral bases other than calcium. The very high content of calcium in the soil/water solution makes any added magnesium and potassium in fertilizers less available owing to competition among ions for sites on soil particles. Thus, those elements must be supplied in larger amounts than on acid soils. Moreover, all the micronutrients except molybdenum are only very slightly available at the high pH values of these calcareous soils. They are usually not available to the tree through soil applications but must be applied as foliar sprays to be taken up in sufficient quantities. An exception is iron, which is not satisfactorily absorbed from foliar spray application. Molybdenum is more available in alkaline than in acidic soils. Sulfur may be used to decrease the soil pH of calcareous soils (as it can be in acidic soils), but very large quantities of sulfur are required over a long period of time. Regular use of fertilizers that acidify the soil, such as ammonium sulfate, also reduce pH.

In bedded flatwoods soils, the normal depth profile can be redistributed in the ditching and bedding process. Root growth can be modified by drainage and water table management. Some citrus groves are on soils with an acidic topsoil, but the subsoil is calcareous. Most of the fibrous roots are in the topsoil, and such groves usually need the same treatment as groves without alkaline subsoil, except that the treatment will be modified to take into account the depth of the acid layer.

The well-drained, sandy soils are naturally acidic (sometimes too acidic for citrus tree growth), and they may increase in acidity under normal fertilizer regimes. This is due partly to the loss of basic elements from the soil through absorption by roots and through leaching by heavy rainfall. The greatest cause of increasing acidity is the use of acidifying fertilizers and the accumulation of sulfur if used for rust mite control. However, the use of sulfur for rust-mite control has declined recently because its use results in many undesirable side effects. As soils become more acid, the micronutrients leach out more readily, increasing deficiencies, and the basic phosphates that keep magnesium and calcium available are also leached. The nitrification process, by which ammonia is converted to the more readily available nitrate form, is inhibited in sandy soils when the pH is below 5.5. Within a pH range of 5.5–6.5 (a suggested target of 6.0 may be optimal), there is maximum availability of most mineral nutrients to citrus trees, as well as active nitrification. It is very desirable that the grove soil be kept within this pH range. At pH values above 6.5, magnesium, potassium, and the micronutrients are increasingly unavailable to the trees. The soil pH should not be allowed to exceed pH 7.0, except where the pH of the soil in its natural state is that high. When high copper or manganese content in the soil makes iron unavailable, however, it is advisable to keep the soil pH at 6.5 or higher. An increase of soil pH is normally accomplished by use of finely ground dolomitic or calcitic limestone and is termed "liming."

ESSENTIAL NUTRIENT ELEMENTS

Nitrogen

Nitrogen is the most important nutrient for citrus production because it is used in large amounts for production of new plant cells, enzymes, and tissues; when nitrogen is in short supply, all new growth and fruiting are limited. Thus, nitrogen is the element whose availability level will, in general, determine the yield throughout the normal range of tree response. When nitrogen is supplied in slightly suboptimal amounts over a long period of time (chronic mild deficiency), the tree responds by reduced growth and fruit production which are not accompanied by any specific foliar deficiency symptoms. If the nitrogen supply is cut off or greatly reduced for trees that have previously had an adequate supply, the tree responds rather quickly (depending on residual supplies of organic nitrogen) with a marked visible leaf chlorosis. The older leaves become yellow and drop earlier than normal. Later the new leaves also do not develop a dark green color, followed in cases of persistent severe nitrogen shortage by defoliation, fruit drop, and shoot death. Above the levels of visible deficiency, shoot growth and yield increase steadily with nitrogen supply up to a maximum value unless some other element is so limiting that the tree cannot utilize the increased nitrogen. Since nitrogen level is more closely correlated with growth and fruiting than any other element, the nitrogen percentage is always listed first in fertilizer formulas or analyses.

When nitrogen is limiting, higher levels of available nitrogen increase vegetative growth, flowering, fruit yield, and percentage of acidity in the juice. Nitrogen can cause a slight increase in percentage of soluble solids in the juice, even though fruit production may decrease at excessive rates. This is because excessive vegetative growth can compete with and reduce flowering. Nitrogen level has no effect on fruit rind thickness on mature trees but may produce a puffy peel for tangerines and other fruit of young trees. A high nitrogen level may delay coloring of the rind somewhat. High nitrogen may also decrease tree uptake of phosphorus, potassium, copper, and boron, making it necessary to provide higher levels of these elements. Excess levels of nitrogen are of little economic value and may be detrimental, as previously discussed.

In the past, the most important considerations in choosing a source of nitrogen was availability and leachability, but it is now evident that there are other factors involved of perhaps greater significance. Natural organic forms seem less satisfactory as fertilizers, even apart from their much greater cost per unit of nutrients, than inorganic forms because they have such a low elemental analysis. Only if micronutrients are not supplied would organic nitrogen sources have any advantage over inorganic forms of N. The availability of organic forms of nitrogen over a longer period of time, as compared with the readily leached inorganic forms, is often cited as an advantage. However, citrus trees are able to take in and store nitrogen during rather short periods of absorption. If groundwater contamination in vulnerable soils is an issue, organic or controlled-release nitrogen forms may be more suitable to use than inorganic forms.

Choice of the two principal inorganic nitrogen forms, nitrate and ammoniacal, is largely a matter of cost. Citrus trees do not distinguish between the various sources, so the cheapest per-unit cost is usually the best option. Exclusive use of ammoniacal forms, however, requires more frequent measurement of soil pH and subsequent correction, since tree vigor may decline markedly if the soil pH becomes too acidic. A mixture of nitrate and ammoniacal forms is a very satisfactory compromise; synthetic ammonium nitrate—half nitrate and half ammonium nitrogen—is a regular and low-cost source material. The differences in tree response are insignificant, however, if pH is controlled. Nitrogen in either form is readily leached from sandy soil, so that it is not possible to build up a supply in the soil for future use except for rather short periods by use of green manures.

Urea is another high-analysis source of nitrogen which can be applied to the soil or as a foliar spray. In the soil, urea is readily broken down by urease (an enzyme) into ammonium. Urea has rarely been applied as a foliar spray in the past, but interest in this method increased in the 1990s as a potential way of reducing nitrate contamination of groundwater. However, spray application of urea is not as commonly used in Florida as soil application of other nitrogen sources. Again, fertilizer price is important, and as costs for ammonium nitrate increase, there is more interest in using soil applied urea.

Nitrogen is applied in traditional dry formulations or as a solution or suspension. Where liquid forms are used, they may be applied directly to the soil, usually as a 21–32–percent nitrogen solution or injected into the irrigation system. This mixture of water and liquid fertilizer is known as "fertigation" and is becoming a popular method of fertilizing in Florida, especially for young trees (see chapter 8).

Nitrogen is absorbed by citrus trees in Florida at all times of year. However, the rate of uptake in winter is about half of that in summer due to lower soil temperatures. Leaching is also much reduced in winter due to lower average rainfall than in summer. Thus, nitrogen applications are about as effective during one time of the year as at another.

Phosphorus

This element is available in fertilizer as P_2O_5, usually termed "available phosphoric acid." In spite of fairly low natural levels in many of Florida's sandy soils, it has been possible to demonstrate phosphorus deficiency symptoms in citrus trees grown without added phosphate on the well-drained soils of central Florida. Some of the soils used for new citrus plantings in poorly drained areas have very low natural phosphorus content and may require phosphates regularly in the fertilizer. Phosphorus application is recommended in all virgin areas for young trees up to seven years of age. However, phosphorus does not leach readily where soil pH is 6.0 or higher, and the fruit removes very little phosphorus, about 1.5 to 2.0 pounds per 100 boxes of fruit. Therefore, regular phosphorous applications are usually not necessary in mature groves which have accumulated adequate phosphorus. Phosphorus levels can be easily checked with soil or leaf tissue analysis.

Approximately 0.05 pound of available phosphoric acid is required per box of fruit. A high phosphorus level actually decreases the percentages of soluble solids and acidity in juice and delays coloring of Valencia oranges. Groves that have been bearing for a dozen years and have been supplied with phosphorus regularly will likely show no economic benefit from further applications.

Potassium

Potassium (K) or potash is calculated and guaranteed on the fertilizer tag as the equivalent K_2O, although potassium is never present in this form in fertilizers. Until 1945, it was standard practice to include more than twice as much potash as nitrogen in the fertilizer mixture. High potash levels were believed to produce fruit with better external quality as well as to cause early maturity and improved freeze hardiness of stem tissues in the winter. Now it is known that for oranges, such high levels of potash actually produce large fruit size, coarse peel texture, and slow loss of green peel color as fruit matures. These are conditions which were long supposed to result from excess nitrogen. "Sheepnosing"(the presence of an unusually pointed stem end of the fruit) of grapefruit can be caused by high levels of nitrogen and potassium especially when there is a small crop load. In addition, potassium has no effect on tree freeze hardiness. Ratios of nitrogen to potash today are usually 1:1, although ratios of 1:1.25 are common and may be needed on calcareous soils. Potassium deficiencies are seen from time to time and are usually the result of high nitrogen rates and high fruit production, resulting in rapid potassium depletion.

Several usual source materials supply available potash about equally well. Potassium chloride, however, will increase salt levels and when used in combination with high salinity irrigation water may cause salinity damage. There is practically no absorption of potassium by citrus trees at low soil temperatures, but the element is readily stored in leaves for future use by trees.

Magnesium

This element assumed an important place in fertilization of citrus trees in Florida about 1930. After the discovery that bronzing of grapefruit leaves (a chlorotic pattern in which green was gradually replaced by a dark yellow color) was due to magnesium deficiency, growers recognized that magnesium deficiency often limited yields and made trees more susceptible to cold injury. Today, magnesium compounds are regularly supplied to augment the very low natural amounts in Florida soils. Dolomitic limestone, a naturally occurring mixture of calcium and magnesium carbonates, serves as well as ordinary calcic limestone to decrease acidity and also supplies large amounts of slowly available magnesium into the soil. Magnesium is needed at about 30 percent of the nitrogen rate for seeded cultivars and at about 15 percent of the nitrogen rate for seedless types. In general, most magnesium will be provided when dolomite is used as an amendment to increase soil pH. The balance of the magnesium—or all of it when calcic limestone is used—is supplied by a readily available form in the fertilizer mixture. Where native soil pH is naturally high, calcium is usually abundant and is an antagonist to magnesium uptake. In such situations, foliar applications of magnesium nitrate are effective. Seedy cultivars respond to magnesium applications to a greater extent than seedless cultivars because seeds store large amounts of magnesium and remove it from the leaves where it is needed to produce chlorophyll. So long as magnesium is provided in amounts that prevent deficiency symptoms, there is no increase in growth or yield as magnesium levels increase. Absorption of magnesium, like that of potassium, slows dramatically in winter when soil temperatures are low.

Calcium

Calcium is a major nutrient component in citrus trees and the supply in the soil is usually adequate in Florida. This is true because large amounts of calcium carbonate are applied to control soil acidity in addition to large amounts of calcium supplied in superphosphate. Furthermore, alkaline soils found in the flatwoods regions of Florida usually have an abundance of natural calcium.

Sulfur

As with calcium, sulfur deficiency has never been found in citrus plantings in Florida. There are two good reasons for this. Large quantities of combined sulfur in available form are regularly supplied in the materials used primarily to provide other elements, especially in superphosphate but also as sulfates of the heavy metals and ammonia. In addition, sulfur is sometimes used to control rust mite, eventually being washed to the ground by rain and undergoing changes in the soil that make it available as a nutrient. Sulfur deficiency symptoms in the laboratory are similar to those of nitrogen deficiency.

Iron

Chlorotic patterns due to iron deficiency have been recognized far longer than any others and always appear first on the young shoots. The soil may have high iron content, yet plants may either not be able to absorb it or not be able to utilize it after absorption. Iron deficiency occurs most often in plants growing in high pH, in waterlogged soils, or in soils very low in organic matter—the so-called sand-soaks. Susceptibility to iron deficiency also varies with rootstock, as trifoliate orange and its hybrids, Swingle citrumelo and Carrizo citrange, are most susceptible; sour orange and Cleopatra mandarin are the least susceptible. Iron deficiency problems have also occurred where high copper levels are present. High soil copper residues, sometimes as much as 1,000 pounds per acre, result from the combination of copper fungicidal sprays and copper added in fertilizer mixtures. When this problem occurs, it is most likely in an old citrus grove or former vegetable planting. Accumulations of either manganese or zinc also depress iron availability, but neither of these metals has yet reached the level in citrus groves where their effects are as serious as those of high copper. Iron is not readily absorbed by leaves, and spraying has not proven satisfactory to correct iron deficiencies on citrus trees. Furthermore, soil applications only increase the already large amount of unavailable iron in the soil. The solution to the problem has been the use of chelated forms of iron, which put iron into the soil in an organic combination that prevents it from combining with other elements and becoming insoluble. On acidic sandy soils with pH less than 6.5, chelated iron gives prompt recovery for iron deficiency symptoms. On calcareous soils, it is more difficult to overcome iron deficiency. Special forms of chelated iron are available for high pH soils and are usually effective in correcting iron chlorosis. Chelates are typically much more costly than spray application of other microelements.

Copper

Conditions known as "red rust," "exanthema," "dieback," or "ammoniation" were noted in Florida citrus trees as early as 1875 and were attributed to various causes, especially the use of ammonia or manure. About 1912 it was discovered that application of copper sulfate to the soil under citrus trees remedied this problem, although no one then thought copper could be essential to plants. It was not until the late 1930s that regular use of copper eliminated dieback as a serious problem. Growers and research workers alike tended to overlook the fact that when copper sprays were regularly used for fungus control, no other measures were needed to prevent dieback. Copper was routinely included in fertilizer mixtures without regard to spray residues, and after some 10–15 years of overuse, the excessive copper accumulation in soils brought about the iron deficiencies discussed earlier. Consequently, no copper should be included in fertilizers if copper sprays are regularly applied or if a soil test shows over 50 pounds of copper per acre in the top six to eight inches of the soil. In young, nonbearing groves or in groves where copper sprays are not used, about 1⁄40 as much copper as the amount of nitrogen should be included in the fertilizer mixture. Where soil tests show copper levels in excess of 50 pounds per acre, the pH should be maintained above 6.5 to prevent toxicity.

Zinc

The effects of zinc deficiency are most often expressed in citrus trees as "frenching" of foliage, a severe chlorosis in which leaf tissue becomes nearly or entirely white except for green veins. At the same time, leaf size becomes progressively smaller as the deficiency becomes more severe, and shoot internodes become shorter, giving a rosette effect. Recognition of zinc deficiency about 1934 led to realization that other metallic elements might also be deficient. Since zinc deficiency severely decreases the leaf area for photosynthesis, extremely zinc-deficient trees have limited growth and are unproductive. Citrus trees are not able to obtain adequate supplies of zinc from soil applications, but zinc is readily absorbed from foliar sprays. After visible deficiency symptoms have been corrected, smaller quantities of zinc are applied to prevent recurrence of the deficiency.

Manganese

A mild form of chlorosis has been associated with manganese deficiency on acid, sandy soils since about 1932. The marl chlorosis or marl frenching found on calcareous soils is the result of combined manganese, zinc, and sometimes iron deficiencies. In the 1930s, serious cases of manganese deficiency were found in many groves on acid soils because the pH had been allowed to drop well below 5.0 and manganese had been leached out. With maintenance of pH above 5.5, manganese accumulates in the soil. Because of the widespread manganese deficiencies, this element was commonly added in fertilizers beginning in 1934. But by 1950 excessive accumulation was becoming apparent. A temporary mild deficiency pattern on new shoots does not affect growth or fruiting of citrus trees. Only in the case of persistent deficiency symptoms should manganese be applied

Boron

Florida growers first became aware of boron because of its toxic effect on the tree due to boron impurities found in certain potash salts. Occasionally, boron toxicity occurred in trees where borax-dipped field boxes from packinghouses had been stacked. Once the source of the boron injury was identified and further boron application was stopped, it could be eliminated from the soil by liming and irrigation or, more slowly, by leaching due to normal rainfall. Boron deficiency is not often found in Florida, but it does occur sometimes when growers use only high-analysis fertilizers or following a prolonged drought. Boron is readily available to citrus trees from applications of a soluble borate either to the soil or as a foliar spray, which results in a more rapid response than soil applications. The difference between amounts of boron that are beneficial and those that are toxic is very small, so unusual care is needed in applying this element. It should never be applied in the fertilizer mixture and as a foliar spray in the same year.

Molybdenum

Ever since the turn of the century, an unusual leaf chlorosis has been observed. Later called "yellow spot," it did not respond to any of the treatments that corrected other chlorotic symptoms. Only since 1950 has it been realized that yellow spot is a symptom of molybdenum deficiency. This element is less available in acid than in slightly alkaline soils, unlike all other heavy metals. Since molybdenum deficiency symptoms in leaves usually occur on soils that have been allowed to become undesirably acid, soil applications of molybdenum compounds do not correct visible deficiency symptoms. Groves with a soil pH above 6.0 are not likely to exhibit molybdenum deficiency.

THE CITRUS FERTILIZER PROGRAM

A fertilizer program should provide all the necessary nutrients that are not obtainable under natural conditions. Some of these elements may be present in concentrations that are too low for plants to thrive, some may be present and available in quantities that severely limit growth and yield, and some even may be present in sufficient quantities but are unavailable to the plant. The optimum fertilizer program must either augment the existing supply until it reaches adequate levels or increase the availability of previously unavailable elements. A sound citrus fertilizer program accomplishes both of these objectives. Such a program is designed not only to meet present needs to enable trees to produce a current crop of desired size and quality but also to maintain healthy, vigorous, productive trees in the future. The effectiveness of a fertilizer program, therefore, should never be judged by the results it produces for a single year because these are most often the results of previous fertility programs. Only by following a fertilizer program for several years is a grower in a position to assess its worth with confidence.

The optimum citrus fertilizer program has three components: the application of soil amendments, primarily for control of soil pH; the application of nutrient elements to the soil using liquid or dry applications; and the application of certain nutritional elements as foliar sprays.

Soil Amendments and pH

The term "soil amendments" covers materials that are incorporated into the soil to improve soil characteristics, making them more favorable for plant growth. These include mixing organic matter with soil to improve water and nutrient holding capacity and to increase aeration and tilth of clay soils. Organic sources of nitrogen, such as chicken and cow manures and sewage sludge, have some effect as amendments in addition to their primary function of supplying nitrogen. However, the amount of nitrogen added is minimal unless large quantities of these materials are applied. Transportation of these low-analysis materials is also expensive and may cause problems with odors and flies. Soil organic matter is usually increased most efficiently by growing and discing-in green manure crops. But even this practice only slowly adds organic matter to the soil due to the high rate of breakdown in Florida's hot, humid climate. Moreover, application of organic matter is expensive, time consuming, and not widely done in Florida. Repeated discing can even lead to losses of organic matter.

The most widely used soil amendments in Florida citrus groves are those that influence soil pH, namely, dolomitic and calcitic limestones. Annually up to 2,000 pounds per acre of one of these materials may be required to maintain the desired pH range on acid sands. Since each pound of sulfur applied in some pest-control programs requires slightly over three pounds of limestone to neutralize its effect on lowering soil pH, it is obvious that as little sulfur should be used as possible that is consistent with good pest control. The present trend toward rust mite control by chemicals other than sulfur has reduced the need for frequent grove liming and is making it possible to avoid previously common fluctuations in soil acidity during the year. The widespread use of irrigation has also resulted in a reduction in the requirement for lime since most irrigation water in Florida contains substantial quantities of calcium and magnesium carbonates.

Soil pH should be checked annually by taking a number of soil samples in each grove to a depth of about eight inches from locations under the outer edge of the tree canopy (dripline). This is where the largest numbers of roots are concentrated and where maximum nutrient leaching occurs. A composite sample combining individual samples from 15–20 trees is very desirable, and the area represented by a single composite sample should not exceed 10 acres. In contrast, site-specific precision agriculture may require many single soil pH samples per acre.

The amount of lime needed will depend not only on the pH but also on soil properties such as texture and organic matter content. The lower the pH value, the more lime is required. The finer the texture and the higher the organic matter content, the more lime is required for a given pH. Under average conditions in sandy Florida soils, about 200 pounds per acre of limestone would be required to raise the pH value by 0.1 unit—that is, from 5.0 to 5.1. A customary practice is to raise the pH by liming to 6.5 whenever it reaches 5.5. Then 10 × 200, or 2,000 pounds (one ton) per acre of limestone will be required to make the correction. Lime applications may be made with equal effectiveness at any time of year; usually it is applied at a time when other grove work is not occurring. As indicated previously, dolomitic limestone is usually preferred because it supplies both magnesium and calcium at the same time it decreases soil acidity.

Application of Nutrient Elements to the Soil

Research has shown that the number and timing of fertilizer applications are of little importance in maintaining sufficient nutrient levels for mature trees provided a sufficient amount and balance of nutrients is applied each year. The number of split applications required to apply fertilizer is important, however, when considering nutrient leaching losses. Fertilizer is usually applied from October to June, when heavy rains are less likely to leach nutrients from the soil. The traditional practice has been to make three dry fertilizer applications per season: one in the fall (October or November), one in the late winter (January or February), and one in late spring or early summer (May or June). The first of these replenishes nutrients that are lost in harvested fruit and from leaching by summer rains. The second application is important for adequate flowering, fruit set, and new vegetative growth. The third provides nutrients for rapidly developing fruit and for the summer growth flush. This application method is commonly used, but there are fertilizer expenses, costs of labor, and potential leaching losses to consider. Fewer split applications are equally effective but are not commonly used in current fertilizer programs. Some growers apply liquid suspension fertilizer mixed with herbicides in a boom applicator. This system places both materials under the canopy directly above tree roots. The fertilizer salts reduce weed growth and help to make the herbicide more effective. There are obvious cost savings with such applications. Citrus grown on vulnerable soils where nitrate leaching could be a problem may require more frequent, lighter applications of fertilizer to minimize leaching. Trees growing in impoverished sites may also benefit from more frequent applications. When irrigation is used to deliver fertilizer (fertigation), the frequency can be increased considerably, up to 10–20 times per year. Some current programs such as the open hydroponic system fertigate with dilute solutions every time irrigation is required. With such a system, fertilizer could be delivered almost any time at reduced labor costs, although material costs may increase.

Fertilizer may be provided from many source materials and in various types of mixtures. Regardless of the sources and combinations, the amounts and ratios of the elements are the important factors for a successful program. The commercial grower must vary this program to account for variations in rootstock-scion combinations, in soil and environmental conditions, in cost of fertilizer materials, and in market demands and prices.

Historical Fertilizer Practices

The amount of nitrogen required for a bearing citrus tree has in the past been based on potential annual yield. Although the amount varied with soil type, in general, the amount averaged about 0.4 pounds of nitrogen per box (90 pounds) of fruit potentially produced for oranges and 0.3 pounds per box for grapefruit (85 pounds). This method of determining nitrogen rates often resulted in excessive fertilization of even highly productive groves based on more recently developed Best Management Practices (BMP) recommendations. For example, under good management, yields in excess of 800 boxes per acre of oranges and 1,000 boxes per acre of grapefruit are not uncommon. Using the base values of 0.4 pounds of nitrogen per box for oranges and 0.3 pounds of nitrogen per box for grapefruit, the amount of nitrogen applied, 320 and 300 pounds per acre, respectively, would be well over the maximum recommended rate of 250 pounds of nitrogen per acre. Nitrogen rates above 250 pounds per acre have not been shown to greatly increase production, are not economically justifiable, and could lead to potential groundwater contamination.

Current Fertilizer Practices

Although the amount of nitrogen that a grove requires may vary somewhat due to crop load, cultivar, tree age, soil type, and other factors, research has shown that 120–250 pounds of nitrogen per acre per year should provide high yields of good quality fruit for mature orange trees. Figure 9.1 provides a range of nitrogen rates based on projected yields for oranges. In contrast, a

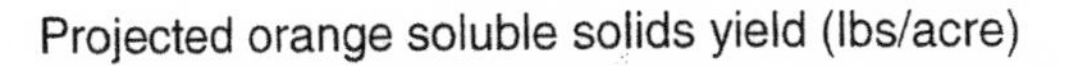

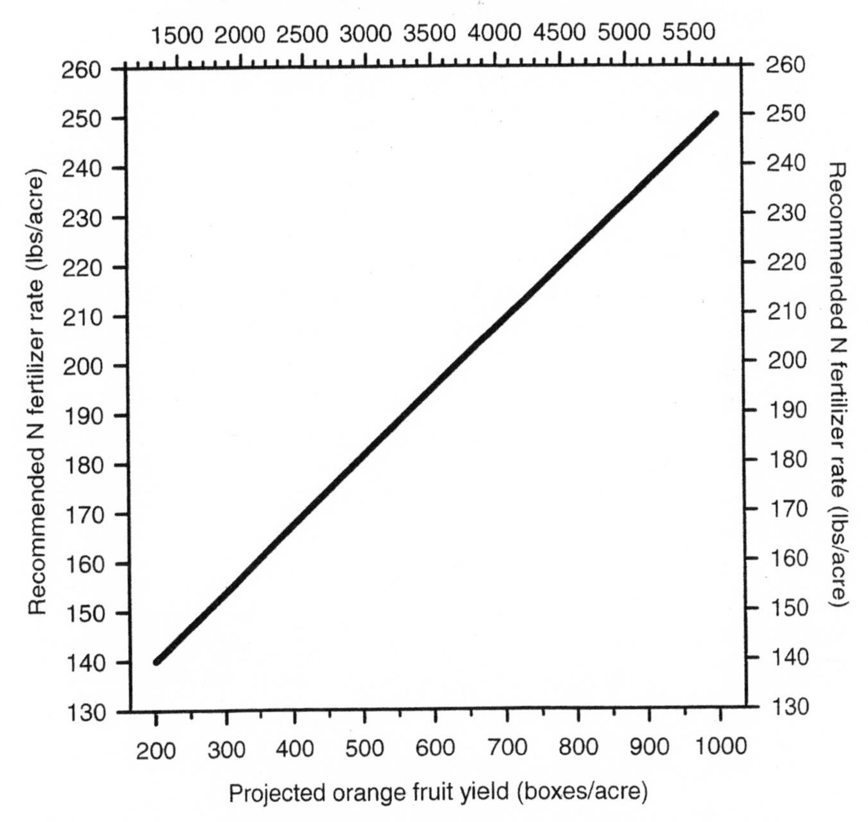

Figure 9.1. Production-based nitrogen fertilization rates for Florida oranges. Source: *Nutrition of Florida Citrus Trees*, edited by T. Obreza and K. Morgan, University of Florida Cooperative Extension, SL 252, 2008.

maximum of 160 pounds of nitrogen should be adequate for grapefruit, with lower rates used for fresh fruit in flatwoods areas. Tangerines, tangelos, and other specialty fruits require nitrogen at the rate suggested for oranges, with particular attention paid to crop load. Higher rates (up to 250–300 pounds of nitrogen) may be necessary in some years, especially in seasons with large crops, for Orlando tangelos and Honey tangerines (Murcott), respectively.

One of the problems associated with the new nitrogen rate ranges is that it is difficult to choose a specific rate for a particular situation. This is especially true if you cannot accurately estimate your projected yield (fig. 9.1). An alternative method is to use a baseline nitrogen rate and adjust it based on the actual average yields for the past four seasons. Yields may be expressed as boxes per acre or as pounds-solids per acre. [pounds-solids, p.s. = juice content (weight) × soluble solids (%)]. A sample calculation is described in "Nutrition of Florida Citrus Trees" edited by T. A. Obreza and K. T. Morgan, SL 253 University of Florida, IFAS Extension 2008. This method coupled with leaf and soil analysis gives a more precise nitrogen rate for a particular situation.

Potash is usually supplied in equal quantity with nitrogen, magnesium at about 20 percent of this rate (if a need is indicated); phosphoric acid is included at 25 percent of the nitrogen amount (again, if needed). Micronutrients such as manganese, copper, zinc, iron, and boron are usually supplied only if a need has been indicated. For groves between 10 and 20 years old, micronutrients are not included unless deficiency symptoms or tissue analysis indicate their need. When deficiency symptoms show iron is needed, iron in the chelated form can be applied, either mixed in fertilizer or as a separate application.

Complete fertilizer mixtures usually contain all of the required elements; by convention, the successive numbers represent the percentage of available nutrients in the order: N, P_2O_5, K_2O, Mn, Cu, Mn. In the past, mixtures with 4 percent nitrogen were common; currently the minimum nitrogen percentage is usually 6, but more commonly the mixture may contain 8, 10, or even 33 percent or more because of decreased cost of applying smaller quantities of higher-analysis materials.

Spray Application of Nutrients

Certain micronutrient elements are taken up better as foliar sprays than via soil application. These elements may be applied separately but are often incorporated into pest-control sprays to reduce application costs. Nutritional sprays for Florida citrus trees date from the 1930s, when zinc was found to be required and to be poorly available from soil applications but more readily available in foliar sprays. In addition, there is considerable interest in the application of elements other than micronutrients by spraying. Soluble nitrogen sources (usually urea) are often applied around bloom time to increase growth. Foliar applications of urea may decrease leaching losses and groundwater contamination that may occur with soil-applied nitrates. Some research suggests that urea may increase fruit set and yields and supplement soil application, especially in flatwoods soils where limited root growth may be a problem. Urea should be applied at high relative humidity, at moderate air temperature (77–88°F), and in spray solutions with a pH between 7 and

8. Potassium sprays are also used to increase fruit size, but low volume, high concentration application (< 100 gallons per acre) may cause spray burn.

Micronutrient nutritional sprays are costly to apply by themselves. Although these sprays are often beneficial to the tree, undesirable or harmful plant reactions may result from their indiscriminate use. Also, as the number of materials in the spray tank is increased, the potential for adverse tank reactions from incompatible materials and foliar spray burn increases. Micronutrients should be used only when their benefits can definitely be realized. The two situations that always warrant their use to correct observed deficiencies are (1) when applications to the soil are not effective; and (2) when a rapid response is desirable and cannot be obtained by soil applications. Nutritional sprays are generally used only for supplying elements needed in very small amounts (the micronutrient elements), or for supplying magnesium (as magnesium nitrate) to trees on alkaline soils. Nutritional sprays may be used to correct deficiencies or as maintenance sprays, intended to maintain nutrient levels above the deficiency range by regular application of small amounts of materials.

Citrus trees absorb zinc compounds from soil applications only to a very limited extent. Zinc is often deficient in Florida soils and is usually applied when needed as a nutritional spray. For such maintenance spraying, three to five pounds of metallic zinc per acre normally satisfies the annual requirement. The spray application is often made during the period when growth is not occurring in spring or at postbloom and may be applied with pesticide sprays to save on application costs. If zinc deficiency symptoms have developed in the grove before any nutritional spray has been used, this indicates that a corrective spray will be needed. This spray should contain twice as much zinc as the maintenance spray and is usually applied only once per season.

Yellow spot, which results from molybdenum deficiency, is quite uncommon but may be observed in summer as small, water-soaked areas on the leaves. Mild symptoms can be corrected by applying five ounces of sodium molybdate per acre, whereas severe symptoms require twice this amount. Spraying is required only when deficiencies are visible.

Manganese and boron are often poorly available from ground applications on alkaline soils, so their deficiencies must be corrected by foliar spraying. The same is true of copper if no fungicidal copper sprays are used. Nutritional sprays of copper usually contain a more soluble form than fungicidal sprays, which are intended to remain on leaf surfaces. On calcareous soils, manganese sulfate may be applied as a foliar spray, but the effect is only visible for one season. If sprays are used, three to five pounds of manganese or copper equivalent per acre is recommended. Boron should be supplied at 1⁄200th the rate of nitrogen. In all cases, avoid excess applications of nutritional sprays and do not use them unless there is a need as indicated by leaf symptoms or leaf analysis.

In summary, the complete nutritional program for commercial citrus production includes liming to adjust the soil pH range to 5.5 to 6.5, soil applications of sufficient amounts in proper proportions of the elements not naturally present or available in the soil, and nutritional sprays as needed to supply elements that cannot be adequately furnished by soil applications. Nitrogen and potassium are usually applied to the soil, but they also may be sprayed.

Soil and Leaf Analyses

Testing the soil and evaluating the nutritional status of leaf tissues may provide growers an opportunity to refine their fertilizer programs. These tests are diagnostic and may help in solving nutritional problems as well. As mentioned previously, soil tests are used to determine soil pH and whether liming is needed. This is the principal value of soil tests, but they may also be used to assess phosphorus, magnesium, and calcium levels and to detect toxic levels of copper. Testing for other elements is of little value due to inherent sampling errors that arise from taking comparatively few and small samples to represent a large volume of soil that is usually not uniform. In addition, many elements, such as nitrates and potassium, are readily leached from the soil, and thus soil analysis of these elements does not correlate well with tree performance.

SOIL SAMPLING

Soil samples should be taken once per year, usually during leaf sampling. A soil core, about eight inches deep, is taken at the dripline of 15–20 trees per acre. The dripline, as the name implies, is located at the outer extent below the tree canopy. It is important to collect samples from a relatively uniform block of trees and with a similar soil type. Soil and tree uniformity are often a problem in many flatwoods citrus groves. The soil cores are combined, mixed, air-dried, and sent to a diagnostic lab for analysis. A printout is returned containing values for soil pH, phosphorus, calcium, and magnesium. Levels of copper and organic matter also may be requested. The lab report gives relative levels of each factor, extraction method (which is important), and recommended options. It is important to consult current University of Florida, Institute of Food and Agricultural Sciences (IFAS) soil sampling guidelines before making adjustments to the current fertilizer program.

Traditionally, soil samples were taken yearly from the same areas of a grove to reduce sample-to-sample variation. Grid sampling is now being used in some groves to create maps that show the variation in soil/leaf readings within a particular area. A grid is then superimposed on a grove map and various areas are identified using global positioning systems (GPS). The new method allows for more precise placement of fertilizers and lime and can pinpoint problem areas that may require better irrigation, drainage, or special soil amendments. In addition, site-specific prescription maps can reduce the number of material applications to areas where they are not needed.

LEAF ANALYSIS

Leaf tissue analysis provides a very accurate method of determining the actual mineral content of the leaf. Proper sampling is extremely important. Therefore, both sampling and analysis are often left in the hands of the analytical laboratory staff or trained personnel.

A standard leaf sample consists of 100 or more mature spring-flush leaves with the petiole attached, usually collected in late July and August. These leaves are typically four to six months old. The leaves should be taken from nonfruiting twigs of at least 15–20 uniform trees to assure a representative sample. The leaves must not have been sprayed with nutritional sprays, since this will cause surface contamination that must be washed off with a non-nutrient soap and water to prevent inaccurate readings. Leaves should also be free of pest and disease damage. Clean dry leaves can be stored in paper bags and sent to a lab for analysis. After chemical analyses have been run, the values obtained can be compared against leaf analysis standards as shown in table 9.1. Ideally, values should be maintained in the optimum range for each element. Since cropload can affect leaf nutrient levels, the range is very wide, especially for microelements, so some degree of interpretation may be required.

PLANT GROWTH REGULATORS

Plant growth regulators (PGRs) are naturally occurring or synthetic materials that alter the growth, development, or fruit quality of citrus trees. PGRs are commonly used in many fresh fruit-producing citrus regions including California, South Africa, Spain, and Australia. They have been tested extensively for use in Florida but have only sporadically been used, probably because Florida's industry is geared toward processing, and the effectiveness of PGRs is often variable; they may even cause fruit or tree damage. PGRs can be used to improve fruit set, reduce floral initiation, extend the harvest sea-

Table 9.1. Guidelines for interpretation of orange tree leaf analysis based on 4-to-6-month-old spring flush leaves from nonfruiting twigs

Element	Unit of measure	Deficient	Low	Optimum	High	Excess
N	%	< 2.2	2.2–2.4	2.5–2.7	2.8–3.0	> 3.0
P	%	< 0.09	0.09–0.11	0.12–0.16	0.17–0.30	> 0.30
K	%	< 0.7	0.7–1.1	1.2–1.7	1.8–2.4	> 2.4
Ca	%	< 1.5	1.5–2.9	3.0–0.49	5.0–7.0	> 7.0
Mg	%	<0.20	0.20–0.29	0.30–0.49	0.50–0.70	> 0.70
Cl	%	?	?	< 0.2	0.20–0.70	> 0.70[a]
Na	%	?	?	?	0.15–0.25	> 0.25
Mn	mg/kg or ppm[b]	< 18	18–24	25–100	101–300	> 300
Zn	mg/kg or ppm[b]	< 18	18–24	25–100	101–300	> 300
Cu	mg/kg or ppm[b]	< 3	3–4	5–6	17–20	> 20
Fe	mg/kg or ppm[b]	< 35	35–59	60–120	121–200	> 200
B	mg/kg or ppm[b]	< 20	20–35	36–100	101–200	> 200
Mo	mg/kg or ppm[b]	< 0.05	0.06–0.09	0.10–2.0	2.0–5.0	> 5.0

a. Leaf burn and defoliation can occur at Cl concentration >1.0%.
b. ppm = parts per million.

Source: Koo et al., 1984. Recommended fertilizers and nutritional sprays for citrus. Florida Cooperative Extension Service Bulletin 536D.

son by delaying fruit drop, improve fruit color and firmness, increase juice content, and thin excessive crops.

Increasing Fruit Set and Yields

Several citrus hybrids such as Orlando and Minneola tangelos, along with Sunburst and Robinson tangerines, usually require cross-pollination with a compatible pollinizer to produce adequate yields. Alternatively, gibberellic acid (GA_3), a naturally occurring PGR, can be sprayed on the tree during bloom. The GA_3 replaces the natural PGR produced by seeds, increases fruit set and yields, and produces seedless fruit. Seedlessness is very desirable for all fresh citrus markets. Sometimes fruit size is decreased by GA_3 application, however.

Reducing Floral Initiation

Some citrus cultivars like navel oranges and the Ambersweet hybrid produce excessive numbers of flowers. Spray application of GA3 in the winter reduces floral initiation and flower number, and in some cases it increases yields. GA3 should not be applied to trees producing average numbers of flowers as this may reduce yields.

Extending the Harvest Season

Citrus growers sometimes want to extend the harvest season especially for grapefruit and navel oranges to take advantage of higher prices later in the season. If fruit are simply stored on the tree, fruit peel quality deteriorates and excessive fruit drop can occur. Spray application of a combination of GA3 and 2,4–D, a synthetic PGR, delays peel color development and fruit softening (GA3) and reduces fruit drop (2,4–D). This combination has been used successfully in many citrus-growing regions for more than 50 years. It has been used for grapefruit in Florida, but problems with peel spray burn and liability issues with 2,4–D (damage to other adjacent crops) have greatly limited this practice. However, spray application of GA3 alone often decreases stem rind deterioration for Minneola tangelo.

Increasing Juice Content

Researchers working with GA3 to extend the harvest season of grapefruit also noted that juice content increased. Further studies suggested that spray application of GA3 at color break in the fall increased juice content of oranges by 2–15 percent. There was considerable interest in the practice in the 1990s but it is rarely used today (2009) due to cost of application and variable results.

Crop Thinning

Some mandarin types, Sunburst, Dancy, and especially Murcott, are prone to alternate bearing. They produce many small fruit in the "on" year and few large fruit in the "off" year. This is undesirable from a price and marketing standpoint. Napthaleneacetic acid (NAA) is sprayed on fruit in the spring about six to eight weeks after bloom or just as natural physiological drop occurs in May. NAA applied in the "on" years reduces fruit number, increases fruit size, and balances alternate bearing. As with other PGRs, the effectiveness of NAA varies seasonally and with cultivar.

IRRIGATION

Irrigation is not always an absolutely necessary operation in Florida citrus groves if trees are widely spaced on deep, well-drained soils. Many groves have been productive with only rainfall, even though periods of water stress have occurred. Nevertheless, with sound management, irrigation usually has increased yields and has been profitable in many instances. In bedded groves or those on shallow soils, where drainage is required during rainy months, irrigation is necessary during droughts because of the limited rooting depth. To understand what factors determine the need for irrigation, it is important to consider plant water use patterns.

Water Requirements

Protoplasm, the living substance in plant (and animal) cells, is able to carry out its functions only when it contains adequate amounts of water. In a dry seed, the protoplasm is alive but almost completely inactive. The first step in germination is absorption of water until the liquefied protoplasm is able to carry on metabolic activities, divide, and grow. Most plant tissues die long before the protoplasm of their cells ever reaches the low moisture content of a dormant seed.

Water has many functions in plant tissues in addition to hydrating protoplasm. Mineral nutrients cannot enter plant cells unless dissolved in water, and organic compounds cannot move from one cell to another except in solution. The universal solvent, water, is one of the raw materials of photosynthesis by which green cells use light energy to manufacture sugars and other necessary metabolites. Mineral elements move as ions from root tips up through trees to the leaves. Tender new shoots and young leaves remain turgid because their cells contain adequate amounts of water. Only when plant cells are turgid can they grow or divide, and only when leaves are fully expanded can they make the most effective use of light. In times of water stress, turgidity is decreased, leaves become wilted, and growth ceases.

The amount of water required for metabolic functions, however, is very small. For terrestrial plants, one of the biggest problems is the constant loss of water from their tissues in a process called "transpiration." The amount of water required for all other plant functions is infinitesimal compared with the amount transpired through stomatal pores (openings) in leaves and fruits of citrus trees. As long as the atmosphere surrounding the plant is not saturated with water vapor, the plant loses water. In general, the drier the air, the faster the rate of water loss. Unless this water can be replaced promptly, leaf tissues begin to dry out and the plant becomes stressed. Practically all the water the trees use is absorbed by their roots from the soil, so the ability of soils to store and supply water is very important.

Soil-Water Relations

In earlier discussions of soil characteristics, it was mentioned that the amount of available water a soil can hold is determined by the difference between the amount present at field capacity and the amount present at the permanent wilting point. A typical Astatula fine sand at the permanent wilting point contains about 1 percent water, at field capacity about 5 percent; it is therefore able to hold (by difference) a maximum of about 4 percent available water. By comparison, soil of the Arredondo series (a phosphatic soil with slightly more organic matter) holds about 2 percent at the wilting point and 8 percent at field capacity, and thus contains 6 percent available water at maximum capacity. Each soil has its own characteristic water-holding capacity.

It appears initially that the soil with the highest percentage of available water would be the one that could supply the most water to a citrus tree. But the total amount of water available to the tree is dependent also upon the volume of soil permeated by its roots, and this means primarily the depth of the soil to either a water table or a hardpan. A deep, well-drained soil with an effective rooting depth of five feet and a maximum of 4 percent available water holds nearly twice as much water as a shallow soil with only one and a half feet to a hardpan with 7 percent available water at field capacity. That is why shallow soils often have greater irrigation requirements than deep, sandy soils. During a drought, the shallow soil must be irrigated more frequently and can absorb and store less water at each irrigation than is the case with a deep soil.

Absorption of water by plants is not the only way that available water is removed from the soil. Evaporation from the soil surface removes considerable water, especially from the top foot of soil (there is little loss by evaporation below this depth). Collectively, evaporation and transpiration are termed "evapotranspiration" (ET). Shallow soils, therefore, lose a larger percentage of their available water through evaporation than deep soils and consequently must be irrigated more frequently than deep soils during dry periods.

Plant-Water Relations

The leaves of a tree with adequate moisture are turgid, but as the water content of the tree decreases, water stress occurs and the leaves gradually lose turgidity and become wilted. The first indications of this condition, called "incipient wilting," are found in early afternoon when transpiration exceeds water absorption from the soil. Incipient wilting is not visible but can be measured using a pressure chamber. This is a specialized piece of equipment that indirectly measures the tension in the xylem (water-conducting tissues). Such a slight moisture deficit in the leaves occurs almost daily, except in rainy weather, and seems to have no adverse effect on growth or yield since trees recover from such deficits at night when transpiration decreases.

As the moisture deficiency within the tree becomes greater resulting from a succession of days when transpiration exceeds absorption, the leaves begin to show a visible temporary wilting for a short but increasing interval each afternoon. During this time, the tree adjusts to the decrease in soil moisture by stomatal closure which decreases transpiration. Stomates are pores located on the underside of citrus leaves which regulate water flow out of the leaf and carbon dioxide flow into it. During photosynthesis, leaf stomates are open and water vapor is inevitably lost from leaf tissues. In addition, the tree can remove water from fruit and the wood, which temporarily reduces stress. Tree trunks and fruit typically shrink during the day and swell at night, and the change in their diameter over time can be a good indicator of tree stress. Soil moisture levels eventually reach the permanent wilting point and the leaves will remain wilted permanently unless water is added to the soil. This condition is seldom reached in Florida for mature trees but growth and yields can be decreased if the soil moisture is not replenished by rain or irrigation.

Irrigation Requirements

About once in every 10 years, Florida experiences a drought of serious economic consequences. About one year in three, the spring rainfall pattern does not provide sufficient soil moisture for adequate spring flush growth and flowering even on the best sites. The critical period for irrigation is usually February to June, during fruit set and rapid cell division. Most growers who irrigate keep a close watch on trees and weather records during this period. Sometimes, growers have waited until leaves are badly wilted before starting irrigation, which can cause extensive flower or fruit drop.

Fall irrigation is usually done less frequently than in the spring due to lower temperatures and evapotranspiration. In addition, excessive irrigation in fall may reduce total soluble solids in the fruit. On the other hand, water-stressed trees may experience some leaf loss after harvesting.

Proper irrigation in the spring increases yields by increasing fruit number and possibly size; it is an economically justified operation. Even then, trees should not be over-irrigated due to cost and the fact that water is being wasted. Leaching of fertilizers and other chemicals into groundwater may be yet another problem associated with over-irrigation.

The Irrigation Program

The irrigation program should be developed with proper understanding of water, labor, power, and equipment availabilities and the basic operating conditions. An economic feasibility study should be made. In the past, citrus growers employed various means of water distribution requiring high labor inputs and often simply used flood irrigation in bedded groves. The gasketed, lockjoint, aluminum portable pipe system later came into vogue and was used during the 1940s. Surface and deep-well water sources were generally available; gasoline and electricity were the power sources, and centrifugal and turbine pumps were used. As grove acreage increased, water availability became more of a problem, and labor for the portable irrigation lines was unavailable. Systems that required higher capitalization but far less labor were developed. Volume guns with underground feeder lines that water large areas, and self-propelled mobile guns that travel down the middle and require labor only when being moved from one middle to another were used. Later, out of economic and practical necessity, a system was developed that consisted of a solid-set of permanent risers with impact sprinklers set along each tree row to cover the entire ground area. In the 1960s, drip, trickle, or microsprinkler systems were developed which distributed water through polyvinylchloride (PVC) and polyethylene (PE) lines and discharged water through multiple emitters or single microsprinklers under each tree. As water supplies become more limited, drip and microsprinkler systems are probably the only types that will be allowed because of strict water use regulations mandated by the water management districts.

Microsprinklers are more effective than drippers in most Florida soils because they provide a wider coverage pattern and higher water volumes (5–25 gallons per hour (gph) vs. 1 gph for most drippers). Microsprinklers cover more of the root zone and if properly used cause less leaching of nutrients than drippers. They are also used to apply fertilizer (fertigation) and for cold protection for young trees. All low volume systems are subject to clogging by particulate matter (pebbles, sand, precipitates), or living matter (insects, spiders, bacteria, algae, or fungi). In addition, they are easily damaged by grove personnel or equipment, and routine maintenance is needed.

Flood irrigation is still used in some bedded groves that are located near canals which provide irrigation and drainage. Water is pumped or moved by gravity via a change in slope onto the grove for a few days and then is removed by the same method. Water should not remain standing for more than three days in the summer and seven days in the winter or root damage may occur. At first glance, flooding appears to use very high volumes of water, but since the water is removed, unlike in other citrus growing regions where it percolates downward, the process is actually quite water and energy efficient.

In recent years there has been interest in using the open hydroponic system (OHS) which was developed in South Africa and Spain. This intensively managed drip fertigation system is predicated on the use of carefully scheduled fertigation which applies water and nutrients continuously throughout the day at low levels. The system is used in conjunction with high-density plantings and with a rigorous pruning regime. OHS has proven successful in arid climates where the root zone is confined to the wetted area. It is being tested in Florida and has the potential to reduce water and fertilizer use while increasing yields and controlling growth and flowering. However, it is also costly to establish and operate, and further research is warranted.

IRRIGATION SCHEDULING METHODS

Methods and instruments are available for directly measuring available soil water content, including tensiometers, capacitance probes, neutron probes, resistance blocks, and time domain refractometry. Generally, irrigation is needed when tensiometer readings reach 10–15 centibars in the spring and 15–30 centibars in the fall and winter. A tensiometer consists of a plastic tube filled with water and attached in the soil to a ceramic cup. The other end of the tube is attached to a vacuum gauge. As soil water is taken up by the tree, water exits the tube via the ceramic cup and a reading in centibars registers on the gauge. The higher the centibar reading, the more tightly water is held in the soil. Another technique uses neutron scattering to measure soil moisture, but this instrument is used primarily for research because it is expensive and has a radioactive source. The probe sends out energized neutrons that lose energy as they hit hydrogen ions in the soil. The probe calculates the ratio between the number of energized and nonenergized neutrons and converts this to percentage soil water. Capacitance probes give real-time water use by the tree at several depths. These probes relate the amount of water in the soil to the capacitance values of sensors in the soil and convert these values to soil water content. They are expensive and must be calibrated for variations in soil type. A water budget system is used occasionally that subtracts daily water losses by evapotranspiration, which accounts for the greatest losses of soil moisture, and adds water supplied by rainfall and irrigation. Such an accounting system must allow for the type and depth of the soil (water-holding capacity) in the individual grove. A similar system is based on soil water depletion from a given root zone. The soil water holding capacity and rooting depth must be known. The soil water depletion can be measured or estimated and the amount of water lost to evapotranspiration is added back to the soil via irrigation. Trees are irrigated when soil water depletion reaches 25–33 percent in the spring and 55–66 percent in the fall and winter. Both systems must be diligently managed, but they do take some guesswork out of the irrigation program. Many growers irrigate based on the calendar system. Irrigation is supplied every 7 to 10 days in winter and 5 to 7 days in summer. This is a convenient but not very accurate means of scheduling irrigation. Ongoing research is investigating several tree-based techniques that can be used to evaluate tree water requirements on a limited number of representative trees in a grove.

PRUNING AND TRAINING

Pruning and training—the removal of plant parts in order to develop a desired shape (training) or to remove dead, diseased, or poorly positioned branches to improve yields or fruit quality (pruning)—have been practiced with many kinds of fruit trees for centuries. Many of the early leaders of the Florida citrus industry were farmers with previous experience in growing deciduous fruit trees in the northern United States and Canada. They tried at first to use the same types of pruning and training for evergreen citrus trees that they had learned to use on deciduous apple or peach trees. Gradually they realized that there was no need for training citrus trees since they have naturally strong branch crotches and limbs seldom split. While citrus trees grew well with different types of training, they were apparently about as vigorous and productive with minimum compared to considerable training. Since these practices added to production costs, growers in Florida largely stopped detailed pruning and training. Today, mature, bearing trees are pruned for size control, to improve fruit quality, for rejuvenation, and to remove freeze-damaged limbs.

Pruning and training during the first four years in the field were discussed in chapter 8. In the early bearing years (from the fifth to the tenth year) limited pruning and training is needed except for sprout removal. For trees older than 10 years (and in the case of closely spaced trees, even younger), more pruning may be needed. This is especially true for the high-density plantings described previously. The following pruning practices are fairly standard in mature, bearing groves.

Sprouting

Once a year, as time permits, sprouts should be removed from the rootstock, trunk, and scaffold branches of trees that tend to have many competing branches arise in the main framework area. Failure to remove such sprouts may result in trees with less-than-desirable shape. In addition, if rootstock sprouts are allowed to develop, they often compete with the scion, thereby allowing the rootstock to comprise a major portion of the tree. Since the rootstock is not the desired cultivar and the fruit is usually inedible, judicious removal of rootstock sprouts is important.

Scion shoots should be pruned if they will not become desirable permanent branches. For example, large, upright water sprouts usually are nonproductive. Sprout removal is a horticulturally sound practice, but in most citrus groves it is rarely performed except on young trees, since it is labor-intensive and expensive. Regular pruning of rootstock sprouts, however, must not be ignored and should be conducted regularly as a part of the total production program.

Deadwood Pruning

There are various reasons for the presence of dead twigs and branches in a citrus tree—among them, cold damage, waterlogging, insect and mite infestations, and fungal infections. Simply removing such deadwood without determining and correcting the cause is poor grove management. In addition, hurricanes, freezes, and droughts (which are largely beyond human control) may cause dieback of large as well as small limbs. Even without these external causes, some twigs regularly die because growth of other branches decreases light penetration into the canopy. Others may become weakened and die due to a reallocation of metabolites to developing fruit. Dead twigs are also an important source of melanose (a fungus) inoculum of leaves and fruit. It is not feasible to prune all of the fine twigs to control melanose, but removal of some larger infected branches is quite helpful as a supplement to spraying. This is especially true for grapefruit, which are very susceptible to melanose infection. Furthermore, as larger limbs die back following freezes or fungal infections, decay may move into the trunk or main limbs.

Hedging

Hedging was developed in the 1950s as a remedy for the situation created in many groves where trees had been planted in rows 20–25 feet apart and had grown so large that they had become crowded. Several circular saws (fig. 9.2) are used to hedge tree sides vertically or at a 5 to 15 degree angle so as to leave a clear middle 7 to 9 feet wide between the outer canopies of the trees. This tree shape admits light to the lower part of the canopy, increasing the bearing area and improving fruit color on the lower branches. Hedging may also reduce costs of spraying and picking.

Any one of several hedging systems can be used, depending on how many sides of each tree need trimming and the frequency of the operation. One of the most satisfactory systems is to start hedging when the trees first begin to interfere with equipment movement in the operating middle by opening either each middle or alternate middles to a good operating width. Hedging is done in one direction only, since most contemporary groves are planted in a rectangular pattern as hedgerows.

Many different hedging programs are used commercially. Hedging frequency is related to tree vigor (a combination of many factors) and distance between trees. In general, frequent light pruning (usually annually) is best because the least amount of leaves, wood, and flowers are removed each time. This helps to keep yields at an optimum level and keeps costs relatively low.

For early and midseason cultivars, hedging often coincides with harvest completion, so many groves are hedged just after the crop is picked. Of course, if harvesting is done in winter, the hedged grove with more open canopies is more subject to possible cold damage to new shoots that emerge after hedging. This is especially true in northern citrus-growing regions. Hedging should be completed, however, before bloom if possible, or at least very shortly after fruit set, to minimize yield reductions.

A special case exists for Valencia orange, since there is a crop on the tree much of the year. Growers seem to be equally divided on timing, some choosing to prune immediately after harvest and others electing to prune carefully

Figure 9.2. Commercial citrus hedging and topping machine. Source: Steve Futch, University of Florida.

in late winter prior to harvest. In either case, pruning will have to be light so as not to remove a large portion the current or future crop. Frequent, light pruning is especially important for the late-maturing cultivars.

Topping

Partial removal of tree tops is often necessary to reduce shading and subsequent fruit loss in the lower canopy. Moreover, tall trees are difficult to spray and harvest.

Topping is now a regular part of most pruning programs and is performed at regular intervals of one to four years. Although many configurations are possible, the most widely accepted shape is rooftop-like with the peak at about 15–18 feet and shoulders at 12–16 feet. Trees are also pruned in the shape of a cube, or a tapered Christmas tree in some groves.

Skirting

Raising the bottom of the canopy of trees, called "lifting" or "skirting," is accomplished by pruning the branches that are touching the ground or very close to it. Skirting is usually done mechanically and removes all limbs that are within two to three feet of the soil surface. Skirting prevents fruit from touching the ground which makes them susceptible to soil-borne diseases such as brown rot. Skirting also allows easier and less destructive movement of herbicide booms and mowers under the tree and permits easier access to service micro-irrigation emitters. Skirting may also improve air movement in the tree canopy and reduce disease severity, although this has not been documented scientifically. Skirting can also allow for better water and fertilizer distribution under the tree canopy

Other Pruning Considerations

If, for any reason, large areas of trunk or limbs are exposed to the sun as the result of heavy pruning or of defoliation by cold, storms, or through top-working (which may be illegal under current nursery restrictions), the exposed surfaces may need a coat of whitewash or white latex paint to prevent sunscald. In spite of their thick bark, these surfaces had been previously shaded by foliage and will be damaged by sudden exposure to direct sun. Sunscald can occur in winter as well as summer, but whenever it takes place, the trunk may be injured so seriously that the tree must be removed. Severe heading back of the top so that all foliage is removed is best done in spring when new shoots regrow most quickly, with temporary whitewash protection of the bark in the meantime.

To paint or not to paint pruning cuts is not nearly as important as whether to make clean cuts that heal readily with no projecting stubs. Thousands of small cuts are made yearly that are not painted without adverse effects. Stubs can be hidden by new growth and are potentially dangerous to operators and equipment. While stub removal is horticulturally sound, it is rarely done today due to the high labor costs.

Efficient mechanical harvesting by machines can require that trees are regularly topped to avoid dropping fruit from overly tall trees. Skirting also facilitates access to fruit pick-up machines. Therefore, as interest in mechanical harvesting increases, growers will have to modify their pruning programs.

COLD PROTECTION

As mentioned in earlier chapters, low winter temperature is an important factor limiting citrus production in Florida. Other aspects of low temperature effects on citrus are discussed in chapters 2, 4, and 7.

Citrus growers should initially choose a rootstock and scion combination suited to the general area to avoid cold injury. Trees that are free from nutritional deficiencies and pest or disease damage will usually withstand low temperatures better than trees that are unhealthy for any reason. In the section on weed control and cultivation in chapter 10, the effect of a cover crop on cold damage and the importance of clean cultivation in winter are discussed. Finally, the grower should avoid practices that cause trees to grow late in the fall. Such trees are more easily injured by cold than trees that have stopped growing.

When growers have done all they can by careful site selection and sound cultural practices to help trees avoid low temperature damage, active cold-protection measures may be needed. During times of plentiful labor and cheap fuel, most cold protection was accomplished by heating or "firing" the groves. Other methods included use of wind machines and, more recently, the use of water via overhead and under-tree sprinklers. Currently, most cold protection methods, except micro-irrigation, are not used in Florida because many citrus plantings moved to southern or coastal parts of the state following the severe freezes of the 1980s.

Grove Heaters (Firing)

Until the freeze of 1894–95, citrus growers in Florida relied solely on choosing a favorable location for cold protection. The first grove heating was done during the freeze of 1899, although far more growers then were in favor of covering trees with tents or lath (perhaps with supplementary heat). The cumbersome nature of covers and the favorable effects obtained with wood fires led to extensive grove heating after 1900. Since pine wood was abundant, resinous heartwood of slash pine (called "lighter wood" or "fat pine") was commonly used for many years.

The last abundant pine wood supplies were used up many years ago, and growers turned to tires and diesel fuel as a source of heat. At one time, the black smoke produced by incomplete combustion was considered desirable and beneficial. It is now well recognized that it is heat, radiant and convective, not smoke, that provides cold protection. This serious source of air pollution can thus be avoided. Stack heaters of various designs and pressurized oil systems to assure more complete combustion, gradually replaced less-efficient smudge pots. Central systems became popular because they were clean, more efficient, easy to ignite, and required no labor for refueling. However, the costs of firing, fuel, and labor have become so great (along with environmental concerns) that grove heaters are rarely used today.

Irrigation

Over-row, high-volume sprinklers are sometimes used for cold protection, especially to protect field nurseries or very rare specimen trees. Even though ice forms on twigs, the internal temperature of the plant tissues cannot fall below 32°F as long as continuous and adequate volumes of water are applied. If the temperature falls far below freezing for a long duration, a thick coat of ice may form, and its weight may cause considerable limb breakage. Nursery trees are very flexible, planted densely, and are close to the ground, so that ice usually forms a framework resting on the ground and puts little load on the branches.

With the development of permanent irrigation systems with risers spaced regularly through groves, it seemed logical to adapt the systems for cold protection. Overhead sprinklers were used during the freezes of December 1962 with disastrous results. Applying water at a typical rate of 0.1 inch per hour allowed it to freeze immediately, subjecting the wood and foliage to lower temperatures than would have been experienced with no irrigation. Severe damage occurred to irrigated trees while adjacent nonsprinkled trees often showed little or no damage. Overhead irrigation applied too little water and sprinkler rotation rate was too slow to provide adequate cold protection. In addition, forming too much ice on the tree caused limb breakage or even splitting of the entire trunk. The events of 1962 discouraged the use of overhead sprinkling for cold protection in bearing groves.

Recent experience with under-tree microsprinkler irrigation systems has shown that they can provide some effective cold protection. In this situation, the water is directed to the lower tree trunk or central canopy. Of course, the fruit and part of the upper canopy may be damaged, but the main portion of the trunk can be saved, and such trees usually regrow quite rapidly the following season. In-tree sprinklers have also provided adequate cold protection for moderate-sized trees in Florida and Louisiana during major freezes in the 1980s. Thermometers, thermocouples, or other temperature sensing devices are necessary so that growers know when to begin and end the use of any cold protection practice.

Wind Machines

Wind machines have been used for cold protection in the past. Like oil heaters, they were used for many years in California before being tried in Florida. Wind machines consist of one or two large propellers mounted on a tower that can be rotated in a circle. They are operated by electric, gasoline, or diesel motors. The principle behind wind machine operation is based on the presence of an inversion layer in which colder air is closest to the ground and warmer air is present up to a height of 40–200 feet, above which temperature falls again with height. The height at which this change occurs is called the "inversion point," and the increase in temperature with height up to this point is called a "temperature inversion," since air temperatures are normally colder with increasing altitude. If the inversion is large (that is, if the air temperature 40 feet above the ground is several degrees warmer than the temperature near the ground), then a wind machine effectively mixes this warmer air with the cold air. If there is a breeze or an advective (windy) freeze, or if the night is quite cloudy, there is likely to be little or no inversion and wind machines are ineffective. Wind machines are not widely used in Florida because many of the recent damaging freezes have been advective. In addition, they are often vandalized and become expensive to operate and maintain.

Cold Damage to Fruit and Trees

Citrus fruit is initially damaged when the temperature falls to 26°F and remains there for four or more hours. The first evidence of damage is the presence of ice crystals in the fruit. When the ice has melted, tiny white crystals of hesperidin are formed along the walls of the fruit segments, especially in sweet oranges. These crystals do not indicate that the fruit is harmful to eat but do indicate in most cases that ice crystals have formed. If the freezing ruptures the juice sacs near the stem end, fermentation is likely to start in a few days; then several days later, the tissues at the stem end begin to dry out as water is lost through the peel. The frozen fruit can still be used for processing, although it contains less juice than undamaged fruit. Following severe freeze damage, processing fruits are harvested as soon as possible to reduce losses in juice content and before fruit drop becomes extensive.

The Florida Department of Citrus, in cooperation with the Fruit and Vegetable Inspection Division of the State Department of Agriculture, monitors fruit condition during and following any freeze to determine the extent of damage and to make sure that only sound fruit is sent to the fresh market. Shipping restrictions are also imposed following a severe freeze so that cold damage will be obvious before fruit is shipped. When shipping is resumed, fruit inspectors carefully examine fruit samples packed for shipment to eliminate any that show freeze damage.

Severe freezes kill leaves, twigs, or even large branches. Leaves initially form watersoaked areas, become curled, and then turn brown before dropping. Stems and large limbs usually turn brown and die back. Growers often prune out this dead or damaged wood soon after a freeze, but experience has shown that such pruning should not be attempted until after the spring growth flush has matured. Generally, fertilizer rates should be cut back in proportion to the amount of canopy that is lost and root death that may also occur. Excessive fertilization of severely damaged trees is costly and will not increase regrowth. Similarly, irrigation rates should be reduced in accordance with canopy loss. However, the tree should not be allowed to become water stressed as new leaves emerge.

GROVE REHABILITATION AND TREE RESETTING

To obtain maximum efficiency of a production unit, or grove, it is essential that every tree location be occupied by a healthy, productive tree. Only then can the grower utilize proven production practices with maximum benefit. For example, if a grove is established with 200 trees per acre but actually has only 180 trees due to losses, it can never reach more than 90 percent of its theoretically possible production. A similar loss of production efficiency occurs at any planting density; however, in cases where trees were closely spaced originally, which is becoming a more frequent trend in the industry, some trees may be removed without decreasing the grove's permanent production capacity. Prompt replacement of dead and declining trees means higher average, long-term returns from the grove.

Dead trees are an obvious problem, but so are trees that are marginally productive or declining. If these trees remain in the grove, growing steadily weaker and yielding less fruit each year, the potential production capacity for the grove declines annually even though production costs remain the same or even increase. In short, potential profits decline slowly but steadily.

It is sometimes possible to rehabilitate an unproductive tree. Growers should carefully consider the cost of such remedial work to determine whether it will be economically profitable. Often, restoration will be more expensive in the long run than replacement.

Replacement of a tree should be considered only if the rootstock is unsuitable (see chapter 5) or the tree is affected by an incurable disease such as blight, CTV, or particularly greening. It is better to think not of "tree rehabilitation" but of "grove rehabilitation." This term embraces the practices necessary to maintain the full complement of trees and theoretical productive capacity in the grove. This program should be conducted regularly rather than being deferred until serious losses in production have occurred.

Grove rehabilitation programs are usually much less important in the early life of the grove than in later years, although problems sometimes arise even in the first or second year that may affect subsequent production. Climatic factors may be more damaging to young trees than to older ones, and poor site selection and systemic diseases such as CTV may cause poor early tree growth. Administration of a grove rehabilitation program involves diagnosis of the problem, appraisal of the extent of current damage and the likelihood of additional future damage, selection of an appropriate course of action, and taking that action. Citrus trees may become unproductive, may decline in vigor, or may even die from many diverse causes including climatic factors such as freezes, droughts, hurricanes, floods, lightning; soil conditions, especially improper drainage or excessive salt concentrations; nutritional disorders resulting from mineral deficiency or toxicity; pathogens, including bacterial, fungal, or viral diseases; insects, mites, and nematodes; pests such as rabbits, gophers, termites, or other animals; and even mechanical injuries caused by grove machinery. Any or all of these may result in lowered citrus tree vigor and production.

In some cases, identification of the problem is a very simple matter; in others, it may prove quite difficult. A tree may respond to several very different causes which are manifested as a decline, dieback, or gummosis. Such general symptoms are of little or no diagnostic value since they indicate only that the tree is not functioning normally.

Once the problem has been identified, it becomes possible to appraise current damage and its effect on the productive capacity of the grove as well as to determine future damage to the tree. Such appraisals are necessary as a basis for deciding what corrective measures will be economically feasible. Furthermore, in the course of an appraisal, additional information may be brought out (such as the interrelationship of certain factors) that may indicate the problem is more or less severe than it first appeared.

The final step in the rehabilitation program consists of remedial treatment, if feasible, or the removal of declining or dead trees and resetting of new trees after any underlying causes have been corrected. Declining trees should be taken out just as soon as they fail to return costs of production and when appraisal shows that they cannot be rejuvenated.

Tree removal is often done in fall but may be delayed until winter in the case of an early-bearing cultivar so that a last small crop can be harvested.

The operation can be conducted at any time if the crop is not a factor. Trees are usually removed by a specially designed front-end loader, after which the area is carefully raked to remove roots. The hole is then refilled to its previous level. Some growers use giant clippers to snip the tree off at ground level to limit the disruption of the grove floor and irrigation system. Soil fumigation or fallowing before replanting should be considered, although the economic value of fumigation has not been documented scientifically and there may be environmental problems associated with it.

Resets should be planted in winter or early spring, preferably with the same cultivar already in the block unless the entire block is being converted to another, more profitable cultivar. Resets of various ages, cultivars, and rootstocks may occur within the same grove, but this practice is not advisable.

Resets planted in an old grove should be given the same or even greater attention as young trees planted in a new grove. Sometimes growers fail to pay particular attention to resets, assuming that the operations used for the mature trees will be adequate for the resets. Resets need much more attention, however, because of competition with the older, well-established trees around them. The program for young trees discussed in chapter 8 should be followed for replacements during their first three or four years in the grove. Trees should be cared for independently of the mature tree program. For example, broadcast fertilizer applications adequate for established trees may not provide adequate nutrients in the root zone of resets, and growth may be reduced unless trees are fertilized individually by hand. Although soil moisture may be adequate for the mature trees, more frequent irrigation may be needed by resets because of their much smaller root systems. Again, precision agriculture with site-specific management practices can be a powerful tool for efficiently managing resets.

TEN

Pest, Disease, and Weed Management for the Bearing Grove

PEST MANAGEMENT

Many insects, mites, and diseases can attack citrus. While these pests will vary from one part of the world to another, their presence and the amount of damage they can cause are generally functions of climatic conditions in the area. The climate in Florida often creates pest problems greater than those in many other citrus-producing areas. While a subtropical or tropical climate is necessary for the production of citrus, the high temperature, rainfall, and relative humidity in Florida create excellent conditions for many pests and diseases. There are areas in the tropics where the climate is even more ideal for pests and diseases, but citrus is usually not grown on a large, commercial scale in those locations.

Strategies

There are many approaches to citrus pest management in Florida, but three emerge as most important. Nature has provided ecological systems of checks and balances that maintain some degree of order in the world. In citrus management, this method of not applying sprays and encouraging nature to help with pest control is called a biological control program. Citrus in Florida was once entirely grown under such a program, and in some areas, such programs are still reasonably successful. Biological control programs do best today in isolated areas away from the influence of pesticides used by other growers. Pesticides can upset the delicate ecological balance of a grove that is under biological pest management. Even chemical sprays in nearby groves can upset populations of parasites or predators that might be needed to help maintain biological control. Fruit produced in groves under a biological pest control system is rarely blemish free, and leaves are often damaged, since such a system usually operates with a low pest population at all times. Where fruit with superior external quality, maximum productivity, and tree vigor are not essential, this program may be satisfactory. Further complicating the situation today are two bacterial diseases—citrus canker, which has been a problem several times this century, and citrus greening, which was identified in Florida in 2005. The severity of both diseases is impacted by the presence of insects. Canker can be harbored in the tunnels of citrus leafminers and greening is transmitted by Asian citrus psyllids. Since even low populations of either of these pests could bring serious disease problems which threaten the productivity and vitality of the tree, biological control of pests will likely be practiced less than it has been in the past for other pests.

A second approach to pest control represents the other end of the spectrum from biological control. This is total control of all pests by chemical sprays. Intensive pest control may keep fruit and trees in the best possible condition, but such a program would be costly and inefficient, and it could contribute to environmental pollution. Further complications would also arise because sprays to control one pest often destroy biological control agents that help control other pests. The scenario becomes very complicated and is certainly less than ideal. Control measures for citrus psyllid to combat greening have caused increases in some other pests that previously were under biological control, such as scale insects.

The third approach is a combination of the first two and is usually referred to as "integrated pest management" (IPM). This program uses biological control where possible, supplementing it with pesticides only when biological agents are unable to do the job alone. Another important aspect of this integrated pest management approach is the judicious selection of pesticides with specificity for the target organism and without adverse effects on friendly biological control agents. Most Florida citrus has historically been grown under the integrated pest management program because it combines the most favorable elements of chemical and biological control programs.

Philosophy

Pest management strategies should be developed with several factors in mind, but the overriding consideration must be the intended destination of the fruit in the marketplace. Fruit destined for processing simply does not have to have the high external quality of fruit going into fresh market channels, and pest management programs will differ substantially among groves producing fruit for different types of markets. Some groves producing the highest quality fruit for the gift fruit market may receive four to five (or even more) sprays per year to achieve the highest external quality possible. Other groves that routinely produce fruit destined for processing may receive only one or two sprays each year or, in some cases, none. The foregoing scenario assumes freedom from the threat of canker and/or greening diseases. When these diseases are present, or even in the area, judicious growers will use preventive sprays to lessen the probability of the diseases or vectors attacking their trees. This, in turn, may significantly increase the number of sprays applied.

Not only the intended market determines pest management strategies, even though this is a dominant consideration; current and historical levels of both insect and disease populations must also be considered. Careful inspection and recordkeeping are very important in all pest management operations. Some pest problems may cause extensive tree or fruit damage and thus require routine control measures. Other problems may cause seasonal or cosmetic damage and do not require routine control.

Inspection

Pest and disease control programs should not be undertaken without the benefit of careful inspection. These inspections (often referred to as "scouting") will form the cornerstone of the pest management program. Most growers utilize some sort of inspection program whereby trees and fruit are examined regularly and results recorded on a form designed for the purpose. These inspections and records will aid decision making when control measures are deemed necessary. Subsequent information collected regarding spray application dates, quantity of chemical and carrier volume used, and other pertinent data should become part of the record. Follow-up inspections should then provide data as to the efficacy of the control measure taken. The records accumulated should be valuable not only for pest control data but for regulatory matters as well.

Many growers are utilizing strong, science-based scouting and data analysis programs to refine the inspection process. These types of scouting programs are especially important in monitoring the spread of greening. Sampling in such a program would integrate pest population and disease data with climate databases, tree-growth response curves, and fruit yield and quality data, and would utilize statistical programming to more accurately forecast consequences of pest management actions. In many cases, sampling methods are modified to better facilitate the sort of in-depth procedures described above. A combination of several factors will promote adoption of increasingly sophisticated procedures and programs to interpret these data and influence pest control actions: (1) economics of the citrus production program and its relationship to pest management costs; (2) regulatory concerns over environmental matters, worker safety, and pesticide residue; (3) restrictions on the use of and potential loss of chemicals currently used for citrus pest management; and (4) necessity to determine more accurately the optimum chemical, application rate, timing of application, and conditions for pest management strategies to increase efficacy and minimize potential undesirable side effects.

While the level of sophistication is left to the individual producer, inspection and recordkeeping are important components of the pest management program. Growers are urged to keep at least the fundamental records and invest in the study of the more sophisticated scouting, recordkeeping, and management technology to increase efficiency. Professional scouting services are available and can provide high levels of service.

Spray Application

Spray application may be made in several ways when it is required. The most frequent application method is through the use of air carrier sprayers; in this system a solution of water and appropriate chemicals is injected into a large volume of high-velocity air which carries the material into the tree canopy. The air volume within the canopy is replaced by very fine spray droplets suspended in air, providing excellent coverage. Hydraulic spraying is another application method and is accomplished by spraying a water/chemical mixture directly on the tree through a spray gun, thereby washing the tree down with the spray solution. This was the standard method used for many years in Florida from the 1940s to the 1960s. While very effective when properly done, this method is very slow, uses large amounts of water, may have excessive runoff, and is not economically suitable to large-scale operations.

Aircraft provide other methods for the application of chemicals for pest and disease management. Both fixed-wing aircraft and helicopters are sometimes used, especially when rapid application is important. Although aircraft application is the quickest way to spray, it can be one of the most superficial. Because the volume of material applied is quite small and the application speed of the aircraft is high, most of the spray is applied only to the outer tree canopy areas. Consequently, some pests and diseases are not adequately controlled. Fixed-wing aircraft, in addition, may not be suitable in areas where groves are intermixed with residential areas. Helicopters are slower moving than fixed-wing aircraft and the propellers provide a forceful downdraft that may increase coverage. However, they are noisy and expensive and also may not be suitable near residential areas.

The mechanics of spray application is a science and technology in its own right. A great deal of information is available on the subject from a variety of sources. Since comprehensive details would be beyond the scope of this book, this chapter will offer only a brief discussion of spray application. The best information available is probably found in the current Florida Citrus Pest Management Guide, a publication that is revised annually by the University of Florida IFAS Cooperative Extension Service.

CITRUS PESTS AND DISEASES AND METHODS OF CONTROL

The management of citrus pests and diseases has been greatly complicated by the introduction of canker and greening bacterial diseases. In fact, there may be little resemblance between spray programs in groves free of the disease(s) and groves which have the disease(s) or are threatened by them. Spread of both canker and greening appears to be inevitable and it is quite likely that all citrus in Florida will be exposed to both diseases. Meanwhile, traditional pest management philosophies prevail in many areas and these methods will be discussed first, followed by a section which deals with pest management where canker or greening may be endemic or at least threatening.

Pest Management in Areas without Canker and/or Greening

FUNGAL DISEASE CONTROL

Traditional pest management for Florida citrus in the absence of greening and canker necessarily revolves around disease control. This is because the five major fungal diseases affecting citrus must be controlled by prophylaxis, or prevention. After the organism has infected the plant material, the damage is done and is usually irreversible. Therefore, timing of disease control sprays is of paramount importance. Since each disease has a prime (and slightly different) time of maximum infection in citrus, the diseases are best discussed individually, as that is how they will have to be handled in the spray program.

Scab (Elsinoe fawcetti)

The fungal disease called "scab" causes problems on only a few different types of citrus (plate 1). It is often most serious on Temple, Murcott, and many types of lemons. It also affects certain types of the tangelos and may be a problem on grapefruit, especially in southern and coastal areas of Florida.

Since only young tissues (fruit, leaves, and twigs) are susceptible and peak spore release occurs in early spring, control is essential during the spring-flush period. Sprays may be needed to protect the young fruit when about

⅔ of the flower petals have fallen and again in late April to early May, when disease pressure is heavy. Proper timing and thorough coverage are essential to success.

Melanose (Diaporthe citri)

The fungus associated with melanose spends part of its life cycle in dead citrus wood. Therefore, groves on a regular hedging and topping program often have less trouble with this disease than their nonpruned counterparts. Melanose incidence is also high following severe freezes, which increase the amount of deadwood in the tree.

Unlike scab, melanose affects all citrus cultivars and is often the greatest problem on grapefruit. The disease, which can blemish leaves (plate 2), fruit (plate 3), and twigs, is usually not severe enough to require control on trees where fruit is destined for processing. On fruit grown for the fresh market, however, blemishes on the fruit peel can make it unmarketable, and melanose must be controlled for maximum pack-out.

Infection from melanose spores is most prevalent at warm temperatures (75°–80°F) and abundant moisture (9–12 hours of continuous wetness). Favorable conditions for the development of melanose occur in the spring and coincide with the period of susceptibility for both fruit and leaves. Where melanose is to be controlled, spray application should be made in late April to early May (postbloom spray). Sprays applied earlier to control scab, even though the same materials may be used, will be of little benefit in controlling melanose, especially if disease pressures are heavy.

Greasy Spot (Mycosphaerella citri)

This fungal disease affects leaves (plate 4) and occasionally fruit (greasy spot rind blotch, plate 5) and can affect all types of citrus, although it is historically most severe on grapefruit. Rind blemishes on the fruit can reduce fresh fruit grade, but most problems from greasy spot result from leaf damage, subsequent leaf drop in the fall and winter, and reduced yields the following year.

Warm temperatures and long periods of wetness or very high relative humidity are essential for spore germination. Spores over-winter on leaf litter from the previous season. Optimum conditions for infection occur in late May and June most years in Florida, and this is also a period when summer-flush leaves are expanding and are most susceptible to infection. The fungus sporulates on the underside of the leaf and enters through the stomata (pores for water and carbon dioxide exchange). Once it has entered the leaf, spray controls are ineffective. Since spore germination and infection proceed rather slowly with greasy spot, sprays are best applied in late May–July (summer spray) for maximum effectiveness. In areas of heavy infestations or in south Florida, a second spray in August may be beneficial.

Alternaria Brown Spot

This fungus disease has become quite serious on certain citrus cultivars in the last two decades. Minneola tangelo and Dancy tangerine are most seriously affected, but Orlando tangelo, Murcott (Honey), and Lee tangerines are also sometimes affected. Spores are produced on recently fallen, infected leaves (plate 6) and also infect fruit (plate 7), but little is known about conditions necessary for infection. Spray applications are timed to coincide with emergence of new growth, and several sprays may be necessary to control the disease.

Postbloom Fruit Drop (PFD) (Colletotrichum acutatum)

This disease, which occurs sporadically, became a problem in the 1990s in Florida. PFD affects all cultivars, but severity is related to time of bloom in relation to rainfall. Cultivars that bloom over long periods of time as well as young trees are especially vulnerable.

The disease is manifested as splotches on and distortion of flower petals and is disseminated by rainfall. When flowers are infected, the young fruit often abscises, leaving the calyx (button) on the stem for a protracted period of time (plate 8). Heavy infections can cause substantial reductions in fruit set and subsequent yield. A model has been developed at the University of Florida to aid growers in determining when control measures are necessary. The model considers the number of diseased flowers and rainfall amounts as a guide to predicting severity.

PFD was primarily a problem for limes and navel oranges but recently has been found on several other orange cultivars. Control is achieved by timely application of fungicides during bloom. In some years with protracted bloom periods, control can be very expensive followed by a significant yield reduction in the next crop year.

Foot Rot/Root Rot (Phytophthora spp.)

Foot rot was one of the first diseases to attract attention in Florida. As early as 1876 it was observed to infect sweet orange seedling trees, most of which were grown on poorly drained sites. Resistance to foot rot shown by sour orange trees under the same conditions was an important factor in the change from sweet orange seedling trees to sweet orange budded on sour rootstocks.

As the name implies, this is a root- and trunk- rotting disease caused by *Phytophthora nicotinanae* and *P. palmivora*. These organisms were considered to be fungi for many years but have been reclassified as algae due to their life cycle and cell wall characteristics. Foot rot lesions occur on trunks or roots, usually at or near ground level, which can debilitate or kill susceptible trees (plate 9). In addition, leaves often show midvein yellowing. The use of resistant rootstocks, such as Swingle citrumelo, offers best control, and trees should be planted at the proper depth to ensure that the bud union is considerably above the ground level. Pruning of low branches and, most important, providing good soil drainage also help to lessen severity of the problem. Prolonged wetting of tree trunks should also be avoided. In addition, chemical control measures are available as soil drenches or foliar sprays.

Root rot is caused by the same organisms as foot rot. While the organism is largely confined to roots, leaf symptoms are similar to those of foot rot. In some cases, there is an interaction between *Phytophthora*, particularly *P. palmivora*, and the *Diaprepes* weevil feeding on the roots termed the Phytophthora/Diaprepes (PD) complex. This complex may cause damage even to rootstocks that are normally tolerant of root rot such as Swingle citrumelo and Carrizo citrange. Control measures for root rot are the same as for foot rot. Recently, sampling techniques have been developed to determine the relative levels of *P. nicontianae* in the soil. Samples are collected from March to November and the relative number of propagules per cm^3 is determined to assess disease severity. Threshold levels above 15 to 20 may cause damage, and appropriate chemical controls are then implemented.

Brown Rot (Phytophthora spp.)

Phytophthora brown rot of citrus fruit has become increasingly more serious in recent years. It is most severe in coastal areas but also occurs in other regions in years of high rainfall. Early and midseason cultivars may have rotted fruit on the tree from August to October due to infection by two species of the same genus that causes foot and root rot. Fruit in the lower canopy, especially those touching the ground, are more likely to exhibit infection. Pruning low-hanging branches (skirting) and applying approved fungicides in advance of infection may prove beneficial. Use of microsprinklers instead of overhead sprinklers also reduces spread of the disease in the tree canopy as it reduces splashing of the spores onto the fruit.

Insect and Mite Control

As stated earlier, proper timing for disease control is usually more critical than for insects and mites. The three traditional major spray periods—bloom

(⅔ petal-fall), spring (postbloom), and a summer spray usually applied in June or July—are discussed in the following section. The only other major period of possible concern in some years is the fall. Spraying frequency will vary with year and pest levels.

This is not to say that some pest problems cannot occur at odd times. Flare-ups of certain insects and mites often occur when least expected and must be controlled. Therefore, the value of regular inspection and monitoring of pest levels becomes increasingly obvious.

MAJOR SPRAY PERIODS—TRADITIONAL SPRAY PROGRAMS

For purposes of convenience, it is best to discuss the major spray periods and the insects and mites that are controlled at these times, and to assume that any other problems can be taken care of on an as-needed basis.

Bloom Spray

This spray period is principally for control of scab (on susceptible varieties) and/or postbloom fruit drop (PFD), but inspection may show other problems are present. If so, they may be controlled by including additional chemical(s) in the scab spray. Tank mixing of two or more pesticides may result in some incompatibility problems, so be sure to read specific label instructions carefully.

Spider mites may be a problem in the spring, and some scales, whiteflies, psyllids, and aphids may be present and require control. The postbloom or spring spray may be a better time to control these pests, but psyllids should be controlled early.

Postbloom or Spring Spray

This is the time period for effective melanose control. If disease pressures have not been heavy, and especially if the fruit is to be processed, the melanose spray may not be necessary. Sprays for *Alternaria* brown spot may be needed for susceptible cultivars. Arthropod pests, such as rust mites, scales, whiteflies, aphids, and mealy bugs, may be troublesome and require control. Careful monitoring of all pest levels will help make this judgment more effective.

Summer Spray

The summer spray is very important in Florida because greasy spot may have a serious adverse effect on tree vigor and production. This spray is best timed in late May–July when the new growth flush leaves are expanding to maximize greasy spot control. Spray timing will vary within the state, with south Florida requiring sprays earlier than the more northern areas. Populations of other pests, especially rust mites, also peak at this time of year, so additional chemicals other than those used for greasy spot may be necessary. Careful inspection should dictate the choice of chemicals based upon population pressures. Consideration should also be given to adverse effects of sprays on non-target friendly organisms that could assist with biological control.

Fall Spray

The fall months are inactive times for control of most diseases but often are busy ones for control of mites. Both rust and spider mites often build up in the cool, dry, fall months. Rust mites cause fruit blemishes of special concern to fresh fruit growers, while high populations of spider mites can cause damage to leaves during the fall that, if left unchecked, may defoliate areas within the trees.

The foregoing text is a rather brief discussion of a very complex subject. Many factors should be considered before any spray is applied. Professional consultants, extension agents, and fertilizer and pesticide representatives can help with questions and should be consulted. Local extension offices have many publications on the complex subject of pest management. The remainder of this chapter is devoted to a more detailed discussion of some of the major pests of citrus and other factors to consider in a spray program. The chapter concludes with information relating to spray programs for groves with citrus canker and/or greening or near groves with these problems.

Major Citrus Pests

SCALES, WHITEFLIES, AND MEALYBUGS

This group of closely related insects comprises a larger number of active pests of citrus trees than any other in Florida. Although individual species sometimes may require particular control measures, usually control for one member of this group will act on other members also. Therefore, a single spray application of may be effective for several different pests.

Purple Scale (Lepidosaphes beckii)

This armored scale was for many years the most destructive insect pest of citrus in Florida and used to require regular spraying for its control. The situation has changed greatly since the 1958 introduction of a parasitic wasp (*Aphytis lepidosaphes*). With the wasp's establishment throughout the citrus belt, purple scale populations have been reduced to such low levels that little or no spraying is necessary. Weather conditions appear to have little relation to the effectiveness of *Aphytis* species. If sulfur is used repeatedly, its harmful effect on the wasp will probably make it necessary to include a scalicide in the spray program to control purple scale.

The slender, dark purple male scale is usually found on the upper surface of leaves, while the much larger female scale (just under 1⁄16 inch, club-shaped, and brownish) mostly prefers the lower sides of leaves (plate 10). These scales often congregate in groups, and their sucking of cell sap causes yellow-brown spots on the leaves. Where they have fed on fruit, green spots persist on the fruit even after the rind develops its mature color. The scale also feeds on twigs and even limbs. The long scale, very similar to purple scale in appearance but longer, is often found mixed in these colonies.

Glover (Long) Scale (Lepidosaphes gloverii)

This relative of purple scale, often referred to as "long scale," does the same damage and is controlled similarly. This scale has been in the state for many years, but it occurred in only small numbers. It was thought to be held in check by parasites. Surprisingly enough, as purple scale became less important, Glover scale populations increased. It can be distinguished from purple scale by the longer, narrower armor, particularly of the female (plate 11).

Florida Red Scale (Chrysomphalus aonidum)

Although not distributed as widely over the citrus belt as purple scale, Florida red scale formerly required equal effort for control. Sometimes, especially in the southern part of the state, it superseded purple scale as the major insect pest in some groves. Especially after mild winters, it occurred in unusually heavy infestations (plate 12). The armor of this scale is circular with a prominent central nipple, is less than 1⁄16 inch across, and varies in color from reddish-brown to reddish-purple (plate 13). Control measures are exactly the same as for purple scale. However, due to the activity of parasites (principally another *Aphytis* wasp species [*Aphytis holoxanthus*]), serious problems with this pest no longer occur.

Citrus Snow Scale (Unaspis citri)

The name is derived from the appearance of the clustered white male scales, which look like flakes of snow on the trunks or limbs, where they are mostly found (plate 14). By 1969 snow scale had spread throughout the state, and by 1971 it had become a serious pest; it is still considered today to be the most damaging of the armored scale insects. Snow scale feeds primarily on trunks and limbs, with heavy populations spreading onto the leaves and fruit. In most areas, parasite (another *Aphytis linganensis* wasp) populations have been established, which today help keep this pest in check.

Yellow Scale (Aonidiella citrina)
This pest produces a round armor much like that of the Florida red scale but lighter in color. No eggs are found under the armor, since the females give birth to live young. Like the Florida red scale, it feeds only on leaves and fruit. In recent years, this scale has spread throughout the state and on occasion has required control.

Chaff Scale (Parlatoria pergandii)
These scales are brownish-gray, nearly circular, and often clustered densely, especially on branches, like overlapping pieces of wheat chaff. The most distinguishing feature of this scale is the purple color of the female, eggs, and crawlers. With heavy infestations, the leaves and fruit develop considerable populations of this scale. The persistent green spots left on the fruit by its feeding are particularly serious problems on tangerines intended for the fresh market.

Other Scale Insects
While the above-mentioned scales belong to the armored scale group, whose representatives are major pests in all parts of the citrus world, there are other scale pests that may at times also cause damage.

The cottony-cushion scale (*Icerya purchasi*) is a soft-bodied scale (plate 15). The adult female is conspicuous because of its greatly expanded, fluted, white, cottony abdomen filled with red eggs. It is rarely a serious problem in bearing groves because the Vedalia ladybeetle (introduced to Florida from Australia via California in 1899) keeps the populations low. Control of this scale by the Vedalia ladybeetle is the classic example of biological control.

Other soft scales, whose armor is part of the actual insect body, includes the Caribbean black scale (*Saissetia negleca*), brown soft scale (*Coccus hesperidum*), and green scale (*Coccus viridis*). These rarely appear in damaging numbers unless parasite activity is disrupted, although Caribbean black scale (*Saissetia neglecta*) does require chemical control at times. Ridges on the upper part of the body of the female black scale portray the letter H. Caribbean black scales often are found feeding near stem nodes (plate 16).

CITRUS MEALYBUG (*PLANOCCOCCUS CITRI*)

Usually not a pest requiring control, this insect may at times become numerous enough to cause serious injury, especially on grapefruit. It is most readily recognized by its white masses, enveloped in cottony frass (insect waste), which are seen on tree trunks in very early spring and later between clusters of fruit. In heavy infestations, mealybugs may cause fruit drop and lowering of grade, especially with grapefruit, because of the hard lumps (feeding injury) on the fruit rind, which becomes yellow prematurely. Chemical control is best applied before bloom or immediately thereafter before the crawlers have become established under the calyx (button) of the fruit.

WHITEFLIES (*DIALEURODES CITRI*)

The common citrus whitefly, which lays yellow eggs, and the cloudy-wing whitefly (*Dialeurodes citrifolii*), whose eggs are black, are both serious pests at times. Often they occur together, but the former is more abundant in the northern area and the latter in the central and southern areas of Florida. The woolly whitefly (*Aleurothrixis floccosus*) is widely distributed but rarely troublesome. The name "whitefly" comes from the adult form, which is a small, two-winged, flying, harmless white insect (plate 17). It is not a true fly. The damage is done by the feeding of the larval stages (nymphs; plate 18), which are much like scale crawlers; they suck nutrients from tender leaves, especially those of vigorous sprouts. They are found mostly on the lower surface of leaves, rarely feeding on the fruit. Whitefly larvae, along with aphids, mealybugs, and soft scales, produce a sweet secretion, called "honeydew," which provides nutrients for the sooty mold fungus, often seen as a black film coating leaves or fruit. Besides making fruit unattractive, honeydew delays the natural loss of green color as the fruit matures and is said to serve as protection from various other pests (but this has not been proven).

Natural enemies (especially the friendly fungi) and the sprays for control of scale insects have long since reduced populations to a low level. Control is easily obtained with the same materials used for scales; usually sprays applied for scales also reduce whitefly populations, so these insects rarely require special control.

OTHER INSECT PESTS

Aphids

Aphids feed on succulent growth, especially very young developing shoots. Several species of aphids are found generally throughout the citrus belt. Although winged forms may occasionally occur, aphids are usually observed as colonies of wingless forms densely clustered on tender shoots. The young leaves from which they suck nutrients become curled inward toward the feeding area and remain permanently dwarfed and twisted. The green citrus aphid (*Aphis citricola*) is perhaps the most destructive species (plate 19), but the melon (*Aphis gossypii*) and black citrus aphids (*Toxoptera auranti*) do very similar injury to new growth. All these species are able to transmit the citrus tristeza virus (CTV). The brown citrus aphid (*Toxoptera citricola*), which migrated from Central America to Florida in the 1990s, is especially effective at transmitting CTV and also feeds on tender growth flushes. There is concern that its presence in Florida increases the severity of CTV problems.

Control of aphids is usually unnecessary in mature citrus groves. Populations are usually controlled naturally by ladybeetles, lacewings, and other predators. Trees that produce a succession of succulent growth flushes or trees that have unusually vigorous and succulent growth because of heavy pruning or freeze damage may warrant spray applications. Young trees, which have a high percentage of young versus mature leaves, often require aphid control.

Aphids reproduce rapidly and damage new growth quickly. Chemical control must be anticipated in order to kill the pests before deformation of leaves has occurred, and for ideal control, repeat applications may be required. As vectors of CTV, all aphids, especially the brown citrus aphid, must be considered in areas of active spread of this virus where susceptible scion-rootstock combinations are used. However, it is impractical and expensive to control aphid populations in most situations using chemicals alone.

Plant Bugs

The true bugs found as pests in citrus groves include the citron bug (*Leptoglossus gonagra*), which feeds on the seeds of the wild citron (related to watermelon) that grows on the sand hills of central Florida, and the leaf-footed plant bug (*Leptoglossus phyllopus*) and the stinkbugs (*Nezara viridula*), which feed on the immature pods of crotalarias and other legumes. After their usual food supplies have dwindled in the fall, these bugs often fly to nearby trees and feed on mature citrus fruits, causing them to drop. Since they puncture the rind with their piercing-sucking mouth parts, it is almost impossible to see where feeding has occurred until the rind around the feeding site changes color. The best control measure is to remove the primary plant hosts from the grove in September, before citrus fruits have become attractive to these bugs, or while the bugs are still in an immature, wingless stage.

Grasshoppers
These pests breed in low-lying moist areas or even in citrus groves at times and may develop in large numbers when the grasses they feed on are eliminated by cultivation or are killed by drought or frost in late fall. Grasshoppers then feed on citrus leaves and may cause partial or complete defoliation of a grove in a short time. Damage may be especially severe on young trees. Two species are common in Florida: the large American or bird grasshopper (*Schistocera americana*), which is sometimes a very serious pest, and the even larger but less active lubber grasshopper (*Romalea microptera*), which is only of minor importance. If they are developing in fields adjacent to a grove, chemical control may be necessary after the grove cover crop has been chopped, disced, or mowed. It will be much easier to kill nymphs than adult grasshoppers. When the infestation develops within the grove, however, proper management of the cover crop is very helpful in control. Clean cultivation, where feasible, during the early spring (February to May) and after the middle of August will reduce grasshopper infestations by preventing egg laying in the soil.

Citrus Leafminer (CLM) (Phyllocnistis citrella)
This very small nocturnal moth lays eggs on newly expanding leaves. The eggs hatch into larvae, which tunnel into the leaf tissue leaving characteristic serpentine "mines" as they move just under the cuticle layer of the leaf (plate 20). CLM pupates in the leaf margins, causing them to roll up. Management of this pest requires a seasonal approach (timed to new flushes), which is incorporated into the regular pest management program.

Damage is most severe on young trees in the nursery or newly planted trees in the field. Chemical control of CLM is not often needed for mature trees. This pest first appeared in Florida in 1994 and became fairly widespread, but it is now under a fair degree of natural and introduced biological control *Aegeniapsis* sp., which was introduced from Australia, parasitizes the larvae. Control of CLM is important in areas with canker to prevent the formation of tunnels which are easily infected by the canker bacterium.

Asian Citrus Psyllid (Diaphorina citri)
This insect was first detected in Florida in 1998. Shortly thereafter it was found to be distributed throughout the state, making eradication impossible. Direct damage from the 1/10-inch pest is confined to young and tender leaves (plate 21). The damage, which causes the leaves to twist and curl, resembles that from aphids. As one would expect, damage is often greater on

young trees since they have a greater proportion of young leaves throughout the year. The damage itself may reduce tree growth, but the real threat from the psyllid is that it is the vector of citrus greening disease—a very serious bacterial disease which can affect all species of citrus without regard to rootstock. Citrus greening affects tree productivity and causes stunting, decline, and ultimately, tree death. No totally effective greening control program is known at this time (2009).

Chemical control of psyllids may minimize the risk of trees from becoming infected with greening but this is not guaranteed. Most growers will likely adopt vigorous psyllid control programs to prevent or slow the spread of greening resulting in radical changes to the spray programs as previously outlined. Extensive, widespread psyllid control programs also may upset the balance of pests and predators, causing an increase in levels of other pests such as scales.

Root Weevils

Several species of weevils (often referred to as "beetles") cause serious damage to citrus trees, particularly in the larval stage which feeds upon the roots. The adult stage feeds upon the leaves, but damage is not serious (plate 22). The grayish-brown Fuller rose beetle (*Asynonychus godmani*) has been known for many years in Florida. In 1952 its white, legless larvae were first found feeding on the citrus roots in the Indian River area; since then it has been reported in other citrus-growing areas of the state. The adult causes some injury by feeding on the leaves, making characteristic notches in the margins. The citrus root weevil (plate 23), known as the "blue-green beetle" (*Pachnaeus opalus and P. litus*), is a serious problem and may be found throughout most citrus regions. The sugarcane rootstalk borer weevil (*Diaprepes abbreviatus*) was inadvertently introduced into Florida and found near Apopka (Orange County) in 1964 (plate 24). This weevil has a wide range of host plants, is found in many locations in Florida, and constitutes a serious threat to the citrus industry. Root damage is often difficult to assess until tree vigor is reduced. Often the larvae form channels in large roots. Chemical control is somewhat difficult for adults, with few chemicals registered for control of larvae. Complicating the weevil damage, *Phytophthora* infections may also be present.

MITES

Mites are related to spiders, often spinning webs, and some species feed on citrus fruit, leaves, and stems. However, there are many species that are beneficial and do not feed on citrus. They differ from insects most obviously in having eight legs normally (whereas insects have six), but they are so small that the grower is not likely to try leg counting. By far the most important economically is the citrus rust mite.

Citrus Rust Mite and Pink Citrus Rust Mite
(Phyllocoptruta oleivora and *Aculus pelekassi)*
The citrus rust mite (CRM) was one of the first pests noticed by early citrus growers in Florida because it causes oranges to be russetted (plate 25) and grapefruit, lemons, and limes to show the silvering of rind sometimes referred to as "sharkskin." Citrus rust mites are difficult to detect until peel injury is observed. They are exceedingly small, about 1⁄200 of an inch long. The yellow, conical, segmented body, broadened at the front end, has only four legs and can be detected only with the aid of a good hand lens. Rust mites feed on leaves, fruit, and green twigs and are capable of causing serious injury at high populations. The population peak is usually reached in midsummer, and fruit russetting may occur in the period from postbloom to September. The pink citrus rust mite generally occurs along with the CRM and causes similar damage to fruit, green stems and leaves as CRM.

Citrus Red Mite (Panonychus citri)
Although "citrus red mite" is the approved name for this pest, the name "purple mite" is unfortunately firmly established in Florida. It is one of a group of mites often called "red spiders"; it is about 1⁄70-inch long, rose-red to purple in color, with the typical eight legs except in the first stage of development, when it has only six. The tiny, round, red eggs can be seen with a hand lens along the midrib near the base of leaves (plate 26). This mite feeds on leaves, fruits, and green twigs, but the principal feeding is done on the upper leaf surface.

The citrus red mite has been in Florida for many years, sometimes producing damage that was ascribed to other causes. Damage began to increase in severity in about 1935, and red mite became one of the major citrus pests. In recent years, its importance has declined due to effective chemical control. Red mite feeding causes leaves to become grayish and produces mesophyll collapse, characterized by translucent, or necrotic, areas of the leaf due to death of mesophyll cells. Firing, drying, and browning of leaves while still attached to the twigs and defoliation often occur in the winter and spring as a result of heavy infestations. The mite also feeds on the green wood and fruit; it causes a scratched appearance on the green rind of the fruit, but the final grade is usually not affected.

Texas Citrus Mite (Eutetranychus banksi)

This mite, found for the first time in Florida in 1951 in Brevard County, had become generally distributed over the state and was a major pest by 1966. It is tan to brownish-green with greenish spots along the sides. Like the citrus red mite, it is found chiefly on the upper side of leaves, and its tiny, flat, disk-shaped eggs, light tan in color, are laid along the midrib. It causes the same sort of injury as the citrus red mite and is controlled by the same chemicals.

Six-Spotted Mite (Eotetranychus sexmaculatus)

This long-time pest in Florida is a broad-bodied, yellow mite with six dark spots on its back; it lives in colonies only on the lower surface of leaves and is very widely distributed throughout the citrus groves in Florida. Its appearance in individual groves is sporadic, requiring vigilance on the part of the grower. During the summer, it is rarely found in groves. Like the citrus red mite, this pest increases in numbers rapidly during the winter and spring if not controlled. It prefers grapefruit as hosts but may also be found on other kinds of citrus trees. It seems to prefer low-humidity areas in the grove and is often most abundant in trees next to a road.

Feeding greatly decreases ability of leaves to produce carbohydrates, and leaves may drop in severe cases. Heavily infested leaves are characteristically crinkled, with large yellowed areas. Fruit drop may be an indirect result of decreased leaf metabolism.

Other Mites

The leprosis mite (*Brevipalpus californicus*), which can carry a virus that causes leprosis of branches (Florida scaly bark) and fruit (nail-head rust) of citrus trees, was considered a serious pest in the first quarter of this century but rarely is seen today.

The broad mite (*Brevipalpus phoenicis*) feeds on young fruit and immature leaves. It is sometimes observed causing sharkskin of fruit or leaf rolling, but it rarely causes much damage.

POTENTIALLY DEBILITATING DISEASES

Diseases may be caused by bacteria, fungi, algae, or viruses. Melanose, greasy spot, scab, postbloom fruit drop, Alternaria, and foot/root rot were discussed previously in the section on traditional spray programs. Most of these diseases can be controlled with traditional spray or biological control programs. However, recently the traditional program has had to be modified significantly with the introduction and spread of two very serious bacterial diseases, canker and greening.

Citrus Canker (*Xanthomonas axonopodis*)

Citrus canker was the only citrus bacterial disease of consequence that had been found in Florida until greening was discovered in 2005. Canker was first observed in 1913 following its accidental introduction in 1910. An intensive program of eradication, which involved burning of infected trees, was instituted at once, resulting in complete eradication after several years. This eradication program was the primary factor in establishing the Department of Agriculture's Division of Plant Industry (formerly called the "State Plant Board"). Citrus canker had not been found in Florida since 1926 until it was allegedly discovered in a nursery in 1984. Millions of nursery trees were destroyed before it was discovered that the disease was not canker but a related disease (citrus bacterial spot), caused by a slightly different strain of the canker organism. The disease emerged again in 1986 in several west central Florida counties and was later thought to be eradicated by 1994. Once again the disease was found in late 1995 near Miami, later moving into Broward and Palm Beach counties. By 1997–98, canker was found again in Manatee and Collier counties. It spread slowly throughout many of the southern counties and was widely dispersed following three hurricanes in 2004 and one in 2005. Eradication attempts were discontinued shortly thereafter.

Canker causes lesions of fruit, stems, and leaves and affects most types of citrus—with grapefruit and Mexican limes being most susceptible and tangerines least susceptible. The disease causes leaf drop, shoot dieback, and fruit drop but generally does not kill the tree. Corky lesions develop on the leaves surrounded by a yellow halo (plate 27) and pustules form on stems that are a source of future inoculum (plate 28). Many regions of the world, such as Argentina, have had canker for many years but have developed management strategies to help control the problem.

The canker discoveries of 1995 and 1997 were addressed for over a decade while sporadic outbreaks occurred in many areas of the Florida citrus industry. Trees were removed and burned and rigid quarantines only slowed the spread. Regulated control measures were ended after it became clear that the three hurricanes which moved through the state's citrus areas spread the disease over much of the citrus-producing area of the state. The state/federal control program ended in 2006 and canker control was left up to the growers.

Canker infections or even proximity to groves with canker infection will usually trigger a response from affected growers. Quarantines have been tightened and traditional spray programs changed to control or moderate the effects of canker. The canker spray program usually takes the form of regular and frequent copper sprays—usually three or more each year. These sprays will also often help control fungus diseases, although timing of the sprays for the various diseases is usually less than optimal. The presence of canker may trigger additional sprays for citrus leafminers as well. Since canker has a direct effect on productivity, tree vigor, and fruit marketability, the spray program should emphasize canker control first, with secondary consideration of other pests and diseases.

Citrus Greening (Huanglongbing HLB) (*Candidatus liberibacter*)

Citrus greening disease has joined citrus canker as the second major bacterial disease to affect Florida citrus. The disease was first diagnosed in Florida in 2005, but the pathogen was no doubt present prior to that time. Since greening is transmitted by the Asian citrus psyllid (which was first found in Florida in 1998), it is quite possible that the disease came along with the insects but tree symptoms occurred much later.

Greening is a serious disease problem because it can infect all citrus cultivars and is not related to rootstock selection and tolerance like tristeza virus or foot rot are. The disease reduces productivity and growth, and eventually kills trees which are severely infected.

Symptoms of greening include vein yellowing and blotchy mottling of tree leaves (plate 29). Other symptoms include appearance of yellow shoots, twig dieback, abnormal flowering, and stunting of affected trees. Fruit from affected trees are small and poorly colored, and they may be lopsided. Fruit set is reduced as the disease affects more and more of the tree. Fruit flavors have been described as bitter, salty, medicinal, and sour.

Control of greening is difficult at best. There is no chemical or biological control agent available. Prevention of the disease by controlling psyllids may help reduce the impact of greening, but total psyllid control is economically unfeasible. Rapid removal of infected trees can help to reduce inoculum levels, but there is no guarantee that all trees will become symptomatic as soon as they are infected. Major research efforts are under way (2009) to solve the greening problem. At this time, greening control measures include use of clean budwood and certified nursery trees, psyllid control, removal of alternate hosts for the bacterium such as *Severinia* and *Murraya*, and rapid removal of symptomatic trees as the best strategies for mitigating the impact of this disease. Many growers have implemented rigorous scouting programs to detect greening symptoms and remove affected trees before the disease spreads.

BIOLOGICAL CONTROL OF PESTS

By no means are all of the organisms found on citrus trees harmful; in fact, some of them play an important part in natural, biological control of harmful pests. Often they greatly decrease the need for applying sprays. Therefore, the choice of a material for pest control should be made with due consideration for any possible effect on these beneficial organisms.

Beneficial Insects

Predatory insects contribute greatly to the control of some insect pests—sometimes giving a very high degree of control and sometimes being only partially effective. The Vedalia ladybeetle is so effective in maintaining the population of cottony-cushion scale that spraying is almost never needed in bearing groves. The Chinese and blood-red ladybeetles feed on aphids, and the Australian (Crypt) mealybug ladybeetle feeds on mealybugs, thus greatly decreasing the need for spraying. These ladybeetles have preferred hosts, but several other ladybeetles and various lacewings, mealywings, and thrips feed on a variety of insect and mite pests.

Parasitic insects, whose larvae feed on the host from the inside after hatching from eggs deposited in the host, also help greatly in control of certain insects. As mentioned previously, an outstanding example of biological control is the *Aphytis* wasp, which controls purple scale, and two related species that control Florida red and citrus snow scales. Certain insecticides have proven undesirable for use in spraying citrus trees because they kill natural predators and parasites and increase pest populations more than if no spray had been applied.

The beneficial insects have received increased attention, especially in areas where a regular spray program is not required. These beneficials may be a major factor in controlling insect pests. Furthermore, their activity often reduces or eliminates the use of chemicals that may have detrimental side effects on plants, animals, or humans. Recent biological control successes include citrus black fly and to a lesser extent citrus leafminer. Levels of both pests have been reduced due to introduction of biological control agents. Biological control of psyllids remains as a target for further research. However, biocontrol of an insect that vectors plant diseases has not been successful to date.

Beneficial Mites

A few species of mites that are pest predators are often found on citrus leaves. To a large extent, they live on the remains of dead insects and mites, but some species consume scale eggs and crawlers, and others feed on six-spotted mites. The extent of their influence on biological control is not well understood.

Beneficial Fungi

Many kinds of fungi have long been known to control various insect and mite pests, especially purple scale, mealybugs, whiteflies, aphids, and rust mites. Although some of the most conspicuous of these friendly fungi, once thought to be of great value in pest control, have been found only on dead insects (as is true of the redheaded scale fungus), others are very effective. An outstanding example is the considerable degree of control of whiteflies by the red and yellow *Aschersonia* (friendly fungus), which makes spraying for this pest rarely necessary so long as spraying for scales is done. The high rainfall in Florida and the resultant high humidity, which are so conducive to development of diseases, are also very favorable for growth of beneficial fungi. Typically rust mite populations decline after summer rainfall begins due to parasitism by the fungus, *Hirsutella thompsonii*.

Summarizing the insect and disease control programs used on Florida citrus is a difficult task. The traditional program provides a starting place, with modifications made as necessary where canker and greening are cause for concern. As always, scouting will remain an important grove management practice and will only become more important as growers battle additional pests and diseases.

DISEASES OFTEN MAKE GROVE REHABILITATION NECESSARY

There are many causes for a decline in tree growth and production. The roles of cold weather, soil moisture, fertilization, and pest management on citrus tree growth have been previously discussed. There are some serious diseases, however, that are not controlled by chemicals and that are often responsible for a progressive decline in tree health and productivity. Among the most important of these are viral, bacterial, and nematode infestations as well as problems of unknown causes such as citrus blight.

Citrus Blight

In 1964 the first symptoms of a new and unknown malady appeared in flatwood and marsh areas of Florida. The problem affected trees beginning at approximately five years of age and symptoms included sparse foliage and premature leaf drop, zinc and manganese deficiency patterns on leaves, and small, misshapen fruits. The name "young tree decline" was given to this condition. Unlike the normal tree losses of 1 to 2 percent per year, much larger random losses and clustered losses rapidly became a serious threat to the industry. Initially the problem occurred primarily for oranges budded on rough lemon rootstock, but other rootstocks and cultivars also showed symptoms. Similar symptoms had been present for many years in groves in well-drained areas where the problem was called "sand hill decline." The name "lemon root decline" suggests the high incidence of the problem on rough lemon rootstock. No definite cause has as yet been identified after more than 100 years of research, although soil-related abiotic factors, viruses, and bacteria have been suggested as possible causes. We now recognize the above to be symptoms of a "disease" that apparently occurs on citrus trees of all types and rootstocks. The problem seems most acute, however, for sweet oranges on the more vigorous rootstocks such as rough lemon and Carrizo citrange. The name "citrus blight" is now used to identify this problem.

Scion-rooting, whereby a mound of soil is placed around the tree trunk usually by mechanical means, to eliminate the rootstock as roots developed from the scion has been a practice in some affected groves. Even this practice is ineffective, however, after the trees have become weakened; total tree replacement is the best treatment. Afraid to continue the use of rough lemon rootstock, growers turned to sour orange, Carrizo citrange, Swingle citrumelo, trifoliate orange, Cleopatra mandarin, macrophylla (*Citrus macrophylla*), or other rootstocks, even though these show varying degrees of susceptibility to blight and some have their own unique problems (see chapter 5). As with Milam (a rough lemon type that is tolerant of burrowing nematode), perhaps a variant type eventually may be found that is blight-tolerant. It is hoped that the efforts of numerous scientists and managers will determine the causative agent and its correction; in the meantime, the grove manager must be aware of the problem and plan accordingly.

Root Rot

Root rot caused by the fungus *Clitocybe* often produces a decline similar to that of foot rot (refer to earlier discussion in this chapter). The roots are infected first, and the infection may spread to the base of the trunk. Small mushrooms (fruiting structures) appearing from the trunk are characteristic of the disease, although they do not always develop. The fungus is usually present in pieces of oak or hickory roots remaining in the soil after land clearing, and the disease can be largely prevented if all such roots are thoroughly removed from the soil before planting citrus trees. This is a disease usually found on high, sandy soils (not on low, heavy soils as foot rot is), and it is not related to rootstock selection. Infected trees should be removed as soon as production decreases and the site should be cleaned of root debris prior to resetting. There are no known chemical controls.

A root rot caused by *Diplodia* fungus may also occur in citrus groves and is often the cause of tree decline in heavy, wet soil after flooding by very heavy rains or excessive irrigation. The root bark decays significantly before the tree suddenly goes into a marked wilting, and decline follows rapidly. By this time, the trunk has usually been girdled also. A characteristic feature is the blackening of wood beneath diseased bark. Wilting is more likely to occur during a drought, because even badly diseased roots may still supply sufficient water while it is available. No cure is known for the disease, but it may largely be prevented by careful irrigation and improved drainage. After diseased trees are removed, it is desirable to fumigate the soil before replanting.

Viral Diseases

Viruses and virus-like organisms are very small parasitic agents consisting of nucleic acid (DNA or RNA) and a protein coat and are capable of reproduction only within a living host. Viroids have similar structure but are usually much smaller and lack the protein coat. Both are responsible for many serious diseases of plants and animals. Citrus trees are susceptible to several viral and viroid diseases. Some cause a very rapid decline, while others cause a slow decline or merely a permanent dwarfing. The principal viral diseases of citrus trees are psorosis, citrus tristeza virus (CTV), citrus exocortis viroid (CEV), and cachexia.

Psorosis, also called "scaly bark" in California, was the first recognized viral disease of citrus. Since 1896 a disease characterized by severe trunk and large branch bark scaling had been known; for many years it was thought to be caused by a fungus, though none could be identified. As the scaly patches become larger, the underlying wood and bark turn brown because of gums and resins produced by diseased cells, and eventually the top begins to die back. Usually trees do not show scaly bark symptoms until they are mature and they may decline slowly for many years. This disease, now called "psorosis A," was shown by Fawcett in 1934 to be virally caused. It is most serious on sweet orange and tangerine, but it can be a problem on all types of citrus worldwide, and incidence of the disease is not unique to any particular rootstock or scion. Any combination is susceptible if infected budwood is used.

Several other forms of psorosis—psorosis B, blind-pocket psorosis, and concave gum psorosis—are known. Caused by strains of the same virus, they are much less common than psorosis A. Not all kinds of citrus trees show bark scaling or gum exudation as symptoms of infection, but all do show a characteristic chlorotic pattern in the developing leaves. This pattern is a symmetrical one of light yellow areas, either small flecks or large blotches, which are translucent when the leaves are held up to the light.

There is no known cure for trees infected with psorosis. The virus is apparently transmitted almost entirely by budwood, not by insects or pruning tools, so there is no hazard to other trees from leaving the diseased trees in place. Trees should be removed and replaced when declining yield makes them unprofitable. This disease is easily prevented with careful budwood selection and is rarely found in Florida.

In 1952, at the request of progressive growers, a citrus budwood registration program was inaugurated to prevent the extreme losses that often were caused by using psorosis-infected budwood. Now the Citrus Budwood Registration Bureau of the Division of Plant Industry of the Florida Department of Agriculture and Consumer Services, the program has been expanded to include other viral problems and other types of problems of a propagable nature. This program now provides the industry seed and budwood sources free of many of these disorders. It has added vastly to our knowledge concerning such disorders.

Tristeza (also known as "CTV," the abbreviation for citrus tristeza virus) first came to serious attention in Florida in the 1940s after it had wiped out the Brazilian citrus industry almost completely. It had appeared initially in Argentine citrus groves between 1925 and 1930, where it rapidly killed trees on sour orange rootstock, and had spread to Brazil by 1937. At first the cause was unknown, but eventually it was shown to be a viral infection. The quick decline disease of California was soon found to be due to the same virus. In 1951 the virus was found in Florida, but in a mild form that had not caused major damage. In various areas of the state, CTV has since been found in more virulent forms and has been a source of concern. It is transmitted by budding and grafting and by aphids feeding on citrus, including the green citrus aphid, the cotton aphid, the black citrus aphid, and most important, *Toxoptera citricidus*, the brown citrus aphid. This aphid, which vectored the disease in Brazil, wiping out millions of citrus trees on sour orange rootstock, was discovered in Florida in 1994 and is now widely distributed. The presence of severe CTV strains and the threat of transmission by the brown citrus aphid have completely discouraged the use of sour orange and other susceptible rootstocks in Florida.

In Florida, CTV causes problems only when certain stock/scion combinations are involved. In some citrus areas of the world including Florida, there are stem-pitting strains which can affect all cultivars and rootstocks. Sweet orange and other species used as scions may be infected without any injury or evidence of infection unless they are budded on sour orange or other susceptible rootstocks. Susceptible rootstocks, other than sour orange, include *Citrus macrophylla*, Mexican lime, grapefruit, some tangelos, and shaddocks. The virus creates a barrier where the rootstock and scion tissues join, preventing movement of metabolites from the top of the tree to the roots. Before long, the rootstock dies, followed by the scion. There is no easy way for growers to identify CTV decline in the field until symptoms become acute; indicator plants and an immunological test called ELISA are used in the budwood registration program (see chapter 6). However, since rootstocks other than sour orange used in the commercial citrus belt do not generally manifest the symptoms, a decline limited to sour orange rootstock may be suspected of being caused by CTV if no other cause is obvious. The decline begins with leaf yellowing and progresses through various chlorotic patterns, to leaf abscission and twig dieback. Usually a piece of the inner bark of sour orange rootstock, taken just below the bud union, will show a honeycomb pattern of fine holes, readily visible under a hand lens. However, this is not always a reliable indicator for CTV. No cure is known for CTV, and affected trees should be taken out as soon as identification is positive because the virus may be carried by aphids to other susceptible trees. Use of tolerant rootstocks (see chapter 5) will avoid damage in new plantings.

Exocortis (citrus exocortis viroid, CEV) is a viroid that affects trifoliate orange and some of its hybrids (citranges), Rangpur lime, sweet lime, and citron. Like CTV, this viroid is present in the twigs of many kinds of citrus all over the state. Injury occurs when CEV is transmitted to a susceptible rootstock by use of infected budwood or when CEV-contaminated pruning tools carry the viroid to a susceptible rootstock-scion combination. Susceptible and nonsusceptible rootstocks should not be mixed in a commercial grove in order to avoid the latter type of transmission. Exocortis-free budwood should be used regardless of the type of rootstock. The disease has been of minor importance in the Florida citrus industry only because susceptible rootstocks were used to such a limited extent. However, several of the newer rootstocks and many in experimental plantings are CEV-susceptible.

The symptoms of CEV include bark scaling on a susceptible rootstock below the union and/or a stunting of the scion. Stunting may be the only symptom and is often extremely severe; when bark scaling occurs, stunting is also present. Soon after 1900 there were reports of dwarfing of some sweet orange cultivars on trifoliate rootstock, while other cultivars had normal growth. It is now recognized that CEV was absent from the budwood of the group growing normally. No cure for infected trees is known. The disease should be avoided at planting time by using certified budwood and proper selection of rootstocks and scions.

Cachexia (xyloporosis) is a viroid disease apparently transmitted only by budwood. The viroid occurs without symptoms in tolerant rootstock-scion combinations, but symptoms develop when cachexia-infected budwood is propagated on *Citrus macrophylla*, Rangpur lime, or Palestine sweet lime. Furthermore, Orlando and Minneola tangelos, Murcott, satsuma, Robinson, Osceola, and Lee (and probably Page and Nova) are susceptible as scion cultivars, regardless of the rootstock used. The characteristic symptom found on the trunk of sweet lime rootstock or tangelo scion is the development of small pits in the outer surface of the wood into which little projections from the inner side of the bark fit. This is easily seen when a flap of bark is cut and lifted just below the bud union (sweet lime) or just above it (tangelo).

Decline occurs in infected trees. Symptoms include bark scaling, gum infiltration, reduction in leaf size, chlorotic mineral-deficiency patterns, leaf drop, tree stunting (often with flattened tops), small fruit, and poor yields. Infected trees are an economic loss in a few years and must be replaced. To avoid this problem, cachexia-free budwood (which must, as a nursery requirement, also be free from psorosis) should be used; it is now available for every citrus cultivar propagated in Florida. Sweet lime (once eliminated for use as a rootstock in Florida because of this viroid) along with other susceptible rootstocks can be used with this clean budwood.

Nematodes

Nematodes are unsegmented worms with cylindrical bodies; they are of economic importance in plants and animals. For many years, Florida citrus growers, like other horticulturists, were familiar only with the root knot nematode (to which citrus trees are immune). This outlook radically changed in 1946 when the citrus nematode was reported widely spread in Florida groves and was strongly suspected of causing spreading decline, a mysterious disease that had baffled researchers for years. In 1953, however, it was definitely shown to be another species, the burrowing nematode, which was the cause of spreading decline. The citrus nematode, *Tylenchulus semipenetrans,* was associated with slow decline and was especially a problem in flatwoods groves with limited root systems.

The burrowing nematode, *Radopholus similis,* widely distributed in tropical areas of Asia and the Americas, feeds on a large number of plant species. As the name implies, spreading decline is characterized by gradual tree decline that slowly but steadily spreads through a grove at an average rate of 50 feet per year from the point or points of its first appearance. The symptoms shown by the above-ground portion of the tree are very general and common to decline from many other causes: a steady decrease in vigor, fruit size, yield, and amount of foliage. But root infestation by the nematode and the progressive spread from tree to adjacent tree are distinctive characteristics. Nematode infection potential is greatest in the subsoil immediately below the topsoil, several inches below the surface because burrowing nematode does not reproduce well at high soil temperatures. Therefore, greatest damage occurs at depths usually below 10 inches.

The nematode moves through the soil in search of new roots after the one it has been feeding on becomes diseased and damaged. Since citrus roots extend out into the soil, the pest can move from the roots of one tree to another. If any of its numerous other hosts (particularly among the weeds) are in the intervening area, they can also become infected and facilitate further spread. The nematode is widely distributed in the soil throughout the citrus area of Florida, but it is only in the deep, well-drained sands, such as on the Ridge, that heaviest damage occurs and the typical symptoms of spreading decline are found.

Control involves preventing introduction to new groves and elimination from infested ones. Citrus nursery stock can be obtained that is certified to be free of nematode infestation. Commercial nursery sites must be certified free from nematodes by the Division of Plant Industry to be approved by the division. In home citrus plantings, there is danger of planting ornamental shrubs or herbaceous perennials that may be infested with nematodes too near citrus trees, thus providing a source of infestation.

Citrus rootstocks that show resistance to burrowing and citrus nematodes have been actively sought, with some success. Milam, a citrus hybrid of unknown parentage, was discovered in 1954 in a Polk County grove that had become infested with burrowing nematodes. The individual tree from which this cultivar originated showed no symptoms of spreading decline. It has been used to a limited extent as a rootstock for over 40 years, but the current number of nursery propagations is quite small. It is particularly recommended as a biological barrier to prevent spread of the nematode. Two other discoveries in infested groves have not had widespread acceptance. Estes, a tolerant rough lemon type, has never become popular, while Ridge Pineapple, though resistant, has (like other sweet seedlings) been unpopular because of foot rot and drought susceptibilities in the past. Carrizo and Kuharske citranges owe some of their current popularity to the fact that they are tolerant, although not resistant, to the burrowing nematode.

Buffer zones used to be established to prevent the spread of the burrowing nematode from one area to another in a grove. Buffers were maintained around the known infested areas, treated with fumigants, and kept free of grass and weeds by using herbicides. This program sought to limit the movement of the nematode through the grove but no preplant fumigants are currently available to control nematodes in Florida.

The citrus nematode, *Tylenchulus semipenetrans,* occurs throughout the world wherever citrus has been grown for any length of time. It was considered a minor problem after its discovery in California in 1912 and in Florida about 1913. Its widespread distribution in Florida in 1946 caused much concern, but because citrus trees can tolerate quite heavy infestations without showing symptoms so long as growing conditions are good, it maintained its status as only a minor pest. There is reliable evidence, however, that the citrus nematode is causing considerably more damage than was thought earlier, even though the symptoms are not as identifiable and devastating as those of burrowing nematode infestations. All rootstocks are susceptible to damage, with the exception of the trifoliate orange and Swingle citumelo, which appear to have resistance. Badly infested trees appear undernourished, with small, slightly yellow leaves, small and often unmarketable fruit, and general stunting and deterioration. For planting new groves or resetting trees in old groves, nursery trees certified free of nematodes should be used. Soil fumigation of reset areas prior to replanting has been used in the past but is not currently recommended. Proper grove management and tree stress reduction help to reduce citrus nematode damage.

Two species of the root-lesion or meadow nematode, *Pratylenchus coffeae and P. brachyurus,* and some species of sting nematode (*Belonolaimus longicaudatus*) have caused damage to citrus in a few cases. These pests are limited in distribution in Florida citrus and may not become a serious problem if nurseries are kept free from infestation.

Nematode monitoring also helps to determine the extent of the problem and can be conducted by contacting the Division of Plant Industry. Growers must be vigilant and well informed through educational programs conducted by the University of Florida, Institute of Food and Agricultural Sciences (UF/IFAS) and by studying relevant publications.

WEED CONTROL AND CULTIVATION

History

Weed control is essential under Florida conditions since weeds compete with citrus trees for water, nutrients, and even sunlight. If weeds are not controlled, they will eventually overgrow the entire grove, making grove operations difficult and costly, and lowering productivity dramatically. In addition, weed control may consume up to 25 percent of the grove production budget.

During the first part of the 20th century, weed control was achieved by hand hoeing. Later, as implements were developed for work with horses, and then with tractors, shallow cultivation was practiced as the principal weed control method. Growers often planted cover crops to add humus to the soil after plants were cultivated.

Recommendations as late as the 1970s for bearing groves suggested doing three or four hoeings per year. All cover crops were to be turned under in the fall to improve organic matter, lessen fire hazard, and increase grove air temperature during freezes. Labor shortages and costs reduced the amount of hand hoeing that could be done, and this was replaced by shallow cultivation using mechanical tree hoes.

Weed control strategies began to change in the 1960s with the advent of chemical herbicides. These materials offered an alternative to mechanical cultivation and provided weed control without tree damage and root pruning which often accompanied mechanical cultivation. Contemporary citriculture depends on chemical herbicides for a major portion of all citrus weed control.

Systems

Three soil management systems have been used in Florida citrus groves: clean cultivation, continuous cover crops, and a combination of these two. The term "cover crop" refers to plants grown as a soil covering to prevent erosion by wind or rain. Various herbaceous crops called "green manures" are also grown in groves and fields for the organic matter they add to the soil when turned under before they reach maturity. In most of central Florida, cover crops are grown for the soil organic matter they add to the soil rather than for erosion control. The sandy soils of Florida are very low in natural organic matter and may benefit from regular additions of it. Adding organic matter improves the water- and nutrient-holding capacity of the soil. Groves planted on raised beds (usually located in coastal and flatwood areas) are usually kept under a permanent sod cover to ensure that the beds do not become eroded. The only vegetation management used in such groves is mowing of the middles and chemical weed control in the tree row.

At one time, clean cultivation throughout the year was advocated by some growers, who thought that no living plants should ever exist in citrus groves except the trees. This resulted in the disappearance of practically all organic matter from the soil, severely reducing its ability to hold water and nutrients. Constant cultivation was expensive and damaged tree roots, and the trees decreased rather than increased in vigor; consequently, this system was abandoned. A new application of the concept of clean cultivation has now emerged in which chemical control methods are utilized, in most cases completely eliminating mechanical cultivation and mowing. Tree growth has been very good under this system, probably because the nonmechanical techniques used do not damage the tree roots as constant cultivation did. This system of clean cultivation using herbicides is often referred to as a "trunk-to-trunk weed-control management system." However, the use of trunk-to-trunk is discouraged in today's systems because of the high rates of herbicides they require and the concern for potential groundwater contamination.

The system most commonly used in Florida citrus groves is to grow cover crops in the row middle which are then mechanically or chemically mowed. Where this system is used in bedded groves, cultivation is usually limited to mowing. Low growing vegetation is encouraged to grow in the summer months; it does not affect tree growth or yields since the rainfall normally is ample for both, and it does provide organic matter (and extra nitrogen, in the case of legumes) for improving the soil. After the rainy season is past, the cover crop is closely mowed to conserve soil water, to eliminate fire hazard from dry plant material, and to prevent possible cold injury due to interference by cover crop plants with radiation of heat from the ground and with flow of cold air. Bedded groves are usually mowed as close to the ground as possible.

Two general types of cover crops are used, which include leguminous and nonleguminous plants. Perennial peanuts are a favored legume and are used by some growers as a cover crop. While it is true that legumes increase soil nitrogen, the amount of nitrogen they add is small; no distinction is made in calculating fertilizer needs of groves with leguminous cover crops and those with nonleguminous cover crops. Moreover, legumes are expensive to propagate, or they require special cultural practices to maximize growth. Some legumes, especially the crotalarias, create a problem in connection with plant bugs, as described earlier in this chapter. Because the disadvantages of using legumes as a cover crop often outweigh the small advantages, many growers prefer not to plant a cover crop and simply allow the native or naturalized weeds to cover the ground. They calculate that the amount of organic matter produced for a given expenditure is the more important measure of a cover crop's worth. The soil vegetation cover will consist of various grasses, especially the pink-plumed Natal grass but even including sandspurs, Spanish needle, and other broad-leaved weeds. Because a considerable amount of organic matter is produced at no cost, this system is most widely used.

As has been suggested, the cover crop, while it is standing, may directly affect tree health, quite apart from its contribution to the soil. It may increase the chance of cold injury on frost nights, or encourage the development of certain insect pests, or fuel a fire set by a carelessly tossed match or cigarette or by lightning. It is important to eliminate weeds around buildings and young trees and along fence rows. These operations formerly required labor-intensive hoeing. It is now possible to use recommended herbicides with considerable savings in both money and human energy. Using herbicides may also be easier than hoeing in particular situations within the grove, such as in and along ditches, and in spot eradication of serious weeds.

The contemporary citrus grower now uses a combination of chemical and mechanical methods in a weed control program. With bedded groves, tree rows are usually cleaned year-round with appropriate chemicals, and middles and ditches are mowed as needed. Nonbedded groves, especially those in colder locations, are maintained in a similar manner but are usually clean-cultivated during the winter months to provide cold protection and reduce fire hazard.

Strategies

Weed control depends first upon proper identification of weed species present in the grove and a knowledge of their biology. Some weeds are innocuous and provide little competition for citrus trees while others are very competitive. Clearly, some weed species will need more attention than others.

IDENTIFICATION

Basically, three types of weeds need to be considered in a control program: broadleaf weeds (including vines), grasses, and sedges. Within these groups, there may be further subdivisions based upon life cycle: perennial, annual, or biennial. Perennial plants are long-lived whereas annual and biennial plants complete their life cycle in one and two years, respectively. The longevity of the plant and methods of reproduction can figure prominently in weed management programs. For example, it would not be logical or cost effective to control an annual weed that is near the end of its life cycle. However, it would be important to quickly control a weed that was about to go into seed production so as to reduce reinfestation. Another example is a weed that reproduces by underground rhizomes or tubers; to attempt to control such a weed by cultivation would be a poor decision because it would likely increase the population by cutting and redistributing the plant.

For purposes of identification, a monitoring program should be established. Just as one scouts for insects, mites, diseases, or nutritional disorders, observations of weed populations and information about species and their growth should be a part of any complete record made for grove management decisions. Weed surveys should be conducted several times each year, especially during periods of rapid growth in the spring, summer, and early fall. Not only should tree rows and middles be monitored, but ditches, roadways, fences, and adjacent properties should be inspected since these may be the site of future problems that may later invade your grove.

Weed species need to be accurately identified and noted as well as their stage of growth and area of cover. Special attention should be paid to particularly troublesome weed species which might represent threats to the citrus trees or may be difficult to control. Identification should be made from several leaves or plant parts taken from several plants in the area to guard against misidentification. Commonly found weed species in Florida and their scientific names are given in table 10.1.

All survey observations are best made on a form prepared specially for the task. A grove plat is also important so that it can be used to pinpoint areas of concern within the planting. The survey forms and marked plats should become a part of the grove records and are fundamental items to be considered when management decisions are made.

CULTIVATION

Cultivation is an effective method of controlling broadleaf weeds, but it is required frequently as new weeds are constantly emerging. In fact, the very

Table 10.1. Weed species that commonly cause problems in Florida citrus groves

	Common name	Scientific name
Grasses	bahiagrass	*Paspalum notatum*
	bermudagrass	*Cynodon dactylon*
	goosegrass	*Eleusine indica*
	guineagrass	*Panicum maximum*
	natalgrass	*Rhynchelytrum repens*
	papragrass	*Panicum purpurascens*
	vaseygrass	*Paspalum urvillei*
	torpedograss	*Panicum repens*
Broadleaf weeds	arrowleaf sida (teaweed, or Mexican tea)	*Sia rhombifolia*
	Brazilian pusley	*Richardia brasiliensis*
	citron melon	*Citrullus vulgaris*
	cudweed	*Gnaphalium pensylvanicum*
	Florida pusley	*Richardia sabra*
	goatweed	*Scoparia dulcis*
	Jerusalem oak	*Chenopodium ambrosioides*
	lambsquarter	*Chenopodium album*
	lantana	*Lantana camara*
	morningglory	*Ipomoea* spp.
	pigweed	*Amaranthus* spp.
	purslane (common)	*Portulaca oleracea*
	ragweed (common)	*Ambrosia artemisifolia*
	spurge	*Euphorbia* spp.
	Virginia pepperweed	*Lepidium virginicum*
	wandering jew	*Tradescantia albiflora*
Vines	balsam apple	*Momordica charantia*
	cats claw	*Bignonia unguis-cati*
	maypop	*Passiflora incarnate*
	milkweed	*Morrenia odorata*
	peppervine	*Ampelopis arborea*
	Virginia creeper	*Parthenocissus quinquefolia*

Source: Dr. D. P. H. Tucker (retired), University of Florida.

act of cultivation often enhances weed propagation by dividing underground reproductive structures of many grasses and bringing buried seeds to the surface, where they can germinate easily. As previously discussed, cultivation is not widely used today; it is principally used for improving grove appearance in most northern, well-drained soils during the fall or cutting weeds that could pose a fire or cold hazard during the winter. Even then, cultivation is basically limited to groves that are not planted on raised beds, since cultivation of beds could contribute to erosion problems and damage shallow roots. Excessive or deep cultivation in any grove can cause tree root damage, loss of soil organic matter through oxidation, or soil compaction and can result in the development of difficult to control weed species in certain situations.

MOWING

Mowing could be used as the total weed management system. However, mowing under trees (especially where microsprinklers are present) is difficult, time-consuming, and expensive. Therefore, most groves are mowed between the rows, and some sort of chemical weed management measures are used within the tree rows. Fortunately, the shading produced by the citrus tree often suppresses some weed species under the canopy, so control is not required as much as it is between the rows.

It is important that mowing be timed to occur before weed seedheads are produced to minimize further infestation. Where rotary mowers are used, caution should be exercised so as to not distribute seeds or plant parts beneath the canopy where they might rapidly infest that area.

Typically, mowing is carried out by a large rotary or flail mower, but chemical mowing has found widespread acceptance. Application of specific chemicals with specialized equipment can retard weed growth in middles and greatly reduce the need for frequent, costly mechanical mowing (especially with high fuel prices). Consequently, many growers are using an integrated program of chemical and mechanical mowing with considerable success. Such a program is often referred to as a "middles management" or a chemical mowing program.

CHEMICAL

The use of weed control chemicals (herbicides) has found widespread acceptance since the 1970s within the citrus industry. It represents an economical tool for the grower to use as a part of the weed management system. Some herbicides are broad spectrum and control many weed species, while others are much more specific. The selectivity of the compounds can be used to advantage in some applications and points to the need for good records, especially proper identification of weed species. An example would be using a selective herbicide to eliminate the taller broadleaf weeds from middles so that the lower growing grasses that remain can be maintained effectively with a chemical mowing program.

Herbicides are classified in a variety of ways. They may be pre-emergence/post-emergence, systemic, or contact. All materials would either be foliar or soil applied. Pre-emergence, as the name implies, means that the major chemical activity occurs as weed seeds germinate and usually before weeds emerge from the ground. Post-emergence chemicals are applied at later stages of plant growth and after the weeds have emerged from the soil. Post-emergence herbicides lack any pre-emergence activity. Systemic chemicals are absorbed by plant roots, leaves, and/or stems and are translocated within the plant or disrupt the photosynthetic system. Foliar- or soil-applied chemicals are terms referring only to the method of application.

Chemical application rates and timing may be difficult to determine since so many variables need to be considered. The weed species, soil type, stage of weed growth, tree age, cultivar, weather patterns, and other variables are all important considerations, as is the choice of chemicals to use. In many cases, a mixture of one or more pre-emergence products is applied in conjunction with a post-emergence contact or systemic herbicide to ensure satisfactory control of a broad spectrum of weed species. Chemical adjuvants are also often added to the herbicide spray tank to increase herbicide uptake and efficacy. Timing is arrived at by maximizing effectiveness in considering as many variables as possible. Under typical Florida conditions, two to three herbicide applications per year may be necessary. The actual number depends on the weed species present, the degree of control desired, the amount of rainfall/irrigation, the choice of chemicals used, and the actual weed pressure present at the time of application.

Improper use of herbicides causes several problems. Over-application may damage citrus trees and contribute to groundwater pollution; also, it is costly. Under-application is also costly because weed species will not be controlled, requiring additional applications.

Obviously, chemical weed control requires a great deal of knowledge, experience, and expertise, and further detailed discussion is beyond the scope of this book. Much advice is available in a wide variety of publications, and professional help is available from consultants, weed- and pest-control professionals, and the University of Florida IFAS Extension Service.

ELEVEN

Harvesting, Maturity, and Grade Standards and Marketing

HARVESTING

Harvesting is a labor-intensive, costly process that can have a significant effect on subsequent fruit quality, especially for fresh fruit. Poorly harvested and handled fruit is susceptible to decay and has a shorter shelf life than properly handled fruit. Some citrus growers have their own harvesting crews, but many opt to harvest on a contract basis to reduce liability and personnel problems.

Prior to 1900, most citrus fruits were pulled from the twigs, often carelessly, and sent to market after a few days of curing in the packinghouse. A great deal of the fruit was injured by plugging (removal of rind tissue when the fruit is pulled from the stem). Such fruit often decayed on its way to market. Because of this problem, the industry turned almost entirely to clipping each fruit from its stem, a slower process, but one that gave a much higher percentage of sound fruits. From 1900 to 1940, fruit clippers were used regularly, but a shortage of labor during World War II brought about a return to pulling oranges and grapefruit for faster harvesting. It was found that by leaving the button (calyx) on the fruit, pulling caused no more fruit injury than usually occurred from clipping or stem punctures. Certain mandarins (tangerines) plug easily when pulled because of their fragile rinds and should be clipped. By the time oranges and grapefruit are ready to harvest, the fruit usually separate readily when pulled properly. Proper harvesting means turning the fruit at right angles to its stem and exerting a sideways motion without pulling straight away from the twig, which may cause plugging even with fully mature fruit.

Another problem associated with harvesting is oleocellosis (also called "oil spotting"). This occurs when fruits are harvested when they are very turgid or wet. As the fruit is placed in the harvesting container, the oil glands in the peel are broken. This causes discoloration of portions of the peel, making the fruit unsuitable for the fresh market. Rough handling of grapefruit may also cause blossom-end clearing where internal bruising and juice leakage from juice vesicles results in a water-soaked appearance at the blossom end of the fruit.

Harvesting Methods

Citrus trees are harvested in two ways: spot-picking, when only certain fruits or portions of a tree are harvested; and clean-picking, when all of the fruits except those that are obviously immature are picked.

SPOT-PICKING

Early in the season, some cultivars may be spot-picked for size, color, or maturity to obtain higher prices for the fruit. Specialty fruits, such as tangerines, Temple oranges, and cultivars for the gift trade, may be spot-picked if labor is available. Oranges, grapefruit, and other cultivars going to the processing plant are clean-picked. Therefore, samples should be harvested knowing that fruit from the top outside and outside portions—especially on the south and southwest sides of the tree—will be the most highly colored and have the highest total soluble solids (TSS) and ratios of TSS to acid. (Fruit quality factors are discussed in the next section.) Fruit from inside portions of the canopy will be less well colored and lower in total soluble solids and ratio than the average for the entire tree. Spot-picking is not widely used except for high value fruits due to labor costs.

CLEAN HARVESTING

In the 1700s and early 1800s citrus fruit were hand-picked into wooden boxes and barrels for shipment to market. As early as 1763, oranges were transported from Florida to England by ship in this manner. In 1873, E. Bean developed the standard, two-compartmented wooden field box which had a volume of 2.23 bushels and held 1 $\frac{3}{5}$ bushels of citrus fruit. Citrus growers are still paid on a standard box equivalent even though these containers have not been used commercially for many years. Most other citrus-growing regions in the world pay for citrus based on weight. Most citrus fruit in Florida are hand-picked into canvas or plastic picking sacks (fig. 11.1) and then dumped into large bins which contain from 9 to 11 standard boxes (fig. 11.2). Bins are loaded onto stripped-down trucks (goats; fig.11.3) with a versatile hydraulic hoist, carried to the roadside, and the bins transferred to flatbed semitrailers (fresh fruit), or the fruit are loaded into large, open trailers which hold

Figure 11.1. Hand harvesting of citrus into a picking sack using a ladder.

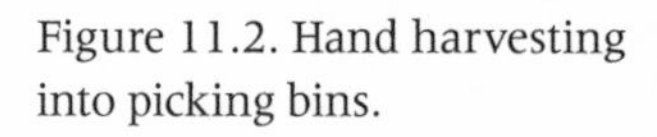

Figure 11.2. Hand harvesting into picking bins.

Figure 11.3. Loading citrus bins into a modified truck called a "goat." Fruit is then transferred to a large trailer.

about 500 boxes of oranges and 400 boxes of grapefruit when full and based on highway weight limits. Fruit is then transported to the packinghouse or processing plant.

MECHANICAL HARVESTING

Mechanical harvesting research began in the late 1950s and early 1960s and was funded by the Florida Department of Citrus and private companies. Citrus fruit, unlike cherries and apples, require a relatively large amount of energy to remove them from the stem, called fruit removal force. Therefore, early research involved finding chemicals that would effectively remove fruit without removing leaves, or in the case of Valencia oranges, small fruit which may be present when mature fruit are harvested. Research was also done on developing effective and economical harvesting machines. Trunk shakers, water cannons, and large, tractor-mounted fans were tested for fruit removal following the application of the abscission-inducing chemicals. The inability to find an effective, consistent chemical; the cost of chemical development; lack of interest by growers due to adequate labor supplies; and most important the severe freezes of the 1980s sidetracked mechanical harvesting efforts by the 1990s.

However, the labor force for harvesting the citrus crop dwindled following the hurricanes of 2004–05. Labor often moved to construction-related industries such as roofing, which paid high salaries. In addition, changes in immigration policies after the September 11, 2001, attacks on the United States further limited the potential labor pool, and labor costs and fringe benefits for harvesting labor have increased over the years. This has led to renewed interest in mechanical harvesting systems by universities, the U.S. Department of Agriculture, commercial companies, and grower organizations. Studies have been conducted along three lines: development of optimum tree sizes, shapes, and spacings to facilitate effective equipment use; discovery and testing of abscission-inducing compounds to reduce the fruit removal force necessary to separate the fruit from a branch; and invention and refinement of equipment to harvest the fruit. Progress has been made with more than 31,000 acres (8 million boxes) of processed citrus fruit being mechanically harvested in 2005, although mechanical harvesting is still not widely used in Florida. Nevertheless, some new plantings have been spaced and pruned to facilitate mechanical harvesting, and a few abscission-inducing chemicals have been cleared for experimental use with the possibility of at least one of them being registered in the near future. Moreover, improved harvesting equipment including trunk and canopy shakers is currently being tested

Figure 11.4. Citrus mechanical harvester. Source: Barbara Hyman, University of Florida.

(fig.11.4). There is also interest in developing harvester assist systems and robotic harvesters. The effort continues today (2009), funded by a Florida Department of Citrus program. Since harvesting costs often exceed production costs, there is little doubt that work will continue. Hand harvesting is still the method of choice not only in Florida but worldwide. As long as harvesting is affordable and suitable labor is available, it will likely continue to be so. However, the economics of the situation dictate that new methods of harvesting will likely be used in the future.

QUALITY STANDARDS (MATURITY)

Knowing when to harvest citrus fruit involves defining fruit maturity. Technically, there is no ripening process in citrus fruit and no such thing as tree-ripe fruit. In bananas, peaches, and apples, as in most other familiar fruits, there is a stage of maturity at which softening takes place rapidly, called the "climacteric" stage. In a few days a hard, starchy fruit ripens and becomes softer, sweeter, and in time overripe. In contrast, citrus fruit are nonclimacteric. They transform from immature to mature and finally to over-mature while on the tree, but the changes are slow and occur over several months. When picked at any developmental stage of maturity, the fruit quality changes slowly after picking unless fruit is infected by fungi or insects and spoils, or it dries out. There is a considerable period of time during which oranges and especially grapefruit have desirable fruit quality, although some mandarins may have a short time between attaining legal maturity and becoming unmarketable due to internal drying.

Citrus growers want not only a high yield but also a crop that will bring a good market price, which is based on both internal and external fruit quality. Internal quality is concerned with juice quantity and flavor, while external quality reflects eye appeal. Although fruit for processing does not need to have the high external quality of fresh fruit, the internal quality is very important.

For homeowners who have only a few trees, it is easy to determine when citrus fruits have sufficiently matured to give the grower a high-quality harvested product. No chemical tests are needed to measure the fruit's maturity, and no regulations assuring that only fruit of acceptable quality is harvested need be considered. Fruit can simply be sampled at intervals until the crop has reached prime eating condition. There is also no pressure for getting the crop off the trees, as there is for climacteric crops like peaches.

In the past, too many growers were willing to ship immature fruit to distant markets, and as early as 1911 legislation was initiated to protect the reputation of Florida citrus by implementing uniform fruit maturity standards. As the Florida citrus industry has increased in size and complexity, more and more laws and regulations have been necessary to control the operations of growers, packers, and shippers for the best interests of the whole industry. These regulations assure the marketing only of a quality product that the consumer can buy with confidence. These regulations pertain to fruit maturity, sizes, and grades that can be handled commercially. They are rather extensive, since it is necessary to establish different standards for each of the different kinds of citrus fruits. Each shipment of fresh fruit and each lot of fruit delivered to the cannery or concentrate plant is inspected to see that it meets the applicable legal and regulatory standards.

What constitutes maturity in citrus fruit is defined by state law in each citrus-growing state. In addition, federal statutes also apply to interstate commerce. Florida citrus fruit cannot be shipped commercially until it is mature as defined by Florida laws. The Florida Citrus Commission of the Department of Citrus is authorized under these laws to issue regulations for determining maturity, sizes, and grades. However, not all fruit that meets maturity requirements and is properly classified by size and grade can be shipped. To further ensure orderly marketing, the citrus industry operates under voluntary Federal Marketing Order 905. The Citrus Administrative Committee is authorized under this marketing order to issue regulations determining the limits of the various sizes and U.S. grades for interstate and export shipments (subject to the approval of the U.S. Secretary of Agriculture). These regulations are enforced by the Division of Fruit and Vegetables of the Florida Department of Agriculture and Consumer Services. Federal regulations are somewhat flexible, so that requirements can be made more stringent or relaxed somewhat in accordance with the nature of the crop being marketed or due to year-to-year variations in fruit quality factors.

Minimum Maturity Standards

The first attempts to frame a legal definition of a mature citrus fruit were made in 1912. A single fruit character was sought that could be readily measured by physical or chemical methods and used to assure acceptance by the average consumer. Reliance on a single fruit quality factor proved impractical for most citrus fruit. Only lemons and limes can be marketed as fresh fruit based on a single character, the percentage of juice by volume (limes), or fruit diameter (lemons). These factors are not related to degree of maturity or edibility. Lemons and limes develop their maximum acidity early in the season while they are quite immature. They may be picked as soon as they meet minimum size and juice requirements, which are dictated primarily by market demand.

Lime fruit are usually picked when immature. The size requirement is to assure adequate juice content and may change at times, but usually no Tahiti limes less than 1 ¾ inches to 1 ⅞ in diameter may be shipped depending on the time of year. The general juice requirement is 42 percent by volume, with a size restriction at or slightly below 1 ¾ inches as set by marketing agreement regulations each season. There is no size requirement for Key limes, but they also must contain 42 percent juice by volume. These minimum requirements are currently irrelevant because most of the Florida lime acreage was eliminated in the early 2000s due to citrus canker.

No single character has been found to measure consumer acceptance of the low-acid citrus fruit; a group of five interrelated characters is therefore used. These are color break, juice content, total soluble solids (TSS) content, acidity, and TSS-to-acid ratio (see tables 11.1 and 11.2).

Table 11.1. Factors used in minimum requirements for quality of Florida fresh citrus fruit as of May 2007

Fruit	Color break	Juice content	Brix	Acid	Brix/acid ratio	
					Required[a]	Minimum
Oranges	Yes	Yes[b]	Yes	Yes	Yes	Yes
Grapefruit	Yes	Yes[c]	Yes[d]	No	Yes	Yes
Tangerines	Yes	No	Yes	No[e]	Yes	Yes
Temples	Yes	No	Yes	Yes	Yes	Yes
Tangelos	Yes	No	Yes	Yes	Yes	Yes
Lemons[f]	No	Yes[g]	No	No	No	No
Limes[h]	No	Yes[g]	No	No	No	No

Source: Citrus Reference Book, Florida Department of Citrus, May 2007.

a. For appropriate Brix.
b. Gallons per 1 3/5-bushel box.
c. Cubic centimeters per fruit.
d. Separate standards for seeded, white seedless, and pink and red seedless cultivars.
e. Honey tangerines maximum acid 1 percent or minimum ratio of 12:1.
f. No specific Florida standards.
g. Volume basis.
h. Restrictions on size of fruit.

Table 11.2. Factors used in minimum requirements for quality of Florida processing fruit as of May 2007[a]

Fruit	Color break	Juice content	Brix	Acid	Brix/acid ratio	
					Required[b]	Minimum
Oranges	No	No	Yes	No	Yes	Yes
Grapefruit	No	No	Yes	No	No	Yes
Tangerines	No[b]	No[b]	Yes	No[b]	Yes	Yes
Temples	No[b]	No[b]	Yes	No[b]	Yes	Yes
Tangelos	No[b]	No[b]	No	No[b]	No	Yes
Lemons[f]	No	No	No	No	No	No
Limes[h]	No	No	No	No	No	No

Source: Citrus Reference Book, Florida Department of Citrus, May 2007.

Notes: a. From August 1 through November 30, processing fruit must meet fresh-fruit standards.
b. December 1 through July 31, no juice, minimum acid, or color break requirements.

COLOR BREAK

Immature citrus fruits are green because their rinds contain large amounts of chlorophyll. When temperatures fall below 55°F, chlorophyll levels begin to decrease as the citrus fruits begin to mature so that the orange pigments (carotenoids) which are also present in green fruit (and leaves) become expressed. Eventually, the green color disappears entirely in fully matured fruit and is replaced by yellow in lemons and grapefruit, orange in oranges, and orange-red in some mandarins and tangelos. This breakdown of chlorophyll takes place only when the temperature is relatively low, as in fall and winter in Florida. In the lowland tropics, oranges remain green when fully mature internally. Moreover, Valencia oranges and grapefruit that have lost their green color in early spring may regreen if left on the trees into April or later in Florida.

Normally, citrus fruit exhibits a change in rind color along with changes in internal fruit quality factors; and the first indication of change from green to yellow-green, if it is developed naturally, is termed "color break." Florida law requires that citrus fruits (other than limes and lemons) show color break to some extent before they can be shipped as fresh fruit. Treatment with ethylene gas, a naturally occurring plant hormone, at 1–5 parts per million and 90–95 percent relative humidity in the packinghouse degreening room may be used to accelerate the breakdown of chlorophyll if the process has already begun naturally. Degreening for 12–72 hours improves peel color and can be used for oranges, grapefruit, tangerines, and Temple oranges. Oranges, grapefruit, tangerines, Temples, and tangelos have a minimum color break requirement all through the season for fresh fruit, but there is no such requirement after November 30 for processing fruit. Because color break requirements are applied after any degreening treatments, fruit are rarely disqualified because of this factor.

Temperature is not the only factor involved in color break, for light also plays a part. Fruits in the outside canopy will show color break earlier than those in the more shaded parts of the tree. Early in the season, it is therefore sometimes necessary to spot-pick exposed fruit. Sometimes, especially in a warm fall, the peel of shaded fruit may remain quite green after the fruit becomes satisfactorily mature internally. Satsuma mandarins, some tangerines, and tangerine hybrids, which are more marketable when they have developed a deep red or orange color, may become somewhat overmature before shaded fruit show much peel color. Parson Brown, Hamlin, and some navel orange cultivars may also be fully mature internally when still exhibiting some green rind color. Warm weather in the fall causes slow development of color break and less intense color when the fruit is fully matured.

The grower may also influence the development of color break by cultural practices. For example, oil sprays after mid-July or excessive nitrogen fertilization may delay color break, while properly balanced nitrogen with potash fertilization may hasten it slightly.

The brightest citrus fruit rind color develops under high sunlight, low humidity, and low temperature during the maturation period. Climatic conditions in Florida normally do not produce intense peel color for oranges, although internal fruit quality may be excellent. Therefore, after degreening, some oranges, Temples, and tangelos are exposed to a dye solution to enhance their eye appeal (color-added). This process is rigidly supervised to prevent inferior quality fruit from being made to appear properly mature. Later in the season, as natural color develops in the rind, these fruits are usually shipped without artificial coloring. Color-added fruit must be labeled as such and they represent only a small percentage of the total fresh fruit packed in Florida.

JUICE CONTENT

Citrus fruit are essentially all peel tissue early in development. Juice sacs then develop and become filled with juice as the fruit matures. Laws regulating minimum juice content were first introduced in 1933 to prevent dry grapefruit from reaching the market. High-quality fruit have a sufficient amount of readily released juice, and maturity laws specify the minimum quantities that must be present. Juice content is easily determined, but minimum levels are presented in different ways for different fruits. The required juice content of oranges is stated as gallons of juice that can be squeezed from the fruit in a standard Florida field (90–pound) box, with the minimum requirement of 4 ½ gallons per box. Juice content of grapefruit is given as cubic centimeters per fruit, with a minimum quantity for each standard size. An alternative juice measurement for grapefruit is based on percentage by weight. The minimum juice requirement changes during the season, with stricter standards early in the season to prevent dry fruit from reaching the market. No minimum juice content is needed for mandarins, Temples, tangelos, or other tangerine hybrids because they typically have a high juice percentage. As mentioned previously, the juice requirement for limes and lemons is given as a percentage of their volume.

The high juice content of Florida citrus fruit, especially of oranges, contributes to Florida's strong processing industry. Early or late in the season, some fruit may have less than the minimum, but as a general rule Florida citrus fruit runs far above the minimum requirement for juice. Sometimes pink and red seedless grapefruit have difficulty in meeting the requirement early in the season.

TOTAL SOLUBLE SOLIDS CONTENT

The earliest minimum maturity standards in Florida were based on percentage of sugar, the direct measurement of which required considerable time and elaborate chemical laboratory equipment. Later it was found that sugars constituted about 75–80 percent of the total weight of the compounds, organic and inorganic, dissolved in the juice of nonacid citrus. These collectively are called "total soluble solids (TSS)." The percentage of TSS can be determined by measuring the density of the solution, and thus the sugar content is easily determined indirectly. There is a steady increase in TSS, and proportionally of sugars, throughout the season until the fruit is well past acceptable maturity. In addition to sugars, the TSS includes certain organic acids (largely citric acid), vitamins, potassium, and very small amounts of inorganic salts and organic compounds that are responsible for the flavors characteristic of the various citrus fruit. The off-flavors that sometimes develop in fruit because of improper handling or overmaturity are usually due to changes in these flavor components. Microorganisms may also cause off-flavors to occur.

Total soluble solids can be measured in two ways. The hydrometer measures juice density in degrees Brix, which is converted by tables, after correction for temperature, into percent TSS. The Brix scale measures the weight of sugar (pure sucrose) per volume of solution for a certain temperature. "Brix" is always capitalized because the technique is named after the German chemist, A. F. W. Brix. Total soluble solids may also be measured using a refractometer, which measures the diffraction of light through a prism. The amount of diffraction is then read on a scale and is proportional to the level of TSS in the juice. This value is also affected by the amount of acidity in the juice and the TSS must be corrected for acid content using standardized tables. Refractometer measurements are easier to make than those from the officially recognized hydrometer method, but they usually result in slightly higher values than those obtained from the hydrometer. Although late in the season of any cultivar the values may be higher, the following figures give the usual range of TSS during the normal harvesting season for the important citrus fruits: oranges, tangerines, Temples, and tangelos, 9–14 percent; Honey tangerines (Murcotts), 12–16 percent; grapefruit, 7–12 percent. Minimum values of TSS are required as part of the maturity standards.

ACID CONTENT

Citric acid is the organic acid chiefly found in citrus fruits, but small amounts of malic, succinic, and tartaric acids are also present. Acid percentage is highest in immature fruit, decreasing steadily as the fruit matures. There is a gradual change in taste from sour to pleasantly tart to sweet. Very late in the season, fruit becomes insipidly sweet due to very low acid levels.

There is no maximum permissible acidity specified for citrus fruit, but the percentage must always be determined in order to calculate the ratio of TSS to acid. Minimum acid content required for fresh oranges and tangelos is 0.4 percent. There is no minimum for other citrus fruit. The acidity is determined by titration of the juice with a standard solution of sodium hydroxide; the equivalent acidity as anhydrous citric acid can be calculated easily from the quantity of base used. The amounts usually found in nonacid citrus fruit during their normal harvesting season range from 2.0 percent early in the season to 0.5 percent late in the season. Limes and lemons, which are highly acidic, have 4–7 percent acid.

RATIO OF TOTAL SOLUBLE SOLIDS TO ACIDS

The flavors of the different citrus fruits are primarily due to small amounts of certain organic compounds, as explained previously; the taste (sweet or sour) is dependent on the relative amounts of sugar (determined as TSS) and acid in the juice. Extensive research into consumer preferences for several citrus fruits using taste panels has shown that no particular sugar content alone will assure acceptable taste. The average consumer wants a pleasantly tart or subacid fruit, and this requires certain proportions of sugar to acid. The percentage of TSS divided by the percentage of total acid gives the ratio of TSS to acid, always stated on the basis of acid as unity, for example, 10:1. Data from thousands of taste tests with fruit of all degrees of maturity made possible the charting of what constituted satisfactory ratios for a wide range of TSS and acids. The present maturity laws were established from these findings.

Early in the season, the TSS are low and acid is high, giving a low ratio. As the season advances, the TSS increase and the acid decreases, giving progressively higher ratios. The laws require that fruits have a certain minimum TSS content and minimum ratios of TSS to acid. A higher ratio is necessary to provide acceptable taste when the sugar content is low than when it is high. Therefore, if fruit exceeds the minimum percent TSS requirement, it is permitted to have a lower ratio. Because the percentage of sugar (and hence of TSS) increases as the fruit becomes more mature, the ratio required by law decreases as the season advances. Oranges with 11 percent TSS, for example, need only have a ratio of 9.1, whereas oranges with 8 percent TSS are required to have a ratio of 10.5:1. Percent acid decreases with increasing maturity so that the same ratio is attained with lower TSS, and the minimum TSS requirement decreases as the season advances. Although naturally colored oranges must have 9.0 percent TSS until the end of October to be shipped, they need have only 8.7 percent TSS from November 1 to November 15 and 8.5 percent TSS after November 15.

Citrus fruits just approaching satisfactory maturity often have an unpleasant, raw taste that disappears as they become fully matured. This unpleasant taste cannot be measured by simple chemical tests. Higher ratios early in the season act to decrease the possibility that immature-tasting fruit will reach the market. On the other hand, fruit with very high sugar content at the end of the season may be insipid. While there is no maximum ratio, the minimum acid requirement effectively limits ratios at around 20.5:1. Ratios this high are not preferred by processors, who must blend this juice with lower-ratio juice to produce a marketable product.

A certain ratio is not in itself a guarantee of quality unless it is tied to a TSS percentage. For example, oranges with 6.0 percent TSS (a very low value for oranges) but only 0.5 percent acid would have a 12:1 ratio but be of very poor eating quality.

Maturity Testing

Maturity testing is no more accurate than the sample that is taken from a group of trees or lot of fruit. To be useful, a sample must be representative; that is, it must reflect as nearly as possible actual conditions on the trees or among the lot brought to the packinghouse or processing plant. Some years ago, all of the fruits on a large, productive Valencia orange tree at the Lake Alfred Citrus Research Center were picked and tested individually. This study showed that fruits on the south side of a tree were higher in TSS and had a higher TSS-to-acid ratio than those on the north side of the tree. There was an increase in TSS and TSS-to-acid ratio as fruits were picked from successively greater heights. Fruits on the outside portion of the tree had superior peel color and higher TSS and TSS-to-acid ratio than those partially shaded in the canopy or fully shaded near the trunk. Fruits in the top of the tree, whether in full sun or partially shaded, had superior peel color and were higher in TSS and TSS-to-acid ratio than either outside or interior canopy fruits. Total acid tended to be lowest in fruits from the northeast side of the tree and to increase for inside fruits picked from successively greater heights. A systematic pattern of juice content was not observed, although large, coarse-textured, poorly colored inside fruits tended to have a lower percentage juice. A sample taken from the outside on all sides of the tree at a height of three to six feet is considered representative of the entire tree, and this technique also is the most convenient for the person sampling the trees. However, this sampling method may be inaccurate for large trees with a high percentage of the fruit in the top of the tree.

Enough samples should be taken to ensure that variations within a grove are included; 20–50 fruits properly chosen will be more representative than several times that number casually picked from the trees. Systematic sampling will help to determine whether the fruit should be harvested at any particular moment. When clean-picking, or when checking the changes in internal quality is the objective of testing, samples of outside fruit taken from all sides of the trees at a height of three to six feet from the ground should be representative of the entire crop as discussed above. When maturity is marginal and any error may mean that the harvested fruit must be destroyed, it is wise to weight the sample by deliberately picking fruit low in the canopy and reaching into the canopy. A sample representative of the block should be taken from the trees, taking into consideration cultivar, rootstock, soil type, drainage, cultural factors, and fruit size among the various blocks. All these factors may affect fruit quality and thus sample validity.

FRUIT SIZES AND CONTAINERS

The market sizes of citrus fruits correspond to the number of uniform fruits that can be packed by set patterns into a standard 4/5-bushel shipping carton. Each size is defined by the maximum and minimum diameters of the fruit in it, since there is a small range of diameters for each size. Oranges, grapefruit, and most tangerines, which are more or less spherical, are easily and readily sized in the packinghouse using a roller system. The fruit move along slightly tapered rollers until they reach an opening equal to their diameter and drop between the rollers onto a belt or into a bin with others of the same size. Some packinghouses separate sizes using different hole sizes in the packing line. More modern packinghouses have singulators, which sort fruits individually by diameter or weight into sizes as determined by high-speed computers. Limes and lemons are longer than they are wide and are not so easily sized by rollers; more often they are sized by weight as they move over a series of counterbalanced trapdoors, the heaviest fruit dropping out first, or photometrically by color for fresh fruit.

In citrus fruits, there is no relation between size and grade as there is, for example, in apples. However, certain sizes are more popular with consumers than others, especially fruit in the middle size ranges. Extra large sizes may be less juicy than average, are more difficult to squeeze for juice, and cost more per fruit; small fruit are usually higher in juice and sugar than average, but more of them must be squeezed to get the same volume of juice. Medium sizes are also preferred for fresh fruit except for gift fruit, where larger sizes—especially of grapefruit and navel oranges—are preferred. The preference for medium-sized fruit is well known to fruit handlers and buyers and is recognized by the marketing agreement committees, which may restrict the sizes as well as the grades of any citrus fruit. When the fruit size is small within a given season, more small sizes will be permitted to go to market than when the bulk of the fruit is medium or large size.

Legal market sizes vary for the different kinds of citrus fruits and are defined in U.S. standards as regulated and amended by the Florida Department of Citrus. The standard nailed 1 ⅗ -bushel box, for which sizes were originally established, was formerly used almost exclusively for shipping citrus fruits but is no longer in use. Instead, commercial shipping sizes are based on the number of fruit in a ⅘-bushel carton as follows: 48, 64, 80, 100, 125, and 163 for oranges, Temples, and tangelos; and 18, 23, 32, 40, 48, and 56 for grapefruit. Sizes for tangerines are 100, 120, 150, 176, 210, 246, and 294. The Citrus Administrative Committee may set further limits on the shipment of small fruit within a given season. Cartons containing ⅖ bushel (based on weight of each fruit size) and 3, 5, 8, 10, and 20–pound polyethylene and plastic mesh bags are also widely used for shipping. Fruit in bags is sold by weight; hence, bags may include more than one fruit size. Limes are not packed by standard sizes, although fruit in any one container must be uniform in size within a range of ¼-inch diameter.

GRADES

United States standards for grades of Florida citrus fruits are defined by the Consumer and Marketing Service of the U.S. Department of Agriculture. Florida grade standards, which apply to citrus fruit shipped within the state only, are similar in most respects to the federal standards. They are established under the Florida Citrus Commission Regulations and the several federal marketing agreements issued by the U.S. Secretary of Agriculture. Also considered are the recommendations of the Growers Administrative Committee that specify what grades and sizes of citrus fruits may be shipped outside of the production area (south of Georgia and east of the Suwannee River). These regulations are established at the beginning of each shipping season and are revised or amended as crop and market conditions warrant during the season. Citrus fruit grades are based mostly on external appearance and represent the degree of eye appeal to the consumer. However, no fruit may be placed in a grade unless it meets minimum maturity requirements.

Four fresh fruit grades are commonly recognized: U.S. Fancy, U.S. No. 1, U.S. No. 2, and U.S. No. 3. Fruit that has not yet been graded is termed "Unclassified." The U.S. Fancy grade is seldom packed since it represents fruit with near flawless external appearance, which is very difficult to produce under Florida's humid conditions. Most citrus fruit is packed as U.S. No. 1, and growers use stringent production practices to achieve this grade category. Occasionally U.S. No. 2 fruit were shipped in the past but this is not common practice today. Fruit that are U.S. No. 3 are rarely shipped to fresh fruit markets but are sent instead to processing plants.

To be classified in any particular grade, citrus fruit must be mature, similar in cultivar characteristics, and free from unhealed cuts and bruises. The Fancy grade requires that fruit be well colored, well shaped, firm, fairly smooth in rind texture, and free from rind blemishes due to pests or mechanical causes (except that up to an average of 10 percent of the surface of any lot of fruit may be discolored by rust mites). These represent external characteristics of the fruit that the purchaser considers as highest quality. The various lower grades specify the extent to which fruit may progressively vary from these characteristics.

The U.S. No. 1 and U.S. No. 2 grades are further divided into subclasses based on discoloration due to rust mites feeding on the peel. Some people believe that russetted fruit represents a different cultivar from undamaged fruit, and others believe that it is sweeter than undamaged fruit. As mentioned in chapter 10, rust mites affect only the appearance of the rind and do not have any effect on the internal quality of fruits unless damage is extreme. Russetted fruit is just as edible as brightly colored, undamaged fruit. Some markets even pay a premium for it, although it is usually discounted somewhat for poor external appearance and may not store as long. The subclasses are further separated on the basis of the proportion of the rind that is discolored and the percentage of fruit showing russetting. The U.S. No. 1 grade has five subclasses: U.S. No. 1 Bright, U.S. No. 1 (no subclass name), U.S. No. 1 Golden, U.S. No. 1 Bronze, and U.S. No. 1 Russet. The U.S. No. 2 grade has Bright and Russet subclasses. All these subclasses have the same grade characters except for rind color and represent progressively greater amounts and intensity of discoloration. The U.S. No. 1 Bright grade may contain no fruit with more than ⅕ of the surface area discolored, while U.S. No. 1 Russet must have at least ⅓ of the fruit (by count) with more than ⅓ of the surface discolored. The discoloration becomes darker as the affected portion of the rind becomes larger. Sometimes the discolored area is dull and will not shine when waxed and polished (early season damage), which is undesirable. In other cases, this area is glossy, although discolored, and will become shiny when brushed, which is desirable from a marketing standpoint.

The U.S. grades for Tahiti limes include U.S. No. 1, U.S. No. 2, and U.S. Combination, which contains at least 60 percent of U.S. No. 1 fruit. Each grade requires fruit to be mostly green, but has subclasses, Turning and Mixed Color, for fruit that is becoming yellow. The minimum grade that is usually shipped is U.S. Combination Mixed Color. This requirement was subject to change upon the recommendation of the Lime and Avocado Administrative Committee. However, the Lime committee has now been disbanded. Key limes must meet standards for U.S. No. 2 Tahiti limes except there is no color specification.

PROCESSING FRUIT

Fruit size and grade standards are mainly applicable for the fresh fruit market but also apply to processing fruit from August 1 until November 30. Fruit quality factors are much more important economically for processing fruit since nearly 90 percent of the total citrus crop in Florida goes to processing plants. In the 2005–06 season, over 95 percent of sweet oranges were processed (mostly to chilled juice), 64 percent of the grapefruit went to processing plants, and 34 percent of the tangerine crop was processed. Temples (70%), Murcotts (36%), and tangelos (61%) were also processed to varying degrees. These fruit are produced primarily for the fresh market and high percentages of processing fruit represent large economic losses to growers compared to prices received for similar fruit in the fresh market.

Internal fruit quality is of greater importance to citrus processors than external appearance. When fruit is in short supply, processors overlook external appearance unless there are rind blemishes that will flake off into the juice, or the fruit is so abnormal in size or shape that it cannot be handled by the processing machinery. As fruit begins to dry out, either through natural processes as it becomes overmature or because of freeze injury, the value to the processor decreases sharply.

Fruit is sold to processors on the basis of pounds-solids (pounds of juice multiplied by percentage of TSS). At one time, some fruit for single-strength juice was sold by the pounds of juice per box or load, but this measure is not in use today. Random samples are taken for fruit quality (maturity) tests from each load upon arrival at the processing plant. Information furnished the processor includes the usual information on Brix (TSS), acids, and TSS-to-acid ratio obtained from a regular maturity test to assure compliance with legal requirements, as well as the number of box equivalents or weight of fruit in the load and the weight of the juice per box equivalent. These data enable processors to calculate the probable yield of final product, to determine how the load should be blended with others to meet buyer specifications, and to calculate the fruit price.

Citrus-processing plants in Florida produce a multitude of products in six principal forms: frozen concentrated (FCOJ); chilled; not-from-concentrate (NFC); reconstituted; canned juices; and frozen, chilled, and canned sections and salads. NFC juice is either nonpasteurized, which requires careful processing and handling, or pasteurized, which has a longer shelf life. There are also numerous valuable by-products, including peel oils, molasses, dried pulp, and even hand cleaner. About 80 percent of all the fruit delivered to processors goes into two products: frozen concentrated orange juice (FCOJ) and chilled juice including orange-grapefruit and other fruit blends. United States grade A pasteurized orange juice must have a minimum Brix of 11.0:1 and reconstituted juice a minimum Brix of 11.8. Minimum ratio ranges from 12.5:1 to 20.5:1 under U.S. and Florida standards for frozen concentrated and pasteurized orange juice. The frozen concentrate product is manufactured principally as either a 3-plus-1 or 4-plus-1 concentrate (These values refer to the ratio of water to FCOJ.). A can of the former has a Brix of from 45° to 47° as purchased in the grocery store. Fresh juice is concentrated in a multistage high-vacuum process to about 55° Brix, after which fresh juice is added back to reduce the Brix to the proper range. Addition of fresh juice (and oil flashed off during concentration) is essential to produce a high-quality product. Add-back was a key component to the development of FCOJ by the Florida Department of Citrus and the University of Florida in the mid-1940s. Other frozen concentrated juices include grapefruit, tangerine, lemon, and lime.

An increasing proportion of the oranges and grapefruit sent to processing plants are being processed as chilled juice. For example, in the 2003–04 season, 38 percent of orange juice was processed as FCOJ and 72 percent as chilled juice compared to 13 percent FCOJ and 87 percent chilled juice in the 2005–06 season. Chilled juice is extracted from fruit of concentrate quality, flash pasteurized, chilled to near freezing, and put into containers for shipment to markets and sale to consumers. Widespread acceptance of frozen concentrated and chilled juices has also led to an increasing number of various types of sections and salads being made available in either frozen or chilled form. A decreasing proportion of fruit is being processed into canned products, with consequent lower returns to the grower whose fruit does not meet concentrate and chilled, ready-to-serve juice standards.

Today, knowledgeable growers recognize that virtually every phase of the cultural program from site selection to initial choices of a rootstock and scion, to the fertilizer, spray, irrigation and other programs influence fruit quality and yield. Moreover, growers of processing fruit have a convenient yardstick in pounds-solids per acre against which they may measure the overall efficiency of their operations. A viable citrus industry is assured as long as citrus consumers continue to identify reconstituted or chilled juice with freshly squeezed juice or fresh fruit. There are many synthetic citrus or citrus-like products that compete with authentic juice for a piece of the consumer market.

PACKINGHOUSE OPERATIONS

In the early Florida citrus industry, fruit were simply harvested, placed in containers, and shipped without any type of sorting or treatment. Of course, storage life was very short and spoilage very great. Primitive packinghouses were established in the early 1900s and fruit were washed and sorted for color. It was observed that washed fruit became desiccated and softened faster than nonwashed fruit due to the loss of natural waxes on the peel. Waxing was introduced into the citrus packinghouse in the 1920s beginning with the use of hot paraffin wax and natural waxes like carnauba. In the mid-1980s, the industry switched from using petroleum-based solvent waxes to water-based waxes that may often include a fungicide.

Basic packinghouse operations have remained unchanged for many years with the exception of computerized sorting and sizing systems available in modernized packinghouses. However, there are many possible configurations of the components in a particular packinghouse. Basically, fruit are brought from the field in bins and drenched in water containing fungicide and a sanitizer. Some fruit are placed in degreening rooms early in the season to improve their natural color as described previously. Degreening does not affect internal quality or fruit maturity. Initially, fruit are dumped onto the packing line, leaves and twigs removed, and the fruit are washed. Fruit are then pre-sized and pregraded and either move to the color-add process (described previously) or are treated with fungicide, dried, waxed, and dried a second time. Fungicide may also be applied in the wax. The fruit undergo a final grading, usually by hand, although sophisticated computerized grading systems are available and sometimes used to supplement hand grading. The fruit are then sized by the methods previously described, labeled, packed in 4⁄5-bushel cartons or bags, palletized, and transported primarily via truck or ship. Fruit are packed by hand or using packing-assist machines. Some packinghouses use totally automated packing systems that rely on suctions cups for lifting and arranging the fruit in the carton.

POSTHARVEST FRUIT DISORDERS

Citrus fruit are susceptible to several important and damaging postharvest problems. The major problems caused by fungi include green mold, Phomopsis, Diplodia, Alternaria, sour rot, and brown rot (alga). All of these can be controlled, but not eliminated, using fungicides and by refrigeration. Canker is a very important bacterial problem that prevents fruit from being shipped to other citrus-growing states like California that currently do not have the disease. Protocols for controlling canker are currently being developed and may reopen some of the previously closed markets to fresh fruit export. Important nonpathogenic disorders include chilling injury, aging, and desiccation. Chilling injury occurs for grapefruit, lemons, and limes when they are stored below 50°F. Florida oranges are not susceptible to chilling injury and can be stored at lower temperatures near 32°F. Tangerines can be held at an intermediate temperature of 40°F.

DETERMINING THE VALUE OF FRUIT

Citrus crops may be sold in a variety of ways. Some are sold in bulk, and the grower is paid for the crop based on an agreed-upon estimate of its value. For example, a buyer may agree to purchase a grower's crop of Valencia oranges in a certain five-acre block for $10,000. The grower would then be paid the $10,000 as a flat sum. Frequently the crop is sold on some sort of unit basis, such as price per standard box (a box of oranges being 90 pounds, grapefruit 85 pounds, and tangerines 95 pounds). For example, a grower may sell a crop on-tree for $5.00 per box. The fruit is harvested, the numbers of bins counted and converted to boxes, and the grower is paid at the $5.00 per box rate. Fruit destined for the fresh fruit market is usually sold this way, and the price is usually strongly related to the external quality of the fruit. Fruit of superior quality will bring a greater price since more of the fruit can be packed as a higher grade and less will have to be sent to the processing due to rind blemish problems. Fruit sold for processing may be priced in one of two ways: either on the basis of pounds-solids per box or pounds of juice per box. Most, if not all, fruit destined for processing is priced on the basis of pounds-solids per box. To arrive at the box value, the grower needs to determine the percent total soluble solids (degrees Brix), multiply by the pounds of juice per box, then multiply that figure by the price per pound of solids. For example, assume 12 percent total soluble solids and 45 pounds of juice per box with a pounds-solids price of $2.00 per pound. Then multiply 0.12 times 45 to determine pounds-solids per box (5.4 ps). Multiply 5.4 times the pounds-solids price of $2.00 to yield the per-box price of $10.80. Some prices are understood to be delivered-in prices. In those cases, the grower must pay the picking and hauling costs out of the proceeds. Other prices are quoted as on-tree prices, in which cases the buyer pays for picking and hauling. A careful analysis of the various selling options will usually result in the best method for selling the crop.

MARKETING THE CROP

Fresh or Processed?

The first step a grower should make is to determine whether the crop should be sold as fresh or processed. Blocks with the highest probability of a high packout (packout is the number of fruit packed as a percentage of the total number of fruit delivered to the packinghouse) with minimum eliminations may be selected for the fresh market. Fruit with low packout and low external quality may be best suited for processing. The earlier this decision is made in the production year, the better, since long before a crop is harvested, decisions must be made that can affect how it will be sold. Production practices used from bloom through harvest can have a profound effect on the fruit—especially on the external quality. Fruit external quality and appearance will have to be very good for the fresh market. Consumers demand fruit with excellent eye appeal and will pay highest prices for the most attractive fruit. This, of course, will usually be reflected in higher returns to the producers of highest quality fruit. Moreover, Florida fruit must compete in the marketplace with fruit from traditionally fresh fruit-producing regions such as California and Spain.

In contrast, fruit destined for processing does not need to have the superior external quality of fresh fruit. After all, juice is the primary product, and the peel is used only as a by-product (citrus pulp) for feeding livestock or for the peel oil.

With these facts in mind, the grower will need to make production decisions that will later influence whether the crop goes to the fresh market or to the processor. Such decisions are never simple. If the fresh market route is chosen, there are usually higher returns than for processing. However, these increased returns come at a cost of additional sprays to maintain high external fruit quality. Other production practices may require extra attention as well, further increasing costs of production. Furthermore, there is always the risk that a poorly timed spray later in the season could result in pest damage that would make the crop unmarketable as a fresh fruit. The grower then will have some very expensive fruit that will have to go for processing. The disparity between on-tree prices per box for fresh and processing fruit varies greatly with the type of fruit. For example, in the 2006–07 season, fresh oranges brought $13.27 per box and processing oranges brought $8.97 per box. In contrast, fresh grapefruit brought $9.51 per box compared with only $0.34 for processing fruit!

The simplest decision is the one that is made to produce the crop for processing and pays little or no attention to factors that might compromise external fruit quality. Since sprays for pest management and other cultural practices are minimized, such fruit is always produced at costs lower than fresh fruit. Harvesting is usually simpler also, since the entire grove can be rapidly harvested at one time and transported in bulk to the processing plant. Contrast this with the fresh fruit harvesting, which requires careful picking, transportation in containers to minimize possible damage, and the occasional need to spot-pick for size and/or color.

Therefore, the decision to choose either fresh or processed market channels is influenced first, by determining early season production practices and second, by an assessment of crop quality at harvest. The second assessment takes into account the alternatives available to the grower at harvest time. If the external quality is good and fresh fruit prices are high, the fresh market is the logical choice. Likewise, if the external quality is poor, the processed market is clearly the only choice. Only when fresh and processed returns are nearly equal does the decision become complicated.

While the fundamental issue of fresh versus processed channels may be determined rather easily, the grower has an extensive array of marketing choices that will still need to be made. Basically, these choices are made on the basis of how the grower will be paid for the fruit.

SELLING THE CROP

Priced or Nonpriced?

When the crop is sold, another series of decisions will need to be made. The first is a choice of how long the grower wants to retain ownership of the fruit. When an outright sale is negotiated, the grower usually gives up all rights to ownership and the fruit becomes the property of the buyer upon completion of the terms of sale. Such fruit is said to be "priced fruit," and the grower has several options regarding how the fruit is priced.

If the grower wishes to retain financial interest in the fruit and any products or by-products from the fruit through the entire marketing process, another option is available. First payment for the crop is made in this case after sale of the end product occurs and expenses have been deducted. This is known as "deferred pricing," and fruit sold in this manner is referred to as "nonpriced fruit." This method of sale is used primarily in cooperatives and participation plans. Several options may be available for the grower.

Priced Fruit

There are several priced fruit options, including pricing at delivery or forward pricing. With forward pricing, several mechanisms will ultimately determine the price the grower receives for the fruit. Since pricing at delivery is the least complicated, it will be examined first.

PRICING AT DELIVERY

If a crop is sold for a price that is determined at harvest or delivery, it has been sold in the spot market. Fruit is sold by the box, by pounds-solids, or in bulk. Buyers may be packers, processors, or independent handlers (known in the industry as "bird dogs"). The price of the fruit is usually negotiated between the buyer and seller and is influenced strongly not only by supply and demand but by the quality and quantity of the crop. Harvesting conditions such as tree size and ease of picking can also influence the price. It is obviously very important for growers to be well informed at the time of sale since they will be negotiating with buyers who are extremely knowledgeable. While negotiations may be difficult, payment for the crop is usually made at completion of harvest or upon delivery.

FORWARD PRICING

In this case, a price is either negotiated prior to harvest at an agreed-upon price (forward contract), or a contract is executed in which the futures price at harvest time will determine the ultimate delivered-in price. With a forward contract, grower and buyer negotiate a price prior to harvest that will be paid for the fruit when it is picked. This can give growers some security in knowing that their fruit will be picked and knowing the value of the crop in advance. Again, the grower needs to be very well informed before entering into contract negotiations.

Growers producing fruit for the processing market may enter into a contract (usually with a processor) that will set the price based upon the price of orange juice futures at the time of delivery. This would then be a contract based on futures price. The grower may then use the futures market either to gamble a little on the price or to hedge to provide some insurance. Such operations are complicated and require considerable expertise. Detailed explanations of the process are beyond the scope of this book.

Nonpriced Fruit

With nonpriced fruit (often referred to as "deferred pricing"), the grower retains at least some ownership of the crop until the final product is ultimately sold. Payment is often made in installments, the last installment coming shortly after the end product is sold. In some cases, however, the grower receives no money until the end product is sold. In this case, the grower must have cash reserves to stay in business.

There are basically two alternatives for nonpriced fruit. One is a pool system in which growers' fruit is combined with fruit from other growers, producing a large volume, and the price is averaged throughout the pool. Cooperative members make up many pools (usually by cultivar), and participating growers retain ownership of the fruit until ultimate disposition. The growers are then paid for the fruit, less the expenses of the cooperative. Since cooperatives are owned by the members, all profits are distributed to members based on the size and quality of their crop.

The other system of pooling operates through a participation plan. These plans are usually offered by a private corporation but may also be offered by cooperatives. In a participation plan, growers pool their fruit and the grower turns over ownership of the product when it is delivered. After the packer or processor sells the product and the price is determined, the manufacturing costs are deducted and the grower receives the residual funds. In this case (unlike a cooperative), the grower has no control of marketing and has to accept the price received when the end product is sold.

With any pooling system there are advantages and disadvantages. Advantages include some assurance that the crop will be harvested and have a "home" and that deliveries will not need to be timed to get the best price. Disadvantages arise from loss of control, having to participate in a pool (especially troublesome if you are an above-average grower with excellent quality fruit), and having to wait several months for the pool to close before receiving final payment for the crop.

Growers may also choose another method of selling with the price deferred through consignment. In this case, the grower has the crop packed or processed but retains ownership until the product is actually sold. The grower is then paid the price received, less expenses for processing.

TWELVE

Citrus Fruits for Home Use

SCION CULTIVARS AND ROOTSTOCKS

Many types of citrus can readily be grown by the urban or dooryard grower in central and south Florida. Some cultivars, such as satsuma mandarins, Nagami and Meiwa kumquats, and Meyer lemons, may be grown in warm locations of north and west Florida. The most popular cultivars in order of season of ripening are presented in table 12.1. This table shows a grower which cultivar combinations to select in order to produce fruit over an entire season in warm locations.

Sweet oranges and grapefruits are eaten regularly by many people. Specialty fruits (tangerines and tangerine hybrids) are excellent for holiday or dessert uses, and acid fruits (lemons, limes, and others with high citric acid content) are used in drinks, garnishes, and ingredients for pies and cakes. Fruit trees have additional ornamental value and enhance the beauty of the surroundings for home landscapes. Specimen trees of exotic types add interest as conversation pieces. "Cocktail" trees, which may consist of several different cultivars on the same tree, are practical and fun to grow. If the area around the home is large enough, several cultivars of each type may be selected.

Sweet oranges are usually the first choice of home growers because they can be eaten fresh or juiced. Hamlin or navel oranges mature before January and are good choices except in very cold locations where it may be necessary to substitute satsuma mandarin. If there is room for more than one or two sweet orange trees, a midseason cultivar (matures in January to February) such as Midsweet could be considered. The late season (matures from March to July) Valencia orange is another possible choice due to its high fruit quality and because it holds well on the tree for several months. In the area north of Ocala, however, Valencia is not recommended because the fruit is too often damaged by cold. Where Valencia can be grown successfully, these three

Table 12.1. Citrus cultivars for use in dooryard plantings

Fruit	Season	Seeds/fruit	Relative fruit size
Oranges			
Navel[a]	Oct.-Jan.	0–6	Large
Hamlin	Oct.-Jan.	0–6	Small
Earlygold	Oct.-Jan.	0–6	Medium
Pineapple	Dec.-Jan.	10–20	Medium
Midsweet	Jan.-Mar.	6–24	Medium
Valencia	Mar.-July	0–6	Medium
Grapefruit			
Marsh, white	Oct.-July	0–6	Large
Duncan, white	Oct.-July	40–65	Large
Redblush, red flesh	Oct.-June	0–6	Large
Rio Red, Ray Ruby, very red flesh	Oct.-June	0–6	Large
Star Ruby, Flame very red flesh	Oct.-June	0–6	Large
Specialty			
Satsuma	Oct.-Nov.	0–6	Medium
Robinson	Oct.-Nov.	Varies[b]	Small-medium
Sunburst	Oct.-Nov.	Varies	Small-medium
Orlando	Nov.-Jan.	Varies	Medium-large
Fallglo	Oct.-Dec.	Varies	Medium-large
Minneola	Feb.-Apr.	Varies	Large
Temple	Dec.-Jan.	15–20	Large
Dancy	Nov.-Dec	7–20	Small
Murcott	Feb.-Apr.	18–24	Medium
Kumquat	Early mid season	0–4	Very small
Acid			
Calamondin	Year-round[c]	0–4	Very small
Tahiti lime	Year-round	0	Medium
Key lime	Year-round	10–15	Small
Lemon[d]	Year-round	0–9	Varies
Limequat	Year-round	0–4	Small

a. Low yields for old-line cultivars

b. These cultivars vary in yield, size, and number of seeds per fruit depending on degree of cross-pollination. For best results, a compatible pollinizer tree should be nearby.

c. Acid citrus bears largest crops in late summer, but some fruit mature all year.

d. Many cultivars can be grown such as Lisbon, Bearss, Villafranca, Meyer, and Ponderosa. Meyer is a lemon hybrid that has become widely planted.

cultivars (Hamlin, Midsweet, and Valencia) will supply fresh fruit continuously from early November to August.

Grapefruit is usually the second choice in dooryard plantings. Some people prefer a white-fleshed cultivar such as Marsh. However, others may prefer a red-fleshed cultivar like Redblush, Star Ruby, Ray Ruby, Rio Red, or Flame because of the combination of flesh color and seedlessness. Many people also believe that red-fleshed fruit taste better than white-fleshed fruit but there is no scientific evidence to support this contention. Usually a single grapefruit tree is all that is needed for home use because trees are very productive and fruit hold well on the tree; they may be harvested throughout most of the year.

Among the mandarins, satsumas (often a substitute for sweet oranges in colder locations) mature during late November but the peel often remains green at maturity and internal fruit quality diminishes very rapidly in about three to four weeks. The mandarin hybrids (Robinson, and Sunburst) are very good early-maturing tangerine hybrids. The homeowner may decide to substitute a tangor or tangelo for a tangerine. Temple tangor has very good fruit quality and appearance and matures from January to March, and Minneola tangelo produces excellent quality fruit with deep red-orange peel color and a distinct neck on the fruit. Minneola is the latest maturing tangelo, February to April, and fruit may be damaged by cold in some locations. Some of the cultivars (especially the hybrids) may be difficult to find at local nurseries. In addition, Minneola and Orlando tangelos and Sunburst tangerine may require cross-pollination with a different cultivar to produce a reasonable crop (see chapter 3).

Acid fruits include lemons and limes, but the true lemon is not recommended for a home fruit in Florida because of its susceptibility to cold damage and to the fungus disease, scab. In recent years, the Meyer lemon, which is likely a hybrid and not a true lemon, has become quite popular for home use. The fruit is very similar in appearance to true lemons and it is used in the same way, but the trees are more cold hardy and heat resistant. For southern Florida, either the Tahiti or Key lime makes a satisfactory acid fruit for home plantings, and fruit can be picked throughout the year. Eustis or Lakeland limequat can be grown wherever sweet oranges can grow, and calamondins produce a suitable acid fruit anywhere in the state. However, all of the above selections are sometimes difficult to find in nurseries.

Dual-purpose fruits are those that combine ornamental value with edible fruit. The kumquats and calamondins are notable examples. Fruits that provide conversational value include Ponderosa lemon, Ruby blood orange, or shaddock (pummelo). Ponderosa lemon produces a very large fruit but has all of the disadvantages of other true lemons. Blood oranges are well suited to Mediterranean-type climates but often do not achieve bright red peel and flesh color under Florida growing conditions. Shaddock fruit look like grapefruit (shaddock is one of the presumed parents of grapefruit) but are much larger and have a different flavor from grapefruit.

Rootstocks should be chosen for freeze hardiness in northern areas, longevity, and influence on fruit quality rather than primarily for yield. Rootstock choices will likely be limited for dooryard growers, but if there is a choice, it is good to be informed (see chapter 5). Satsuma mandarins should be budded onto trifoliate orange rootstock to achieve maximum cold hardiness and good fruit quality. In northern Florida, other citrus fruits may also be grown on trifoliate orange. In central Florida, Swingle citrumelo, Kuharske citrange, and Cleopatra mandarin will probably give the highest fruit quality—that is, the best balance of sugars and acids.

SITE SELECTION AND TREE CARE

Any well-drained site that is suitable for building a home and having a garden is likely to be satisfactory for growing citrus trees if low winter temperatures are not limiting. Citrus trees will tolerate moderate shade but will be more productive and have better fruit quality if they are not shaded by other trees. They should also not be planted so close together that they shade the lower branches of neighboring trees. For home plantings, the spacing recommended between trees for commercial plantings (see chapter 7) should be increased because citrus trees become very large if left unpruned.

Vigorous one-year-old nursery trees growing in containers should be obtained from a reputable commercial nursery. It is very important to purchase citrus trees that are certified free of the common virus diseases and nematodes (microscopic worms that feed on roots). The homeowner who wants to grow citrus trees should also be very careful not to plant ornamentals that may be a source of nematode infection. Trees may be planted at any time of the year but it may be risky to plant before February 15 in southern Florida or March 15 in northern Florida because of late-season freezes.

Trees should be planted slightly higher than the soil (media) level in the container. This allows for some settling of the tree after planting and assures that the rootstock will be well above the soil surface. It is advisable to mud-in the root system while backfilling the planting hole to remove air pockets and to provide good root-soil contact. A basin should be made around each tree that holds about 5–10 gallons of water and is periodically filled to soak the root area of the young tree. Some of the artificial media should be removed from the root mass when the tree is planted. This will expose many of the outer roots and allow them to grow quickly into the new planting area provided. This media removal is also important since the difference in composition between the potting mix and the planting site can result in difficulty in wetting the soil and decrease subsequent root growth. If subfreezing temperatures are forecast after trees are planted, they should be protected as described in chapter 8. Small trees also may be covered with a blanket or cover (tarp) that should reach all the way to the ground. The blanket or cover reduces heat loss from the soil and is only effective alone during light to moderate freezes. A light bulb or continuously running sprinkler should be placed under the cover to provide heat in severe, hard freezes.

Water is the most important requirement initially for young tree growth, and it will be used (transpired) rapidly by trees with a dense canopy, which healthy container-grown trees should have. Let a hose run slowly in each basin for about 10 minutes several times each week for long enough to fill the basin during the first month after planting. Thereafter, trees should be watered about every five days in the winter and every three days in the summer if there has been no appreciable rainfall. The need for frequent irrigation will be less as the tree's root system increases in size. Some trees may require watering through the fourth year after planting. Alternatively, a microsprinkler or sprinkler system can be used for irrigation. Frequency and amount of irrigation is described in chapters 8 and 9.

Fertilization should commence when new growth begins. In the first year, fertilizer should be applied about every six weeks from early April until early November in north Florida. In southern areas of Florida, trees can be fertilized throughout the year. A 6N-6P-6K or 8N-8P-8K analysis fertilizer may be used at rates from ½ to 1 cup for the first application to 1 ½ pints in November, increasing the quantity steadily all season. For the second, third, and fourth years, the fertilizer schedule given in chapter 8 can be followed. Fertilizer should be scattered within the basin area the first year. In succeeding years, a good rule of thumb is to spread the fertilizer as many feet beyond the dripline of the canopy as the age of the tree in years (up to 10). For the home planting, it is convenient to remember that one pint of mixed fertilizer weighs about one pound depending on the fertilizer analysis. Be sure not to use a one-pound capacity coffee can or similar container since such a can may actually hold nearly three pounds of mixed fertilizer. It is preferable to weigh the fertilizer before application to avoid damaging the tree.

Citrus trees in home plantings are less likely to show mineral element deficiencies than those in commercial groves because of the greater amount of organic matter usually present in most garden soils. As a precaution against the possible development of deficiencies, however, it is wise to use fertilizer mixtures containing microelements during the first few years after planting. These formulations are readily available at garden stores and nurseries and the amount of each micronutrient will be expressed on the label. If zinc deficiency symptoms appear, it will be necessary to apply a foliar spray as described in chapter 9. Zinc application is rarely needed in plantings on acidic, sandy soils. On alkaline soils, however, zinc and manganese deficiencies may appear unless annual nutritional sprays are applied. Iron deficiency may also occur and should be corrected with soil applications of chelated iron (a special formulation of iron) which is available to the tree even in alkaline soils.

Many home gardeners have grown citrus trees successfully without measuring soil pH. If the tree's foliage is unhealthy or yellowish, however, and the cause is not obviously an insect or disease infestation, there may be a nutrient deficiency. In this case, the homeowner can take a soil sample to the county extension office or local garden center to check the pH. If the soil pH is not in a 5.5–6.5 range, the situation should be corrected as outlined in chapter 9 (liming). If the soil is naturally basic (7.5 or above), it will be very difficult to change the pH, and this should not be attempted.

Fertilization of bearing citrus trees for home plantings should be based on age rather than yield because of the inconvenience of keeping yield records. The recommended fertilizer ranges given in chapter 9 will also work for home plantings. However, these commercial recommendations are for trees growing without competition from lawns. Citrus trees in home plantings will grow best if they are cleanly cultivated, but sometimes they are planted within the lawn area. In such cases, the turf should be kept from growing closer than three feet from the tree trunk. In the first 10 years after planting, the amount of fertilizer should be increased about 25 percent at each application to provide enough for both the grass and the tree. In addition, fertilizer applied around the tree may injure the grass if it is not promptly irrigated into the soil. Weed growth around the base of the tree can be controlled using mulch, approved herbicides, or hand cultivation. The mulch should be kept at least six inches away from the trunk to avoid potential foot rot problems.

Pest control should be done only as needed. Citrus trees in home plantings may live for years with little widespread damage from pests. Often a few leaves show signs of pest or disease damage but these represent only a small percentage of the overall leaf number. When pest damage does become a problem, homeowners will often find it easier and more economical to have the spraying done by a pest control operator or a neighboring commercial grower rather than to spray the trees themselves. A homeowner may opt to spray trees up to bearing size, but larger trees require large and expensive spray equipment to achieve adequate coverage.

The discussion of pests, diseases, and their control in chapter 10 is just as valid for home as for commercial citrus growers. However, some pesticides require the use of special equipment and training for applicators and are not suitable for home use. Also, certain pesticides can only be sold to and used by persons holding restricted pesticide licenses.

Some pest and disease problems commonly found on citrus trees are easy to diagnose and treat; others are difficult to identify, even for experienced citriculturists. Tables 12.2, 12.3, 12.4, and 12.5 provide useful guidelines for identifying problems associated with fruit and vegetative portions of citrus trees. The tables also refer to color photographs in the center of the book for some of the more common problems. Additional information can be obtained from University of Florida/IFAS publications and through county extension offices or local garden shops and nurseries. It is important to remember that mature citrus trees are quite tolerant of environmental, disease, and pest-related stresses. Trees can remain productive for many years by following basic cultural programs.

Table 12.2. Problems associated with citrus fruit

Symptoms and causes	When usually observed	Time of development	Control
Fruit drop prematurely: physiological disorder. Various causes, usually stress related. Some early fruit drop is normal since trees often overbear.	Most acute at bloom, during May and June, and near harvest.	Throughout season.	Follow recommended cultural practices with emphasis on adequate irrigation.
Fruit is rusty or brown with smooth texture: citrus rust mite(Plate 25) Primarily a cosmetic problem	When fruit size and begins to mature.	Usually during fruit development.	Check for rust mites and spray only if populations are very high.
Splitting of fruit: physiological disorder. Not reversible. Follow recommended practices next season.	Prior to and during harvest season.	Just prior to fruit maturity.	Follow fertilizer recommendations.
Smooth brown or tan irregular blemishes on peel: mechanical damage from abrasion. A cosmetic problem.	Near harvest.	Usually when fruit is small.	No control available.
Black, sooty covering on fruit: sooty mold fungus (Plate 32). The fungus grows on exudate from scales, whiteflies, and aphids.	Prior to and during harvest.	Summer.	Control insects that cause the problem.
Granulation, drying of juice sacs: physiological disorder. Problem is worse on young or vigorous trees.	At harvest.	Near harvest.	Regular irrigation, good cultural practices.
Thick peel: physiological disorder. Worse on trees that are young or vigorous trees.	At harvest.	Throughout fruit development.	Follow good cultural practices; do not use excessive fertilizer.

Table 12.2 *(continued)*

Symptoms and causes	When usually observed	Time of development	Control
Raised scabby bumps on fruit: citrus scab fungus (Plate 1). Affects only certain cultivars such as Temple orange and grapefruit. Also see Table 12.5; citrus canker.	Any time fruit is on tree.	Shortly after bloom.	Preventive spray program early in the season.
Brown, pinhead-sized raised lesions on fruit about 1/16-inch diameter: melanose fungus (Plate 3). Sprays will not reverse damage; must be prevented.	Summer until fruit harvest.	Late spring, usually after May 1.	Preventive spray program; remove deadwood that harbors the fungus.
Grapefruit rind pitting: greasy spot fungus (Plate 5). A cosmetic problem.	At or near harvest.	Summer.	Preventive spray program.
Raised specks or spots of various colors, usually less than 1/10 inch, removable: scale insects (Plates 10–16). (There are many types.) High populations affect tree vigor, but usually cosmetic on fruit.	Anytime.	Summer, fall.	Spray if population warrants control. There are also several natural biological controls.
Premature coloring, black decay in fruit core: black fungus. An infrequent problem.	Late fall.	Early fall.	Remove affected fruit. Sprays are usually not justifiable for control.
Premature coloring, very small puncture wounds, some fruit drop: plant/stink bug injury. An infrequent problem.	Late summer, early fall.	Late summer, early fall.	Watch for insects and spray if necessary.

(continued on next page)

Table 12.2 *(continued)*

Symptoms and causes	When usually observed	Time of development	Control
Tan, leathery decay with foul odor: brown rot fungus. An infrequent problem. Spray will not reverse existing damage.	Near harvest and post-harvest.	Just prior to harvest.	Preventive spray may be helpful if problem is recurrent. Usually occurs only in very wet years.
Crease-like, depressed lines in fruit peel: physiological disorder (creasing). An occasional problem of certain cultivars.	Prior to or during harvest.	Spring and early summer.	Follow good cultural practices, particularly fertilization practices and potassium application.
Discolored, often necrotic, sunken areas of damage to peel: chemical burn. Sprays must be applied properly, observing all label instructions.	Any time.	Any time.	Usually results from improper spray rates or application.
White, cottony masses usually located around stem: mealybugs or cottony-cushion scale insects (Plate 15). Only an occasional problem. High populations may case fruit drop.	Any time.	Summer.	Spray should be applied if population warrants.
Black, raised bumps on fruit stem about 1/8 inch in diameter: black scale insects. High populations can cause fruit drop.	Near harvest.	Late spring.	Spray should be applied if population warrants.
Holes in fruit of various sizes, depths: chewing insects, birds, or rodents. Control measures will depend on nature of problem.	Near harvest.	Near harvest.	No economical control available.

Table 12.3. Problems associated with citrus leaves and twigs

Symptoms and causes	When usually observed	Time of development	Control
Black, sootlike covering on leaves: sooty mold fungus (Plate 31). The fungus grows on exudates of scales, white flies, and aphids.	Anytime.	Mainly spring, summer.	Need to control pests that cause problem.
Distortion, curling, and/or cupping of leaves: aphids (Plate 19) or psyllids(Plate 21). Controlling the insects will prevent the problem, but will not repair existing damage.	Anytime.	Mainly spring, summer when leaves are tender.	Spray if populations are high. Aphids are under biological control. Psyllids difficult to control and may vector citrus greening disease.
Scratched, silvery appearance to leaves: spider mites, usually citrus red mite (Plate 26). May cause leaf drop in fall with severe infestations under dry conditions.	Anytime.	Mainly spring, fall.	Spray when mites are observed.
Leaves cut, chewed, or notched: chewing insects such as grasshoppers, crickets, weevils. Usually not a serious problem except on young trees.	Anytime.	Anytime, but usually during warmer months.	Difficult; probably best removed by hand.
Leaves distorted, often with bumpy scabby lesions: scab fungus (Plate 1). Must be prevented; spray will not correct. May be on fruit and twigs also.	Anytime.	During spring growth flush period.	Requires well-timed sprays or use resistant cultivars.

(continued on next page)

Table 12.3 *(continued)*

Symptoms and causes	When usually observed	Time of development	Control
Dark brown pinhead-sized pustules with sandpaper texture on leaf surface: melanose fungus (Plates 2). Spray will not correct. May attack fruit and twigs also.	Anytime.	Late spring, early summer	Preventive spray program. Remove deadwood which harbors the fungus.
Dark tarlike spots on leaves usually less than ⅛ inch in diameter: greasy spot fungus (Plate 4). Sprays will prevent, not correct, existing problem. Severe cases will cause defoliation.	Anytime, but becomes most apparent in the fall.	Summer.	Preventive spray program. Remove fallen leaves under tree.
Raised specks or spots of various colors, usually less than 1/10 inch, removable: scale insects (Plates 10–16). There are many types. High populations affect tree vigor. May also attack fruit and twigs.	Anytime.	Summer, fall.	Spray when population becomes high.
White, cottony masses on leaves, twigs, fruit: mealybugs or cottony-cushion scale insects (Plate 15). Usually not serious except on small plants, but may cause fruit drop.	Anytime.	Spring, summer.	Hand removal or spray if warranted.
Small, 1/16-inch, translucent discs under leaves with white flying insects present: whiteflies (Plates 17,18). Sooty mold may be more of a problem than whiteflies.	Summer.	Late spring, summer.	Sprays are usually not necessary except under extreme conditions. Natural controls keep populations in check.

Table 12.3 *(continued)*

Symptoms and causes	When usually observed	Time of development	Control
Appearance of bird droppings feeding on leaves: orange dog (Plate 30). Caused by larvae of swallowtail butterfly. Raised red horns when prodded.	Spring, fall.	Spring, fall.	Remove with a stick or by hand, and crush.
Leaf necrosis at tip or margin; may be some drop, twig death: salt accumulation from excess water or fertilizer. Irrigate with fresh water to leach salts from roots.	Anytime.	Anytime.	Check water supply for salts; reduce fertilizer amounts.
Red or yellow raised spots about 1/12 inch on underside of leaves: *Aschersonia*, friendly fungus which attacks immature whiteflies.	Anytime.	Summer.	Do not control, as this is a friendly fungus.
Leaf drop, often followed by twig dieback: root damage, most often due to flooding. If flooding has occurred, tree may need to be removed or replaced.	Anytime.	Anytime.	Avoid excess water by providing for adequate drainage at planting.
Pale green leaves with darker veins: nutrient deficiency—usually microelements. Follow fertilizer and nutritional spray recommendations carefully.	Anytime.	Anytime.	Nutritional spray needed.
Leaf yellowing, necrosis, some leaf drop: spray burn—may be salt or chemical. Salt spray is often a problem near the ocean.	Anytime.	Anytime.	Wash off excessive residues. Avoid excessive spray rates.

(continued on next page)

Table 12.3. Problems associated with citrus leaves and twigs *(continued)*

Symptoms and causes	When usually observed	Time of development	Control
Tunnel-like lesions on leaves: citrus leafminer (Plate 20).	Late spring-summer but may be evident anytime of year.	Worse when young tender growth is on tree.	Preventive spray problems.

Table 12.4. Problems associated with citrus limbs, trunk, or entire tree

Symptoms and causes	When usually observed	Time of development	Control
White, snowlike specks on bark of limbs, trunk: citrus snow scale (Plate 14). Severe infestations may cause limb loss, fruit drop.	Anytime.	Anytime.	Remove mechanically or use chemical spray.
Tree declines, small yellowish leaves, bark lesions near soil often present: foot rot alga (Plate 9). Cure will be difficult. May need to contact County Extension Office or local nursery.	Anytime.	Anytime.	Remove any organic material near trunk, disinfect area. Allow trunk to dry out. Apply chemical.
Tree declines without outward obvious symptoms: blight, virus disease, nematodes, or weevils. Diagnosis of actual problem will likely be difficult. Contact County Extension office. Also see Table 12.5; citrus greening.	Anytime.	Anytime.	No easy control available for these problems.

Table 12.5. Problems that may involve fruit, leaves, and twigs, and may affect tree vitality

Symptoms and causes	When usually observed	Time of development	Control
Dark brown to black raised lesions which may be surrounded by yellow halos on infected fruit and leaves: citrus canker (Plates 27, 28).	Any time, but often appears after leafminer infestations.	Any time, but spread is worst following rainy season.	No totally effective control measure available. Copper sprays may be helpful.
Blotchy leaf chlorosis of affected limbs. Twig dieback, stunting of tree. Fruit may be lopsided and have bad taste: citrus greening (Plate 29).	Any time, often worse during the growing season.	Any time; spread by Asian citrus psyllid insect (Plate 21).	No control once tree is infected. Control psyllids as preventive measure.

Appendix

2006–2007 Citrus Budgets and Reset-Replacement Tree Costs

Ronald P. Muraro
Professor of Food and Resource Economics
University of Florida
Institute of Food and Agricultural Sciences
Citrus Research and Education Center
Lake Alfred, Florida

CITRUS BUDGETS

Annually, citrus budgets are tabulated for the Central Florida (Ridge), Southwest Florida, and Indian River citrus production regions. The 2006–2007 citrus budgets can be found in this appendix.

These costs may not represent your particular grove situation. However, they represent the most current comparative cost estimates for Florida citrus. The budget cost items for the Central Florida (Ridge) and Indian River represent a custom managed operation. The Southwest Florida costs are more representative of an owner-managed operation.

Budget analysis provides the basis for many grower decisions. Budgets can be used to calculate potential profits from an operation, determine cash requirements, and determine break-even prices. The budget costs presented will serve as a format for growers to analyze their own individual records. The cost data were developed by surveying citrus caretakers-custom operators, suppliers, growers, colleagues with UF/IFAS, and County Extension Agents in each production region.

The 2006–2007 summary budgets are shown in Tables A.1, 2, and 3 for the Central Florida (Ridge), Southwest Florida (both represent a processed juice market program) and Indian River (represents a fresh market grapefruit program), respectively. Each budget is presented in two scenarios. Scenario one represents costs of typical grove care practices that do not include citrus canker and greening management control programs.. Scenario two is same cultural program as scenario one but expanded to include the additional costs for managing citrus canker and greening summarized in Table A.4.

With the introduction of citrus greening in 2005, Florida citrus growers have had to develop new management strategies to identify and remove infected trees along with adding new spray programs to control the insect vector, the Asian citrus psyllid. Likewise, with the end of the citrus canker eradication program in 2006, to reduce the impact of canker infestations on new tree flushes and to reduce fruit drop, copper spray material is being added with each spray tank mix. For fruit grown for the fresh fruit market, additional costs are incurred by growers to assure that the production blocks and fruit can be certified "canker free" for shipments to the U.S. domestic and European markets. Table A.4 presents the estimated additional costs required to manage citrus greening and canker and were based on the cultural programs being implemented in UF/IFAS CREC research groves and information obtained from citrus growers. The citrus greening costs were incorporated into Tables A.1, 2, 3, and 5.

The budgets shown in the tables list the costs of individual grove care practices normally performed in a citrus grove. These costs reflect current grove practices being performed by growers. The estimated costs are for a mature grove (10+ years old); the grove care costs for a specific grove site may differ depending upon the tree age, tree density, and the grove practices actually performed. For example, tree losses due to blight or citrus greening could double, if not increase more, the tree replacement costs. Travel and set-up costs may vary due to the size of a citrus grove and the distance from the equipment barn. Citrus canker and greening control costs will also vary between individual blocks due to variety and fresh or processed market destination.

The budget costs with greening are shown as an expanded "delivered-in" format in Table A.5. The delivered-in costs include cultural/production, management, regulatory, and harvesting costs. The costs are presented in per acre and per box cost units as well as cost units in per pound solids for oranges and per carton for grapefruit. The per acre yields used in Table A.5 represent an above average production and reflect an additional 2.3% average annual tree loss for all age trees due to citrus greening.

Delivered-in break-even prices are shown in Tables A.6-a, 6-b, and 6-c for the three respective production regions. For oranges the yield ranges from

300 to 600 boxes per acre and are presented on a price per pound solids basis. For grapefruit the yield ranges from 350 to 650 boxes per acre and these are presented on price per Florida field box basis. Comparative break-even prices for both with and without the additional citrus greening management costs are shown in the cost tables.

RESET-REPLACEMENT TREE COSTS

Replacement of dead or diseased trees is an important part of the cultural program in a Florida citrus grove. Growers view an empty tree space as costly and nonproductive since the cost of weed management must still be incurred and equipment for fertilization and spraying must continue to pass by each tree space in a grove. Also, a successful resetting or tree replacement program gives perpetual life to a citrus grove and does not require the investment of capital and lost income of total replanting.

Average annual tree loss in Florida citrus groves has ranged between 3% and 4% but can vary markedly in individual groves. Typical causes of tree loss have been disease such as blight and tristeza, root rot, and occasionally lightning strikes. However, with the introduction of citrus greening in 2005, an additional 2% to 5% probably should be added to the annual historic tree loss rates.

Florida citrus growers have had to develop new management strategies to control citrus greening. A currently recommended program is scouting up to four times per year to identify and remove infected trees along with additional insecticide sprays to control the Asian citrus psyllid insect vector. For young reset trees, this includes Admire systemic soil drenches. Before citrus greening, the additional cost for a reset-tree replacement program would add about 10% to the total grove care costs. With citrus greening, the percentage of total grove care costs has risen to at least 14%.

The costs for tree removal, planting, and cultural maintenance through three years of age are shown in Table A.7 and include the additional citrus greening costs required to manage resets with citrus greening. The additional citrus greening costs were based on the cultural programs being implemented in UF/IFAS CREC research groves and information obtained from citrus growers. The costs are presented as a cost per tree ranging from 1–2 trees per acre up to 26+ trees per acre. As shown in Table A.7, the per tree costs decrease as the number of reset trees per acre increases.

KEY POINTS TO TABLES

a. Central Florida production area refers to Polk and Highlands counties. However, the costs presented in this report are applicable to other counties such as Hardee, Hillsborough, Lake-Orange, Osceola, and Pasco counties.

Southwest Florida production area refers to those counties in the Florida Agricultural Statistics Service "Southern Production Area." However, the costs shown are applicable to other South Central Florida counties such as DeSoto and Sarasota counties.

Indian River production area refers to the citrus producing counties on Florida's east coast.

Where *equipment use* or *application* is listed (mowing, spray and herbicide application, etc.), the costs include a charge for equipment repairs, maintenance, labor, and overhead management charges/costs. The exceptions are costs of items such as hedging and topping where average custom charges are used. A *management charge* for equipment supervision and fruit marketing is not included. Management charges/costs could be based on a monthly charge ($3 to $6/acre) or percentage of gross sales. In addition to these charges, a harvesting supervision cost (10¢ to 20¢/box) for overseeing and coordinating harvesting may be charged. Other cost items which are not included in the budget are ad valorem taxes and interest on grove investment. In addition to these cost items, overhead and administrative costs, such as water drainage/district taxes, crop insurance, and other grower assessments, can add up to 12 percent to the total grove care costs. These costs vary from grove to grove depending on age, location, and time of purchase or establishment and are estimated in the expanded tables A.4 and A.5.

Included in the materials expense for the Central Florida and Indian River budget costs is a supervision (or handling) charge of 10 percent of cost/price of the materials. The budget costs for Southwest Florida represent an *owner-managed operation* for the production of oranges for processing and grapefruit for the fresh market. Therefore, the *10 percent handling and supervision charge* added to the material cost for a custom-managed operation is *not included* in the costs.

The budget cost items have been revised to reflect current grove practices being used by growers such as chemical mowing, different spray materials, rates of fertilization, microsprinkler irrigation, more reset trees, and hedging and topping practices. Therefore, the revised costs for each grove practice shown may be higher, or lower, than previously reported.

Although the estimated annual per-acre grove costs listed are representative for a mature citrus grove (10+ years old), the grove care costs for a specific grove site may differ depending upon the tree age, tree density, and grove practices performed; for example, spot herbicide for grass/brush regrowth under trees could add an additional $10.26 per acre; Diaprepes control could add $93.18 per acre for each foliar application; extensive tree loss due to blight, tristeza, or citrus greening could substantially increase the tree replacement and care costs; travel and set-up costs may vary due to size of the citrus grove and distance from grove equipment barn and could add $36.08 per acre.

b. Spray materials include copper (Cu), oil, and nutritionals.

c. Per-acre costs shown in parenthesis are for 2007.

d. Irrigation Expense includes the following:

	Microsprinkler	*Drip*
Variable Operating Expense (Diesel)*	$ 73.24	$ 70.08
Fixed-Variable Expense (annual maintenance repairs to system)	56.90	49.70
Total Cash Expenses**	$130.14	$119.78
Fixed-Depreciation Expense	56.56	45.25
Total Cash and Fixed Expense	$186.70	$165.03

* Adjusted for higher fuel costs.

** Where applies, there may be an additional cost of $16.72 per acre for water control in/out of ditches and canals plus $18.56 per acre for ditch and canal maintenance plus $17.48 for weed control in ditches and canals.

Table A.1. A listing of estimated comparative *Central Florida (Ridge)* production costs per acre for *Processed Oranges*, 2006–2007*

Costs represent a mature (10+ years old)		
	Processed Cultural Program	
Central Florida (Ridge) Orange Grove. PRODUCTION/CULTURAL COSTS[a]	Without Canker-Greening	With Canker-Greening
Weed Management/Control:		
Mechanical Mow Middles (4 times per year)	$ 43.68	$ 43.68
Chemical Mow Middles (2 times per year)	17.02	17.02
General Grove Work (2 labor hours per acre)	30.88	30.88
Herbicide (1/2 tree acre treated):		
Application (2 residual applications)	28.86	28.86
Material	73.90	73.90
Spot Treatment (material and application)	25.46	25.46
Total Herbicide Cost	128.22	128.22
Spray/Pest Management:		
Temik (33 lbs):		
Application	—	14.93
Material	—	108.54
Total Temik Cost	—	123.47
Winter-Spring #1 (February):		
Application (125 GPA)	—	29.20
Material	—	34.09
Total Spring #1 Cost	—	63.29
Winter-Spring #2:		
Application (125 GPA)	—	29.20
Material	—	38.41[b]
Total Spring #2 Cost	—	67.61
Summer Oil #1:		
Application (125 GPA)	29.20	29.20
Material	44.59	35.09
Total Summer Oil #1 Cost	73.79	64.29
Summer Oil #2:		
Application (125 GPA)	29.20	29.20
Material	73.34[b]	55.59
Total Summer Oil #2 Cost	102.54	84.79

Table A.1 *(continued)*

Central Florida (Ridge) Orange Grove. PRODUCTION/CULTURAL COSTS[d]	Processed Cultural Program Without Canker-Greening	With Canker-Greening
Fall Insecticide:		
Application (125 GPA)	—	29.20
Material	—	14.33
Total Fall Insecticide Cost	—	43.53
Total Spray/Pest Management Costs	176.33	446.98
Field Inspections for Citrus Greening (4 inspections @ $22.73)	—	90.92
Mandatory Citrus Canker Decontamination Costs	29.35	29.35
Fertilizer (Bulk):		
4 Applications	37.44	37.44
Material (16-2-16-3MgO @ 220 lbs N)	197.04	197.04
Total Fertilizer Cost	234.48	234.48
Dolomite (one ton applied every 3 yrs), Material/ Application	11.45	11.45
[c]Pruning[c]:		
Topping ($28.92/A _ 2.5 yrs)	11.97	11.97
Hedging ($27.33/A _ 2 yrs)	13.67	13.67
Chop/Mow Brush after Hedging ($11.99/A _ 2 yrs)	6.00	6.00
Total Pruning Cost	31.64	31.64
Tree Replacement, 1 thru 3 years of age (3 trees/acre without greening; 6 trees/acre with greening)		
Remove Trees:		
Pull, Stack & Burn (Clip-Shear & Front-End Loader)	19.08	31.80
Prepare Site and Plant Tree (includes reset trees)	43.29	81.00
Supplemental Fertilizer, Sprays, Sprout, etc. (Trees 1–3 years old)	37.62	98.40
Total Tree Replacement Cost	99.99	211.20
Irrigation: Microsprinkler System[d]	186.70	186.70
IRRIGATED PROCESSED FRUIT PRODUCTION COSTS	$989.74	$1,462.52

* The listed estimated comparative costs are for the example grove situation described in the Economic Information Report Series entitled "Budgeting Costs and Returns for Central Florida Citrus Production" and may not represent your particular grove situation in Central Florida.

Table A.2. A listing of estimated comparative *Southwest Florida* production costs per acre for *Processed Oranges*, 2006–2007*

Costs represent a mature (10+ years old)		
	Processed Cultural Program	
Southwest Florida Orange Grove. PRODUCTION/CULTURAL COSTS[a]	Without Canker-Greening	With Canker-Greening
Weed Management/Control:		
Mechanical Mow Middles (3 times per year)	$ 25.09	$ 25.09
Chemical Mow Middles (3 times per year)	24.77	24.77
General Grove Work (2 labor hours per acre)	30.88	30.88
Herbicide (1/2 tree acre treated):		
Application (3 residual applications)	30.75	30.75
Material	82.15	82.15
Total Herbicide Cost	112.90	112.90
Spray/Pest Management:		
Temik (33 lbs):		
Application	—	14.93
Material	—	98.67
Total Temik Cost	—	113.60
Winter-Spring #1 (February):		
Application (125 GPA)	—	23.94
Material	—	30.99
Total Spring #1 Cost	—	54.93
Spring #1:		
Application (125 GPA)	—	23.94
Material	—	34.92[b]
Total Spring #2 Cost	—	58.86
Summer Oil #1:		
Application (125 GPA)	23.94	23.94
Material	40.54	73.50
Total Summer Oil #1 Cost	64.48	97.44
Summer Oil #2:		
Application (125 GPA)	23.94	23.94
Material	66.67[b]	50.54
Total Summer Oil #2 Cost	90.61	74.48
Fall Insecticide:		
Aerial Application (10 GPA)	—	8.00
Material	—	13.03
Total Fall Insecticide Cost	—	21.03
Total Spray/Pest Management Costs	155.09	420.34

Table A.2 *(continued)*

Southwest Florida Orange Grove. PRODUCTION/CULTURAL COSTS[a]	Processed Cultural Program	
	Without Canker-Greening	With Canker-Greening
Field Inspections for Citrus Greening (4 inspections @ $22.73)	—	90.92
Mandatory Citrus Canker Decontamination Costs	29.35	29.35
Fertilizer (Bulk):		
4 Applications	26.28	26.28
Material (17-4-17-2.4MgO @ 220 lbs N)	209.32	209.32
Total Fertilizer Cost	235.60	235.60
Dolomite (one ton applied every 3 yrs) –Material/ Application	14.76	14.76
Pruning[c]:		
Topping ($28.92/A ÷ 2.5 yrs)	11.97	11.97
Hedging ($27.33/A ÷ 2 yrs)	13.67	13.67
Chop/Mow Brush after Hedging ($11.99/A _ 2 yrs)	6.00	6.00
Total Pruning Cost	31.64	31.64
Tree Replacement, 1 thru 3 years of age (4 trees/acre without greening; 7 trees/acre with greening)		
Remove Trees:		
Pull, Stack & Burn (Clip-Shear & Front End Loader)	25.44	37.10
Prepare Site and Plant Tree (includes reset trees)	57.72	94.50
Supplemental Fertilizer, Sprays, Sprout, etc. (Trees 1–3 years old)	50.16	114.80
Total Tree Replacement Cost	133.32	246.40
Irrigation:		
Microsprinkler System[d]	186.70	186.70
Clean Ditches (Weed Control)	18.56	18.56
Ditch and Canal Maintenance	17.48	17.48
Water Control (Pump water in/out of Ditches and Canals)	16.72	16.72
Total Irrigation Cost	239.46	239.46
IRRIGATED PROCESSED FRUIT PRODUCTION COSTS	$1,032.86	$1,502.11

* The listed estimated comparative costs are for the example grove situation described in the Economic Information Report Series entitled "Budgeting Costs and Returns for Southwest Florida Citrus Production" and may not represent your particular grove situation in Southwest Florida.

Table A.3. A listing of estimated comparative *Indian River* production costs per acre for *Fresh Market Grapefruit,* 2006–2007*

Costs represent a mature (10+ years old)		
	Fresh Market Cultural Program	
Indian River White Grapefruit Grove. PRODUCTION/CULTURAL COSTS[a]	Without Canker-Greening	With Canker-Greening
Weed Management/Control:		
Mechanical Mow Middles (3times per year)	$ 32.76	$ 32.76
Chemical Mow Middles (3 times per year)	25.53	25.53
General Grove Work (2 labor hours per acre)	30.88	30.88
Herbicide (1/2 tree acre treated):		
Application (3 residual applications)	42.28	42.28
Material	90.37	90.37
Total Herbicide Cost	132.65	132.65
Spray/Pest Management:		
Temik (33 lbs):		
Application	—	14.93
Material	—	108.54
Total Temik Cost	—	123.47
Winter-Spring #1 (February):		
Application (125 GPA)	—	29.20
Material	—	34.09
Total Spring #1 Cost	—	63.29
Winter-Spring #2 (Nutritional):		
Application (125 GPA)	29.20	29.20
Material	38.41	38.41
Total Spring #2 Cost	67.61	67.61
Winter-Spring #3:		
Application (125 GPA) @ 2 Times & 3 Times	58.40	87.60
Material	31.68	47.52
Total Spring #3 Cost	90.08	135.12
Late Spring #4 (Nutritional):		
Application (125 GPA)	—	29.20
Material	—	49.51
Total Late Spring #4 Cost	—	78.71
Summer Oil #1:		
Application (125 GPA)	29.20	29.20
Material	49.51	80.85
Total Summer Oil #1 Cost	78.71	110.05

Table A.3 *(continued)*

Indian River White Grapefruit Grove. PRODUCTION/CULTURAL COSTS[a]	Fresh Market Cultural Program: Without Canker-Greening	Fresh Market Cultural Program: With Canker-Greening
Summer Oil #2:		
Application (PTO – 125 GPA)	29.20	29.20
Material	55.59[b]	55.59
Total Summer Oil #2 Cost	84.79	84.79
Fall Insecticide:		
Aerial Application (10 GPA)	8.00	8.00
Material	36.28	
Total Fall Insecticide Cost	44.28	58.61
Total Spray/Pest Management Costs	365.47	721.65
Field Inspections for Citrus Greening (4 inspections @ $22.73)	—	90.92
Clean Blocks Before Certification and Harvesting	—	31.08
Inspections Before "Canker Free" Certification (2 inspections @ $22.73)	—	45.46
Mandatory Citrus Canker Decontamination Costs	29.35	29.35
Fertilizer (Bulk):		
4 Applications	37.44	37.44
Material (12-2-12-2.4MgO @ 120 lbs N)	129.88	129.88
Total Fertilizer Cost	167.32	167.32
Dolomite (one ton applied every 3 yrs): Material/Application	14.76	14.76
Pruning[b]:		
Topping ($28.92/A ÷ 2.5 yrs)	11.97	11.97
Hedging ($27.33/A ÷ 2 yrs)	13.67	13.67
Chop/Mow Brush after Hedging ($11.99/A ÷ 2 yrs)	6.00	6.00
Raise Skirts of Trees ($15.75/A ÷ 2 yrs)	7.88	7.88
Total Pruning Cost	39.52	39.52
Tree Replacement, 1 thru 3 years of age (4 trees/acre without greening; 7 trees/acre with greening)		
Remove Trees:		
Pull, Stack & Burn (Clip-Shear & Front-End Loader)	25.44	37.10
Prepare Site and Plant Tree (includes reset trees)	57.72	94.50
Supplemental Fertilizer, Sprays, Sprout, etc. (Trees 1–3 years old)	50.16	114.80
Total Tree Replacement Cost	133.32	246.40

(Continued on the next page)

Table A.3 *(continued)*

	Fresh Market Cultural Program	
Indian River White Grapefruit Grove. PRODUCTION/CULTURAL COSTS[a]	Without Canker-Greening	With Canker-Greening
Irrigation		
Microsprinkler System[c]	186.70	186.70
Clean Ditches (Weed Control)	18.56	18.56
Ditch and Canal Maintenance	17.48	17.48
Water Control (Pump water in/out of Ditches and Canals)	16.72	16.72
Total Irrigation Cost	239.46	239.46
IRRIGATED FRESH FRUIT PRODUCTION COSTS	$1,211.02	$1,847.74

* The listed estimated comparative costs are for the example grove situation described in the Economic Information Report Series entitled "Budgeting Costs and Returns for Indian River Citrus Production" and may not represent your particular grove situation in the Indian River Production Area.

Table A.4. Additional costs for managing citrus canker and citrus greening—2007

	Oranges for Juice Processing	Grapefruit for Fresh Market
	$/Acre	
Citrus Canker		
Copper Spray Costs (Additional Materials)	15.12	45.36
Clean Blocks Before Certification and Harvesting	—	31.08
Grove Inspections Before "Canker Free" Certification (2 inspections)	—	45.46
Mandatory Citrus Canker Decontamination Costs	29.35	29.35
Total Additional Costs for Citrus Canker	44.47	151.25
Citrus Greening (control psylla)		
Temik (Application & Materials)	118.54	118.54
Spray Costs (Application & Materials)	132.10	172.87
Field Inspections for Identifying Trees with Greening (4 inspections)	90.92	90.92
Total Additional Costs for Citrus Greening	341.55	382.32
Total Additional Costs for Citrus Canker and Greening	386.02	533.57

Table A.5. Estimated total delivered-in cost for Central (Ridge) Florida Valencia Oranges, Southwest Florida Hamlin Oranges and Indian River Fresh-Packed grapefruit with citrus canker and greening, 2006–2007

Represents a mature (10+ years old) citrus grove	Central Florida Valencia Oranges Processed Cultural Program With Canker-Greening and Resetting-Tree Replacement			Southwest Florida Hamlin Oranges Processed Cultural Program With Canker-Greening and Resetting-Tree Replacement			Indian River Grapefruit Fresh Market Cultural Program With Canker-Greening and Resetting-Tree Replacement		
	$/Acre	$/Box	$/P.S.	$/Acre	$/Box	$/P.S.	$/Acre	$/Box	$/Carton
Total Production/Cultural Costs	$1,462.52	$3.499	$0.5145	$1,502.11	$3.280	$0.5466	$1,847.74	$4.307	$2.1535
Interest on Operating (Cultural) Costs	73.13	0.175	0.0257	75.11	0.164	0.0273	92.39	0.215	0.1077
Management Costs	48.00	0.115	0.0169	48.00	0.105	0.0175	48.00	0.112	0.0559
Taxes/Regulatory Costs:									
Property Tax/Water Management Tax	61.00	0.146	0.0215	61.00	0.133	0.0222	61.00	0.142	0.0711
Fly Protocol Cost	—	—	—	—	—	—	56.65	0.132	0.0660
Water Drainage District Tax	—	—	—	—	—	—	65.21	0.152	0.0760
Total Direct Grower Costs	$1,644.65	$3.935	$0.5786	$1,686.22	$3.682	$0.6136	$2,170.98	$5.061	$2.5303
Interest on Average Capital Investment Costs	321.22	0.768	0.1130	321.22	0.701	0.1169	321.22	0.749	0.3744
Total Grower Costs	$1,965.86	$4.703	$0.6916	$2,007.43	$4.383	$0.7305	$2,492.19	$5.809	$2.9047
Harvesting and Assessment Costs:									
Pick/Spot Pick, Roadside & Haul and Canker Decontamination	1,070.50	2.561	0.3766	1,177.06	2.570	0.4283	1,003.00	2.338	1.1690
DOC Assessment	91.96	0.220	0.0324	100.76	0.220	0.0367	150.15	0.350	0.1750
Total Harvesting and Assessment Costs	1,162.46	2.781	0.4090	1,277.82	2.790	0.4650	1,153.15	2.688	1.3440

(Continued on the next page)

Table A.5 *(continued)*

Represents a mature (10+ years old) citrus grove	Central Florida Valencia Oranges Processed Cultural Program With Canker-Greening and Resetting-Tree Replacement			Southwest Florida Hamlin Oranges Processed Cultural Program With Canker-Greening and Resetting-Tree Replacement			Indian River Grapefruit Fresh Market Cultural Program With Canker-Greening and Resetting-Tree Replacement		
	$/Acre	$/Box	$/P.S.	$/Acre	$/Box	$/P.S.	$/Acre	$/Box	$/Carton
Total Delivered-In Cost	$3,128.32	$7.484	$1.1006	$3,285.25	$7.173	$1.1955	$3,645.34	$8.497	$4.2487
Yield	418 boxes/acre; 6.8 P.S. per box for Valencia Oranges			458 boxes/acre; 6.0 P.S. per box for Hamlin Oranges			429 boxes/acre Assumes 100% packout		
Average Tree Density	120 Trees per acre			145 Trees per acre			145 Trees per acre		

P.S. = pound solids. Two cartons per box

Table A.6a. Break-even price for processed Valencia Oranges in Central Florida (Ridge), 2006–2007

	Boxes Per Acre						
	300	350	400	450	500	550	600
	Delivered-in Price Per Pound Solids*						
Without Canker-Greening	$1.129	$1.026	$0.949	$0.889	$0.841	$0.802	$0.769
With Canker-Greening	$1.373	$1.235	$1.132	$1.051	$0.987	$0.935	$0.891

* Assumes 6.8 pounds solids per box.

Table A.6b. Break-even price for processed Hamlin Oranges in Southwest Florida, 2006–2007

	Boxes Per Acre						
	300	350	400	450	500	550	600
	Delivered-in Price Per Pound Solids*						
Without Canker-Greening	$1.307	$1.186	$1.096	$1.026	$0.970	$0.924	$0.886
With Canker-Greening	$1.580	$1.421	$1.301	$1.208	$1.134	$1.073	$1.023

* Assumes 6.0 pounds solids per box.

Table A.6c. Break-even price for fresh market grapefruit in Indian River Florida, 2006–2007

	Boxes Per Acre						
	350	400	450	500	550	600	650
	Delivered-in Price Per Box						
Without Canker-Greening	$7.90	$7.25	$6.74	$6.34	$6.00	$5.73	$5.49
With Canker-Greening	$9.81	$8.92	$8.23	$7.67	$7.22	$6.84	$6.52

Table A.7. Estimated cost of planting and maintaining a reset citrus tree through three years of age, 2007—*With Greening*

	Resets/Replacement Trees Per Acre				
	1–2	3–5	6–10	11–25	26+
	$ Cost Per Tree				
Tree Removal					
(clip-shear trees; remove with front-end loader)	7.95	6.36	5.30	4.24	3.18
Site Preparation					
Disk Tree Site	1.33	1.15	0.98	0.83	0.71
Rotovate Tree Site	1.46	1.27	1.08	0.92	0.78
Repair-Rebuild Beds	1.68	1.47	1.25	1.06	0.90
Total Site Preparation	4.47	3.89	3.31	2.81	2.39
Planting Cost					
Tree Cost (Container Tree)	8.25	8.25	8.25	8.25	8.25
Plant Tree and First Watering (Custom Charge)	2.64	2.30	1.95	1.66	1.41
Total Planting Cost	10.89	10.55	10.20	9.91	9.66
Total Site Preparation and Planting Costs	15.36	14.44	13.51	12.72	12.05
Supplemental Maintenance					
Year #1	6.60	6.09	5.74	5.41	5.09
Year #2 (Trees 1–3 years old)	7.86	7.03	6.15	5.38	4.71
Year #3 (Fertilizer, Tree Wraps, Sprout, etc.)	5.91	5.25	4.51	3.87	3.33
Total Supplemental Maintenance Costs	20.37	18.37	16.40	14.66	13.13
Summary of Tree Replacement Costs					
Tree Removal Costs	7.95	6.36	5.30	4.24	3.18
Site Preparation and Planting Costs	15.36	14.44	13.51	12.72	12.05
Supplemental Maintenance Costs (Years 1 thru 3)	20.37	18.37	16.40	14.66	13.13
Total Three-Year Cumulative Costs	43.68	39.17	35.21	31.62	28.36

Index